Anonymous

Annual Report of the Sanitary Commissioner With the Government of India

With appendices and returns of sickness and mortality among European troops,

native troops, and prisoners, in India, for the year

Anonymous

Annual Report of the Sanitary Commissioner With the Government of India
With appendices and returns of sickness and mortality among European troops, native troops, and prisoners, in India, for the year

ISBN/EAN: 9783337148751

Printed in Europe, USA, Canada, Australia, Japan

Cover: Foto ©berggeist007 / pixelio.de

More available books at **www.hansebooks.com**

ANNUAL REPORT

OF THE

SANITARY COMMISSIONER WITH THE GOVERNMENT OF INDIA

FOR

1905,

WITH

APPENDICES AND RETURNS OF SICKNESS AND MORTALITY AMONG EUROPEAN TROOPS, NATIVE TROOPS, AND PRISONERS IN INDIA, FOR THE YEA

CALCUTTA:
OFFICE OF THE SUPERINTENDENT OF GOVERNMENT PRINTING, INDIA.

1906.

TABLE OF CONTENTS.

SECTION I.

METEOROLOGY OF THE YEAR.

SECTION II.

EUROPEAN ARMY OF INDIA.

SECTION III.

NATIVE ARMY OF INDIA.

CONTENTS.

SECTION IV.

JAILS OF INDIA.

SECTION V.

VITAL STATISTICS OF THE GENERAL POPULATION.

SECTION VI.

GENERAL POULATION—HISTORY OF THE CHIEF DISEASES.

SECTION VII.

GENERAL HISTORY OF VACCINATION.

CONTENTS.

SECTION VIII.

SANITARY WORKS.

SECTION IX.

GENERAL REMARKS.

APPENDICES.

ANNUAL SANITARY REPORT FOR 1905.

SECTION I.

METEOROLOGY OF THE YEAR.

The following report on the Meteorology of India has been kindly furnished

Summary of the meteorological phenomena of the year. by the Meteorological Department of the Government of India.

The chief meteorological features of the year 1905 were—

(1) The severity and persistency of the cold weather conditions as shown by the amount of snow and rain in northern India and the lowness of the temperature over the greater part of the country. This low temperature lasted until the end of April.

(2) The lateness and scantiness of the monsoon rains, especially in north-west India and the consequent high temperature.

(3) The extreme weakness of the retreating monsoon in the Peninsula except in Malabar and the extreme south of the Madras presidency.

During the first twenty days of January weather was feebly unsettled in northern India with occasional showers in the plains and snow in the hill districts. Temperature there was accordingly very unsteady but the departures from the normal were at no time very large or important. In the Peninsula on the other hand weather was generally dry and temperature moderately high during this period. On the 21st a fairly deep depression entered northern India from the west and occasioned heavy precipitation in north-west India and showers in the north-east. It was followed by a very strongly marked cold wave which appeared on the north-west frontier on the 21st and advancing across northern India during the next five days reached Upper Burma on the 26th. Maximum or day temperatures were from 12° to 32° and the night temperatures from 8° to 26° below the normal during this period and severe frost was experienced both on the hills and plains, a number of stations registering the lowest night temperatures on record. As usual during the cold season these low temperature conditions did not extend to Madras. A second depression appeared in Baluchistan towards the close of the month and gave showers in north-west India.

February opened with unusually low temperature over the whole country except in the extreme south and Burma. These low temperature conditions lasted throughout the month but were on the whole most marked during the first three days during which period a large number of stations experienced much lower temperatures than any previously on record for February. Sharp frost occurred in many parts of northern and central India. On the mean of the whole month temperature was in very large defect in northern India and the Deccan, the defect ranging from 5° in the Deccan to 9° in the Upper Sub-Himalayas. The day temperature was more affected than the night temperature and in consequence the daily range was much smaller than usual. Weather on the other hand was warmer than usual in Madras.

The cold weather which ordinarily does not outlast the end of February was in the present year prolonged into March and the greater part of April. Five depressions in all affected the weather in March. Two of them were important and gave fairly heavy rain over upper India and very heavy falls of snow in Kashmir. Practically no rain occurred during the month in lower Sind, Gujarat, western Rajputana, Bombay and the west of the Central Provinces and of the Deccan, as well as within small areas over the south Madras coast and the coast of Lower Burma.

Temperature in March was as markedly in defect as during the two preceding months ; the only exceptions occurred in South Madras, Mysore and the Tenas-serim coast.

In April conditions were on the whole less disturbed than they had been since the beginning of the year. The precipitation of the month exceeded the normal in Bengal, the Central Provinces and Madras but was normal or in defect in the remainder of the country : the defect was most pronounced in Burma where the usual thunderstorms were almost entirely absent. The most noteworthy feature of the meteorology of the month however was the abnormally low temperature over nearly the whole of the Indian region. The defect was considerable to large over northern India where both day and night temperatures averaged from $4°$ to $7°$ below the normal.

In May a considerable change occurred in the conditions that had prevailed so far. Little or no rain fell over the dry area in the north-west and the precipitation of the month was less than the average in all districts excepting Upper Burma, Bengal, Orissa, Ganjam, the east and centre of the Central Provinces and parts of the south and south-west of the Peninsula. Accordingly the low temperatures of the previous four months disappeared and were replaced by an excess except in north-east India, the Gangetic plain and the coast districts of the Peninsula. The excess ranged up to $5\frac{1}{2}°$ and the defect up to $3\frac{1}{2}°$. The highest temperatures of the month were recorded generally in the second or third week but were by no means remarkable, though at Jacobabad the thermometer rose to $124°$ on the 13th, $1°$ above the highest reading previously recorded for May.

June was unusually dry over by far the greater part of the country. The monsoon rains were greatly retarded on both sides of India, not setting in till the 10th on the Bombay coast and the 29th in Bengal. The Arabian Sea current was weak and unsteady and it was not until the 27th that it penetrated into the interior. The rainfall was in excess of the average in Burma, Assam and the south of India and in those regions the temperature was in very slight defect. On the other hand in the remainder of the country where the rainfall was less than the average the temperature was much above the normal, the excess amounting to $8°$ in Chota Nagpur.

In July the rainfall of the month was lighter than usual and the temperature above the normal over a considerable part of the country. At the commence-ment of the month a fair amount of rain was falling over the Peninsula, the central parts of the country and north-east India, but very little elsewhere. There was, in consequence, a considerable excess of temperature over northern India, a slight excess over the Peninsula, and a considerable deficiency over the central parts of the country across the head of the Peninsula. This distribution of temperature continued till the 5th; between the 6th and the 10th the tempera-ture conditions were largely controlled by the rainfall of a depression which passed westward from Bengal through the central parts of the country and disappeared on the 10th. During this period the mean temperature was steadily high over north-west India and over the Peninsula, and low elsewhere—more particularly at the central stations. A great change then occurred. A shallow storm appeared at the head of the Bay on the 11th and passing through west Bengal disappeared over the United Provinces on the 15th. A strong flow of monsoon winds into northern India occurred during this period occasioning fairly heavy general rain and bringing about a great reduction of temperature. The greatest deficiency was shown in the day temperature which was from 15° to 27° below the normal at several stations in the Punjab and Kashmir. During this time there was no rain of importance over the Peninsula, and the mean temperature ranged above the normal. As the rainfall grew lighter and contracted in the north this area of excessive temperature spread northward and on the 16th, 17th and 18th the mean temperature exceeded the average in nearly all parts of the country except Burma and north-east India. Between the 19th and 25th another depression passed from the head of the Bay westward, as far as lower Sind, occasioning rain over the central parts of the country and the Peninsula. The rainfall was heaviest over Gujarat and the greatest depression of temperature was reported from that area. Showers fell over the north-west Himalayas and neighbouring districts on the 23rd and the mean temperature was lower than usual over the North-West Frontier Province for some days. In other parts of the Indian land area the mean temperature was generally higher than usual between the 20th and the 28th. On the 27th a fresh depression appeared over Bengal and marched northward through that province. Rain fell over Burma and in the north-east, and the mean temperature became lower than usual over these areas.

August was even drier than July, the total rainfall of the month being below normal in all parts of the country with the exception of Assam, Bengal, Bihar and Madras. The deficiency was greatest in north-west India which obtained only 20 per cent. of its normal supply. Temperature in consequence was more or less above the normal over the whole of the Peninsula and north-west and central India ; the excess was most marked in Rajputana and the adjacent districts of the Punjab where it averaged nearly 6° in amount. The heat in the north-west was most intense between the 19th and 21st when maximum temperatures ranging up to 111° and from 10° to 15° above the normal were recorded at several stations.

During September rain occurred frequently in Burma and north-east India and the mean temperature there was below the normal ; over the remainder of the country however rain was infrequent or light and the heat was greater than usual. The excess of temperature was not so great as in August. The most striking features of the temperature conditions were (1) a large excess in north-west India during the first ten days of the month and (2) a large defect over the whole of northern India from the 11th to 17th. Conditions were very steady over the greater part of the Peninsula where on most days temperature was above the normal.

Weather was less disturbed than usual in October except in Burma, Assam, parts of Bengal and the south of the Peninsula where the retreating monsoon current was somewhat more active than usual. During the first four days of the month the weather was very quiet and there was little rain beyond light showers in the north-east. In consequence the mean temperature, except in parts of Assam and Bengal, was generally above the normal, the excess being greatest and amounting to $5°$ over Rajputana and the neighbouring parts of the United Provinces and over the Bombay Deccan and Khandesh. On the 5th rain commenced over the southern half of the Peninsula and fairly heavy daily rain continued over that region from the 5th to the 21st, when, owing to the formation of a depression over the Bay, and the consequent diversion of the moist current from the south of the Peninsula to the depression area, the fall somewhat abruptly ceased. In consequence of the above rainfall the mean temperature of the Peninsula was generally lower than usual. During this period temperature over northern India was unsteady and varied by small amounts from the normal. On the 11th, 12th, and 13th there was some snow and rain over Kashmir and light rain over the Punjab. In consequence a decided fall of temperature took place in the north-west and in some divisions the mean daily temperature was reduced to below the normal. On the 14th and 15th it became rapidly hotter in Sind and these high temperatures spread eastward over Gujarat, Rajputana and Central India. On the 21st a large shallow low pressure area lay over the Bay of Bengal and rain was falling over Burma, deltaic Bengal, Orissa and the Circars : the low pressure area gradually concentrated and on the 23rd a depression was shown off the coast near Vizagapatam. This depression advanced north-north-eastward and north-eastward up the coast, entered Bengal on the 25th and then passed eastward through Upper Burma and Cachar during the 26th. The storm occasioned a considerable fall of rain in the north-east and the temperature over Burma, Assam and Bengal was somewhat lower than usual. During this period the mean temperature was very excessive in the north-west. From the 27th to the end of the month the weather was practically rainless throughout the country and the mean temperature was generally above the normal—more especially in the north-west where the average excess ranged between $2°$ and $10.°$

Quiet weather prevailed over the Indian land area during November. From the 1st to the 9th the only rainfall reported was some scattered showers, mainly over Burma, Assam and the extreme south of the Peninsula. During this period the most marked feature of the weather was the excessive heat prevailing over north-west India, where the mean temperature varied between $4°$ and $10°$ above the normal. Temperature was also higher than usual—though to a less extent than in the north-west—over north-east India, the showers in that area having been less frequent and less heavy than usual. On the 10th moderate to heavy rain set in over the southern half of Madras. The rainfall diminished both in amount and extent between the 12th and the 16th, extended again on the 17th and 18th, was light between the 19th and 23rd, but became heavier and more general on the 24th and subsequently extended along the west coast reaching as far north as Bombay, Poona and Ahmednagar on the 27th and 28th. During the period from the 11th to the 15th the temperature was unsteady over northern India but was very high over north-west and central India. On the 16th a fall took place, but, though the mean temperature in the north-west was lowered, the fall was not sufficiently large to bring it below the normal in that area. In the north, east and centre, on the contrary, the mean temperature fell to below the normal and remained normal or below till the 24th. On the 25th a decided increase of temperature occurred over the Peninsula, and though the rainfall over the south of the Peninsula, the west coast and the Bombay Deccan temporarily made the air cooler than usual in those regions, the interior of the Peninsula remained warmer than the average until the close of the month. In north-west India the temperature remained high during this interval while in north-east India it was generally normal or below the normal.

Quiet weather prevailed over the Indian region during the first week of December, and the only rainfall reported was a few showers over the south of the Peninsula. The mean temperature was at the commencement of the week generally higher than usual except over the north-west and the north-east but temperature gradually fell and on the 6th, 7th and 8th was below the average except in Burma. The deficiency of temperature was especially marked over the centre of the country. On the 8th a barometric depression, which had apparently advanced in a northerly or north-easterly direction from the south of the Bay, was shown over the east of the Bay off the west of the Andamans. It advanced in a north-easterly direction during the next two days and apparently broke up near the Arakan coast. It occasioned rainy, unsettled weather over the Andamans and Burma between the 6th and the 10th and brought about a considerable fall of temperature in the Andamans on the 6th and 7th and in Burma between the 8th and 10th. Between the 8th and the 11th temperatures lower than usual prevailed over the greater part of the Indian area, but on the 12th a rise set in and by the 14th the mean temperature generally exceeded the normal except over the central districts and along the west coast. Light showers of snow occurred on the north-west Himalayas between the 13th and 15th, and temperature was unsteady in the extreme north-west, but on the 16th, 17th and 18th a rise took place and on the 18th immediately before the appearance of the second storm of the month, the mean temperature was very high over the whole of north-west India. On the 19th, a shallow depression was shown over the Punjab and neighbouring regions and moderately heavy rain was falling over the North-West Frontier Province and the Punjab. A slight cold wave succeeded this precipitation and passed slowly across northern India between the 20th and the 25th. It was succeeded by very high temperatures in the north-west which in their turn extended eastward over the central parts of the country. On the 29th, 30th and 31st light showers fell in the north-west and north-east and temperature was reduced to below the normal in both areas; but over the remainder of India the mean temperature was generally in excess of the normal during this period and in parts of the Deccan, the central districts and Bengal the excess was considerable.

The mean temperature of the whole Indian land area for the year was only 0·4° in defect ; the defect was shown entirely in extra tropical India, temperature being 0·1° in excess in tropical India. It was above the normal in all months of the year with the exception of the first four and December. The defect was unimportant in December but equalled 4·7° in February and 3·0° in March and April. The excess on the other hand was greatest in June, the month of greatest deficiency of rainfall.

The mean pressure of the year was ·009″ higher than usual. The excess was much more marked in the Peninsula than in northern India, averaging ·014″ for the former and ·003″ for the latter area. The most striking feature of the pressure conditions during the year was the large excess in April and November, associated in the former month with a considerable depression and in the latter with a moderate elevation of temperature.

The year was drier than usual. The mean relative humidity of the air was slightly below the normal. This dryness was solely a result of a large deficiency of aqueous vapour, the pressure of which was ·012″ below the normal.

The mean cloud amount was above the normal from January to March and below it during the remainder of the year. The cloudiest month relative to the normal was March ($+1\cdot2$), and the skies were least overcast in June when the mean cloud amount was $1\cdot1$ below the normal.

The rainfall of the year exceeded the normal over Burma, Assam, Bengal, Bihar, the east submontane districts of the United Provinces, the central and west Punjab, the North-West Frontier Province, the west of the Central Provinces and Baluchistan, but was lighter than usual in all other parts of the country. The deficiency was largest in Rajputana, Sind, the south-east Punjab and the western and central districts of the United Provinces where it ranged between 48 and 62 per cent. The greatest excess on the other hand occurred in north Bihar, central and deltaic Bengal and Baluchistan which received 23 to 32 per cent. more than their normal amounts.

The deficiency of rainfall was not distributed over the whole year but occurred almost solely in the period June to September.

SECTION II.

EUROPEAN ARMY OF INDIA.

2. The average strength of the European Army of India during 1905, India. Appendices A and B to Section II, Tables I, III, and LIII. excluding officers, was 71,343 as compared with 71,083 during 1904, the number of new arrivals in the country during the corresponding trooping seasons having been 15,178 and 16,366 respectively. From the marginal table it will be seen that the health of the British troops during the year has been exceptionally good, the rates of admission into hospital, of constantly sick,

India		All causes. Ratios per 1,000.		
		1895-1903.	1904.	1905.
Admissions	...	1100·6	900·4	834·3
Constantly sick	...	68·05	57·0	52·3
Deaths	... ...	13·46	10·83	10·05
Invalids	... ...	33·99	35·27	21·24

of mortality and of invaliding all being the lowest on record; and this is the more satisfactory as 52 per cent. of the strength were under 25 years of age and 25 per cent. had less than one year's service in India. Taking India as a whole, the healthiest months were March, April, May and June, and the most unhealthy August, September and January.

The chief causes of sickness were venereal diseases, ague and simple continued fever, which accounted for 18·42, 13·35 and 5·74 per cent, respectively, of the total number of admissions from all causes. The decrease in sickness was due chiefly to the diminution in the number of admissions on account of ague and venereal diseases. The admission rate for ague was 111·4 per mille against 174·0 in 1904, and an average of 311·2 in the preceding decennium; that for venereal diseases was 153·7 against 198·5 in 1904 and 385·5 for the decennium. Enteric and remittent fevers also show diminished admission rates in comparison with those of 1904 and of the decennium, but the rates for simple continued fever, influenza and small-pox rose from 23·7, 4·9 and ·7 per mille, respectively in 1904 to 47·9, 14·2 and 1·4, respectively, in the year under review. How far the diminution in the admission rate of ague is due to a real reduction in the prevalence of the disease as a result of more effective prophylaxis and treatment, and how far it may be explained by the increasing admission rate of simple continued fever, is difficult to gauge exactly, and in this connection a reference may be made to the remarks in the subsequent paragraph upon simple continued fever.

The main causes of death were, as usual, enteric fever and hepatic abscess, the former accounting for 29·7 per cent. and the latter for 11·7 per cent. of the total number of deaths. The mortality from each of the chief diseases with the exception of heat-stroke, pneumonia and dysentery was less than in the previous year.

The main causes of invaliding were, in order, debility, ague, tubercle of the lungs, valvular disease of the heart, disordered action of the heart and syphilis, these causes together accounting for over two-fifths of the total. There was a notable decrease (from 643 in 1904 to 315 in 1905) in the number of men sent home on account of ague, tubercle of the lungs and syphilis. Only 23 men were invalided for *Bilharzia hæmatobia* as compared with 71 in 1904 and 158 in 1903.

A comparison of the marginal tables in the first paragraphs of this and the following section shows that the admission rate of European troops during 1905 was nearly one and a half times as high as that of Native troops, and the constantly sick and invaliding rates were more than twice as high. The death rates of European and Native troops were, respectively, 10·05 and 8·09 per mille.

The average strength of the European troops of the garrison in the Aden Hinterland was 349 and altogether 345 admissions and 3 deaths were reported, giving an admission rate of 988·5 and a death rate of 8·60 per mille. Ague, as usual, was the chief cause of sickness and accounted for 46 per cent. of the total number of admissions. The three deaths were due to pneumonia, hepatic abscess and intestinal obstruction.

3. As in 1904, the Northern Command shows lower constantly sick and admission rates than the Eastern, and the Eastern Command than the Western; and in 1905 these commands stand in that order also as regards mortality, the death rate of the Eastern being considerably higher than that of the Northern. The invaliding rate, as in 1904, was lowest in the Eastern and highest in the Western Command. The admission, constantly sick, death and invaliding rates all show an improvement on the corresponding figures of the previous year in *each* command; but, while the three commands contribute very equally to the improvement in the constantly sick rate, the diminution in the admission and death rates (compared with the figures of 1904) is most marked in the Northern Command. The admission rate for venereal diseases was highest in the Eastern Command (165·8) and was closely followed by that of the Western (155·4); and in the Western alone was the admission rate for venereal diseases exceeded by that for ague—the rate for ague in the Western being 162·8 per mille in comparison with 111·4 amongst European troops in India as a whole. The death rates for enteric fever were approximately the same in the three Commands; those for heat-stroke were about twice as high in the Northern and Eastern as in the Western; those for hepatic abscess were more than three times as high in the Western and Eastern as in the Northern. As regards divisions, the statistics in the 4th (Quetta) Division were the most unfavourable showing as they do admission and constantly sick rates of 1024 and 60 per mille respectively. The 1st (Peshawar) Division had a higher admission rate (1031) but a lower constantly sick rate (54), and the 9th (Secunderabad) Division with the low admission rate of 813 had the high constantly sick rate of 62 per mille. The statistics of the 5th (Mhow), Burma and 3rd (Lahore) Divisions compare most favourably with the corresponding figures of the previous year, but the statistics of the 1st (Peshawar) Division which last year showed the lowest admission and constantly sick rates (740 and 44 per mille, respectively) come next to those of the 4th (Quetta), the most unhealthy of all the divisions. During 1905, the health of the European troops was most satisfactory in the 2nd (Rawalpindi) and 3rd (Lahore) Divisions. As in 1904, tubercle of the lungs was most fatal among troops in the 6th (Poona) Division and abscess of the liver among those in the Aden Brigade. Enteric fever was most fatal among troops in the 1st (Peshawar), 4th (Quetta) and 5th (Mhow) Divisions; heat-stroke among those in the Aden Brigade, 1st (Peshawar), 8th (Lucknow) and 3rd (Lahore) Divisions.

Amongst the largest stations, the highest constantly sick rates were at Lucknow (61·4), Secunderabad (57·9), Quetta (55·4) and Rawalpindi (55); the lowest, at Ambala (48·1). The incidence of enteric fever and hepatic abscess

Commands. Divisions. Stations.
Appendix A, Tables I, III, IV and V.

was greatest at Poona and Lucknow and least at Ambala, Bangalore and Quetta. In regard to the prevalence of dysentery, however, Quetta came next to Lucknow. Bangalore (20·4), Lucknow (18·4) and Secunderabad (18·2) furnished the highest constantly sick rates for venereal diseases, and Quetta (7·9) the lowest. Poona which gave the highest constantly sick rate for ague (5·4) returned the lowest for simple continued fever (·3); the high rates for simple continued fever of 5·3 and 4·6 were furnished by Lucknow and Quetta, respectively.

4. The geographical groups which, judged by their statistics, may be regarded as having been the most unhealthy for

Geographical groups. Appendix B, Table II.

European troops during 1905, were Bengal-Orissa (IV), Indus Valley (VII), Western Coast (X), Burma Coast (I) and Central India (VIII), while at the other end of the scale were the Hill Stations (XIIa), Southern India (XI) and Burma Inland (II). Comparing the relative position, as to healthiness, of these groups for the past three years (1903-05), it is found that Bengal-Orissa (IV), Burma Coast (I) and Central India (VIII) invariably appear amongst the most unhealthy and the Hill Stations (XIIa) and Southern India (XI) amongst the healthiest. Omitting Convalescent Depots (XIIb), the highest constantly sick rates were in Bengal-Orissa (IV) with 77·3 per mille, Western Coast (X) with 58·3, and Indus Valley (VII) with 57·2; the highest admission rates in Burma Coast (I) with 1220·8, Bengal-Orissa (IV) with 1160·4 and Indus Valley (VII) with 1092·7; the highest death rates in Bengal-Orissa (IV) with 17·75, Central India (VIII) with 11·89 and Indus Valley (VII) with 10·66. The constantly sick, admission and death rates of the Hill Stations (XIIa) were 45·4, 754·5 and 8·95, respectively, as compared with 52·3, 834·3 and 10·05, respectively, for all India. In these geographical groups the average strength varied from 1,259 in Burma Coast (I) to 13,724 in Upper Sub-Himalaya (VI) and in all of them, except four, the admission and constantly sick rates show a decrease—most marked in Burma Coast (I), Gangetic Plain (V) and Central India (VIII)—on the corresponding figures of 1904. The four groups whose rates compare unfavourably with the previous year's are Southern India (XI) with an increased constantly sick rate, Western Coast (X) and Hill Stations (XIIa) with increased admission rates and Indus Valley (VII) with a considerable increase in both the constantly sick and admission rates. Ague was most prevalent among the troops located in Bengal-Orissa (IV) and Indus Valley (VII); venereal diseases among those in Bengal-Orissa (IV), Western Coast (X), Burma Coast (I) and Southern India (XI). The mortality from enteric fever was highest in Central India (VIII), Upper Sub-Himalaya (VI), Indus Valley (VII) and Deccan (IX); in the first two of these groups, it amounted to 41 per cent. of deaths from all causes. Hepatic abscess caused about 17 per cent. of all deaths in Deccan (IX), Bengal-Orissa (IV) and Southern India (XI).

Combining the statistics of all the European troops in India and comparing the aggregates with those of the previous year, there were, during the year under review, 4,483 fewer men admitted to hospital, 319 fewer men constantly sick, 53 fewer deaths and 999 fewer men invalided, notwithstanding the fact that the average annual strength of these troops was greater by 260 men. Even if we leave aside the diminished loss from deaths and invaliding, the gain derived solely from the decrease in the number of men constantly sick may be represented as equivalent to the work of 116,435 men for one whole day: this gives an idea of the significance of the above figures.

5. Influenza was especially prevalent during the year among European

Influenza. Appendix B, Table VI.
troops, the admission rate, which was 3·1 per mille in 1903 and 4·9 in 1904, rising to 14·2 in 1905. The disease was prevalent in all but one, Southern India (XI), of the geographical groups, the incidence being greatest in North-Western Frontier and Indus Valley (VII) and Bengal-Orissa (IV), which were affected to the extent of 94 and 60 per mille of their strength, respectively. The only other group with a rate above the mean for India (14·2) was Upper Sub-Himalaya (VI). Some idea of the occasional fluctuation of incidence in the same group from year to year is obtained when it is noted that North-Western Frontier and Indus Valley (VII) which in 1905 was so severely affected by influenza was, in 1904, free from the disease. Against 349 admissions with no fatality in the previous year, there were altogether in the year under review 1,014 admissions, with three deaths, reported from 45 stations, of which Peshawar (428), Fort William (97) and Mian Mir (96) contributed 61 per cent. of the total number of admissions. At Peshawar, which in 1904 furnished no case under this heading, the means by which the disease was introduced could not be discovered. The disease was characterized by the usual symptoms—lassitude, fever, pains in back and limbs, headache, great prostration and debility, and although of a mild type (the case which ended fatally having been complicated by pneumonia) became very prevalent amongst the general population. A careful microscopical examination in each case excluded the possibility of the disease being malaria. Both the intestinal and respiratory varieties were represented: in the former, abdominal pain, nausea, vomiting and diarrhœa were prominent; in the latter, there was well marked coryza. In all cases convalescence was prolonged and marked by great debility. The epidemic occurred during May and June. For the third year in succession Fort William appears on the list of greatest incidence; the months of maximum prevalence have invariably been August and September. Here also any etiological connection with malaria was excluded. At Mian Mir, influenza was widely prevalent during April and May and of a mild type, but followed by great debility and in two cases accompanied by pneumonia.

6. The troops in the Northern Command remained free from cholera during the year. In India as a whole, ten cases with nine deaths were returned during

Cholera. Appendices A and B, Table VII.
1905 as compared with 16 cases and 11 deaths in 1904 and 81 cases and 57 deaths in 1903. Of the ten cases during the year, five occurred in November on board the R. I. M. S. "*Irrawaddy*" which was carrying troops from Mandalay to Rangoon. The cause of infection was attributed to one of these five men having broken bounds and visited a native village which was afterwards found to be cholera-stricken. All five, who were transferred for treatment to Thayetmyo, died of the disease. Of the other cases, two were reported from Jubbulpore, and one each from Madras, Allahabad and Dinapore. Cholera is said to have been prevalent in Madras and to have been introduced by emigrants *en route* to Natal. At Dinapore, the disease was probably contracted in the bazaar, where an outbreak of cholera had occurred after the dispersal of natives from the Sonepore fair. As regards the case at Allahabad, the source of infection was associated with a restaurant outside cantonment limits, but it is not stated if the disease was prevalent in the surrounding district. In the Jubbulpore hospital reports, no reference is made to the two cases mentioned above, which were returned as *choleraic diarrhœa*.

7. Small-pox was twice as prevalent amongst the European troops during
1905 as during the preceding year and was report-
ed from all the geographical groups except two—
Burma Inland (II) and Bengal-Orissa (IV).
The Western Command was chiefly affected,
contributing 77 cases (one fatal) while the Northern furnished six cases and
the Eastern one only.   There were in all 99 cases with one death as compared
with 49 cases with four deaths in 1904 and 68 cases with eight deaths in
1903.   One or more cases were recorded in 23 stations, the largest numbers
occurring in Poona (26), Mhow (19) and Colaba (9).   At Poona, the epidemic
prevailed from January to May—most of the cases occurring in March, and it is
stated that   the disease was contracted, in the first instance at least,
in the bazaar, where a severe epidemic was raging amongst the natives.
All the infected men bore satisfactory marks of vaccination and none of them
had previously suffered from small-pox.   No death occurred amongst the 26
men attacked at Poona, while a woman who had not been vaccinated was attacked
by the disease and died.   The outbreak at Mhow took place in March and fresh
cases continued to occur till June.   The disease was prevalent among the native
population and the source of infection was traced to the *sadar* bazaar which is
situated near the barracks of the cavalry, who suffered most.   None of the
patients had previously suffered from small-pox and all had been vaccinated (one
man had been vaccinated just a week before he became infected).   In the station
of Colaba small-pox prevailed from January to March ; the disease was severely
epidemic in Bombay City but the source of infection, as regards the troops, was
not definitely traced.   All the nine men that were attacked had been vaccinated
and seven of them revaccinated on one or more occasions.

8. During 1905, a year of abnormally small autumn rainfall, there was exceptionally little malaria among the European troops, intermittent fever being the cause
of 13 per cent. only of the total admissions as com-
pared with 19 per cent. during 1904 and 23 per cent.
during 1903. The admission rate for the European
army of India as a whole was 111 against 174 in 1904 and an average for the
preceding decennium of 311, and the death rate fell from ·24 in 1904 to ·11 per
mille.   There was also a coincident diminution in the prevalence of the disease
during 1905 among Native troops and prisoners, the rates for these, (except for
that of prisoners in 1896[*]), being the lowest on record.   To the reduction in the
prevalence of this disease among European troops all three commands contributed, their admission rates falling from 202 in the Northern, 212 in the Western
and 129 in the Eastern to 70, 163 and 113, respectively.   The incidence in the
Northern thus fell to about one-third of what it was in 1904 and that in the
Western, although considerably smaller than in the preceding year, is still the
largest in the three commands.   The great prevalence of malarial fever in the
Western Command was due chiefly to the troops of the 5th (Mhow) Division
being particularly affected.   Of the ten divisions, the admission rates were
highest among the troops in the 5th (Mhow) Division with 184 per mille of
strength, the 1st (Peshawar) Division with 178·9 and the 6th (Poona) Division
with 155·6.   Among the troops of the Aden Brigade the admission rate was
286·2 per mille and the death rate 3·17 against 486·5 and 4·39 per mille,
respectively, in 1904.

Small pox. Appendices A. and B,
Tables III and IV.

Ague (1). Appendix B. Tables X
and LIII.

---

[*] The admission rate of prisoners in 1896 was 278·5 per mille as compared with 325·5 in 1905.

As regards geographical groups, the greatest prevalence was, in order, in Bengal-Orissa (IV) with an admission rate of 255·6 per mille, Indus Valley and Frontier (VII), Central India (VIII), Burma Inland (II) and Western Coast (X), and the least in Burma Coast (I) with an admission rate of 27 per mille, Hill Stations (XIIa) and Upper Sub-Himalaya (VI). Of these, Central India (VIII) and Indus Valley and Frontier (VII) generally take a high place, but the Upper Sub-Himalaya (VI) group which in former years usually accompanied these two (VIII and VII) in the list of groups chiefly affected appears in 1905 at the other end of the scale with an admission rate of only 74·6 per mille as compared with 208·8 in 1904. Excluding the Hill Stations, in which the prevalence of malaria depends on the number of infected men arriving from stations in the plains, and comparing 1905 with the previous year, a decrease in the admission rate is recorded in 49 stations and an increase in 22. If stations in which the annual average strength was below 200 be omitted, the highest rates were recorded at Ahmedabad, Delhi, Thayetmyo and Fort Allahabad, the rates in all being above 400 per mille of strength and being considerably higher than those of the previous year. The admission rate at Dthalla (Aden Hinterland) was 456 per mille of strength.

The seasonal prevalence of ague is most marked in the Northern Command (May to November) and least in the Western, due

Influence of temperature.

doubtless to the effect on the mosquito and the *plasmodium malariæ* of the low temperature during the winter and spring months in the former command. The influence of temperature on the sexual generations of the malarial parasite and on experimental malaria has been recently investigated by Jancso'. Mosquitoes of the species *Anopheles claviger* were infected by being fed on patients suffering from malaria (separate experiments being made for each variety of the malarial parasite), then kept at different degrees of temperature and finally killed, embedded and examined microscopically in sections at stated intervals. If the mosquitoes were kept at 24—30° C., the development of the parasite progressed rapidly and well; at a lower temperature development was prolonged, while at a higher the mosquitoes themselves were unfavourably influenced. If, however, the mosquitoes were kept from the very time of feeding onwards at a temperature below 16° C., no cysts (zygotes) developed; whereas, if, after being kept at a favourable temperature till cysts formed, the temperature was then lowered even much below 16° C., healthy cysts with infective germs were found. The lower limit of temperature permitting the development of the malarial parasites to proceed to the sporocyst stage was 16° C. Jancso also demonstrated that the possibility of infection from the bites of infected mosquitoes is independent of the temperature at which the infected mosquitoes were kept. It would appear, therefore, from these experiments that an infected mosquito even after hibernation may still be infective.

Leaving aside diseases of the nature of relapsing fever, African tick fever and possibly also yellow fever, in which the specific

Immunity and prophylaxis.

causative micro-organisms—*spirochætæ* (the systematic position of which in the animal or vegetable kingdom is still unsettled, Novy and Knapp' having recently brought forward a mass of evidence against *spirochætæ* being classed as protozoa), disappear from the blood with convalescence, immunity in tropical protozoal infections of the human blood, *e.g.*, malaria and trypanosomiasis, is only partial; it is more of the nature of a mutual "tolerance" or symbiosis,' the specific protozoon remaining in the body after convalescence. This persistence of the protozoa in the body is possibly due to these small representatives of the animal kingdom being themselves, like their hosts, possessed of the power of producing immune-bodies for self-defence; and this same power may explain the complete failure of serum therapy[4] in these diseases of man. Assuming, therefore, this very natural hypothesis, the methods by which we can hope to combat these diseases are obvious. There is no hereditary immunity[5] or absolute acquired immunity[5]; we cannot rely on the body tissues or fluids being able to rid themselves of these parasites; we have to find some extraneous agent which, while relatively harmless to the host, has a powerful destructive effect on the parasites. For malaria, such an agent we have in quinine. By means of this drug a relative immunity very similar to the acquired immunity of natives may be attained; but the quinine immunity is preferable to the acquired immunity, because natives with the latter, by harbouring the parasite (latent infection[7]), may remain a source of infection long after convalescence. The methods of administering quinine to obtain immunity are various. Craig[8] states that quinine acts on *every stage* in development and growth of *all varieties* of the *plasmodium*, but especially on free spores and young intracorpuscular forms, and that the best method of prophylaxis is to keep the blood all the time full of quinine by regularly repeated doses. Celli[9] recommends 5 to 8 grains daily, given in the form of sugar-coated pills, and states that this method is better and produces less cinchonism than that of giving larger doses at intervals of some days. Plehn prefers the administration of quinine at long intervals to giving small doses daily and recommends 8 grains every fourth or fifth day[10]; but evidently he is not quite sure of the value of this method for in a more recent publication he adds that, if the danger of infection be great, instead of 8 grains every fourth *or* fifth day, 15 grains preferably in divided doses should be given on two consecutive days[11]—on either the fourth *and* fifth or the fifth *and* sixth days. It will be observed that there is a vast difference between these two methods put forward by Plehn. Koch[12], as is well known, recommends 15 grains every eighth and ninth day and Külz[13] has recently reported good results from this method. Morgenroth[14] advocates 15 grains on two consecutive days each week but admits that 15 grains every eighth and ninth day before breakfast is generally sufficient. According to him, 8 grains are quite an inadequate dose for prophylaxis and even 15 grains every tenth and eleventh day are not protective. The best results are claimed by Morgenroth for quinine hydrochloride in cachets or dissolved in weak hydrochloric acid; tabloids of quinine sulphate were least satisfactory—being often found almost unchanged in the stools after 24 hours; tabloids of quinine hydrochloride varied greatly in solubility according to the method by which they were prepared. He advises that the dose should never be less than 15 grains and in cases of great intolerance euquinine should be given. The same author suggests that attacks of malaria, which occur during this method of prophylaxis, are probably due to defective absorption of the quinine and, therefore, as absorption is most active in the morning quinine should be given before breakfast: such attacks, as occur during quinine prophylaxis are stated to be always very mild[9] and never followed by relapses. The date for beginning quinine prophylaxis at any particular station is determined by the appearance of *anopheles* at that station, taken in conjunction with the prevailing temperature. From the long use of quinine administered in these large doses no permanent injury has been observed. Cases of severe hæmorrhage into skin and mucous membrane during a quinine course of 15 grains every eighth and ninth day have been reported by both Morgenroth and Külz[15], but such results are invariably associated with concurrent scurvy.

---

[4] See paragraph 68 in Section IV.

These various methods of quinine prophylaxis may be divided into two groups, the *rationale* of each group being quite distinct. In the one group, small doses at frequent and regular intervals are given with the object of keeping the blood *always* under the influence of quinine so that the malarial parasite cannot enter the body; in the other group, large doses are administered at intervals of some days in order that quinine may be in such concentration in the blood that the parasites—should they meanwhile have gained an entrance into the body and begun to grow and multiply—are quickly and easily killed, being still in those stages of development most susceptible to the action of quinine. The former is the ideal method and corresponds closely with natural immunity but is not practical, for a time comes when owing to some accident or want of foresight the supply of quinine to the blood is not kept up, immunity is lost for a period during which infection may occur and the administration of small frequently repeated doses, when resumed, may not be sufficient to destroy the parasites that have now established themselves within the human body. Kendall,[15] although he advocates the use of quinine in moderate daily doses and employed this method at Panama, admits that quinine thus administered does not by any means absolutely guarantee freedom from malaria or completely drive organisms from the circulation. Of all the methods referred to above, that of Plehn of giving 8 grains every fifth day is probably the least satisfactory; even supposing the method were scrupulously carried out quinine would not be given sufficiently frequently to prevent the parasite from entering the body, nor would the dose be sufficiently large to destroy the parasite after it had gained an entrance. Attention has already been drawn to the great difference between this, his usual method, and that which he advises under conditions where the danger of infection is great; and it need only be added that his advice to give the 15 grains " preferably in divided doses" detracts much from the value of his method. For practical prophylaxis the methods of Koch and Morgenroth are most probably the best and quinine prophylaxis can hardly be said to have had a fair trial unless the quinine has been administered in doses of 15 grains on two consecutive days in the week. When anti-mosquito and in particular anti-larval measures are impossible or futile, our only hope of success in combating malaria lies in (1) quinine prophylaxis, (2) segregation and mechanical defence (mosquito-nets, etc.) of healthy and infected alike, and (3) thorough treatment with quinine of malarial patients and cases of latent infection. The importance of rendering natives free from the *plasmodium* is shown by Craig[7] who found that 62·2 per cent. of adults and 48·3 per cent. of children among the native population in the Philippines were examples of latent infection; and the effect of mosquito-proof wards has been demonstrated by Gorgas[7] on the Isthmus of Panama. In connection with anti-mosquito measures, the habits of certain fish and possibly also tadpoles, which may prey on mosquito-larvæ, are worthy of attention; in Barbadoes, the absence of anopheles is attributed to the presence of small fish (*Girardines versicolor*) which destroy mosquito-larvæ[15].

Where quinine prophylaxis is extensively employed as among the European troops in India, the demonstration by the microscope of the malarial parasites in true cases of ague becomes proportionately more difficult; in such cases the increase in the relative number of large mononuclear leucocytes,[19] which seems characteristic of most protozoal blood infections, may be of assistance in coming to a conclusion as to the nature of an illness.

The effect of X-rays in the treatment of malaria has been investigated by Demarchi[19], who found that these rays have no effect on the course of an attack or on the parasites but have a favourable action on the chronic large spleen.

The more important points brought out by a study of the special malaria reports of stations are here shortly summarized.

Reports from medical officers.

Of 8,106 cases of malaria among European troops in which the type of fever was diagnosed by clinical methods, 73·78 per cent. were considered by the medical officers to be cases of benign tertian infection, 1·28 per cent. quartan, 5·00 per cent. malignant tertian, ·38 per cent. being cases of " malarial cachexia " and 19·56 per cent. remaining undifferentiated. In 72 stations the microscope was used as an aid to diagnosis and the number of cases in which the diagnosis was thus confirmed was 2,797. In 85·2 per cent. of these, benign tertian parasites were found, in 11·7 per cent. malignant tertian and in 2·5 per cent. quartan. Quartan fever was recorded in 17 stations and at Port Blair was nearly three times as prevalent as benign tertian. The proportion of malignant to benign tertian infection was highest in Port Blair, Jacobabad, Jubbulpore, Kamptee and Allahabad, the malignant type being more prevalent than the benign in the first two stations named. At Allahabad, two cases of mixed infection (malignant tertian with benign tertian) were observed. As there were difficulties, however, in differentiating the types of parasites in some cases, the above figures are only approximate. Some medical officers returned cases as malignant tertian only if crescents were found, so the prevalence of this type of fever is probably underestimated. At Rangoon and some other stations, those cases only in which parasites were found in the blood were returned as malaria, the others being recorded under *simple continued fever ;* and the effect of this procedure is well exemplified in the statistics of Mian Mir, where in 1904 and 1905, the admission rates for malaria were 507·4 and 45·7 per mille, respectively, and those for simple continued fever 76·9 and 349·9 per mille, respectively. That the recorded decrease in the prevalence of malaria is by no means wholly accounted for by this change in diagnostic procedure is at once evident from a glance at the tabular statement (and in particular at the column of totals) given later on in the paragraph on simple continued fever. The influence which the arrival of infected troops may have on the statistics of a station is well seen in the case of Karachi, where the 1st South Wales Borderers, " saturated " with malaria, arriving from Mian Mir in March, caused a rise in the admission rate of the station from 93·6 per mille in 1904 to 245·6 in 1905. At Wellington one man was admitted four times, seven men three times each and eleven men twice each, 19 men thus accounting for 67 admissions. Such results as these show the necessity of segregating and systematically treating with quinine infected troops arriving at a healthy station and also of thoroughly disinfecting, (*i.e.*, freeing from parasites) malarial patients and convalescents. In some stations quinine is administered in doses which are far too small, and this is particularly true of the prophylactic dose which, especially in stations which are reputedly malarious, should never be less than 15 grains and should be given on two consecutive days each week during the malarial season. In about eight stations the prophylactic dose was not more than 5 grains given twice a week. At Taragarh, where convalescent and infected arrivals from stations in the plains underwent a systematic course of quinine, *viz.*, 15 grains daily for seven days and then 15 grains twice a week on two consecutive days for a month, there was a total absence of malaria, as compared with 25 admissions (giving an admission rate of 595 per mille) in 1904 when no such precaution was taken. Quinine prophylaxis (15 grains twice a week) was stopped tentatively for a short period at Ahmedabad and the number of admissions for malaria immediately increased. Lastly, the following striking example of the value of quinine prophylaxis is reported from Kamptee: 500 men of the Royal Scots getting 15 grains of quinine sulphate once a week from 1st September to 15th November gave 17 admissions (equivalent to an admission rate of 34 per mille), while 154 men of the 62nd Battery, Royal Field Artillery, that were not taking quinine regularly gave during the same period 20 admissions (equivalent to 129·8 per mille).

At Dum Dum and Dinapore it was observed that quinine tabloids frequently passed almost unchanged through the alimentary canal; quinine was subsequently administered in solution only. Treatment by hypodermic or intramuscular injections of quinine—chiefly the bihydrochloride, is pretty generally followed and is recognised by many as being more rapid and powerful in its action. This method is found valuable in severe or intractable cases, where the results of administration by the mouth are unsatisfactory, e.g, the desired effect not being produced or, the drug not being tolerated. From Mandalay it is reported that a highly neurotic European woman, six months' pregnant, suffering from malignant tertian fever received an injection of 5 grains of quinine bihydrochloride daily until the fever was checked, without the production of any untoward effect on the uterus. In some cases (Dum Dum), where quinine by the mouth was not retained, good results were obtained by giving quinine in rectal injections.

As regards special anti-mosquito measures, the great decrease in prevalence of malaria at Mian Mir is attributed chiefly to the prevention of canal irrigation water from running through cantonments and that at Jubbulpore to the excellent results of the extensive surface drainage carried out near the barracks by the Royal Engineers in 1904. From Peshawar, on the other hand, it is reported that two regiments whose barracks were very near the military grass farm were especially affected during irrigation of this land and had double the admission rate for malaria of the other regiments not so exposed. The increase of prevalence at Thayetmyo is attributed to the flooding of the country by the Irrawaddy and that at Dum Dum to the neighbouring rice-fields. In Dinapore, however, although anti-mosquito measures were heavily handicapped by the presence of rice-fields all round cantonments, the prevalence of malaria among European troops is less than half, and that among Native troops less than a quarter of what it was in 1904. At this station 10 grains of quinine were given under strict supervision on three consecutive days in the week, at first in tabloid form but latterly, after evidence of the relative insolubility of these preparations, in solution. The decrease of malaria cannot in this instance be accounted for by a change in diagnostic procedure, as admissions on account of simple continued fever have also fallen from 53 in 1904 to 7 in 1905. Consequently, the good results obtained in this station must be attributed principally to quinine prophylaxis and possibly to a less extent to the anti-mosquito measures undertaken. Anti-mosquito operations were carried on methodically and enthusiastically in a number of stations with good results, but in some stations such measures were regarded as impracticable owing to the difficulty of dealing with rice-fields, irrigation canals and other large masses of water in and around cantonments, while in others the efforts of the military authorities were to a great extent rendered futile by the mosquito breeding grounds near cantonments being under the jurisdiction of municipal commissioners who could not be prevailed upon to co-operate. To attain complete success in the campaign against mosquitoes, the systematic concerted action of all residents is necessary. The absence of *anopheles* from Jubbulpore is attributed to the lasting effects of the extensive surface drainage referred to above. In most stations malarial patients were, whenever possible, isolated in special wards and supplied with mosquito curtains; the latter, however, the men did not seem to appreciate. At Ambala, the hospital doors were self-closing and fitted up with wire gauze; at Poona the wards were mosquito-proof. The prevalence of malaria in the Aden Brigade is attributed to the indifference of the men to precautions especially during expeditions into the interior. At Aden itself there is very little malaria because of the high winds and the absence of vegetation: a few *anopheles*, however, are present—due, in the opinion of the Senior Medical Officer, to the larvæ being introduced with the water supply. In the Aden Hinterland the prophylactic use of quinine was continued throughout the year and the camp on the line of communications was removed to a site a mile further away from the river—measures, which have been followed by good results. The effect of methylene blue in malaria was tried at Ahmedabad and Calcutta; in the former station the results are reported as favourable, in the latter as unfavourable. At Kamptee the action of the Italian remedy, *fevrolo*, which is said to have a rapid and powerful lethal effect on the malarial parasite, was tested, but no opinion on its value is expressed. Judging, however, from the temperature charts this powder does not seem to have any very rapid lethal effect on crescents and apparently after three or four days administration was invariably displaced by quinine. *Anopheles* were reported during the year in 52 stations in the plains and in 47 of these identification of the species present was attempted. The presence of *A. rossii* was noted in 36 stations, *A. culicifacies* in 20, *A. stephensi* in 7, *A. fuliginosus* in 6, *A. nigerrimus* in 5 and *A. listoni* in 3. Other species mentioned were *A. maculatus, A. theobaldi, A. barbirostris, A. jamesi, A. fluviatilis,* and *A. lindesayi*. The malarial "endemic index" was given by 14 stations but as in all instances except one, the number of children examined was very small, much value cannot be attached to the results.

9. The admission rate on account of remittent malarial fever was 2·3 per mille as compared with 3·4 in 1904 and 13·5 for
Remittent fever.   Appendix   B, the decennium (1894-1903). This fever was
Tables XI and LIII. more than five times as prevalent in the Western and Eastern Commands as in the Northern. As regards geographical groups the admission rate was highest in Burma Coast (I), where, in contrast with those for India as a whole, the rates are on the increase, being 26·2 per mille as compared with 14·6 in 1904 and 6 for the decennium. The greatest diminution in the admission rate has occurred in Bengal-Orrisa (IV), where the rate has fallen from 34·5 for the decennium and 28·4 in 1904 to 4·7 in the year under review. Southern India (XI) recorded no case during the year. The decrease in the admission rate of European troops during the year has been coincident with a fall in the death rate, which was ·14 per mille in 1903, ·10 in 1904 and ·01 in 1905.

The seasonal incidence for India, taken as a whole, corresponds closely with that of intermittent fever; and it is impossible to say how far the reduction in the reported admission rate of remittent fever can be taken as indicating a diminished prevalence of the severer forms of malarial fever. In this connection it may be noted that several medical officers report that the prophylactic use of quinine, even although occasionally it may not produce any great decrease in the incidence of malaria, has some mitigating influence on the severity of the attacks. A considerable part of the reduction, however, is probably due to more careful diagnosis and the more frequent use of the microscope for diagnostic purposes. It is noteworthy that of the five large stations, above mentioned, in which intermittent fever was most prevalent during the year, four—Ahmedabad, Thayetmyo, Fort Allahabad and Dthalla, have not returned a single case of remittent fever; on the other hand Mhow recorded 39 cases, Port Blair 33, Delhi 20, and Dinapur 12 : such facts as these suggest that the diagnostic procedure followed by the various medical officers is a factor influencing remittent fever returns.

· 10. Since 1902, the admission rate on account of simple continued fever has

Simple continued fever.
Appendix B, Tables IX and LIII.

increased in a striking manner. The accompanying tabular statement gives the admission rates of intermittent and the commoner continued fevers and the total admissions of the less frequent continued fevers.

	ADMISSION RATES PER 1,000 OF STRENGTH.						ACTUAL ADMISSIONS.				
Years.	Intermittent.	Remittent.	S. C. Fever.	Influenza.	Enteric Fever.	Intermittent, remittent, simple continued, influenza and enteric fevers combined.	Dengue.	Cerebro-Spinal Fever.	Malta Fever.	Relapsing Fever.	Dengue, spinal, Malta and relapsing fevers combined.
1891-1900	...	326	15	35	7	34	427	199	2	3	...	204
1901	...	393	6	24	9	13	345	...	1	5	...	6
1902	...	347	7	14	2	17	277	293	1	3	1	308
1903 ...	...	343	4	13	3	20	285	373	...	9	...	312
1904	...	174	3	21	5	20	225	429	...	6	...	415
1905 ...	...	111	2	48	14	16	191	415	2	4	...	421

Amongst the former group of fevers, the progressive diminution of the rates for malarial fevers is worthy of note, as is also the coincident increase during the past four years of those for simple continued fever and influenza ; amongst the latter, the increase during recent years in the prevalence of dengue, which, however, affects only Madras and Burma, is deserving of notice and is probably explained, as pointed out in this report for 1904,[1] by the more accurate diagnosis of malarial fevers. It is also to be noted that the totals of the ratios of intermittent and the commoner continued fevers show a continuous decline. The admission rate of simple continued fever during the year was 47·9 per mille which is double that of 1904 and also considerably greater than that of the decennium 1891-1900, a period during which many other continued fevers, e.g., enteric fever, Malta fever, cerebro-spinal fever, influenza, etc., which were not so well recognised and differentiated as they now are, were doubtless frequently returned under the heading of simple continued fever.

All the commands contribute to this rise in prevalence of simple continued fever, but chiefly the Northern and Western, the admission rates of which in 1905 were about three times as great as those in the preceding year. As regards geo-graphical groups, all show an increase of incidence except Bengal-Orissa (IV) and Sanitaria (XIIb), the highest admission rates being in Burma Coast (I) and Gangetic Plain (V) with 136 and 104 per mille, respectively, and the lowest in Western Coast (X) and Central India (VIII) with 2 and 9 per mille, respec-tively. The stations, with a strength of over 200, at which the highest rates were reported, were Mian Mir (350 per mille), Meiktila (232), Jubbulpore (172), Lucknow (167), Cherat (161), Rangoon (148) and Quetta (146) : those at which no case of simple continued fever was reported were Colaba, Wellington, Madras, Bellary, Thayetmyo, Deolali, Kuldanah, Kasauli, Neemuch, Dum Dum, Ahmedabad and Jutogh. For the European troops of India as a whole, the maximum seasonal prevalence during 1905 was from June to October and the minimum during the first three months of the year, and the same seasonal incidence is roughly applicable to each individual geographical group. The most frequent causes assigned for the attacks were excessive atmospheric heat, exposure to the sun without sufficient protection, chills, errors in diet and constipation.

The difference in the prevalence of simple continued fever at Mian Mir in 1904 and 1905 is said to be due to "different methods of diagnosis ; " in the former year the microscope was utilised in the diagnosis of only 54 cases of fever as compared with 295 cases in 1905, the usual procedure now, apparently, being to return all cases of doubtful or obscure etiology (the microscopical examination of the blood, and possibly also Widal's reaction, being negative) under the heading of simple continued fever. From a station in Burma, which followed the same procedure in diagnosis, and returned 88 cases of simple continued fever it is reported that the fever was "easily controlled by hypodermic injections of quinine" and that characteristic symptoms of the illness thus recorded were headache, pain in limbs, constipation and a temperature ranging from 102° to 105° for a period of three to seven days. It must be borne in mind that inability to demonstrate the malarial parasite (unless after repeated examination of the blood by an expert) especially now-a-days when quinine prophylaxis is so general, is far from being absolute proof of the non-malarial origin of a fever ; and further that even after the exclusion of malaria and enteric fever there are other continued fevers, *e.g.*, influenza, Malta fever, paratyphoid, etc., all endemic in India, which have to be considered in the differential diagnosis before a case is returned under the heading of simple continued fever. At Multan, for instance, some patients who on admission were placed under treat-ment for malaria were subsequently found to be suffering from Malta fever. The more extended use of the microscope for diagnostic purposes is doubtless largely responsible for the great increase of the reported prevalence of simple continued fever, and this must be taken into account in gauging the signi-ficance of the decrease of malarial fevers—intermittent and remittent.

11. There were ascribed to Malta Fever four admissions with no deaths

Malta Fever. Table LIII.

during 1905 against six admissions with three deaths in 1904, and nine admissions with no death in 1903. The reporting stations were Ferozepore (February), Quetta (May), Murree (May), and Rawalpindi (September) ; but the patient at Murree had been admitted from the line of march between Rawalpindi and Ghora Dhakka. The patient at Ferozepore was under treatment for syphilis when the disease developed. He had a total service of six years, four of which had been spent in

India : he had never served in a Mediterranean station. The diagnosis was based on the absence of malarial parasites from the blood and the failure of large doses of quinine to reduce the temperature, on clinical grounds—muscular and articular pains and undulant character of the fever, and on the serum test (carried out at the Pasteur Institute, Kasauli) reported as positive for *M. melitensis* in dilutions of 1 in 160, and 1 in 320, and negative for *B. typhosus*. This patient, who was no doubt infected at Ferezepore, was subsequently transferred to Dalhousie. In the case of the patient at Quetta the diagnosis was made on clinical grounds combined with the history of his having served in Malta and having been previously invalided from there for Malta fever. It is not stated when this invaliding had taken place and whether any relapses had occurred in the interval. The blood reaction was not tested, but the clinical picture seems to have been fairly typical. The patient at Murree was most probably infected at Rawalpindi. He had never served in a Mediterranean station, but had been in South Africa. The diagnosis was arrived at on clinical grounds and on the blood reaction, carried out in the Northern Command laboratory and reported to be " positive for *M. melitensis*, 1 in 50 dilution." The patient at Rawalpindi was at first admitted for malaria (malignant tertian) as the parasite was found in his blood and was treated by intramuscular injections of quinine. As this treatment had little effect on the temperature, which continued very irregular, the serum test with *B. typhosus* was carried out, but with a negative result. The liver also was explored twice with the same result. On the 37th day of illness, the blood tested in the Northern Command laboratory gave a " positive reaction for *M. melitensis* in a dilution of 1 in 200." During his stay in hospital he had a severe hæmorrhage from the bowel, stated to be from an ulcer on the anterior wall on a level with the prostate. Several of the characteristic symptoms and signs of Malta fever were present in this case, but the course of the fever, as presented on the temperature chart, was very irregular. It is not stated if he had served in a Mediterranean station.

12. In this place it is usual to give a summary of recent investigations which advance our knowledge of enteric fever.

Typhoid ,2'.

Before confining our attention to diseases of the typhoid and paratyphoid types, it may not be out of place to refer to a work by Krencker[1] on the biology of the coli-typhoid class of micro-organisms, wherein is given a tabular survey of the differential biological and chemical characters of the bacteria belonging to this class. These organisms Krencker divides into th following five groups : (1) *B. coli* group, (2) the group of food-poisoning bacilli, including *B. typhi murium*, *B. suipestifer* and *B. paratyphi* A and B (this group bridging the gulf between the first and third groups), (3) *B. typhosus*, (4) *B. dysenteriæ* Shiga-Kruse and (5) *B. fæcalis alkaligenes* (a typhoid-like bacillus found in fæces and polluted water). He states that by growth on Drigalski-agar and Endo-agar, the *coli*-group can be distinguished from all the others and that on Rothberger's neutral-red-agar the growth of the large second group resembles that of the *coli*-group. With regard to the last mentioned group, which of late has had considerable interest attached to it owing to Altschüler's statement that he had caused *B. fæcalis alkaligenes* by culture to give the reactions of *B. typhosus*, or, in other words, that *B. fæcalis alkaligenes* could be transformed into *B. typhosus* and *vice versâ*, the mystery has now been cleared up. Döbert[2] working with *alkaligenes* cultures from the same source as Altschüler's found the same extensive near relationship between

B. fœcalis alkaligenes, and *B. typhosus*, but drew attention to the fact that it is only this one strain which corresponds so closely with *B. typhosus*. Then Berghaus[3] pointed out that in this culture there was a mixture of *B. typhosus* and *B. fœcalis alkaligenes*, and that they could be separated by allowing them to grow for several days on gelatine plates. The results of Berghaus, who in a subsequent publication gave a minute description of a large number of *alkaligenes* strains, were confirmed by Trommsdorff.[1]

With regard to the campaign against endemic typhoid on the south-west frontier of Germany, the results not only go far in support of the correctness of Koch's methods of suppressing this disease but throw a flood of light on its etiology. One epidemiologically important fact brought out by these measures and investigations is that the healthy or at least non-specifically sick can harbour typhoid bacilli and excrete them—to the danger of the public. These bacilli-carriers (*Bazillenträger*), according to Klinger[3] may be of either sex and of any age between 18 months and 60 years; they have no specific symptoms of typhoid and are apparently in perfect health. The bacilli are excreted in the fæces and can be demonstrated by the ordinary methods. For demonstrating *B. typhosus* we have not yet got a method with anything like the same certainty and simplicity as the peptone enrichment method for the isolation of the cholera vibrio. The bacilli are sometimes found in the urine, especially of women (seven out of eight cases), but never in great numbers and so may possibly have been accidentally carried into the urine where they multiply. "Carriers" are of two classes: (1) "acute carriers" (*akute Bazillenträger*), who have never before or after the demonstration of the bacillus in their fæces shown any symptoms of sickness, have been in direct contact with the sick and, as in the case of the majority of sick after recovery from the disease, excrete the bacillus for a short time only and in small numbers, being thus, in comparison with the second class, of little significance; (2 " chronic carriers " (*chronische Bazillenträger*), who have, a short or long time before, gone through a regular attack of typhoid and may excrete for months pure cultures of typhoid bacilli. Of 23 ' carriers," detected by Klinger from among 1,700 individuals examined during a period of 1¼ years, 11 (6 men and 5 women) belonged to the first class and 12 (3 men and 9 women) to the second. To get some idea of the number of convalescents that become chronic "carriers," 482 cases of typhoid fever were kept under bacteriological observation during recovery with the result that 63 (*i.e.*, 13·1 per cent.) were found to excrete the bacillus during convalescence, but only 8 (*i.e*, 1·7 per cent.) for longer periods than six weeks after the disappearance of fever; and of these 8 chronic "carriers" one was still excreting the bacillus a year after the attack of typhoid and the other seven excreted for periods of 11, 8, 5 and 3 months. When examining persons who had been in contact with typhoid patients, Klinger discovered four chronic " carriers, " one having had typhoid 30 years and another 10 years before. These four cases, as they were in contact with the sick, may have ingested the bacilli shortly before observation, or they may have been reinfecting themselves again and again from the outer world or from their own dejecta ever since their illness, but Klinger thinks it likely from other observations and investigations that typhoid bacilli under certain not yet fully understood conditions can live a saprophytic existence in the gall-bladder of a person who has recovered from an attack. This persistence of typhoid bacilli in the gall-bladder seems to predispose to the formation of gall-stones ; and, in fact, two of

Klinger's chronic "carriers" had after their attack of typhoid suffered from gall-stones. That these chronic "carriers" may become bacilli-free was only proved with any degree of certainty in one case where the bacilli could not be detected after three months from the date of disappearance of fever ; Klinger states that the others, to avoid trouble and inconvenience, probably sent in for examination the stools of other persons. Particularly in those cases where the bacilli were being excreted for 11 months, was this form of deception likely, for one such woman was persuaded on the occurrence of a fresh case of typhoid in her house to come to hospital for a short stay and in her fæces typhoid bacilli were found almost in pure culture.

Theoretically, as with cholera, diphtheria and cerebro-spinal meningitis, typhoid bacilli from these "carriers" may infect other individuals and cause typhoid, but experimentally this cannot be proved, for although cultures from "carriers" have been shown relatively virulent for guinea-pigs this does not demonstrate their pathogenicity for man. We have, therefore, to fall back on epidemiological observations and these, although few and from their nature affording no absolute proof, are yet fairly convincing. Kayser,[6] in relating his own experience in Strassburg of the danger which these "carriers" are to the public, gives a number of instances of which the following is an example. At a certain bakery a journeyman fell ill of typhoid and died in hospital ; and at the same time the fact was elicited that for the past few years every apprentice in this bakery shortly after his arrival suffered from intestinal catarrh and indigestion. A few days after the journeyman, suffering from typhoid, had been taken to hospital, the proprietrix of the bakery, who had had typhoid ten years before, was found by the bacteriological institute to be a " carrier. " About two months afterwards, this woman temporarily refused to send her stools for examination and the stools subsequently sent in, for three consecutive examinations at weekly intervals, were found free from bacilli. Under the circumstances Kayser suspected that in order to escape the inconvenience of being under bacteriological observation this woman had sent for examination the stools of another person. The following month, however, her tenant, who used the same privy, got a severe attack of typhoid and, although brought at once to hospital, died. As Kayser was unable to discover any source of infection other than this chronic "carrier," and as it was of great importance to obtain for examination material, the source of which was beyond question or doubt, the proprietrix under pressure of the police authorities was brought to hospital and her fæces, obtained from the hospital, were invariably found extremely rich in typhoid bacilli. Her blood serum, in a dilution of 1 in 100, gave a prompt positive agglutination reaction with *B. typhosus.* There was no fever ; and no bacilli, even by using " enrichment " methods, could be isolated from her blood. There were thus, within the space of one year, two fatal cases of typhoid in the household of this chronic "carrier." It is possible, therefore, that these chronic carriers, especially if they are unaware of the potential danger of their dejecta and are not cleanly in their habits, may disseminate the disease ; and in outbreaks where a definite cause cannot be found and where in spite of immediate isolation and thorough disinfection fresh cases still occur, we should bear in mind the possible presence of undetected chronic carriers.

The search for carriers has been chiefly amongst those who have been in contact with typhoid patients. Minelli,[7] however, made an attempt to discover them away from the vicinity of the sick. He examined the stools of

250 inmates of Strassburg prison and found one carrier, who, from all the evidence that could be gathered, had not been in contact with any typhoid patient. The blood of this carrier in a dilution of 1 in 1000 agglutinated the laboratory strain of *B. typhosus*. According to Forster and Kayser, the serum of those who merely harbour the bacillus as a saprophyte in the intestine without having actually suffered from typhoid ("acute carriers") has generally no agglutinative power; so this inmate of Strassburg prison had probably had a slight attack of typhoid and was a chronic carrier.

To add to our difficulties in the campaign against this disease, we are unable, in the present state of our knowledge, to free from bacilli those carriers that have been detected; and measures suggested to minimise the risks of infection, *e.g.*, thorough cleaning and disinfection of hands and anus after defæcation, continuous disinfection of dejecta and their proper and careful disposal, are expensive and cause inconvenience to the carrier and are therefore impracticable. In connection with the (at least apparent) natural immunity of acute "carriers," reference may be made to an article by Wassermann and Citron[8] on the local immunity of tissues and the place of origin of typhoid immune-bodies. These authors maintain that a local immunity of tissues may exist spontaneously (*e.g.*, the intestinal mucous membrane of acute "carriers") as well as after having undergone an attack of typhoid, the bacillus in such individuals living in the intestine without producing any symptoms and without causing any increase of agglutinating bodies in the blood.

That typhoid, primarily and essentially, is not an intestinal disease but a true septicæmia or bacillæmia is now generally accepted. Peyer's patches and the solitary glands are not the channels of infection; the intestinal lesions are secondary to the bacillæmia in the same way that the rose-spots of the skin are. According to Brien and Kayser,[9] the portal of entry is in the alimentary canal, but above the stomach and possibly in the region of the tonsils. Bacilli can as a rule be demonstrated several days earlier in the blood than in the fæces, and in cases of typhoid without any intestinal lesions they may not be demonstrable in the fæces throughout the illness. It is well known that *B. typhosus* may cause diseases other than that clinically and pathologically regarded as classical typhoid fever. Mouisset, Mouriquand and Thévenot[10] have recently described a case of Eberthian septicæmia, which clinically in no respect resembled typhoid fever; there were no abdominal signs, but there were many of the characters of pernicious anæmia; no clue to the nature of the illness was found till *B. typhosus* was demonstrated in the blood of the patient. A case of uncomplicated primary typhoid appendicitis has been reported by Stokes and Amick.[11] Inasmuch as the temperature dropped to normal 48 hours after removal of the appendix, from which *B. typhosus* was isolated, it is assumed that infection was limited to that organ. The patient had had typhoid 13 years before and the authors express the opinion that possibly *B. typhosus* had remained in the gall-bladder during those 13 years; but there is accumulating evidence that the appendix itself and the cæcum (Lentz[12]) as well as the gall-bladder may be places of abode of *B. typhosus* long after convalescence from typhoid. Napier[12] has recorded a case of typhoid nephritis (nephrotyphoid) with the symptoms of an ordinary acute nephritis and few or none of typhoid. *B. typhosus* was found in the urine for about four months after the onset of the illness and the diagnosis of typhoid was corroborated by the fact that two of his attendants were infected, the disease in their case running a normal course.

The lesions of typhoid fever, according to Hirsh[1] in his paper on the bacteriology of the blood in typhoid fever, have no intimate relation to *B. typhosus*: the toxins cause proliferation of endothelial cells (phagocytes) and so give rise to the swollen spleen and lymphoid tissue, the necrosis of Peyer's patches being due to thrombosis. In support of this he states that in *post-mortem* examinations of a fœtus no bacilli are, as a rule, found, yet Widal's reaction is usually present, associated with an enlarged spleen and swollen lymphoid tissue and Peyer's patches. The earliest attempts to isolate *B. typhosus* from the blood were attended with little success, a fact attributed by Stern[2] to the germicidal action of the blood. Subsequently, however, with Castellani's[3] "dilution method" (1899) the results were more successful. Following the dilution method, Hirsh took from the arm, aseptically by means of a hypodermic syringe, 10. c.c. of blood and divided this amount among four flasks, each containing 300 c.c. of bouillon which he then incubated at 37° C., for 24 to 36 hours. In a study of the blood in 100 cases of typhoid fever, *B. typhosus* was found by him in 78 per cent. of all cases, in the first week of illness in 75 per cent. of cases, in the second week in 86 per cent., in the third week in 79 per cent. and in the fourth week in 14 per cent. ; in relapses (cases of reinfection of the blood), the bacillus was isolated in 100 per cent. of cases from the second to the fifth day. The earliest date on which a positive result was obtained by Hirsh was the fourth day of the disease and after the twenty-first day of illness the bacillus was usually absent from the blood. Coleman and Buxton[4] collected the results of the bacteriological examination of the blood in 604 cases of typhoid fever and state that the bacillus was demonstrated in the blood at some stage of the disease in 75 per cent. of cases. The bacilli were found in the blood at the different stages of the disease as follows : in the first week in 93 per cent. of cases (the earliest positive result being obtained on the third day), in the second week in 76 per cent , in the third week in 56 per cent. and in the fourth week in 32 per cent. Of 21 cases of relapses the bacillus was isolated in 18. Schottmüller[5] has been successful in demonstrating the bacillus as early as the second day of illness. Conradi[6] conceiving the idea that bactericidal bodies are the result of coagulation of the blood, tried to isolate the typhoid bacillus by keeping the blood fluid by means of ox-bile. Blood from the ear was received direct into ox-bile-glycerine-peptone and incubated at 37° C. for 10 to 16 hours, when some of this bile culture was poured on to Drigalski-Conradi plates. In this way he examined the blood of 28 patients and succeeded in cultivating the typhoid bacillus in 22 cases and the paratyphoid bacillus in six. Some of these successful results were obtained in the stage of convalescence after fever had disappeared and some in cases of the ambulant type of typhoid (*typhus levis*). Frequently the bacteriological examination of the blood clinched the diagnosis before other tests were of any avail. Sodium oxalate is recommended by Eppenstein and Korte[7] as a substitute for bile for preventing coagulation of the blood and acting as an "enrichment" method for *B. typhosus*. Müller and Gräf,[8] however, maintain that clotting does not destroy the bacillus in the blood and state that by grinding a blood clot on litmus-lactose-agar they were able to isolate *B. typhosus* in all cases of the disease, using only ½ to 1 c.c. of blood. On the other hand, Kayser,[9] availing himself of the fact that *B. typhosus* grows well in bile, makes use of a simple bile tube as an "enrichment" method in the bacteriological examination of the blood for typhoid and paratyphoid bacilli. The patient's blood mixed

The bacilli in blood, urine, fæces, bile, sputum and cerebro-spinal fluid.

with ox-bile is incubated at 37°C. for 14 to 20 hours and typhoid bacilli, if present, having meanwhile increased in numbers are easily recognised on Endo's or Drigalski's medium. In 21 cases of typhoid thus examined, Kayser invariably found the bacillus in the blood in the first week of illness. Kayser's results, in all their details, have been confirmed by Fornet.[10] Pöppelmann[11] recommends stained blood-film preparations, as in the case of malaria, for the provisional diagnosis of typhoid. According to him, typhoid bacilli are frequently found in great numbers in the blood and if, in a case clinically "typhoid," the characteristic short rods be found (using an oil immersion lens) the disease is either typhoid (including paratyphoid) or coli-bacillosis.

Taking a general survey of the above results obtained by a bacteriological examination of the blood, it must be acknowledged that for the early, rapid and comparatively easy diagnosis of typhoid, the method of blood-cultures bids fair to take the first place. Further advantages of this method are its special value in obscure cases and in differentiating typhoid from para-typhoid, also the fact that, *B. typhosus* having been isolated, the diagnosis is assured. In his account of anti-typhoid measures in the district around Strassburg, Olbrich[12] states that of 344 bacteriological examinations undertaken 60 were of blood, 163 of fæces and 121 of urine, with positive results in 33, 19 and 6 cases, respectively, giving the relative success (expressed in percentages) of the examination of blood, fæces and urine as 55, 12 and 5, respectively. These figures, however, must be taken with some reserve as the three series of figures may not be strictly comparable ; for instance, many of the samples of fæces and urine possibly came from contacts, while specimens of blood were probably in all cases derived from persons actually sick.

B. typhosus occurs in the urine in from 25 to 30 per cent. only of ordinary cases of typhoid, and is rarely found before the end of the second week, so the bacteriological examination of urine offers but little aid to the diagnosis of this disease. The possible danger, however, of infection from this source is to be borne in mind. Young[13] has given an account of a case where the bacillus was shown to persist in the urine for seven years after an attack of typhoid fever.

With regard to the examination of fæces, Hammerschmidt,[14] trying to simplify and improve on former methods, recommended the addition of cresol to the media. *B. coli* and other bacteria having been found more sensitive than *B. typhosus* to the action of cresol, it was hoped that media containing this substance would give us an " enrichment " method, for the typhoid bacillus comparable to that which we have for the isolation of the cholera vibrio. The stool of the typhoid patient was received into bouillon containing 2·5 per cent. of liquor cresoli saponatus and this after 24 hours was planted out on Drigalski-agar. Cresol, however, does not seem to have come up to expectations and we are still dependent on the older methods. The relative value of these methods has recently been thoroughly investigated by Klinger[15] who regards Endo's medium as superior to Drigalski-Conradi's for the following reasons : (1) in Endo's the acid-forming colonies produce an interstitial colouration and typhoid colonies could be picked out on an Endo's plate which was quite red, (2) fuchsin inhibits the growth of fæcal bacilli more than litmus does, even when the latter is assisted by crystal-violet, (3) a greater differentiation by colour is possible on Endo's, these finer differences in colour being more marked if the medium be not quite colourless but slightly pinkish, (4) the results, taken collectively, were 33 per cent. better with Endo's than with Drigalski-Conradi's, (5) fuchsin-agar is cheaper and more easily made. Klinger, however, recommends the use of two series of plates—one series of fuchsin-agar and one of litmus-agar, for, although Endo's as a rule gave quicker and more certain results, on two occasions typhoid colonies were found on the litmus series when the fuchsin series was blank. The enrichment methods of Lentz and Tietz (malachite-green-agar) and of Hoffmann and Ficker (caffeine and crystal-violet) were next investigated by Klinger. It was found that a preliminary culture on the former for 24 hours or in the latter for 13 hours greatly increased the chances of success. Of the two enrichment methods, that of Lentz and Tietz is considered by Klinger the better, as it requires less time and is attended with greater success; the technique of Hoffmann-Ficker's. is relatively more difficult. Finally, all four methods were used simultaneously in 43 cases with the following results : litmus gave a positive result in 13 cases (30 per cent.), fuchsin in 16 (37 per cent.), preliminary caffeine in 20 (46·5 per cent.) and preliminary malachite-green in 26 (60·5 per cent.), a positive result being given exclusively by malachite-green in nine cases and exclusively by caffeine in four. By the simultaneous use of all four methods the typhoid bacillus was demonstrated in the fæces in 80 per cent. of cases actually suffering from the disease. Reischauer[16] comes to much the same conclusions as Klinger regarding the relative values of these methods. He states that Drigalski's and Endo's are quicker and cheaper but not so successful if enrichment methods are omitted, and that therefore during convalescence when the bacilli are present in fewer numbers enrichment methods should be used. For enrichment, however, he prefers the addition of caffeine instead of malachite-green to the agar medium, because malachite-green hinders somewhat the growth of *B. typhosus*. Having pointed out that an objection to all these methods is that only a small amount of fæces can be brought under examination, he suggests that the stools be flooded with water and subsequently treated by the Hoffmann-Ficker method in the same way as in isolating the bacillus from water. Nowack, [17] in discussing the limits of applicability of malachite-green-agar in demonstrating the typhoid bacillus in fæces, expresses the opinion that this method is suitable in cases of very rich bacterial mixtures provided the typhoid bacilli are not present in small numbers; where typhoid bacilli are few and perhaps injured, even although the accompanying bacteria be present in small numbers, malachite-green is unsuitable. The hopes which had been entertained of the method of fæcal cultures for the diagnosis of typhoid have not apparently been realised. The method is laborious and yields but poor results compared with those of blood cultures.

The presence of typhoid bacilli in the gall-bladder, during convalescence from typhoid and subsequently, has of late received considerable attention. Lejars[18] has published a case of typhoid cholecystitis which occurred during convalescence from typhoid. Symptoms came on suddenly and were so urgent as to demand laparotomy. The gall-bladder, containing five gall-stones, was found swollen and inflamed. The presence of gall-stones had not been suspected till *B. typhosus* set up inflammation. From the bile of a woman suffering from gall-stones Kisskalt [19] isolated *B. typhosus* in pure culture: in this case the bacillus could not be found in the stools, nor was there any history of her ever having had typhoid, although in the previous year cases of typhoid had occurred at her home. Forster and Kayser [20] investigated the bacterial contents of the gall-bladder in 148 corpses. Out of eight cases who had died of typhoid, *B. typhosus* was found in seven ; of the others, two gave the bacillus and yet had no history of ever having had typhoid. A closer examination showed that in cases of typhoid infection there is always an inflammatory condition of the bile ducts and that the gall-bladder is one source of the typhoid bacilli that occur in fæces. The same two investigators, about eight weeks after the intravenous injection of typhoid or paratyphoid bacilli into rabbits, found the bacilli in the bile and contents of the upper part of the small intestine. A further interesting observation was that the mucous membrane of the gall-bladder frequently contained bacilli when fluid bile showed itself sterile. They lay stress on the relationship of typhoid to gall-stones and the fact that most chronic carriers are women and conclude that the gall-bladder is an important vegetating place of *B. typhosus*—which passing from the blood through the liver into the bile takes up its abode in this organ. These results have in a general way been confirmed by Doerr[20] and also by Müller[21] who found *B. typhosus* in fatal cases of cholecystitis and cholangitis unaccompanied by any typhoid lesions in the intestine. In the case reported by Müller, the patient was jaundiced and his blood serum gave Widal's reaction.

B. typhosus is sometimes found in sputum in cases of typhoid fever ; if the bacillus produces any specific lesion in the lung, that lesion (pneumotyphoid) usually takes the form of a broncho-pneumonia (Robinson[22]) : lobar pneumonia, occurring in typhoid fever, is caused by the pneumococcus. The sputum should be disinfected in all cases of typhoid fever.

Two instances have recently been reported by Schütze[23] in which the typhoid bacillus had been found in cerebro-spinal fluid withdrawn by lumbar puncture. In both cases cephalalgia and rigidity of the neck, were in evidence, and in one case neither clinical symptoms nor bacteriological examinations had fixed a definite diagnosis of typhoid until the bacillus had been found in the cerebro-spinal fluid. Such cases are liable to be wrongly diagnosed.

Numerous examples of water-borne typhoid epidemics, showing their characteristic features, have been reported during the past year from European countries and in not a few was the proof of the mode of spread supported by both epidemiological and bacteriological investigations. Dr. Reece[24] in his report, to the Local Government Board, on the typhoid epidemic in the city of Lincoln, 1904-5, states that the outbreak " owed its origin to the specific pollution of the public water-supply." This conclusion was based chiefly on the following considerations : the absence of evidence implicating the milk-supply, the conditions of drainage, etc., the wide separation locally of the earliest cases, the sudden explosive outburst of the disease followed by an abrupt decline, the diffusion of cases over the whole area served by the city water-supply, the obvious pollution of the river and streams from which water was derived, the inadequacy and irregularity of the filtering arrangements especially at the time of the widespread infection and, lastly, the discovery of *B. typhosus* in an intake to the waterworks. The bacillus was isolated by Dr. Klein and subjected to 15 morphological, chemical and biological tests for identification. Another interesting report to the Local Government Board is that by Copeman[25] on an outbreak of typhoid at Fulbourn asylum, near Cambridge. A vast basin of chalk, in places exhibiting definite fissures, underlies the district in which this asylum is situated. In 1905 an epidemic of typhoid occurred in the asylum and previous to this outbreak it had

Infection by water. Bacillary diagnosis.

been shown that the drainage was defective and leaking in many places. A bacteriological examination of the water of the well of the asylum, however, failed to support the suspicions entertained regarding it as the source of infection. Not satisfied with these results, Copeman sunk a pit to the depth of $6\frac{1}{2}$ feet near the sewage farm of the asylum. Into this pit water was poured till it was half full, a solution of fluorescin and sodium hydrate then added and finally large volumes of water run in for an hour, the water disappearing with great rapidity, as the soil and subsoil at this spot was very pervious to water. (Fluorescin in the presence of an alkali even in a dilution of 1 in 100,000,000 gives a distinct green colour.) The subsoil water was then tested for the presence of the green colour at different places, including the asylum's well, the Cambridge water-works' well, etc. A series of investigations in this manner was made with the result that the green colour was detected in nearly all the places chosen for examination. These fluorescin experiments left no doubt that liquid sewage can be conveyed from the sewage farm through chalk fissures in various directions to distances of at least half a mile ; but of course it does not follow that bacteria would be conveyed to such a distance. Further, these experiments pointed to the existence of currents, in the underground water, subject to modification as to direction and rapidity by circumstances at present unknown—the green colour having been found, for example, in the wells of both the asylum and the Cambridge water-works, but the rate and degree of transmission to these varied widely in the series of experiments. The report contains abundant evidence to show that the water of the asylum's well and that of the Cambridge water-works' wells, although nearly a mile apart, are directly connected with one another and are derived from a " continuous mass of underground water," and that diffusion of contaminating materials throughout this mass may be extensive. With regard to the outbreak in the Rhondda' urban district, the proof is entirely epidemiological : opportunities for the fæcal pollution of the water-supply were obvious and the general behaviour of the outbreak was consistent with the water-borne view. In September, 1905, an epidemic of typhoid, attributed to the dispersal of sewage water over the fields in a district where the soil was mainly limestone, occurred at Pantoise in France.' Fissures were present in the soil and wells were contaminated by soakage. For exhaustive reports on recent water-borne epidemics in Germany, the reader may refer to those by Matthes and Gundlach,' Matthes and Neumann' and Beck and Ohlmüller.' The last mentioned show convincingly how easy it is for wells in places where the soil is very permeable to become infected and they clear up many obscure points in the doubtless water-borne epidemic in Detmold (1904) and in particular the otherwise inexplicable escape of certain quarters from the disease.

Childs' in a comparative study of the Lincoln, Maidstone and Worthing epidemics draws attention to the effect which acceleration of filtration has on the passage of bacteria through filters. He also points out that the most prompt system of notification could not detect the presence of an explosive invasion till two or three weeks after widespread infection had occurred, this interval of time between infection and notification being necessary for incubation and diagnosis. It is after this interval that the explosive outburst of notifications comes. By this time secondary infections (contact) have already begun, so investigations of the cause of an epidemic and the relation of cases (water or contact) to be effective, exhaustive and based on trustworthy data should already be under weigh. It is further pointed out that as the time of the health-officer at this juncture is fully occupied with routine work, a specially trained staff of experts is required to visit the locality and undertake these investigations. An abnormal prevalence of diarrhœa or isolated cases of typhoid have preceded these outbreaks and are regarded as premonitory. It is therefore important to take this warning and make an immediate and exhaustive investigation of such cases to establish their etiology and to determine what preventive measures are necessary. At present, in the absence of such a trained staff of experts, the abnormal prevalence of diarrhœa or isolated cases of typhoid must be regarded as a possible indication of a coming invasion and all channels by which infection may be conveyed must be safeguarded.

The vitality of *B. typhosus* in water and sewage has been investigated by Russell and Fuller,[9] the attempt being made to carry out the experiments under conditions as nearly natural as possible. They made use of three types of sacs (celloidin, parchment and agar), permeable to chlorides, sugar and peptone, and in these sacs they enclosed the bacilli while exposed to the influence of water and sewage bacteria. In the first series of experiments, the bacilli imprisoned in the sacs were exposed in flowing lake water (Mendota) and under these conditions they lived from eight to ten days. When similarly exposed to the direct action of sewage bacteria, the bacilli lived only from three to five days. When, however, the bacilli were exposed merely to the diffusible products of sewage bacteria and not to their direct action they lived almost as long as in the first series of experiments (with flowing lake water). Lastly, in the fourth series of experiments the typhoid bacilli were enclosed with sewage in the sacs and placed in lake water and under these conditions the bacilli were found to live five days, about the same period as when in sewage. It is evident from these experiments that the longevity of *B. typhosus* depends more on actual contact with the sewage microbes than upon contact with by-products capable of diffusion through these membranes and that the duration of life of *B. typhosus* in water is materially affected by the germ contents of its surroundings. Exposed to water highly polluted with saprophytic bacteria, *e.g.*, sewage, the typhoid bacilli cannot survive more than a few days (three to five in the experiment), a period much shorter than that noted in natural unpolluted water. These results, which are said to have been remarkably uniform, strongly support the evidence which is gradually accumulating that *B. typhosus* is a short-lived organism in water under natural conditions. Fehrs[10] reports that it is not bacteria but protozoa that cause the disappearance of typhoid bacilli from water and sewage, flagellata being their natural enemies, and explains the persistence of *B. coli* in water as due to this bacterium being too prolific for protozoa to cope with. Whipple and Mayer[11] state that if a liberal amount of oxygen be present in water, the typhoid bacillus lives longer than under conditions where the amount of oxygen is small, as in summer water in contrast with winter water, in polluted water and in the contents of septic tanks.

To increase the chances of isolating the typhoid bacillus from water, all modern methods aim at bringing as many of the suspicious germs as possible under investigation. Examples of the chief means of gaining this object are—(1) sedimentation or precipitation by chemical agents or by the centrifuge, (2) separation of typhoid bacilli from others by the use of a specific serum, (3) enrichment methods, *e.g.*, caffeine and malachite-green which have little or no deleterious effect on *B. typhosus*, but inhibit the growth and multiplication of other bacteria. To these Drigalski[1] would add taking the sample for examination from the surface of the water ; and judging from the success which he has had with this method, he appears to be justified in his belief that many typhoid bacilli remain near the surface of the water. Instead of mixing all the germs by sedimentation or precipitation, he would separate them by their sinking capacity ; the saprophytic germs being coarse and of high specific gravity tend to fall to the bottom, while the inherent lightness and mobility of typhoid bacilli together with the fact that fermenting fæcal particles tend to float bring them to the surface. The water for the test is therefore taken from the surface and received in sterile cylindrical jars (not too wide) of five to ten litres capacity. The suspected water is kept in the jars at room temperature and in diffuse light for one or two days, when with a sterile pipette water taken from the surface is brought on to Drigalski plates. The plates are incubated at 37° for 18 to 20 hours, when suspicious colonies are subjected to the ordinary tests for identification. In the search for typhoid bacilli in suspected wells, particularly if some weeks have passed since the well may have been polluted, the sediment or mud should also be examined. Experimenting with typhoid bacilli in aquarium water, Hoffmann[3] made the following observations : for three days after pollution of the water, B. typhosus could be demonstrated by direct planting of the water on Drigalski's plates ; for four weeks after pollution, the bacillus could be detected by the aid of Hoffmann-Ficker's enrichment method ; two months after pollution, when the water failed to give a positive result, the bacillus could still be easily found in the bottom mud.

For the isolation of B. typhosus from water the preliminary precipitation method is advocated by Willson[4] and Müller[5]. As regards the two old rival media—Drigalski's litmus-agar and Endo's fuchsin-agar, the praises of the latter have been sounded by Marschall[6] and Herford[7]. Very recently, however, the use of media (agar, gelatine and broth) to which malachite-green has been added has been strongly recommended by Loeffler[8] not only for the isolation but also for the differential diagnosis of typhoid bacilli and nearly allied organisms.

As regards the agglutination test, Sadler[9] reports that the optimum temperature, whether with dead or living cultures, is 55°C. and that at this temperature agglutination is quicker and more intense than at 37°. Of some importance not only with regard to Widal's test but also for identification is the observation of Sehrwald[10] that the rapidity of agglutination of a strain of B. typhosus is increased by growth on a potato which has been sterilised by steam; the quickening of agglutination is lost if grown again on other media. Hirschbruch[11] investigating typhoid bacilli whose agglutinability had been experimentally lowered, e.g., physically by growth at high or low temperatures, chemically by additions to culture media and biologically by symbiotic growth with yeast or B. coli or keeping cultures till they get old, came to the conclusion that variations in agglutinability have no specific significance. With regard to bactericidal sera (Pfeiffer's test), Besserer and Jaffé[12] (also Friedberger and Moreschi[13]) have met with strains of B. typhosus which do not react typically but resist the action of the serum. It seems, therefore, that failure to respond to agglutination and bactericidal tests is no proof that an organism is not the one it is suspected to be : it may be an atypical strain of that species. Those interested in morphological variations of B. typhosus may consult Pèju and Rajat's[14] paper on experimental polymorphism of the typhoid bacillus under the influence of certain salts.

As instances of the successful use of these methods of isolation and identification may be noted that of Terburgh[15] who by Hoffmann-Ficker's "enrichment" method with subsequent plating on Drigalski's medium isolated the bacillus from river water and that of Ströszner[16] who following the same methods isolated the bacillus from the water of a house-well about four or five weeks after the last case of typhoid had occurred in the house : the bacilli were identified by Widal's and Pfeiffer's reactions.

Testing the value of a bacteriological examination of tidal mud as an index of pollution of rivers, Savage[17] found that the typhoid bacillus disappeared from sterilised river-mud in 50 to 60 days, but if other bacteria were introduced as with ebb and flow, the bacillus could be demonstrated for only 35 days. Major MacNaught,[18] R.A.M.C., gives the duration of life of $B.$ $coli$ in water and sewage as follows : in sterile water for several months, in non-sterile water from 15 days to several months depending on the degree of purity of the water, in non-sterile sewage for a few weeks, in sterile sewage for some months; the addition of urine to the water favoured the vitality of $B.$ $coli$. With reference to the significance of $B.$ $coli$ in potable waters, Vincent[19] states that the number of these bacilli is proportionate to the degree of pollution and that if 10 to 50 or more colonies be found per $c.$ $c.$ the water is strongly infected. An "enrichment" method consisting in the use of an acid ($i.e.$, non-neutralised) bouillon for the demonstration of $B.$ $coli$ in water has been brought forward by Venema,[20] who by bringing river-water into acid bouillon and incubating at $37°$ for 16 to 24 hours isolated $B.$ $coli$ in 85 per cent. of the samples. The chief difficulty met with was an excessive growth of $B.$ $proteus$. The cases, however, are too few to enable a definite opinion to be given on the value of this method for detecting excretal contamination ; and, after all, with $B.$ $coli$ present in large numbers (see the remarks below on shell-fish) a topographical survey is usually necessary before delivering judgment.

The incriminating evidence against milk and other foods of spreading typhoid infection has hitherto almost invariably been indirect—by a process of excluding all other possible channels. Konradi,[1] during a typhoid outbreak in the past year, has succeeded in isolating the typhoid bacillus (identified by the usual morphological and biological tests) from suspected milk. Although many epidemics are known to have arisen in this way on only one previous occasion (in Chicago) has the bacillus been demonstrated in milk. In Konradi's case it was found that at the dairy-farm to which the infected milk had been traced, the farmer's son was suffering from a slight attack of typhoid which did not prevent him from milking the cows and so the bacilli from his hand had entered the milk. The same author relates an epidemic, in the Hungarian town of Arad, where the virus was conveyed by whipped cream, the better classes only, and especially women and children, being affected. It was shown that from a house, in which there was a person suffering from typhoid, some whipped cream had been sent to certain confectioners and that most of those who sickened had partaken of this luxury. In a recent epidemic at Melbourne[2] it was found on enquiry that the milk-supply of many of the patients came from one dairy and that at this dairy a child was ill ; this illness proved to be typhoid and further investigations showed that

Infection by milk and other foods.

the original source of infection was at the country-farm from which the milk came. Kayser[3] has traced the cause of some small epidemics to contamination of the milk by bacilli-carriers and advises a search for carriers to be made in dairies in cases where the source of infection is traced to milk. Perhaps the commonest way by which the typhoid bacillus gets into milk is through the use of contaminated water for cleaning milk utensils or for adulteration purposes. Park,[4] when dealing with the milk-supply of the city of New York, called attention to the great danger arising from washing milk-vessels with water obtained from the vicinity of leaky privies. Infection by milk can be effectually prevented by pasteurisation.

The following conclusions have been drawn by Klein[6] from his experiments and observations on the vitality of *B. typhosus* and other sewage microbes in oysters and other shell-fish : infected oysters can clean themselves in nine to twelve days if placed in clean water and if this water be frequently changed (in the dry state they take much longer) ; mussels like oysters can destroy *B. typhosus* ; in the body of the cockle actual growth of the bacillus takes place and to destroy the bacillus steaming under pressure for three or five minutes is required ; oysters taken from clean layings many miles from shore contain large numbers of *B. coli*. The last mentioned fact may be explained by the observation of Houston and Eyre that the faecal matter of birds and fish contains *B. coli*.

Although it is not proved that the typhoid bacillus cannot remain for years in earth and then enter the human body and cause enteric fever, the success of the German methods,

Infection by contact.

which have as their sole object to destroy the specific virus as it leaves man, strongly supports the assumption underlying this working hypothesis—namely, that there is no other source of typhoid infection except man. If, as the exponents of these methods maintain, infection arose in other ways, *e.g.*, from soil to man, their efforts would have been futile. The specific virus is present chiefly in the faeces and urine of infected persons, but also to some extent in the vomit, sputum, saliva, nasal mucus, and sweat—practically in all secretions and excretions. Clothes, bedding, furniture, food, utensils, etc., soiled by these, can easily carry infection to others. Contact infection may be (1) immediate or direct, from the body of the patient to that of attendants, or (2) mediate—through the medium of articles soiled by patients, *e.g.*, clothes, utensils, accoutrements, latrines, etc. In the latter case it is obvious that for contact infection to occur it is not necessary to live in the same room or barrack with or even to have seen or met a typhoid sufferer. Isolated cases separated by intervals of days or weeks, as opposed to the explosive outbreak from a common source of infection such as contaminated water or milk-supplies, are characteristic of infection by contact, where infection occurs in the series of cases at different times and in different ways. Contact infection, however, plays a part in all epidemics : in water-borne epidemics, after the explosive outbreak the contaminated water-supply is cut off, and thereafter all fresh cases of infection are due to contact. Schian[1] who was in medical charge of the German troops in South-West Africa reports that segregation on the march or in camp was often impossible and typhoid infection was spread for the most part by contact ; that disinfecting measures and particular attention to field latrines were most useful in combating the disease. In connection with typhoid in the United States' volunteer encampments (1898), Childs' states that personal infection, direct or indirect, especially from unrecognised and abortive cases was the chief factor in the spread of the disease. Similarly in the official report[2] on the origin and spread of typhoid in the United States' camps in the Spanish war of 1898, it is held that water was not an important factor, that infected material was transported on the person, clothing and tents and that infection by contact was all the more frequent as only about half the cases were correctly diagnosed by the army surgeons. For this reason it is strongly urged that all excreta should be disinfected and carted away to some distance from camps. In civil life, we find the same state of things. Millard[4] gives it as his opinion that water has not been an important agent in the spread of typhoid in New South Wales, that infection from a previous case either directly in the sick-room or indirectly from excreta being imperfectly disinfected or disposed of has been a factor of great importance. Reports from one of the medical societies of Göttingen[5] contain a description of an epidemic in a sanatorium in the neighbourhood. This epidemic was spread by contact and its intensity was aggravated by the occurrence of abortive and atypical cases; persons with the disease undiagnosed, owing to no precautions having been taken, acted as unsuspected sources of infection and created a certain degree of uncertainty in carrying out preventive and suppressive measures.

Leaving aside occasional abortive and incorrectly diagnosed cases of infection, there are two classes of the public, viz., children and typhoid bacilli-carriers, that owing to certain peculiarities in their reaction to the typhoid bacillus are specially dangerous to the community. Children suffer from this disease usually in a mild form, incorrect diagnoses being very frequent, and both from their habits and from their being together in school or at play for a great part of the day are particularly liable to become infected themselves and to infect others: the incidence of the disease is greater among children than among adults. The greater liability of infection from children than from adults has been pointed out by Baginsky[6] who has studied the course of the disease, as it affects children between one and fourteen years of age, in 83 cases. Frequently the disease, appearing as a feverish dyspepsia, was not recognized. This fact together with the greater liability of children to soil their surroundings make it obvious that children are very apt to spread infection by contact. In this connection the account given by Conradi[7] of the part played by children in endemic typhoid in Metz (Lothringen) is particularly interesting. In Metz during the past 25 years there have died from typhoid 45 natives (i.e., residents from birth) and 382 aliens or in-comers, and yet, up to 15 years of age, natives and aliens died in about equal numbers. As regards the age incidence amongst residents from birth, over two-thirds of all the cases of typhoid were among children under 15 years of age. Conradi states that the great incidence of typhoid among children is chiefly responsible for the dissemination of the disease amongst the inhabitants of Metz and that on this latent contact epidemic amongst children (the disease as it affects them often not calling for medical aid) rests essentially the endemicity of typhoid in the city. Further, this contact epidemic amongst the children of Metz going on uninterruptedly from generation to generation explains (1) the adult native's immunity, which is acquired in youth and not inherited, and (2) the high incidence among the newly arrived non-immune alien adults who contract the severe adult type of the disease, the alien children with their native companions being equally liable to the disease in its mild form. It is obvious, therefore, that whenever the floating alien population is increased in such an endemic focus, e.g., for public works or military operations formidable epidemics are almost certain to occur. Wright and Archibald[8] have reported from Glasgow some instances of typhoid in children under 10 years of age, where personal infection seems to have taken place during play. For a detailed description of 63 cases of typhoid fever in children the reader is referred to the original paper by Pater and Halbron[9]. The peculiarities which make carriers specially dangerous to the public are the difficulty of detecting them, the length of time during which unknown to themselves or others they may be disseminating a virulent specific virus, our inability in the present state of our knowledge to free them from the specific virus and the practical difficulty, almost impossibility, of getting such carriers as are detected to remain under surveillance and bacteriological control and to carry out continuous and efficient disinfectant measures in order to prevent infection being conveyed to others. From the researches of Klinger,[10] the proportion of typhoid fever patients that may become chronic carriers is 1·7 per cent., but according to Lentz[11] the proportion may be as high as 4 per cent. : reference has already been made to the instances brought forward by Kayser[12] of the possible danger which these carriers may be to the public. Finally, before leaving this subject, attention may be drawn to a suggestion made by Friedel[13] that the existence of " typhoid houses " may be largely if not entirely due to the presence in them of the carrier, rather than on the presence in dust or crevices of typhoid bacilli, and consequently that in such houses a systematic examination of fæces and urine is a desideratum.

Statements regarding the duration of life of *B. typhosus* outside the human body are most conflicting, some maintaining, that

Soil. Dust. Flies.

the bacillus dies after a short period, others that it survives indefinitely. The latter extreme view has received no further support since the publication by Clauditz[1] of his experiments (referred to in this report for 1904) demonstrating that by gradually accustoming the *B. typhosus* to earth by growing it in bouillon with gradually increasing amounts of earth (*e.g.*, for the growth of the first strain or generation a small quantity of earth being added to the bouillon and for the next strain a somewhat greater amount of earth, and so on) the bacillus could survive for successively longer periods when placed in soil. Investigating the behaviour of *B. typhosus* which he had sown in sterile (1) red river-sand, (2) sifted vegetable soil, and (3) sifted rubbish, Rullmann[2] found after a period of six months numerous living bacilli present in all three substances, and after 18 months none in the sand, a few only in the vegetable soil but many in the rubbish and the only biological change produced by the sojourn in these varieties of earth was a diminution in agglutinability. Almquist[3] reports that the typhoid bacillus grows well in moist sterilised manured earth or pure dung and that its virulence remained undiminished for several weeks. The researches of de Franceschi[4] on the influence of soil on the virulence of *B. typhosus* may be summarised as follows : in dry earth there is a distinct loss of virulence in 24 hours ; in earth moistened with water virulence is unaffected in 24 hours ; in earth made damp with bouillon or fæces, virulence is increased after 24 hours and still more increased by further passages through this earth ; but, in all cases, after a considerable sojourn in earth, virulence declines. Although under exceptional conditions *B. typhosus* may for a time live, multiply and even gain in virulence, evidence is steadily accumulating that in soil as in water after a comparatively short space of time—especially when these conditions return to normal, virulence declines and the bacillus dies.

The typhoid bacillus is said by Heim[5] to withstand drying for at least 213 days, so this possible method of infection must not be forgotten although infection in this way is probably much less frequent than supposed. In the official report[6] on the spread of typhoid in the United States' camps in the Spanish war of 1898, the opinion is expressed that dissemination by dust was probable to some extent, the camping-grounds being grossly polluted, and that " flies undoubtedly served as carriers of infection." Similarly Childs' from a study of the typhoid epidemics in the United States' volunteer encampments (1898) comes to the conclusion that there was strong presumptive evidence to show that infection was spread to a considerable extent from the latrines by the hosts of flies which swarmed in many of the camps. Dr. Nash,' health-officer of Southend-on-Sea, has brought forward some very striking evidence in support of the view that flies play a very important part in carrying the infective agent of epidemic diarrhœa and typhoid ; he recommends the adoption of measures to prevent the breeding of flies and the provision of a destructor for the disposal of refuse. As regards India (see below), flies, although they do not appear to be the most significant factor in the spread of this disease in the majority of geographical groups, no doubt contribute their quota of cases of infection ; and the success which has attended Dr. Nash's measures in his own district should lead to the adop-tion of similar measures elsewhere.

The first clear diagnosis of a case of paratyphoid fever, that is, of a

Paratyphoid and meat-poisoning. bacterial septicæmia clinically like typhoid but associated with a bacillus quite distinct from Eberth's, was that of Gwyn in 1898, where a "paracolon" bacillus was culti-vated from the blood of a patient with all the clinical features of typhoid fever. Cushny (1900) in America and Schottmüller (1900-01) in Germany by a systematic application of the method of blood-culture in cases of "typhoid" showed that paratyphoid was not an uncommon disease in these countries. Dur-ing the past five years cases have been reported from many parts of the world—including India and Japan. Recently Boycott' has published an account of three fully established cases of infection with *B. paratyphosus B*, which were met with in England during 1905 while examining a series of 176 specimens of blood sent in for Widal's test in suspected cases of typhoid. For a detailed description of these cases together with a useful bibliography on paratyphoid infections, the original paper should be consulted. Bacteriologists are by no means in agreement with regard to the systematic grouping of the organisms con-cerned in the production of these fevers ; but the following classification appears to be the one which reflects the most recent views of authorities on this subject : (1) *B. paratyphosus A*, (2) Gärtner's meat-poisoning bacillus, (3) Hog-cholera group, including *B. typhi murium*, etc., (4) *B. paratyphosus B*. The A type of paratyphoid bacillus is easily distinguished from the other three groups by cultural methods, is more nearly related (especially culturally) to Eberth's bacillus and according to universal experience is much less frequently met with than *B. paratyphosus B*. Gärtner's bacillus is differentiated from *B. paratyphosus B* and the hog-cholera group by the agglutination reaction with sera prepared from known cultures. The differences, however, in agglutinability between *B. para-typhosus B* and the hog-cholera group are so slender that at times it is impossible to distinguish them, yet many authorities who merge them in one group acknowledge that there is a certain amount of contrast between the two. Smidt' is of the opinion that hog-cholera, mouse-typhoid and *B paratyphosus B* cannot be differentiated, morphologically, culturally, experimentally or by agglutination, and places them together under the hog-cholera group. Bock' has demonstrated the close relationship between hog-cholera, mouse-typhoid and the Kaen type of meat-poisoning bacillus and although he places the B type of paratyphoid bacillus along with these as showing "no essential differences " culturally or in pathogenicity for animals, he admits that there is a certain amount of "contrast " and that the relationship between the first three mentioned is much more intimate than between these three and *B. paratyphosus B.* Boycott considers the distinction between the B type of paratyphoid bacillus and the hog-cholera group, although slight, is real. In the classification of Böhme' not only is *B. paratyphosus B* but also the Aertryck type of meat-poisoning bacillus and *B. psittacosis* found along with *B. typhimurium* in the hog-cholera group, which is differentiated from Gärtner's bacillus by the immune-serum reactions. In contradiction of Bonhoff's statement, Book also reports that Gärtner's bacillus is easily differentiated from the others by the agglutination test.

In Germany and America it is estimated that about 10 per cent. of cases clinically diagnosed as " typhoid " are in reality paratyphoid. In England, Boycott from the past year's experience puts the proportion at 3 per cent., but admits that this is probably too low. In paratyphoid, intestinal ulceration is more frequently absent and the prognosis is much more favourable, the case-mortality of paratyphoid being 1 to 4 per cent. as compared with 17 per cent. for typhoid (London). It is therefore desirable for the patient's as well as the bacteriologist's satisfaction that an accurate diagnosis should be made. Although for diagnostic purposes a bacteriological examination of the blood is preferable and occasionally absolutely necessary (Kolle, [5] Citron [6] and Brion and Kayser [7]) for differentiation and identification (hence the value of a systematic examination of the blood by blood cultures in cases of " typhoid "), a diagnosis of paratyphoid can generally be made from the agglutination reactions of the serum with known cultures of the various organisms (Boycott). Basing their observations on 40 cases of true typhoid, Grünberg and Rolly [8] report that 70 per cent. of all cases gave the group agglutination reaction with paratyphoid bacilli in a dilution of 1 in 30 or higher and in 33 per cent. the paratyphoid bacilli were agglutinated in a higher dilution than the typhoid bacilli. They therefore maintain that a bacteriological blood examination alone under certain circumstances can show the true nature of an illness. In reply to Grünberg and Rolly, Korte and Steinberg [9] having investigated the group agglutination reaction in 70 cases of true typhoid (for a positive result a dilution limit of 1 in 40 being required) state that in all cases the typhoid bacillus gave the reaction in higher dilution than the paratyphoid bacillus and explain the results of Grünberg and Rolly by (1) their not having estimated the exact agglutination limits, (2) their having used the macroscopic test which is not so exact as the microscopic, especially at the limits of agglutinability and (3) the occurrence of " zones of inhibited agglutination " which are more marked in the macroscopic test. Similarly, Manteufel [10] says that in his experience paratyphoid bacilli seldom gave the group reaction in dilutions over 1 in 50, in all cases higher limits were got with the typhoid bacilli and in no case did agglutination occur with paratyphoid bacilli when absent with typhoid bacilli. Korte and Steinberg observed that occasionally the typhoid bacillus failed to agglutinate with serum in low dilutions (zones of inhibited agglutination), while if the serum were more diluted agglutination appeared ; and the same authors also noted that this phenomenon was materially misleading only in the macroscopic method. According to Falta and Noeggerath [11] the agglutination-preventing bodies (*Hemmungskörper*) which cause these zones of inhibited agglutination, appear generally towards the end of illness and to prevent error it is advisable to use a dense emulsion of bacilli and test the serum to its agglutination limits. Although anomalous results are occasionally obtained, irregularities in agglutination results can generally be attributed to one or other of the following : (1) errors in technique and judgment of results, (2) a tendency to spontaneous agglutination of some strains (therefore use a strain of known agglutinable capacity), (3) the ultimate limits of agglutinability, especially by the microscopic test, of the different organisms not having been attempted, (4) zones of inhibited agglutination, (5) mixed infection. For the diagnosis of mixed infections, Castellani's test of absorption or exhaustion of agglutinins is necessary. Castellani found that an excess of typhoid bacilli would remove not only primary or homologous (*i.e.*, typhoid) but also secondary or heterologous (*e.g.*, coli) agglutinins from the serum of a rabbit inoculated with typhoid bacilli, while from a serum elaborated in response to inoculation with both *B. typhosus* and *B. coli* neither typhoid nor *coli* bacilli alone but only both together—simultaneously or successively, would remove all agglutinins. An instance of mixed infection by typhoid and paratyphoid, established by Widal's, Pfeiffer's and Castellani's tests, has been recorded by Gaehtgens.[11] Ficker's paratyphoid diagnostica are two new bacterial suspensions—substitutes for living bacterial cultures, of *B. paratyphosus A* (Brion-Kayser) and *B. paratyphosus B* (Schottmüller), respectively, specially designed to be of service for the agglutination test to practising physicians. Minelli [13] reports favourably on these proprietary preparations. Stühlinger [14] by killing paratyphoid bacilli with chloroform or better still by aseptic autolysis (in normal saline at 37° for two months) has obtained very good paratyphoid diagnostica.

A case of infection by *B. paratyphosus A* (Brion-Kayser) where the bacillus was isolated from the blood, urine and fæces has been recorded by Kayser [15] : and Brion[16] lays stress on the fact that paratyphoid like typhoid is a bacillæmia rather than an intestinal disease. Meat-poisoning bacilli, which, it will be observed, find a place in almost all groups in the above classification, have been divided by Levy and Fornet[17] into two broad classes—(1) those, like *B. botulinus* (sausage poisoning), that produce nervous symptoms and (2) those that have an accentuated action on the gastro-intestinal tract. They give an account of a family of seven that was infected with *B. paratyphosus B*, the symptoms being very like those of typhoid : the bacillus was isolated from the excreta of all seven and was agglutinated by the serum of the patients in dilutions of 1 in 200 to 1 in 10,000. An instance of sausage-poisoning, caused by *B. paratyphosus B*, has been reported by Krehl.[18] The fever ran a course like that of a moderately severe case of typhoid ; the bacilli were found in great numbers in the stools from the beginning of the illness, and agglutination (1 in 1,000) appeared on the eleventh day of illness and only with *B. paratyphosus B*. Curschmann[19] recognises the gastro-intestinal and nervous forms of meat-poisoning and regards *B. botulinus*, outside man, as a pure saprophyte. Gärtner's bacillus on the other hand generally occurs in the flesh of cattle that have been sick during life. Smidt,[2] Pottevin[20] and Klimenko[21] all draw attention to the significance of sickness among domestic animals at a time when meat-poisoning and typhoid-like illnesses of man occur. Pottevin isolated a bacillus very like Gärtner's from a ham which had infected four members of a family of seven. The blood of the four patients agglutinated this bacillus in dilutions of 1 in 50 to 1 in 500, while tests with the serum of other persons even in a dilution of 1 in 25 were negative.

The bacillus was pathogenic for young cats, producing after ingestion with milk a diarrhœa which persisted for about three weeks. Klimenko isolated *B. paratyphosus B* from a perfectly healthy adult dog and believes that this may throw some light on the occurrence of sporadic cases or even epidemics of paratyphoid fever. The bacillus was pathogenic for mice, rats, guinea-pigs, rabbits and young dogs, causing fever and diarrhœa, and retained its vitality in milk for a long period, being still alive and virulent after one year and four months. ◦

In India, paratyphoid no doubt exists and is possibly not infrequent ; but in no case does the causative organism appear to have been isolated and identified. During the past year, a case which was almost certainly paratyphoid fever has been recorded by Major Morgan, R.A.M.C., at Lebong (see page 44). Accounts of atypical forms of "typhoid" which in all probability were cases of paratyphoid have been described by Lieutenant-Colonel Spencer,[12] I.M.S., (1900), Lieutenant Mackie,[13] I.M.S., and Hospital Assistant Wali Mohammad.[14]

As the general and special measures usually adopted in the preven-
Preventive measures.
tion and suppression of typhoid have been so fully discussed in previous numbers of this report, the following remarks will be confined for the most part to the subject as modified by researches referred to in the preceding paragraphs. It is now generally acknowledged that typhoid is spread by one of the following three ways : (1) contact, (2) water, (3) food (especially milk) and that there is little ground for fearing *B. typhosus* in our surroundings unless it has fallen into water or milk, when we may have mass-infection or epidemics. Water-borne epidemics arise from rivers, water-works or wells being contaminated ; mass-infection by milk is relatively rare and is due usually to the milk-vessels having been washed with contaminated water, less frequently by the hands of the sick (especially ambulant cases), of carriers or of attendants on the sick conveying the bacillus into the milk. To stamp out water-borne and milk epidemics the same measures are required as for the prevention of infection by contact, *viz.*, those which attack the virus as it leaves man. Infection by contact may be (1) immediate from the sick patient—especially if wrongly diagnosed, from the ambulant case or from the healthy bacillus-carrier ; (2) mediate from articles contaminated by them. One frequently hears of intestinal typhoid, pneumotyphoid, nephrotyphoid, etc., but it must be borne in mind that typhoid is not a local infection of the intestine, lungs or kidneys but a bacillæmia and that the bacillus leaves the human body with all secretions and excretions although chiefly in urine and fæces. In fæces the bacilli are usually most numerous during convalescence and in both fæces and urine, although in their physical appearance there may be nothing to suggest infectivity, the bacilli may be excreted for years after convalescence. Other points to be remembered are that infection with *B. typhosus* may result in many atypical forms of illness which are liable to be wrongly diagnosed and have to be cleared up bacteriologically ; that in children typhoid is not an infrequent disease, is usually of a mild type and therefore very often either wrongly diagnosed or left altogether without medical treatment (ambulant). Another interesting, if unusual, source of infection has been described by Dudgeon and Gray,[1] and shows the importance of paying particular attention to bone lesions after typhoid. In this case an abscess appeared in a man's femur two years

after an attack of typhoid and led to a sinus which discharged pus containing typhoid bacilli. The man usually dressed the sinus himself but his wife was in the habit of picking up the soiled dressings and burning them. The wife was infected and died of typhoid. It is obvious that proportionate with the increase in our knowledge of the etiology of typhoid has been the increase in magnitude of the difficulties to be faced in the campaign against the disease.

Prompt diagnosis is essential to success and the value of the method of blood cultures in this connection seems established. The agglutination test, however, is by no means to be laid aside although anomalous results occasionally occur. Even typhoid bacilli isolated from the spleen and blood of patients frequently show abnormalities—culturally and biologically as well as in agglutination, although these irregularities gradually disappear by successive passages through animals. As regards Ficker's diagnosticum, Eichler [1] reports that it remained unaltered during a five months' voyage in the tropics with an average temperature of $30°$—$34°$ C. ; Flatau and Wilke,[2] Meyerhoff,[4] Sadler[5] and Demetrian[6] speak very favourably of the results they obtained by its use and the advantages of being independent of living cultures, incubator and microscope. On the other hand, Güttler[7] and Salter[8] recommend living cultures, stating as their reasons that (1) the dilution limits are always higher with living cultures or, in other words, the reaction with living cultures is more accurate and finer, (2) in true typhoid, paratyphoid bacilli are sometimes agglutinated in higher dilution than Ficker's typhoid diagnosticum and so a wrong diagnosis may be made, (3) agglutination is frequently weaker and slower with dead than with living cultures, *i.e.*, the diagnosticum is less sensitive, and therefore at the beginning of illness living cultures give an earlier reaction, (4) in several abortive cases of typhoid the dead cultures gave much poorer results. To these disadvantages of Ficker's diagnosticum may be added its high price. Those interested in the phenomena of spontaneous agglutination or natural agglomeration and variations in agglutinability according to source and method of culture should consult the original papers by Porges and Prantschoff[9] and Aaser[10]. The use of the bactericidal reaction of blood serum for diagnosis has been investigated by Korte and Steinberg[11], Ulrichs[12] and Elischer and Kentzler.[13] According to Elischer and Kentzler the bactericidal power of typhoid serum is limited to a certain degree of concentration : Ulrichs is of the opinion that Widal's test is superior to the bactericidal test in confirming the diagnosis in suspicious cases of typhoid : Korte and Steinberg state that the agglutination and bactericidal tests are two quite distinct, independent reactions and for diagnosis are of equal value ; that the latter however, as it requires more time and its technique is more complicated, is used only when the former gives a doubtful result ; that the bactericidal reaction can be demonstrated in very high dilution ; that the bactericidal power sinks in convalescence ; and that relapses sometimes occur with the serum highly bactericidal.

If we find ourselves face to face with an outbreak of typhoid our duty is to try and find the source of infection. Having first excluded the possibility of infection having been imported, attention is directed to water and food, which epidemiological evidence can as a rule easily incriminate or exonerate. If water and food are free from suspicion, infection by contact alone remains, but still the exact source of infection may be far to seek. The most natural procedure now is to undertake a critical bacteriological examination of all suspicious cases of sickness in and out of hospital (in search of wrongly diagnosed and ambulant cases), paying particular attention to children ; and, if no success attend such measures, final efforts must be directed to detecting chronic carriers among those who have at some time or another passed through an attack of typhoid. Here it may be well to draw attention to the fact that it is comparatively rare for carriers to excrete the bacillus in their urine especially when, as usually obtains among troops in India, convalescents from typhoid have undergone a thorough course of urotropine : *B. typhosus* is more likely to vegetate in the gall-bladder and vermiform appendix. Consequently, before convalescents from this disease are allowed to rejoin their regiments, their fæces rather than their urine should be examined bacteriologically several times after recovery at weekly intervals to determine whether or not they are capable of carrying infection. It has been shown above that from two to four per cent. of typhoid patients become chronic carriers and that Klinger combining the four best methods of isolating the bacillus from fæces successfully demonstrated the bacillus in only 80 per cent. of cases. The latter fact indicates not only the probability that the proportion of patients that become carriers is under-estimated but also the uncertainty of detecting by the best means at present in our possession *all* those convalescents who, by becoming carriers, remain constant foci of infection. A system of allowing men recovered from typhoid to rejoin their regiments after a course of urotropine is not sufficient to preclude the presence of carriers, who doubtless exist in all large stations in India ;* and latrine disinfection and similar measures by no means safeguard *all* channels and eliminate the danger of infection from this source.

As it is impossible by present methods to detect all sources of infection, for the defence of the susceptible recourse must be had to anti-typhoid inoculation. Vaccines and bacterial extracts have increased in numbers during the past year but most of them are only on trial. Bassenge and Mayer[14] inoculate with a germ-free filtrate of living cultures macerated in distilled water, and report that protection from one injection has lasted six months. Over 5,000 of the German troops going to South-West Africa were inoculated by Kolle and Pfeiffer's method[15], which consists in inoculating large doses, usually between the clavicle and nipple, in increasing amounts two or if possible three times at intervals of eight days. The vaccine is prepared from 24 hours' agar-cultures of a special typhoid strain, the cultures being suspended in equal amount of normal saline and killed by heating at 60° for one hour and then mixed with ·3 per cent. carbolic acid. Morgenroth's[16] report on the results is here shortly summarised. The course of the disease was observed in 424 cases (100 inoculated and 324 non-inoculated). Of the inoculated 30 had been inoculated only once, 52 twice and 18 three times. Those who sickened soon after inoculation had the worst time, the negative phase having not always disappeared in three weeks. Of the non-inoculated 11·1 per cent. died, 25·3 per cent. were severe, 21·3 per cent. of average severity and 42·3 per cent. slight : of the inoculated 4 per cent died (three of the four men who died had been inoculated only once), 10 per cent. were severe, 20 per cent. of average severity and ι6 per cent. slight. Complications occurred in 34·9 per cent. of the non-inoculated and in only 20 per cent. of the inoculated. The medical officers that were in actual charge of the sick, reported that among the inoculated, as compared with non-inoculated, headache was seldom present, the effect on the heart was slight, the fever

---

* From the Central Research Institute, Kasauli, it is reported that two cases of enteric fever at that station were proved to be due to contact with a chronic bacillus carrier.

ran a shorter course and did not reach the same height, and sequelæ were rare. Inoculation is, therefore, especially recommended by Morgenroth for soldiers on service in a country where sanitary measures against typhoid cannot be efficiently carried out. At the same time stress is laid on the importance of not sending the inoculated to an infected locality for at least three weeks after the second inoculation. Major Porter,[17] R.A.M.C., has given the statistics of a regiment sent to this country where only two cases occurred amongst 200 inoculated but 60 among 400 non-inoculated. In its interim report the Anti-typhoid Inoculation Committee[18] states that it is satisfied that " the records which are available up to date furnish proof that the practice of anti-typhoid inoculation in the army has resulted in a substantial reduction in the incidence and death rate from enteric fever among the inoculated ". The statistics available for 1904 show that the incidence of typhoid among the inoculated was reduced to at least half of that among the non-inoculated and a similar reduction was observed in the case-mortality of those who contracted the disease after inoculation compared with that among the non-inoculated.

A case of re-admission for typhoid two months after the end of the first attack, reported by Jürgens,[19] is of interest in connection with the question of immunity after an actual attack of typhoid or anti-typhoid inoculation. In the first attack the typhoid bacillus was isolated from the blood and fæces, and the serum agglutinated *B. typhosus* in a dilution of 1 in 1,000 and was also bactericidal. After re-admission the bacilli were found in the blood, fæces and urine, and the serum agglutinated the bacillus in a dilution of 1 in 200 and, in the second week of the relapse, in a dilution of 1 in 800. This case seems to show that immunity may be absent in spite of the normal formation of agglutinins and bactericidal substances. (A similar instance which occurred in India in 1905 is given in paragraph 11 below.)

Serum therapy has received a considerable amount of attention during the past year. The French were the first to show that it was possible to obtain an anti-endotoxic serum by injecting nothing but typhoid bacilli. Soluble endotoxins (Besredka[20]) are got from the dried dead bacilli by trituration and maceration: they are not secretions of living bacilli but are probably identical with the filtrates of cultures of living bacilli in bouillon. An anti-serum obtained by the injection of dead or living bacilli into animals neutralises endo-toxins *in vitro* or *in vivo*. Macfadyen[21] has obtained an anti-typhoid serum from a goat immunised by the " cell-juices " of bacilli (hence it is an anti-endotoxic serum). The serum has agglutinative and bacteriolytic properties and in a dilution of 1 in 10,000 protected against 10 lethal doses of *B. typhosus*: it was also of use when injected at the onset of toxic symptoms. Chantemesse[22] claims to have reduced, by means of his anti-typhoid serum, a mortality of 15 per cent. to 5 per cent. and Brunon[23] has recently published his statistics which are confirmatory of those of Chantemesse. Both state that all cases where serum therapy was begun in the first week of illness ended in recovery and without any complications.

Impossible as it has been in the vast majority of instances to ascertain the exact source of typhoid infection, station reports contain abundant evidence to show that so far as European troops in India are concerned water-borne epidemics of typhoid are rare. This doubtless is due in a great measure to the care bestowed on

Opinions of medical officers.

the water-supply in cantonments. There are as usual several isolated cases recorded where it is surmised that soldiers have been infected by drinking water from way-side wells and similar sources while on the march, on manœuvres or on shooting expeditions, by drinking aërated waters in bazaars or bazaar milk which has possibly been adulterated with polluted water, or by eating bazaar fruits and vegetables which had been washed in impure water by the native vendors. The chief instance of infection in this manner was that of a draft of 23 men on the line of march to Fatehgarh. Shortly after arrival at that station 12 of these men were admitted to hospital for typhoid. Infection was considered to have been conveyed in the aërated water supplied by natives at one of the camps *en route*. Water of cantonments has been under suspicion in only two or three stations. At Rawalpindi it is said storage (now remedied) was bad, the water having been kept in open-mouthed earthenware *chatties* which were placed on the floor in verandahs; at Poona the water-supply, even after filtration, was regarded with suspicion and recommended to be boiled ; at Ahmednagar *B. coli* was found in the soda water manufactured by a regiment. Apart from milk and other articles of food consumed in or brought from bazaars, the authorised supply of these articles (in canteens, coffee-shops, etc.) to regiments has been as a rule above suspicion. The chief exception was the milk-supply of two batteries at Rawalpindi. These two batteries, which had a common dairy in charge of a contractor who kept no cows himself but procured milk from native villages, furnished several cases of typhoid, while a third battery, occupying lines between the other two, had its own dairy under good management and remained free from the disease. At Aden the consumption of oysters taken from the harbour and sold by natives in the bazaar was held responsible for the infection of some men. In not a few stations (Peshawar, Rawalpindi, Quetta, Mhow, Meerut, Campbellpore, Jhansi and Bangalore) dust and flies have been put forward as possible factors in the spread, particularly as the proximity of latrines to cook-houses, stables to barracks, filth-trenches to cantonments, the dusty nature of the country and the small rainfall, favoured such means of dissemination. In Poona, however, dust was considered unimportant, because the greatest incidence was during the rains and the healthiest time of the year was during the dusty season (see paragraph 14 below).

One striking point brought out by a study of these station reports is the frequency of imported cases. Apparently most of the men admitted for typhoid in the hill-stations have been infected in stations in the plains or on their way to the hills. In the plains, also, importation is not an infrequent occurrence : 18 at Dinapore, 11 at Agra and 11 at Meerut contracted the disease outside their respective stations. In many stations (Peshawar, Quetta, Campbellpore, Mhow and Poona) typhoid is said to be endemic; in some of them the endemicity is attributed to the disease being prevalent amongst the native population, and the proximity of native villages and bazaars to cantonments and the constant intercourse between them are regarded as sources of danger to British troops. At Peshawar the prevalence of typhoid among natives was confirmed by Widal's reaction; from Sialkot the statement comes that typhoid was very prevalent among native children, but no evidence is given in support of this. In other stations, *e.g.*, Campbellpore, the source of infection was held to have been probably in the barracks themselves, although there had been no case of typhoid in the station so far as known for over 18 months before the outbreak : an opinion, however, is not expressed regarding the manner in which the chain of infection had been sustained during that period. Of direct contact infection many instances are given. At Peshawar no less than seven hospital orderlies in attendance on typhoid patients were attacked ; at Rawalpindi three nursing orderlies contracted the disease and they in their turn infected several others in their barrack room. Agra, Landour, Jullundur and Mian Mir also furnished cases of infection among attendants on typhoid patients : the hospital orderly who contracted the disease at Mian Mir was employed in the special ward nursing a case which was afterwards diagnosed as typhoid. At Bangalore the first typhoid patient among the troops contracted the disease in hospital where two officers were under treatment at the time suffering from severe attacks of the disease. The following instances explain to some extent the facility with which infection by contact, immediate or mediate, may take place. At Campbellpore when it appeared that an outbreak was inevitable, all the men of the recently arrived drafts were medically inspected. One man was found suffering from malaise and diarrhœa but not so ill as to report sick. He was admitted to hospital and kept under observation ; a sample of his blood sent to the Sanitary Officer, Northern Command, for examination was reported as giving a positive agglutination reaction in a dilution of 1 in 40 within half an hour with $B.\ typhosus$. This man had suffered from diarrhœa since his arrival in the station but had no fever. Subsequently, he became seriously ill and died after ten days' illness. At the *post mortem* examination healed enteric ulcers were found in the ileum, and the whole intestinal tract was congested. The medical officer of Campbellpore considers it probable that this ambulatory case of typhoid was the origin of the epidemic at that station. At Peshawar it was observed that in many cases symptoms were very slight and the men did not seem very ill although the serum reaction was quite distinct, and that frequently men did not come to hospital for some days after feeling unwell. Such cases, it is pointed out, are powerful agents in spreading infection. The first man that was attacked by typhoid in the regiment which suffered most severely during the epidemic at Peshawar was ill eight days before coming to hospital and his admission was quickly followed by others from the same and neighbouring barrack rooms. At Cherat there was a well marked connection between the first typhoid patient, who was ill for some days before coming to hospital, and the next three men admitted. At Karachi, one soldier had been going about for weeks before he reported sick and when admitted to hospital was at first considered to be suffering from ague : he subsequently had a relapse and died. Similar cases were reported from Delhi, Sialkot, Kasauli, Mhow, etc. Ambulant, abortive and atypical forms of typhoid infection are, therefore, not infrequent in India and men suffering from these types of infection are a real danger to their comrades. At Rawalpindi the arrival of convalescents from the Murree hills was followed by a distinct increase in the number of admissions for typhoid. This led to a careful examination of convalescents with the result that $B.\ typhosus$ was isolated from the urine of several of them. A man, who had been in hospital at Jhansi for some weeks on account of simple continued fever, was sent to Naini Tal for a change. As the history of the case raised suspicions, he was, on arrival at Naini Tal, admitted to hospital for debility, treated with urotropine, kept under observation for 16 days and then segregated with the other typhoid convalescents. Five weeks after arrival he complained of abdominal discomfort and was admitted to hospital. He had a typical relapse ; Widal's reaction was positive. The medical officer at Naini Tal states that he considers all convalescents from simple continued fever arriving from the plains as suspicious if fever had been present for more than two or three days, and that he treats all such cases as typhoid convalescents. Cases like this show the importance, nay, the necessity of viewing all cases of fever as possibly enteric until that disease can be definitely excluded from the diagnosis.

Latrine infection takes a prominent place in many reports. At Peshawar a group of bungalows with their latrines were regarded as infected and on the removal of the troops from this infected area to other barracks the disease ceased: the latrines were of a bad type and situated too near the cook-houses and barracks; the dry-earth system was in vogue and soldiers often omitted to cover their excreta with earth; and flies and dust were plentiful. At Mian Mir infection was considered to have been spread by contact, especially through the medium of infected clothing, bedding and soiled latrine seats: barracks were evacuated; contacts put under canvas, provided with separate latrines and urinals, and their excreta boiled; and clothing, bedding, etc., as well as the evacuated barracks and latrines were disinfected: these measures were followed by good results. At Bangalore the 6th Dragoon Guards were particularly affected, and the only sanitary defect that could be discovered was the proximity of latrines to cook-houses. Barracks and latrines being regarded as infected, the regiment was put under canvas for one month and the vacated barracks and latrines were disinfected. While under canvas there was not a single case of infection, but a couple of weeks after return to barracks the disease broke out again and spread rapidly. The permanent latrines were, therefore, closed and temporary ones constructed on the opposite side of barracks, so as to be as far away from cook-houses as possible, and all excreta were mixed with carbolic acid and boiled. During the six weeks of the year remaining after the adoption of these measures only one man was admitted to hospital for typhoid. In Meerut and Mhow good results seem to have followed the use in latrines of a two per cent. solution of crude carbolic acid instead of dry earth. The following case reported from Naini Tal is of interest. One man had suffered from a severe attack of typhoid at Agra and after being in hospital there for 117 days was sent for a change to Naini Tal where two weeks after arrival he had a relapse.

In most stations diagnosis was confirmed by Widal's test; but in only two stations, apparently, has an attempt been made to isolate the *B. typhosus.* At Sialkot the method of blood cultures was tried in several cases with considerable success; at Karachi bacteriological examinations of blood, urine and fæces were made in each case, but the results are not stated. At Lebong there were only three cases of "typhoid" reported during the year and the three men happened to be in hospital at the same time. Two of these were imported cases; the third case was that of a man who had had no connection with the other two, came from a quite separate part of barracks and had evidently been infected in the station. Widal's test was carried out simultaneously for the three patients who were all suffering from a disease clinically resembling typhoid. The blood serum of the two patients who imported the disease agglutinated *B. typhosus* in all dilutions made, while the serum of the locally infected patient entirely failed to agglutinate *B. typhosus,* but agglutinated *B. enteritidis* (Gärtner) in all dilutions in which the test was carried out. This case, which the medical officer regards as one of paratyphoid, would have been of much greater interest and value had the specific organism been isolated and the exact type of paratyphoid bacillus established. In connection with the possible occasional association of paratyphoid with domestic animals, it is worthy of note that at Ambala a strict limit has been placed on the number of dogs, poultry and pigeons kept in barracks.

13. Besides an account of inspections and routine laboratory work chiefly concerned with microscopical, bacteriological and chemical examinations, the reports of the Sanitary Officers of Commands for the year 1905 contained the results of investigations of the origin of special outbreaks of enteric fever and other epidemic diseases. For the Northern Command, Major Scott, R.A.M.C., made a special investigation of the typhoid epidemics at Peshawar and Campbellpore. At Peshawar, food and water-supplies were exonerated from all blame : the continuance of the epidemic was regarded as due to " latrine infection , using the term in its broadest sense." The affected bungalows were all in one group, with the latrines not far distant and too near the kitchens. Infection quickly became general in these bungalows because latrines were so placed in relation to bungalows that the latrine which a man from any particular bungalow used depended on which side of the bungalow he emerged, the nearest latrine being invariably the one to which he resorted. The evacuation of this group of bungalows, the thorough disinfection of latrines and the use of crude carbolic acid (three per cent.) instead of dry earth for pans, etc., soon checked the epidemic. Between the first four cases at Campbellpore there was probably a personal connection as all four had supper together on the night prior to the departure of one of the four for Multan. Reasoning by a process of exclusion, the Sanitary Officer comes to the conclusion that here again latrine infection was the cause of the spread and this conclusion is supported by the later history of the outbreak ; for, with the exception of a few ordinary sanitary recommendations for the station in general, measures were restricted to disinfection of latrines and bungalows, and the disease immediately ceased. Two cases are specially mentioned to show the extreme difficulty of determining the immediate cause of an outbreak of typhoid in this country. The first was that of a Gunner who was found to be excreting *B. typhosus* in his urine six months after his discharge from hospital. After a course of urotropine, further attempts to isolate this bacillus from the urine proved futile. The second case occurred at Campbellpore and was an example of the ambulant form of typhoid. This case has already been referred to at page 4 . The report of Major Elliot, R.A.M.C., the Sanitary Officer of the Western Command deals specially with the outbreaks of enteric fever at Ahmedabad, Mhow and Poona. At Ahmedabad the source of infection was traced to contaminated mineral waters sold in the bazaar. Samples of the suspected water were analysed and found to contain *B. coli* and *B. enteritidis sporogenes*. The use of waters from this source was discontinued and the outbreak promptly terminated. This epidemic is the only one attributed to contaminated water, which has occurred in the Command during the past five years, the usual method of spread being, in the opinion of the Sanitary Officer, contamination of food by dust and flies which carry the virus from polluted soil. At Mhow all cases had their origin in the station ; all seemed to have arisen independently, and there was no epidemic prevalence. Typhoid is endemic in this station and consequently latrines, urinals and their vicinity were regarded as probably infected. The dry earth system of conservancy was discontinued and carbolic acid extensively used : the provision of a high pressure steam disinfector for use at the

station hospital was recommended. The prevalence of typhoid at Poona was due, in the opinion of Major Elliot, not to causes within regimental lines but to the insanitary condition of their surroundings. Typhoid is said to have been frequent among the natives and also among the civil population of the cantonment. No care was taken to disinfect the excreta of such cases and the night-soil carts from the bazaar and other places passed near or through regimental lines and by droppings specifically fouled the roads. Dust and flies were abundant during the year and were regarded as the carriers of infection. Major Aldridge, R. A. M. C., in the report for the Eastern Command draws attention to the difficulty of differentiating species of bacteria which are of intestinal origin from those which are not. In waters there are *coli-like* bacilli, not necessarily of intestinal origin, which, if reliance be placed on the usual tests, may be mistaken for *B. coli*. Similarly, there appear to be several species of bacilli which, if the milk reaction alone be relied on, might be wrongly identified as *B. enteritidis sporogenes.* It is pointed out that " coincidently with the late winter, the usual spring exacerbations of enteric fever were delayed about a month at most stations in the Eastern Command " and that the enteric seasons correspond with the periods of greatest prevalence of flies. At Lucknow special observations were made on the number of flies in barracks, kitchens, latrines and filth trenches and the increase of flies during the latter part of March corresponded very closely, allowing for the incubation period, with the commencement of enteric fever. The total number of flies was undoubtedly less than usual, and, while those found in the lines were mostly common house flies *(Musca domestica determinata)*, the only species bred from larvæ and pupæ taken from trenches belonged to the genera *Hylemyia* and *Ulidia*. Apparently, therefore, for some reasons at present unknown, the trenching grounds during the first half of the year were not furnishing the enormous number of flies they usually do and, coincidently, the admissions for typhoid were less than half the corresponding average for the last eight years. Outbreaks of enteric fever at Meerut and Dehra Dun were specially investigated. Stress is laid on two factors which under favourable climatic conditions seem to determine the prevalence of enteric fever among British troops in India: (1) a community peculiarly susceptible on account of age, and (2) a system of sewage removal and disposal eminently suited to the dissemination of *B. typhosus*. In both stations, measures directed against the spread of infection by latrines and clothing were attended with success. The report contains an interesting account of an outbreak of peripheral neuritis, regarded at the time as beri-beri, among British troops at Cawnpore. Although the symptoms were compatible with the supposition that the disease was beri-beri, it was difficult to reconcile many facts with this view. The only other satisfactory explanation of such symptoms occurring in epidemic form seemed to be that they were due to chronic arsenical poisoning and this hypothesis is supported by the facts that a considerable quantity of arsenic was found in the hair of one patient and a trace in the urine of two others and that all the men attacked were beer-drinkers. On the other hand, no arsenic could be detected in the beer or other articles of food and drink. Samples of the beer that was being issued at the time of the epidemic were, however, not available (the special investigation of the epidemic having taken place some time after the actual outbreak), so the result of the analysis of the beer which happened to be in stock after the epidemic had ceased detracts little from the assumption that the epidemic was one of chronic arsenical poisoning.

14. The prevalence of enteric fever among European troops in India during

Enteric fever in 1905. Appendices A and B to Section II, and D to Section III, Table IV.

1905 may now be briefly noted. There were 1,146 admissions and 213 deaths during the year as compared with 1,395 admissions and 267 deaths in 1904 and 1,584 admissions and 293 deaths in 1903. The admission rate per mille of strength was 16·1 against 19·6 in the previous year. The average duration of a case in 1905 was about 70 days, in 1904 about 68 days, and in 1903 about 54 days. This progressive increase, year after year, in the stay of a typhoid patient in hospital is doubtless due to a more thorough appreciation by medical officers of the danger which convalescents from this disease are to their comrades, if permitted to return to barracks without undergoing a prolonged course of urotropine. That freeing the urinary system thus from typhoid bacilli is not sufficient and does not necessarily completely disinfect the convalescent has been shown in the preceding paragraphs. The average number of men constantly sick was 220·50 and the total loss of service due to enteric fever alone was 80,482 days against a constantly sick rate of 256·20 and a loss of 93,769 days in 1904. The death rate fell from 4·19 per mille in 1903 and 3·76 in 1904 to 2·99 in 1905. The case-mortality is slightly lower than that of the previous year : among every 1,000 patients five more recovered during 1905 than during 1904 and 26 more than during 1903.

There is every reason for believing that this satisfactory improvement is due to the prophylactic measures now in force, in particular those measures which have as their object the limitation of infection by contact, e.g., the segregation of newly arrived drafts and regiments, the early diagnosis and isolation of cases of enteric fever, the attempts to disinfect those who are convalescent from this disease and the disinfection of latrines, barracks and clothing which may possibly have been contaminated. Much still remains to be done in the way of discovering mild ambulant cases, of recognising atypical forms of the disease and of detecting those convalescents who become carriers in spite of the most thorough administration of urotropine.

15. The disease was equally prevalent in the three commands but was rather

Enteric fever in commands and geographical groups. Appendices A and B, Table VIII.

more fatal in the Western than in either of the other two commands. As regards divisions, the incidence was greatest among troops in the 1st (Peshawar), the 6th (Poona) and the 2nd (Rawalpindi) Divisions and least among troops in the 10th (Burma) Division. Of geographical groups, Deccan (IX), Upper Sub-Himalaya (VI) and Indus Valley and Frontier (VII) showed the greatest prevalence during 1905 and of these Deccan (IX) and Upper Sub-Himalaya (VI) always occupy an unenviable position in this respect. On the other hand, the admission rates in Gangetic Plain (V) and Central India (VIII), which have a very bad record for enteric fever, were only about half those of 1904 or the corresponding decennial means. Hill Stations (XIIa) show a progressive improvement, the admission rate which was 29·4 per mille for the decennium 1894—1903 having fallen to 14·7 in 1904 and 10·9 in 1905. Burma Coast (I) and Western Coast (X) together did not furnish more than 1·1 cases per mille of strength. The following tabular statement, of the monthly incidence, based on the total admissions during the past ten years, in the geographical groups throws a little light on the possible part which dust and flies play in the dissemination of enteric fever in the different groups,

Seasonal incidence of enteric fever in geographical groups during the decennium 1896-1905.

GEOGRAPHICAL GROUPS.	January.	February.	March.	April.	May.	June.	July.	August.	September.	October.	November.	December.	Total numbers of admissions and average annual admission ratios during the decennium 1896-1905.
I—Burma Coast and Bay Islands ...	5	1	4	7	4	13	26	10	4	5	4	3	59
	'7	'1	'3	'6	'4	1'1	2'3	'9	'4	'5	'4	'3	3'2
II—Burma Inland	6	6	8	4	5	5	16	9	7	3	13	2	85
	'3	'3	'4	'2	'3	'3	'6	'4	'3	'1	'5	'1	3'8
IV—Bengal and Orissa	9	28	23	31	10	27	25	15	13	9	11	5	2'9
	'4	1'2	1'1	1'6	'5	1'3	1'2	'9	'7	'5	'6	'2	10'2
V—Gangetic Plain and Chotia Nagpur	157	98	217	204	310	78	45	131	103	97	170	150	1,791
	24	1'4	3'1	4'6	3'3	1'1	'7	1'0	1'6	1'5	2'5	2'8	27'1
VI—Upper Sub-Himalaya ...	246	113	161	295	555	312	132	190	127	131	293	432	3,123
	1'6	'7	1'0	3'1	5'5	3'4	1'4	1'0	1'4	1'2	1'9	2'7	24'5
VII—North-West Frontier, Indus Valley and North-West Rajputana.	53	35	53	52	94	135	95	35	28	43	70	61	803
	1'0	'6	1'0	1'7	2'3	3'4	2'5	'9	'9	1'2	1'1		16'9
VIII—South-East Rajputana, Central India and Gujarat.	140	85	155	258	200	62	69	217	247	127	105	150	1,960
	2'4	1'4	1'7	5'5	3'3	1'0	1'7	3'6	4'2	2'2	1'8	2'7	33'0
IX—Deccan	77	74	140	155	154	121	2'2	327	353	172	105	58	2,010
	'8	'8	1'3	1'6	1'8	1'3	2'1	4'2	3'8	1'8	1'1	'5	21'3
X—Western Coast	5	5	12	9	6	8	12	7	9	4	5	3	85
	'3	'3	'7	'5	'4	'5	'8	'5	'6	'2	'3	'2	5'4
XI—Southern India	25	48	57	43	27	53	65	44	32	37	37	35	535
	'B	'4	1'3	1'4	'9	1'7	1'0	1'9	'9	1'1	1'3	1'0	16'0
XII (a)—Hill Stations	15	6	39	293	340	285	325	311	315	176	147	17	2,351
	'5	'2	1'8	3'1	2'4	1'9	2'2	2'3	2'4	2'2	3'6	'5	26'2
XII (b)—Hill Convalescent Depots and Sanitaria.	1	1	23	125	108	87	47	47	32	25	16	1	511
	'1	'1	1'2	2'0	2'0	1'6	'9	'9	'5	'7	1'1	'1	15'2
INDIA...	543	521	1030	1879	1800	1156	1172	1512	1585	913	1092	1075	14,251
	1'3	'9	1'5	2'8	2'5	1'9	1'7	2'2	2'1	1'4	1'6	1'6	21'3

If dust is a very significant factor in the spread of the disease, the incidence should be relatively low during the rains. Applying this test to the different geographical groups we find that dust can be a possible factor of *significant* importance in only four groups: Gangetic Plain (V), Upper Sub-Himalaya (VI), Indus Valley and Frontier (VII), and Convalescent Depots (XII-*b*): in the first two groups (V and VI), however, where there was a decided increase of incidence during August, dust cannot by any means be the only possible significant factor. As regards flies, it is difficult in so extensive a country as India, which includes regions with vastly different climates, to lay down general tests relative to the prevalence of flies, about the life-history of which we have still much to learn. Precise information (obtained from observations extending over several years) on the relative monthly prevalence of these insects in the stations of India would be of great help in gauging the importance of flies in the dissemination of typhoid. In the present state of our knowledge the easiest test to apply is a comparison of the relative frequency of admissions for enteric fever in the two seasons of the year when flies are (1) most common and (2) least common. Flies are numerous everywhere in the months of August and September, and probably in most stations they are more numerous then than in any other two months, but it is only in groups VIII, IX, XI, and XIIa that the incidence of enteric fever is relatively heavy during these months : in one group (*viz.*, VII) the incidence of enteric fever is lowest in these months and February. Flies are not specially numerous in Northern India in the winter, and yet in group VII (North-West Frontier and Indus Valley) the incidence of enteric fever then is comparatively high. In the accompanying table it will be observed that the winter incidence of this disease in the most northerly Indian stations is far from

Seasonal incidence of enteric fever during the decennium 1896-1905, in stations in the Northern Command and in seven other Indian stations in which the disease is very prevalent.

Stations.	1. Actual numbers of admissions 2. Mean (monthly) admission ratios per 1,000 of strength } During the decennium 1896-1905.												Total numbers of admissions and average annual admission ratios during the decennium 1896-1905.
	January.	February.	March.	April.	May.	June.	July.	August.	September.	October.	November.	December.	
Peshawar	44	19	28	40	44	94	85	30	26	27	58	42	527
	2·2	·9	1·3	2·0	3·5	8·1	7·4	1·7	2·2	1·6	2·9	2·1	21·9
Nowshera	3	1	1	10	22	25	7	7	3	6	1	5	91
	·4	·1	·1	1·3	3·8	4·8	1·6	1·6	·7	1·0	·1	·6	14·2
Rawalpindi	41	15	12	78	130	125	49	50	41	33	47	95	656
	1·1	·4	·3	2·4	7·0	7·6	3·3	3·3	2·6	1·5	1·3	2·5	25·0
Campbellpore	...	...	...	5	25	22	1	1	2	12	1	1	80
	...	...	...	1·9	13·3	8·6	·4	·4	·8	4·8	·4	·4	32·7
Sialkot	15	...	4	12	44	37	9	8	6	12	18	18	183
	1·2	...	·3	1·0	4·6	4·1	1·0	·9	·7	1·2	1·6	1·7	17·3
Mian Mir	12	16	14	17	35	23	18	7	8	4	12	20	186
	1·3	1·7	1·3	2·0	5·0	3·3	2·7	1·1	1·3	·6	1·2	2·1	23·0
Fort Lahore	3	4	1	6	3	3	1	2	...	1	6	5	35
	2·3	3·8	·9	5·8	2·9	2·0	1·0	3·0	...	1·1	5·8	4·6	34·8
Dalhousie	...	...	...	33	50	16	9	10	7	2	2	...	150
	...	...	...	6·6	3·5	1·0	·6	·7	·5	·2	2·2	...	19·0
Multan	1	2	6	29	21	11	1	3	3	8	5	6	94
	·1	·2	·5	3·1	1·7	1 5	·1	·4	·4	1·0	·5	·4	10·4
Ferozepore	5	5	4	31	40	9	8	8	5	5	5	11	136
	·5	·5	·4	3·2	4·8	1·1	1·0	1·0	·6	·6	·5	1·2	14·7
Jullundur	5	1	11	32	31	14	3	8	3	4	4	12	119
	·6	·1	1·1	3·0	5·6	1 6	·6	1·6	·6	·7	·5	1·8	17·8
Ambala	87	31	50	55	76	27	11	22	15	17	95	145	635
	3·3	1·1	1·8	4·0	6·5	2·3	1·0	1·9	1·5	1·1	3·4	5·2	33·6
Kasauli	...	...	6	25	12	19	5	11	5	3	2	...	69
	...	...	3·0	4·6	1·9	2·5	·8	1·8	·9	·9	3·1	...	24·3
Dagshai	1	...	2	38	33	42	33	13	6	1	1	1	174
	·5	...	·8	5·0	3·3	4·6	4·2	1 8	·5	·2	·5	·5	31·1
Subathu	...	...	8	51	28	21	14	36	44	4	...	...	236
	...	...	5 6	8·4	5·2	2·9	2·2	5·1	6·7	·8	...	...	52·8
Quetta	13	...	6	8	10	67	108	157	170	163	127	15	852
	·6	...	·3	·2	·4	2·5	4·2	6·1	6·7	6·7	5·8	·6	25·4
Mhow	44	30	36	104	52	30	79	74	47	18	17	25	527
	2·5	1·8	2·1	6·4	3·3	1·9	4·9	4·6	2·9	1·1	1·1	2·2	35·3
Meerut	45	31	42	111	103	50	19	51	31	31	73	80	661
	2·4	2·0	2·1	6·7	6·6	3·3	1·2	3·3	2·0	1·2	3·5	3·8	37·0
Lucknow	95	45	122	142	99	41	22	52	48	45	82	102	832
	4·0	1·8	4·2	6·6	4·5	1·8	·9	2·3	2·1	1·9	3·3	4·1	37·9
Poona	12	9	23	16	7	9	30	148	141	42	13	8	465
	·6	·4	1·2	·8	·4	·4	1·4	7·1	6·8	1·2	6·3	·4	22·7
Secunderabad	27	23	42	26	17	32	71	111	94	66	42	20	528
	1·0	1·0	1·5	1·0	1·5	1·3	2·8	4·4	3·8	2·4	1·5	·7	22·4
Bangalore	19	35	48	27	24	47	57	50	22	31	16	24	410
	1·1	2·0	2·8	1·8	1·5	2·9	3·4	3·5	1·3	1·8	1·0	1·4	24·4

being insignificant. The same table also shows the seasonal incidence in seven other Indian stations, outside the Northern Command, which generally suffer severely from this scourge. It is hoped that the figures given may be of some assistance to officers in medical charge of these stations in forming an opinion on the extent to which the seasons of greatest prevalence of flies coincide with those of heaviest incidence of enteric fever.

16. From 25 stations as against 28 in 1904 admission rates of over 20 per mille were reported and among these the following may be noted with their admission rates per mille :—

Enteric fever in stations. Appendices C and D to Section II, Tables III, IV and VIII.

In group VI.		In group V.		In group VIII.		In group IX.	
Campbellpore	... 190·3	Fatehgarh	89·2	Indore	44·1	Poona	43·2
Meerut ...	... 57·3	Dinapore	32·8	Nowgong	28·0	Secunderabad	27·3
Lahore ...	... 49·6	Lucknow	27·8	Jhansi	26·8	Kirkee	24·0
Amritsar ...	... 24·5			Mhow	23·0		
Rawalpindi ...	... 23·9						
Mian Mir ...	... 22·1						
Ferozepore ...	... 20·2						

The list also includes Peshawar (4·27), Barrackpore (21·0) and Quetta (20·5) as well as five hill-stations and convalescent depots. There was a total absence of enteric fever from 20 stations including Pachmarhi where in 1904 there were 24 admissions and four deaths among 162 men. All stations in group (VIII) Central India show a decrease (compared with last year's figures) except Indore and Nowgong, and, similarly, all stations in group (V) Gangetic Plain, except Fatehgarh and Allahabad. Table-VIII at the end of this volume sets forth the monthly incidence of enteric fever among European troops and shows how rarely a sudden outburst, so characteristic of an epidemic due to infected water or food, occurs in Indian stations ; Appendix D of Section II in the latter part of this volume yields evidence which is strongly suggestive of " contact infection." The special points of interest in connection with the causation of the more important outbreaks have been summarised in previous paragraphs.

17. In 1905 as in previous years the greatest liability to suffer and die from enteric fever fell upon the age-period 20 to 25 and upon the first year of Indian service. The extent of this liability is indicated by the fact that nearly 73 per cent. of the total admissions at all ages were among men up to 25 years of age and that nearly 64 per cent. of the total admissions were among men during their first two years of Indian service. The tendency to a rise in the admission rates for the higher age-periods continues : in 1905 the admission rate of men

Relation of enteric fever to age and length of residence in India. Tables XV and XVI.

over 35 years of age was 4·2 per mille as compared with 4·0 in 1904 and 3·5 in 1903. On the other hand, there has been a break in the progressive rise of the admission rate for the higher service-periods : the admission rate during 1905 of men with five years' service and upwards was 6·9 per mille as compared with 11·4 in 1904 and 10·2 in 1903.

18. Among European troops in India in 1905 there were recorded thirteen

Plague and some other general dis-
eases.   Table LIII.

cases of plague with four deaths as compared with two cases with no fatality in 1904 and twelve cases with two deaths in 1903. Of the thirteen cases reported during 1905, seven cases with two deaths were at Aden, three with one death at Bombay, one (fatal) at Benares and one case each at Fatehgarh and Port Blair. At Aden all seven cases occurred between 19th and 24th April, the source of infection having been traced to the regimental coffee-shop which was infested with plague-stricken rats. The town, in which the disease was prevalent, had been out of bounds for two months before the outbreak among the troops. At Bombay the exact source of infection was not ascertained, but the disease was prevalent among the natives in the town. The patient at Benares was probably infected in the *Sadar* bazaar which suffered severely from plague and was separated from barracks by a road only. The medical officer at Fatehgarh suggests that the patient at that station was infected by his favourite cat which was known to have killed rats in the infected quarters of hospital servants. The case at Port Blair was imported from Rangoon. There were altogether 13 admissions with one death on account of measles, the majority occurring at Quetta and Lucknow. Erysipelas accounted for 36 cases with three deaths. Of the 46 admissions with five deaths on account of beri-beri, no less than 25 with three deaths were recorded at Aden and ten with one death at Cawnpore. The outbreak at Cawnpore, which occurred between August and October, has already been referred to in paragraph 11 above. At Aden the disease is said to have been unknown among British troops in that station until the arrival of a regiment from Burma about six years ago. The incidence is said to have been high among beer-drinkers but no trace of arsenic was detected on analysis of the beer. The patients were between 24 and 36 years of age and about half of them lived in the upper stories of barracks and about half in the lower. The spread of the disease is attributed to infected buildings under the favouring conditions of heat and moisture. (The native troops at Aden furnished one case of beri-beri.) There were 415 cases of dengue with one death as compared with 439 cases with one death in 1904 and 303 with one death in 1903. The disease was practically confined to two stations in the Secunderabad Division and two in the Burma Division, Fort St. George (Madras) with 48 cases and Rangoon with 354 cases being the chief places affected. At Rangoon no case of dengue was recorded from January to May : most cases occurred in November. In 1902 the epidemic at Rangoon lasted from March to July, so the seasonal incidence varies considerably. With regard to the cases reported in December, the medical officer states that the diagnosis is open to question as the main clinical features of dengue were absent. The blood of a large number of these was carefully examined for malarial parasites but with negative results. *Bilharzia hæmatobia* accounted for 22 admissions during the year against 63 in 1904 and 318 in 1903. The number of stations in which a case or cases were detected was 11 as compared with 26 in the previous year and 33 in 1903. The number of men invalided for this cause fell from 71 in 1904 to 23 during the year under report.

19. The admission rate of European troops in India on account of tubercle of the lungs fell from 2·8 per mille in 1904 to 2·1 in 1905. The admission rate of the Western Command (3·0) has increased while those of the Eastern and Northern Commands have decreased, that of the Northern Command (1·7) being less than half what it was in the previous year. The stations reporting the largest number of admissions were Poona with 14, Rangoon with 10, Quetta with 9 and Lucknow with 8. The death rate fell from ·38 in 1904 to ·28 per mille during 1905 and the number of men invalided for this disease from 223 to 116. The death rates from tubercle of the lungs of European troops, native troops and prisoners during the year were in the proportions of 5, 9 and 65, respectively, as compared with 5, 7 and 48, respectively, in 1904.

Tubercle of the lungs. Appendices A to Section II and E to Section IV, Tables III and IV.

20. In 1905 the admission and death rates for pneumonia were 4·1 and ·63 per mille respectively as compared with 3·4 and ·52 in 1904, the corresponding decennial means being 4·3 and ·62 respectively. There were 296 admissions and 45 deaths from this disease in 1905 against 240 and 37 in 1904. Both admission and death rates were greatest in the Northern Command and least in the Eastern. The three groups in which pneumonia is usually most prevalent among European troops are (VII) Indus Valley and Frontier, (VI) Upper Sub-Himalaya and (XIIa) Hill Stations and in 1905 these same three groups head the list with admission rates of 7·1, 5·4 and 5·8 per mille respectively, (IV) Bengal-Orissa coming next with 4·7 and (VIII) Central India next with 4·4. Compared with the previous year there was an increase in the admission rate in all the geographical groups except (I) Burma Coast, (XI) Southern India and (XII b) Convalescent Depots, and the increase was greatest in (IV) Bengal-Orissa and (VII) Upper Sub-Himalaya. Excluding Fort Lahore and Khandalla where the strength was low, the highest admission rates for the year were recorded at Dum Dum with a rate of 17·8 per mille and an average strength of 281 men, Quetta with 13·1 and a strength of 2,590 men, Nowshera with 12·5 and a strength of 720 men, Ambala with 11·7 and a strength of 2,314 men, Agra with 10·5 and a strength of 1,146 men and Peshawar and Thayetmyo each with 10·2 and strengths of 1,663 and 488, respectively. For India as a whole the months of greatest prevalence are, as a general rule, December, January and February and those of least prevalence July, August and September ; for the year under report most cases occurred in December and fewest in July.

Pneumonia and other respiratory diseases. Appendices A and B to Section II and H to Section III, Tables XII, III and IV.

There was an increase in the admission rate on account of " other respiratory diseases " as compared with the previous year, viz., 23·0 per mille against 21·4 but the death rate was only ·06 per mille as compared with ·10. The increase in the admission rate is shared by all three commands but the greatest rise was in the Eastern Command where the rate rose from 17·3 to 22·5. The highest rate (26·4) was in the Western Command. The geographical group in which these diseases were most frequent during the year was (IV) Bengal-Orissa, (XII b) Convalescent Depots coming next ; and the stations (excluding those with an average strength below 150) which recorded admission rates above 40 per mille were Dum Dum, Quetta, Wellington, Kuldana, Karachi, Kasauli and Maymyo.

21. There has been a slight increase in both the admission and death rates

Dysentery and diarrhœa. Appendices A and B to Section II and E to Section IV, Tables XIII, III and IV.

on account of dysentery, the rates being 13·4 and ·46 per mille as compared with 12·6 and ·42 in 1904. Native troops show a slight increase in the admission rate but a decrease in the death rate; while, as regards prisoners, there was a decrease in the admission rate but an increase in the death rate. Of the three commands the disease was most prevalent and fatal during the year among European troops in the Western and, as regards divisions, it was most prevalent in the 9th (Secunderabad), 4th (Quetta) and 5th (Mhow). The death rates were highest among troops in the 10th (Burma) and 5th (Mhow) Divisions. Among geographical groups, the disease was most prevalent and fatal in (I) Burma Coast. Excluding stations with a strength below 150, the highest admission rates were recorded at Dthalla (Aden Hinterland) with a rate of 65·9 per mille among 349 men, Barrackpore with 55·9 among 286 men, Rangoon with 40·2 among 1,119 men and Aden with 33·8 among 947 men. For the European army as a whole the months of greatest prevalence were July, August and September and those of least prevalence February and March.

While dysentery was the cause of 959 admissions and 33 deaths, diarrhœa accounted for 1,087 admissions with no fatality. The admission rate for diarrhœa was 15·2 per mille as compared with 13·1 in the previous year. The disease was most prevalent in the Eastern Command and least in the Northern. As regards geographical groups, the highest rates were in (XIIa) Hill stations, (I) Burma Coast and (IV) Bengal-Orissa. In the first mentioned are included the two stations with the highest ratios, *viz.*, Ranikhet (108·6 per mille) and Lebong (49·4), as well as Dagshai and Chaubattia which have ratios of 42·2 and 33·8 respectively.

22. The admission and death rates on account of abscess of the liver were

Hepatic abscess. Appendix A to Section II, Tables III and IV.

2·1 and 1·18 per mille as compared with 2·6 and 1·36 in 1904. In the Northern Command the admission rate was only about half and the death rate only about a quarter of the corresponding rates in either the Western or Eastern. The disease was most prevalent among troops in the 6th (Poona), 8th (Lucknow), 5th (Mhow) and 9th (Secunderabad) Divisions and the Aden Brigade. The death rates were highest in the Aden Brigade and 5th (Mhow), 8th (Lucknow) and 9th (Secunderabad) Divisions. As regards geographical groups, (IV) Bengal-Orissa still retains an evil reputation which during the year under report is shared with (X) Western Coast, (XII b) Convalescent Depots and (IX) Deccan. There were in all 153 admissions with 84 deaths recorded during 1905 as compared with 185 admissions and 97 deaths in the previous year. One or more cases were reported from 60 stations and, excluding Poonamallee and Calicut where the average strength was low, the highest admission rates were at Barrackpore (three cases with one death), Dinapore (five cases with four deaths) and Kasauli (three cases with two deaths). At Lucknow, however, there were thirteen admissions with five deaths, at Poona eleven admissions with two deaths, at Secunderabad seven admissions with five deaths, at Wellington six admissions with four deaths and at Jhansi five admissions with four deaths, these being stations with an average strength of about 1,000 and over. There was a history of dysentery in eight of the thirteen cases at Lucknow, in four of the seven cases at Secunderabad, in two of the eleven cases at Poona and in two of the five cases at Jhansi, but in none of the three cases at Barrackpore: in a few instances hepatic abscess is mentioned as having occurred after enteric fever. As in previous years multiple abscesses were found in the majority of the fatal cases.

23. There were 197 admissions with 11 deaths from alcoholism during 1905 against 220 admissions and 9 deaths in the previous year, the corresponding average numbers for the decennium ending 1904 being 250 and 8, respectively; compared with the decennial average rates, there was, in 1905, a fall in the admission rate but a rise in the death rate. The liability to death from alcoholism has a distinct tendency to rise with age and length of residence in India.

Alcoholism. Tables XVI and LIII.

24. The recorded admission rate for all forms of venereal disease in 1905 was 153·7 per mille of strength as compared with 198·5 in 1904, the actual numbers being 10,966 in 1905 and 14,113 in the previous year. The constantly sick rate was 15·3 per mille against 17·5 in 1904 and the average period during which a case remained in hospital was 36·26, which is an increase of four days on the average stay in hospital of each patient during 1904 The total loss of service, however, was 397,584 days as compared with 454,206 days in 1904 and 572,864 days in 1903. The non-efficiency, thus reckoned, due to detention in hospital on account of venereal diseases during 1905 amounts to about five times the loss incurred on account of enteric fever.

Venereal diseases. Tables III and IV.

There were 13 deaths (·18 per mille of strength) attributed to venereal diseases during 1905 as compared with 15 (·21 per mille) in 1904 and 10 (·14 per mille) in 1903 : the number of men invalided for this cause was 83 (1·16 per mille) as compared with 199 (2·80 per mille) in 1904 and 213 (3·02 per mille) in 1903.

All three Commands record a diminution in the admission rates for venereal diseases ; but for 1905, as in the previous year, the Eastern Command heads the list with an admission rate of 165·8 per mille followed by the Western with a rate of 155·4. The admission rate in the Northern Command was only 116·1 per mille while the rate among troops outside this Command was 166·7 ; the corresponding figures for Native troops are 18·3 in the Northern Command and 19·9 outside this Command. As regards divisions, venereal diseases were most prevalent in the 10th (Burma) with an admission rate of 206·9 per mille of strength and the 9th (Secunderabad) with a rate of 194·0, and least in the 1st (Peshawar) with a rate of 74·1 per mille and the 2nd (Rawalpindi) with a rate of 105·7 per mille. Comparing 1905 with the previous year, the divisions which show the greatest reduction in prevalence of these diseases are the 10th (Burma), the 7th (Meerut) and the 4th (Quetta).

Admission rates of over 200 per mille were reported from 21 stations, four of which had an average strength below 100 men; and, if we omit Lower Topa (577·8 per mille) where the average strength was only 45, the highest rate was recorded at Maymyo (338·1). Excluding stations at which the average strength was less than 100, venereal diseases in the Northern Command were most prevalent at Kasauli (with an admission rate of 214·7 per mille), Mian Mir (118·7) and Kuldana (158·6), in the Western Command at Deolali (255·5), Bombay (265·0) and Purandhar (246·2), in the Eastern Command at Fort

William (283·5), Barrackpore (241·3) and Lebong (229·4), in the Secunderabad Division at Madras (266·8) and Bellary (263·7), and in the Burma Division at Maymyo (338·1), Rangoon (253·8) and Shwebo (175·2). The stations with an average strength of over 100 which in a comparison with the previous year showed the greatest improvement were Maymyo, Meiktila, Agra, Muttra and Neemuch: altogether 78 stations out of a total of 104 recorded a diminution in the prevalence of venereal diseases during the year under review.

The admission rate on account of gonorrhœa was 74·8 per mille of strength (the total number of admissions for gonorrhœa being 5,338) and that on account of soft chancre 43·3 per mille as compared with rates of 95·3 and 54·2 per mille, respectively, in 1904. Eight men, as compared with 24 in the previous year, were invalided during 1905 for gonorrhœa.

Syphilis accounted for 2,537 admissions, 13 deaths and 75 invalidings as compared with 3,488 admissions, 15 deaths and 175 invalidings in 1904. The admission rate fell from 49·1 per mille in 1904 to 35·6 per mille in 1905. The figures, however, for 1905 are not strictly comparable with those of 1904 and previous years as in June 1904 changes in the methods of classification and treatment of this disease came into force and these changes considerably affect the statistical returns. Between 1st June 1904, and the end of that year the names of 2,947 men were placed on the " syphilis register " but this number doubtless includes the names of many men who contracted the disease in previous years. The number of men whose names were placed on the register during 1905 was 1,470; this represents the number of men who came under treatment for syphilis for the first time and may be taken as approximately the number of cases of fresh infection during the year and yields a rate of about 21 per mille. A large number of men required admission or re-admission to hospital while undergoing treatment as out-patients, and the number of men who had relapses after being struck off the register was 35. The average number of days each case was under treatment was 261. As regards the methods of treatment, the most common whether in or out of hospital were (1) by injection and (2) by the mouth; of the men in hospital, roughly speaking, for every five treated by injection four were treated by the mouth, while of the men treated out of hospital, for every one treated by the mouth seven were treated by injection.

That there has been in recent years a real reduction in the amount of venereal diseases in the European Army of India, there can be no doubt, although from the changes in classification (for statistical purposes), the introduction of the " syphilis register " and the treatment of some cases as out-patients, it is difficult as yet to gauge the amount of real reduction in prevalence and the relative values of the causes at work in producing this diminution. At first sight it might be thought that, if in consequence of the above changes admission statistics were not comparable, from the combined death and invaliding rates we should be able to get some approximate idea of the real reduction in the prevalence of these diseases. This, however, is not so. There has been a steady and, latterly, fairly rapid fall in the number of invalidings and this has not been accompanied by any increase in the number of deaths in India, but the more prolonged and thorough treatment now undergone by those found suffering from venereal diseases may in a great measure explain these favourable results.

Medical officers attribute the prevalence of venereal diseases to the youth and ignorance of the soldier and the difficulty experienced in controlling infection outside cantonments, particularly on the line of march between stations. For limiting the spread of infection reliance is placed chiefly on (1) medical inspections of troops to detect infected men and place them in hospital, (2) prolonged and thorough treatment, which diminishes the number of sources of infection, (3) the enlightenment of the soldier on the ravages and dangers of venereal diseases, and (4) the encouragement of games, etc., to occupy his leisure hours pleasantly. By such means as these it is reasonable to hope that the prevalence of venereal diseases may be considerably reduced.

25. The death rate from heat-stroke was ·77 per mille against ·49 in 1904
Heat-stroke. Appendix A to Section II, Tables XVI and LIII.
and ·77 in 1903; and as usual the mortality is highest in the Northern Command and lowest in the Western. The total number of deaths recorded under this heading was 55 as compared with 33 in 1904 and 54 in 1903. The stations which furnished the largest numbers during 1905 were Peshawar (6), Mian Mir (5), Agra (5), Rawalpindi (3), Lucknow (3), Bombay (3) and Aden (3).

26. The average annual number of suicides reported in the decade 1891-
Suicide. Tables XVI and LIII.
1900 was 19 and in 1905 the number was 24, of which eleven were by gunshot, four by cut-throat, four by multiple injuries (including fracture of the skull and dislocation of the spine), three by drowning, one by hanging and one by oxalic acid poisoning.

27. From the whole European army in India 1,508 men (21·24 per mille of
Invaliding. Appendix A to Section II, Tables XVII and LIII.
strength) were invalided in 1905 as compared with 2,507 (35·27 per mille) in the previous year; the chief causes of loss by invaliding have been mentioned in paragraph 2 above. The ratio per 1,000 of strength ranged from 28·61 in the Western Command to 14·95 in the Burma Division. Of the 152 men invalided for debility 106 were under 30 years of age and 80 had been less than four years in India; and of the 116 men invalided for tubercle of the lungs 98 were under 30 years of age and 64 had been less than three years in India. Of the total number of men invalided, 47 per cent. were under 25 years of age, an age-period which included 52 per cent. of the total strength; and 82 per cent. of all invalids were under 30 years of age, an age-period which included 85 per cent. of the total strength. As regards length of residence in India, 25 per cent. of all invalids had less than two years' service in this country and 65 per cent. less than five years; the proportion to total strength of men with under two years' service in India was 43 per cent. and of those with under five years' service 84 per cent. The loss from invaliding among men in the lower age-periods and also among men of shorter service in India, was not so great as in the previous year.

28. The average strength of commissioned officers with European troops in
Officers. Appendix E to Section II, Table XVIII.
India in 1905 was 2,323 and of these 1,608 had been under medical treatment, 123 were invalided and 21 died during the year. The admission, invaliding and death rates were lower than in the previous year, while the constantly sick rate remained the same. The admission rates for ague, remittent fever, tubercle of lungs, dysentery, diarrhœa and hepatic affections show a decrease as compared with the figures of 1904, but those for small-pox,

enteric fever, simple continued fever and respiratory diseases an increase. There were six deaths from enteric fever, the chief cause of mortality, giving a rate per mille of 2·58 which is practically the same as in 1904, and is lower than the rate for the troops (2·99). There were 55 cases of dysentery without a fatality against 63 cases with one death in 1904. There was only one case (non-fatal) of each of the diseases, hepatic abscess and plague. As compared with the incidence on troops the following diseases were more prevalent in proportion to strength : influenza, enteric, remittent and simple continued fevers, dysentery, diarrhœa and congestion of the liver, while the officers suffered less than the men from small-pox, ague, hepatic abscess, venereal diseases, tubercle of the lungs and other respiratory diseases. Enteric fever, debility, ague and dysentery were the chief causes of invaliding.

29. There was a distinct improvement in the health of the women during the

Women. Appendix F to Section II, Tables XIX to XXI.

year under review. The average strength was 3,375 which is an increase of 155 on that of 1904. There were in all 2,183 admissions and 37 deaths giving admission and death rates of 646·8 and 10·96 per mille respectively as compared with rates of 697·2 and 12·42 in the previous year ; the constantly sick rate (27·9 per mille) showed an improvement of 2·7 per mille. The average admission, constantly sick and death rates for the decennium 1894—1903 were 873·4, 36·5 and 15·68 per mille respectively. The chief causes of admission to hospital were, as usual, debility, the diseases peculiar to women and ague which together accounted for· 61·88 per cent. of the total number of admissions for all causes. Of the 37 deaths five were due to enteric fever, five to childbirth and abortion, and three each to small-pox, tubercle of the lungs and dysentery. Among the commands and independent divisions, the highest admission rate was recorded in the Eastern Command and the lowest in the Burma Division, the highest death rate in the Western Command and the lowest in the Secunderabad Division.

30. There was less sickness among the children during 1905 than in either

Children. Appendix G to Section II, Tables XXII, XXV and LIII.

of the three previous years, but the constantly sick rate was exceedingly high and the death rate although much lower than that in 1902, was considerably higher than those of the two previous years. In 1905 the average strength was 5,154, the admission rate 405·9 per mille, the constantly sick rate 29·3 and the death rate 38·80. The chief causes of sickness were respiratory diseases, ague and diarrhœa which together accounted for 36 per cent. of the total number of admissions for all causes. There was a considerable diminution in prevalence of ague, measles and eye-diseases as compared with 1904. Out of a total of 200 deaths, 38 were attributed to debility and immaturity at birth, 32 to respiratory diseases, 25 to diarrhœa and 15 each to teething and convulsions, which taken together account for 63 per cent. of the total mortality. There were 26 admissions with two deaths from enteric fever, giving a case-mortality of 8 per cent. ; while for diarrhœa there were 207 admissions and 25 deaths, yielding a case-mortality of 12 per cent. Among the exanthemata to which children are specially liable there came under treatment during the year 46 cases of chicken-pox, 12 of measles, 2 of rubella, 5 of scarlet fever and 9 of small-pox ; there were also 23 cases of whooping cough, 12 of mumps, 8 of diphtheria and 3 of dengue. Sickness was greatest among children in the Secunderabad Division and mortality among those in the Eastern Command ; sickness and mortality were both least in the Burma Division.

The strengths at the different age-periods, the death rates per mille and the relative liability to death at each period are shown in Table XXV. Immaturity at birth was the cause of death in 37 per cent. of the total mortality among children under six months. Of the two children who died of enteric fever, one was between five and ten and the other between ten and fifteen years of age.

Papers and Books referred to in Section II.

Abbreviations used below.

A.H. = Archiv für Hygiene.

A.K.G.A. = Arbeiten aus dem kaiserlichen Gesundheitsamte.

A.P. = Annales de l'Institut Pasteur.

B.I.P. = Bulletin de l'Institut Pasteur.

B.J.H.H. = Bulletin of the Johns Hopkins Hospital.

B.K.W. = Berliner klinische Wochenschrift.

B.M.J. = British Medical Journal.

C.B. = Centralblatt für Bakteriologie.

D.M.W. = Deutsche medizinische Wochenschrift.

H.R. = Hygienische Rundschau.

I.M.G. = Indian Medical Gazette.

I.P.H. = Indian Public Health.

J.A.M.A. = Journal of the American Medical Association.

J.H. = Journal of Hygiene.

J.I.D. = Journal of Infectious Diseases.

J.H.H.R. = Johns Hopkins Hospital Reports.

J.P.B. = Journal of Pathology and Bacteriology.

J.P.P.G. = Journal de Physiologie et de Pathologie Générale.

J.R.A.M.C. = Journal of the Royal Army Medical Corps.

J.T.M. = Journal of Tropical Medicine.

Jahresb. v. W. and P. = Jahresbericht über die gesammte Medicin von Waldeyer und Posner.

L. = Lancet.

L.G.B. = Report of Medical Officer, Local Government Board.

M.M.W. = Münchener medizinische Wochenschrift.

N. = Nature.

P.J.S. = Philippine Journal of Science.

S.C.I. = Annual Report of the Sanitary Commissioner with the Government of India.

Z.H. = Zeitschrift für Hygiene.

(1) MALARIA.—[1]Jancsó in *C. B., Originale*, Bd. 38, 1905, page 650 ; [2]Novy und Knapp, in *J. I. D.* of 18th May, 1906, page 291 ; [3]Plehn, reported in *B. I. P.* of 15th June, 1906, page 518 : [4]Laveran, reported in *B. M. J.* of 2nd June, 1906, page 1287 ; [5]Proceedings of the 77th Congress of German Scientists and Physicians, reported in *H. R.* of 1st February, 1906, page 158 ; [6]Laveran, reported in *L.* of 28th April, 1906, page 1198 ; [7]Craig, in *P. J. S.* of June, 1906, page 523 ; [8]Craig, reported in *J. A. M. A.* of 9th June, 1906, page 1801 ; [9]Celli, reported in *J. T. M.* of 2nd July, 1906, pages 205 and 209 ; [10]Plehn, reported in *B. I.P.* of 15th June, 1906, page 518 ; [11]Plehn, reported on page 21 of *Epitome* in *B. M. J.* of 11th August, 1906 ; [12]See *S. C. I.* for 1902, page 50 ; [13]Külz, reported in *H. R.* of 15th June, 1906, page 660 ; [14]Morgenroth, reported on page 91 of *Epitome* in *B. M. J.* of 23rd June, 1906 ; [15]Külz, reported in *H. R.*

of 15th June, 1906, page 660; [10]Kendall, in *J. A. M. A.* of 28th April, 1906, page 1266; [17]Gorgas, reported in *J. A. M. A.* of 13th January, 1906, page 148 (see also *J. A. M. A.* of 12th May, 1906, page 1416); [18]Reported in *J. T. M.* of 1st May, 1906, page 141; [19]Proceedings of the 77th Congress of German Scientists and Physicians, reported in *H. R.* of 1st February, 1906, page 158; [20]Demarchi, reported in *J. T. M.* of 2nd July, 1906, page 212; [21]See *S. C. I.* for 1904. page 29.

(2) TYPHOID.—[1]Krencker, in *C. B., Originale,* Bd. 39, H. 1., page 14; [2]Döbert, in *Arch. f. Hyg.,* Bd. 52, H. 1., page 70 (see *Jahresb.* v. W. and P., Bd. 1, Ab. 3, 1905, page 628); [3]Berghaus, in *H. R.* Vol. 15, 1905, pages 761 and 1185; [4]Trommsdorff in *M. M. W.,* 1905, page 1667 (see *Jahresb.* v. W. and P., Bd. 1, Ab. 3, 1905, pages 622 and 628); [6]Klinger, in *A. K. G. A.,* Bd. 24, H. 1, 1906, page 91; [6]Kayser, in *A. K. G. A.,* Bd. 24, H. 1, 1906, page 176; [7]Minelli, in *C. B., Originale,* Bd. 41, 1906, page 406; [8]Wassermann and Citron, in *Z. H.* Bd. 50, 1905, page 331; [9]Brion and Kayser, reported in *J. A. M. A.* of 24th March, 1906, page 919; [10]Mcuisset, Mouriquand and Thévenot, in *J. P. P. G.* of 15th May, 1906, page 508; [11]Stokes and Amick, in *J. H. H. B.* of August, 1905, page 284; [12]Lentz, referred to by Minelli in *C. B., Originale,* Bd. 41, 1906, page 406; [13]Napier, in *B. M. J.* of 26th May, 1906, page 1222.

The bacilli in blood, urine, fæces, etc.—[1]Hirsh, in *J. A. M. A.* of 23rd June, 1906, page 1922; [2]Stern, in *Zeits. f. Klin. Med.,* 1890, Bd. 18 (referred to by Hirsh, loc. cit., page 1923; [3]Castellani, in *B. M. J.* of 5th May, 1906, page 1071; [4]Coleman and Buxton, reported in *B. M. J.* of 25th November, 1905, page 1417; [5]Schottmüller, in *D. M. W.* of 9th August, 1900, and in *M. M. W.* of 1902, page 720 (referred to by Hirsh, loc. cit., page 1923); [6]Conradi, in *D. M. W.* of 11th January, 1906, page 58; [7]Eppenstein and Korte, reported in *D. M. W.* of 28th June, 1906. page 1150; [8]Müller and Gräf, reported in *D. M. W.* of 25th January, 1906, page 157; [9]Kayser, reported in *D. M. W.* of 10th May, 1906, page 770; [10]Fornet, reported in *D. M. W.* of 14th June, 1906, page 970; [11]Pöppelmann, in *D. M. W.* of 14th June, 1906, page 947; [12]Olbrich, in *A. K. G. A,* Bd. 24, H. 1. page 159; [13]Young, in *J. H. R.,* Vol. 8, 1900, page 401 (referred to by Hirsh, loc. cit., page 1923); [14]Hammerschmidt, in *C. B., Originale,* Bd. 40, 1906, page 747; [15]Klinger, in *A. K. G. A.,* Bd. 24, H. 1, page 35; [16]Reischauer, in *C. B., Originale,* Bd. 39, 1905, page 116; [17]Nowack, in *Arch. f. Hyg.,* Bd. 53, page 374 (see *Jahresb.* v. W. and P., Bd. 1, Ab. 3, 1905, page 622); [18]Lejars, reported in *J. A M. A.* of 28th July, 1906, page 311; [19]Kisskalt, in *C. B., Originale,* Bd. 41, 1906, page 701; [20]Forster and Kayser, in *M. M. W.,* 1905, page 1472 (see *Jahresb.* v. W. and P., Bd. 1, Ab. 3, 1905, page 623); [21]Doerr, in *C. B., Originale,* Bd. 39, 1905, page 624; [22]Müller, in *Zeits. f. Heilk.,* Bd. 26, H. 7 (see *Jahresb.* v. W. and P., Bd. 1, Ab. 3, 1905, page 623); [23]Robinson, reported in *J. P. P. G.,* Vol. 7, 1905, page 890, also in *B. I. P.,* Vol. 3, 1905, page 678; [24]Schütze, in *B.K. W.* of 20th November, 1905, page 1465.

Infection by water; bacillary diagnosis.—[1]Reece, reported in *B. M. J.* of 31st March, 1906, page 754; [2]Copeman, reported in *B. M. J.* of 12th May, 1906, page 1120; [3]Reported in *L.* of 17th March, 1906, page 770; [4]Reported in *L.,* Vol. 2, 1905, page 862; [5]Matthes and Gundlach, in *A. K. G. A.,* Bd. 24, H. 1, page 83; [6]Matthes and Neumann, in *A. K. G. A.,* Bd. 24, H. 1, page 116; [7]Beck and Ohlmüller, in *A. K. G. A.,* Bd. 24, H. 1, page 138; [8]Childs, reported in *I. P. H.* of July, 1906, page 441; [9]Russell and Fuller, in *J. I. D.,* Supplement No. 2, February, 1906, page 40; [10]Fehrs., in *H. R.* of 1st February, 1906, page 113; [11]Whipple and Mayer, in *J. I. D.,* Supplement No. 2, February, 1906, page 76; [12]Drigalski, in *A. K. G. A.,* Bd. 24, H. 1, 1906, page 68; [13]Hoffmann, reported in *H. R.* of 15th February, 1906, page 180; [14]Willson, in *J. H.* of October, 1905, page 429; [15]Müller, reported in *C. B., Referate,*

Bd. 37, 1906, page 665 ; [16]Marschall, in *C. B., Originale*, Bd. 38, 1905, page 347 ; [17]Herford, in *A. K. G. A.*, Bd. 24, H. 1, page 62 ; [18]Loeffler, in *D. M. W.* of 22nd February, 1906, page 289 ; [19]Sadler, reported in *H. R.* of 1st March, 1906, page 246 ; [20]Schrwald, in *D. M. W.* of 16th February, 1905, page 261 ; [21]Hirschbruch, reported in *D. M. W.* of 24th May, 1906, page 849 ; [22]Besserer and Jaffé, in *D. M. W.* of 21st December, 1905, page 2014 ; [23]Friedberger and Moreschi, in *D. M. W.* of 25th January, 1906, page 149 ; [24]Péju and Rajat, reported in *B. I. P.* of 15th May, 1906, page 399 ; [25]Terburgh, reported in *D. M. W.* of 11th January, 1906, page 74 : [26]Strüszner, in *C. B., Originale*, Bd. 38, 1905, page 19 ; [27]Savage, in *J. H.* of April, 1905, page 146 ; [28]Mac-Naught, in *J. R. A. M. C.* of July, 1905 ; [29]Vincent, in *A. P.*, 1905, page 233 ; [30]Venema, in *C. B., Originale*, Bd. 40, 1906, page 600.

Infection by milk and other foods.—[1]Konrádi, in *C. B., Originale*, Bd. 40, 1906, page 31 ; [2]Reported in *L.* of 19th May, 1906, page 1439 ; [3]Kayser, in *A. K. G. A.*, Bd. 24, H. 1, 1906, page 173 ; [4]Park, reported in *I. P. H.* of August 1906, page 32 ; [5]Klein, reviewed in *L.* of 14th October, 1905, page 1113.

Infection by contact.—[1]Schian, reported in *D. M. W.* of 18th January, 1906, page 122 ; [2]Childs, reported in *B. M. J.* of 17th March, 1906, page 624 ; [3]Reviewed in *J. H. H. B.*, Vol. 16, 1905, page 348 ; [4]Millard, reported in *J. A. M. A.* of 23rd June, 1906, page 1975 ; [5]Reported in *D. M. W.* of 22nd March, 1906, page 486 ; [6]Baginsky, reported in *H. R.*, Vol. 15, 1905, page 21 ; [7]Conradi, in *A. K. G. A.*, Bd. 24, H. 1, page 97 ; [8]Wright and Archibald, in *B. M. J.* of 9th June, 1906, page 1338 ; [9]Pater and Halbron, reported in *B. M. J.* of 7th April, 1906, on page 49 of *Epitome* ; [10]Klinger, in *A. K. G. A.*, Bd. 24, H. 1, page 91 ; [11]Lentz, referred to in *J. A. M. A.* of 3rd February, 1906, page 365 ; [12]Kayser, in *A. K. G. A.*, Bd. 24, H. 1, page 176 ; [13]Friedel, reported in *H.R.* of 1st April, 1906, page 372.

Soil, dust and flies.—[1]Clauditz, in *H. R.*, Bd. 14, 1904, page 865 ; [2]Rullmann, in *C. B., Originale*, Bd. 38, 1905, page 380 ; [3]Almquist, in *Z. H.*, Bd. 52, 1906, page 179 ; [4]De Franceschi, reported in *H. R.* of 15th January, 1906, page 63 ; [5]Heim, in *Z. H.*, Bd. 50, 1905, page 123 ; [6]Reported in *J. H. H. B.*, 1905, page 348 ; [7]Childs, reported in *B. M. J.* of 17th March, 1906, page 624 ; [8]Nash, reported in *L.* of 26th August, 1905, page 619.

Paratyphoid and meat-poisoning.—[1]Boycott in *J. H.* of January 1906, page 33 ; [2]Smidt, in *C. B., Originale*, Bd. 38, 1905, page 24 ; [3]Bock, in *A. K. G. A.*, Bd. 24, H. 2, 1906, page 238 ; [4]Böhme, in *Z. H.*, Bd. 52, 1906, page 97 ; [5]Kolle, in *Z. H.*, Bd. 52, 1906, page 287 ; [6]Citron, in *Z. H.*, Bd. 53, 1906, page 159 ; [7]Brion and Kayser, reported in *J. A. M. A.* of 4th March, 1906, page 919 ; [8]Grünberg and Rolly, in *M. M. W.*, No. 3, 1905, page 105, reported in *H. R.* of 15th February, 1906, page 197, also in *C. B., Referate*, Bd. 37, 1906, page 280 ; [9]Korte and Steinberg, reported in *H. R.* of 1st June, 1906, page 591 ; [10]Manteufel, reported in *H. R.* of 1st June, 1906, page 592 ; [11]Falta and Noeggerath, reported in *H. R.* of 1st June, 1906, page 595 ; [12]Gaehtgens, in *C. B., Originale*, Bd. 40, 1906, page 621 ; [13]Minelli, in *C. B., Originale*, Bd. 41, 1906, page 583 ; [14]Stühlinger, in *A. K. G. A.*, Bd. 24, H. 1, page 54 ; [15]Kayser, reported in *J. P. P. G.* of 15th May, 1906, page 560 ; [16]Brion, reported in *C. B., Referate*, Bd. 37, 1906, page 548 ; [17]Levy and Fornet, in *C. B., Originale*, Bd. 41, 1906, page 161 ; [18]Krehl, reported in *D. M. W.* of 22nd February, 1906, page 326 ; [19]Curschmann, reported in *D. M. W.* of 7th June, 1906, page 940 ; [20]Pottevin, reported in *H. R.* of 15th July, 1906, page 751 ; [21]Klimenko, in *C. B., Originale*, Bd. 41, 1906, page 617 ; [22]Spencer, in *I. M. G.* of April, 1900, page 122 ; [23]Mackie, in *L.* of 23rd September, 1905, page 874 ; [24]Mohammad in *I. M. G.* of June, 1906, page 219.

Preventive Measures.—[1]Dudgeon and Gray, in *L.* of 7th July, 1906, pages 26 and 36; [2]Eichler, reported in *H. R.* of 15th February, 1906, page 196; [3]Flatau and Wilke, reported in *H. R.* of 15th February, 1906, page 196; [4]Meyerhoff, reported in *D. M. W.* of 15th February, 1906, page 275; [5]Sadler, reported in *H. R.* of 1st March, 1906, page 246; [6]Demetrian, reported in *D. M. W.* of 27th April, 1905, page 685; [7]Güttler, reported in *H. R.* of 1st March, 1906, page 246; [8]Selter, reported in *H. R.* of 15th February, 1906, page 196; [9]Porges and Prantschoff, in *C. B., Originale*, Bd. 41, 1906, pages 466, 546 and 658; [10]Aaser, reported in *H. R.* of 1st March, 1906, page 247; [11]Korte and Steinberg, reported in *C. B., Referate*, Bd. 37, 1906, page 671; [12]Ulrichs, in *H. R.* of 1st July, 1906, page 685; [13]Elischer and Kentzler, reported in *H. R.* of 1st June, 1906, page 594; [14]Bassenge and Mayer, in *D. M. W.* of 4th May, 1905, page 697; [15]Kolle, in *D. M. W.* of 23rd March, 1905, page 449 and reported in *H. R.* of 15th February, 1906, page 198; [16]Morgenroth, in *Arch. f. Schiffs u. Trop. Hyg.* of December, 1905, and reported in *B. M. J.* of 3rd March, 1906, on page 35 of *Epitome*; [17]Porter in *B. M. J.* of 11th August, 1906, page 339; [18]Interim Report of the Antityphoid Inoculation Committee, reviewed in *L.* of 19th May, 1906, page 1402; [19]Jürgens, reported in *H. R.* of 1st March, 1906, page 245; [20]Besredka, in *A. P.* of July, 1905, page 477, in *A. P.* of 25th February, 1906, page 149, and in *A. P.* of 25th April, 1906, page 304; [21]Mac Fadyen, in *B. M. J.* of 21st April, 1906, page 905; [22]Chantemesse, reported in *B. M. J.* of 17th March 1906, page 647, also in *D. M. W.* of 29th March, 1906, page 528; [23]Brunon, reported in *L.* of 10th March, 1906, page 716.

SECTION III.

NATIVE ARMY OF INDIA.

31. The health of the Native troops during 1905 was exceptionally good. The average strength, excluding those on duty in China and Somaliland was 123,434 as compared with 124,055 in 1904. The marginal statement gives a tabular view, enabling an easy comparison to be made, of the statistics of sickness, mortality and invaliding during 1905, the previous year, and the quinquennial period from 1897 to 1901. The rate of admission into hospital was slightly higher than in the previous year but the constantly sick and death rates were the lowest on record.

India: Field Forces, etc., Appendices A and B to Section III, Tables XXVI and LIII.

NATIVE TROOPS.		ALL CAUSES. RATIOS PER MILLE.		
		1897-1901.	1904.	1905.
Admissions	...	794·4	603·2	607·1
Constantly sick	...	30·5	23·8	23·2
Deaths	...	11·92	8·46	8·09
Invalids	...	12·66	12·98	9·02

The low death rate is all the more noteworthy as, out of the total number (998) of deaths during the year, 185 were killed at Dharmsala by the collapse of buildings during the earthquake. The slight increase in the admission rate compared with that of the previous year, which was a record figure, was entirely due to the greater prevalence of the minor and less usual causes of sickness; simple continued fever, "other respiratory diseases," dysentery, remittent fever, diarrhœa and enteric fever, alone, showed increased rates, but the increase arising from the combined rates of these diseases amounted to only 18·5 per mille, while the decrease under the heading of intermittent fever alone was 22·9 per mille. A reduced mortality was recorded during the year from all the principal diseases with the exceptions of enteric fever, circulatory diseases and diarrhœa.

The chief causes of sickness in 1905 were intermittent fever, dysentery, "other respiratory diseases," venereal diseases, simple continued fever and pneumonia, in order of their relative prevalence, the first mentioned accounting for 28 per cent. and dysentery for nearly 5½ per cent. of the total number of admissions from all causes. The admission rate for simple continued fever was more than double that of the previous year, with the result that for the year under review this disease occupies a more important place than pneumonia as a cause of sickness. The chief causes of death were pneumonia, remittent fever and tubercle of the lungs, these diseases accounting for 23·5 per cent., 6·9 per cent. and 6·2 per cent., respectively, of the total number of deaths from all causes. The number of men invalided for discharge from the service during 1905 was 1,113 as compared with 1,610 in 1904, the chief causes of this source of loss being, in order, debility, tubercle of the lungs, rheumatism, venereal diseases and intermittent fever.

If Table XXVI is compared with Table I it will be seen that the Native troops suffered less than the European troops from influenza, small-pox, enteric fever, simple continued fever, diarrhœa, hepatic affections and venereal diseases, but that they suffered more from each of the other chief causes of sickness, as also from scurvy.

The average strength of the China Garrison during the year was 1,753. The chief causes of sickness were influenza, "other respiratory diseases," ague, dysentery and venereal diseases, these together accounting for nearly half of the total number of admissions to hospital ; the admission rate fell from 351 per mille in 1904 to 339 per mille in 1905. There were altogether eight deaths, giving a ratio per mille of 4.56 ; five deaths were attributed to tubercle of the lungs. The average annual strength of the Somaliland Field Force was only 525 as compared with 2,521 in 1904 : admissions numbered 596 and deaths two, yielding rates of 1,135 and 3.81 per mille, respectively, against 930 and 17.06, the corresponding figures in 1904. The most prevalent diseases were ague, dysentery and diarrhœa, the admissions for ague alone amounting to 367, i. e., 62 per cent. of the total from all causes. The Aden Column Field Force, the average strength of which was 695, suffered chiefly from ague, dysentery and diarrhœa, these three together accounting for 64 per cent. of all admissions. The actual number of admissions to hospital was 908, which gives a rate of 1,306 per mille ; there were two deaths, one from pneumonia and the other from ague. The Sikkim-Tibet Mission Force, having an average annual strength of 358, reported 293 cases of sickness and 12 deaths during the year, the ratios per mille per annum being 818.4 and 33.52, respectively. Malaria, dysentery and diarrhœa were the chief causes of sickness, and remittent fever was responsible for seven out of the twelve deaths.

32. Among the commands in 1905, the Northern had the highest admission,

Commands and divisions. Appendix A to Section III, Table XXVI. constantly sick and death rates, due as regards the chief causes of sickness to the relatively great prevalence of pneumonia, " other respiratory diseases," simple continued fever, and tubercle of the lungs, and as regards the chief causes of death, to the excessive mortality from pneumonia, tubercle of the lungs and remittent fever, as well as to the results of the earthquake at Dharmsala. The Eastern Command, although showing higher admission, constantly sick and death rates than in the previous year, still ranked first in order of healthiness. Among the ten divisions, if we omit the 3rd (Lahore) Division, in which the constantly sick and death rates were very high—due to the number of men injured and killed at Dharmsala at the time of the earthquake, morbidity and mortality were greatest in the 2nd (Rawalpindi) Division and lowest in the 9th (Secunderabad) Division.

33. Judged by the statistics of sickness during the decennium 1891-

Geographical groups. Appendices B and C to Section III. Table XXVII. 1900, the geographical groups most unfavourable to the health of the Native troops are, in order, Assam (III), Burma Inland (II), Bengal-Orissa (IV), Hill Stations (XII) and North-Western Frontier and Indus Valley (VII), the admission rates in all of these having been over 1,074 per mille, and the constantly sick rate over 35 per mille as compared with 852 and 32, respectively, the corresponding decennial means for India as a whole. During 1905 it is found that these same groups, with the exception of Burma Inland (II), were the most unhealthy. For the past two years Burma Inland (II) has had admission and constantly sick rates lower than the corresponding means for India : in this group the average strength during 1905 was only 2,916 against 6,083, the average for the decennium 1891-1900, so doubtless during the year under review the troops, thus greatly reduced in numbers, were better housed and in more favourable surroundings than formerly ; this may in part explain the fall in the admission rate from 1,165 for the above decennium to 539 in 1905. It should be noted that the average strength in

Assam (III) was only 949 and in Bengal-Orissa (IV) 2,558, while in Hill Stations (XII) and North-Western Frontier and Indus Valley (VII) the average strengths were 20,224 and 17,946, respectively.   The admission rate for intermittent fever was highest in Bengal-Orissa (IV) and Assam (III); that for dysentery in Bengal-Orissa (IV), Burma Coast (I), Assam (III) and North-Western Frontier and Indus Valley (VII) ; that for remittent fever in Bengal-Orissa (IV) and Hill Stations (XII); that for influenza in Bengal-Orissa (IV); that for pneumonia in North-Western Frontier and Indus Valley (VII).   Enteric fever was most prevalent among Native troops in Upper Sub-Himalaya (VI) and Hill Stations (XII).   Excessively high rates for simple continued fever, ranging from 57·3 to 20·7 per mille were return-ed from the five groups Southern India (XI), Burma Coast (I), Hill Stations (XII), Bengal-Orissa (IV) and Deccan (IX) ; in all these groups the rates reported for 1905 were much higher than the corresponding figures for 1904, and this rise in the rates for simple continued fever was almost invariably coincident with a con-siderable fall in the rates for malaria.   The maximum prevalence of tubercle of the lungs was, as usual, in Upper Sub-Himalaya (VI), Hill stations (XII) and North-Western Frontier and Indus Valley (VII).   The highest death rates during 1905 were returned from Hill Stations (XII) and Western Coast (X), in which groups there was a high mortality from each of the three chief causes of death among Native troops—pneumonia, tubercle of the lungs and remittent fever.

In Appendix C to Section III will be found a comparison of the statistics for some of the chief diseases as they affect Native troops located in the plains and in the hills (1) above 5,000 feet and (2) below 5,000 feet (sea-level).   Hill stations below 5,000 feet in height are, as a general rule, the most unfavourable to the health of Native troops as regards both sickness and mortality ; but during 1905 the death rate was exceptionally high (27·13 per mille) in hill stations above 5,000 feet, owing to the number of men killed in the catastrophe at Dharm-sala.   If the deaths at Dharmsala are excluded, the death rate in hill stations above 5,000 feet is only 7·72 per mille as compared with 9·87, the rate in hill stations below 5,000 feet.   It is worthy of note that, as regards the Native army, the health of troops located in hill stations above 5,000 feet in height compared un-favourably, both in 1904 and 1905, with that of troops located in the plains.

34. In 1905 there were 36 stations with an average strength of over 1,000,

Stations. Tables XXVIII to XXX. Regiments. but in only one of these, namely, Dera Ismail Khan, were the admission and death rates excessive, the rates of this station being 967·0 and 16·52, respectively.   The rates of admission for malaria, dysentery, pneumonia, debility and diarrhœa at this station were all high ; and, of the 38 deaths, 22 were due to pneumonia and 5 to remittent fever.   Among the regiments with the most unhealthy records during the year were the 11th Rajputs and the 75th Carnatic Infantry, both stationed in Mauritius, who suffered severely from malaria, as did also the 120th Rajputana Infantry at Baroda ; and the 1-6th Gurkha Rifles at Abbottabad—ague, simple continued fever, mumps and bronchitis being the chief causes of sickness.

35. The admission rate on account of influenza fell from 1·8 per mille in

Influenza. Appendices B and G to Section III, Tables XXVI to XXIX and XXXI. 1904 to 1·5 in 1905, and the death rate from ·02 to nil. This disease was much less prevalent, during 1905, among the Native troops than among the European troops or prisoners.   The largest numbers of admissions were recorded among the troops at Mardan, Karachi, Fort William and Ferozepore, the Northern and Eastern Commands being chiefly affected. The months of greatest prevalence were January, February and September, the high incidence in September being due to all the cases at Fort William having occurred in that month.

36. There were among the Native troops during 1905, only eleven cases of
Cholera. Appendices A and B, Tables XXVI to XXIX and XXXII. cholera with seven deaths, as compared with 27 cases and 22 deaths in 1904, and 135 cases and 80 deaths in 1903. The Northern Command reported during the year only one case (at Abbottabad) which ended fatally ; the Western Command three non-fatal cases (all at Jhansi) ; the Eastern Command four cases (two at Dinapore and two at Allahabad), all fatal; the Secunderabad Division three cases (all at Madras) with two deaths ; the Burma Division returned no case under this heading. The three cases at Madras occurred in the 88th Carnatic Infantry ; the disease was present in epidemic form in the city. At Allahabad both cases occurred among the men of the 9th Bhopal Infantry and the source of infection could not be traced ; one man was attacked while returning from leave and died a few hours after rejoining his regiment. The case at Abbottabad was attributed to the consumption of infected water or food on the line of march ; cholera was prevalent in the surrounding district. Nothing is said by the medical officers regarding the other cases, which occurred in the 12th Pioneers at Jhansi and the 5th Light Infantry at Dinapore.

37. There were 77 admissions and one death from small-pox during the
Small-pox. Appendices A and B, Tables XXVI to XXIX. year as compared with 56 admissions and three deaths in 1904 ; the rates per mille of strength for 1905 were ·6 and ·01, respectively, the former being higher but the latter lower than the corresponding mean rates for the decennial period 1891-1900. The Western Command was chiefly affected, returning 50 cases out of the total of 77 ; the Eastern Command and the Burma Division reported no cases. The largest number of admissions (seven) at any one station was recorded at Bombay, all among the men of the 113th Infantry ; there were six cases at each of the three stations Dera Ismail Khan, Mhow and Kirkee. The six cases at Dera Ismail Khan, occurred during August among the men of the 55th (Coke's) Rifles ; five had been vaccinated and one bore the marks of small-pox. The men attacked at Mhow had not been vaccinated since enlistment, but bore marks of vaccination in childhood. The patients at Kirkee had contracted the disease in Poona ; all had been vaccinated, and one was deeply marked by an attack of the disease which he had undergone many years before. In many reports no remarks are made on the source of infection or on the presence, on the men attacked, of the marks of vaccination or of a previous attack of the disease. In 1905, as in the previous two years, small-pox was less prevalent among the Native than among the European troops ; the disease is much more fatal to the latter.

38. Intermittent fever accounted for over 28 per cent. of the total
Ague, remittent fever and simple continued fever. Appendices A. B and C, Tables XXVI to XXIX and XXXIV to XXXVI. number of admissions from all causes in 1905, but the admission rate, which was 358 per mille in 1901, 274 per mille in 1902, 247 per mille in 1903 and 194 per mille in 1904, fell in 1905 to 171 per mille, which is less than half the mean (349) for the decade 1891-1900. Of the three commands, the Western was chiefly affected during the year, the next place being taken by the Eastern ; all three showed a decrease in

prevalence as compared with the previous year, but the decrease was greatest in the Northern Command. The admission rates recorded in all but two of the twelve geographical groups were lower than in 1904, the two exceptions being Bengal-Orissa (IV) and Southern India (XI); the incidence of the disease was highest in Bengal-Orissa (IV), Assam (III), Central India (VIII) and Western Coast (X). Taking India as a whole, the month of greatest prevalence was November, December coming next, and fewest cases occurred, as usual, in March. October has been, as a general rule, the month of greatest incidence in previous years; the unusual incidence in 1905 may be due to an outburst of infection, or relapses, after the issue of quinine for prophylaxis had been stopped for the season. The stations (excluding those with an average strength under 100) at which the highest admission rates were recorded, were Baroda (1,319 per mille), Sibi (711), Fort William (662), Ahmedabad (630) and Port Blair (605); at Mauritius the rate was 1,715 per mille of strength. The admission rate of the troops composing the Aden Column Field Force fell from 1,823 per mille in 1904, to 688 in 1905, 478 admissions having been recorded in the latter year on an average strength of 695 men. At Baroda the 120th Rajputana Infantry were very severely affected, the admission rate being more than four times as high as in the previous year. The medical officer states that the civil dispensary books show that the disease was also exceptionally prevalent among the civil population, and reasoning by a process of exclusion he has come to the conclusion that the small Vishwamintri river, running in the Gaekwar's territory and coming within a quarter of a mile and to windward of the lines, is the chief breeding place of mosquitoes. There were many stagnant collections of water along the river side, especially at places where cattle were watered, and it was impossible to deal effectually with these breeding grounds of the mosquito. The increase in prevalence of the disease at Mhow was said to be due to many infected men having arrived from Tibet and the Aden Hinterland; the great increase at Meerut (also Deesa, Indore, etc.) was similarly regarded as due to the arrival of infected troops from very malarious stations. The two regiments located in Mauritius suffered severely; out of a total of 2,000 admissions to hospital, from all causes, of the men of the 11th Rajputs, there were 1,749 admissions on account of malaria (1,658 for ague and 91 for remittent fever); and out of a total of 1,170 admissions from all causes of the men of the 75th Carnatic Infantry, there were 848 on account of ague. The heavy incidence is ascribed to the heavy rains and the general unhealthiness of the island during the year; the men were provided with mosquito nets and quinine was issued prophylactically. For further remarks on malaria as it affected troops in various stations, the reader is referred to paragraph 8 of Section II.

Remittent fever was more prevalent during 1905 than in the previous year, but less prevalent than in 1903, the admission rate during the year being 8·4 per mille. The death rate (·56 per mille), however, was less than that (·71) of the previous year; the disease accounted for about 7 per cent. of the total number of deaths from all causes during 1905 as compared with about 8 per cent. in 1904, about 9 per cent. in 1903 and about 10 per cent. in 1902. The incidence of the disease was, as usual, heaviest in the Northern Command, the Eastern Command coming next; the groups most affected were Bengal-Orissa (IV), Hill Stations (XII), Indus Valley and Frontier (VII) and Gangetic Plain (V). The regiments which furnished most admissions were the 11th Rajputs in Mauritius, the 3rd

Brahmans at Dibrugarh, the 62nd Punjabis at Fyzabad, the 45th (Rattray's) Sikhs at Nowshera and the 2-4th Gurkha Rifles at Pathankote. The seasonal incidence of this disease is very inconstant ; during 1905, the prevalence was greatest in July and least in February.

The admission rate on account of simple continued fever, which in 1903 was 3·5 per mille and in 1904 was 6·7, rose in 1905 to 14·3, the actual numbers admitted to hospital being 441 in 1903, 829 in 1904 and 1,768 in 1905. Among geographical groups, the highest rates during the year were recorded in Southern India (XI), Burma Coast (I) and Hill Stations (XII). Most cases were reported in the month of June, and fewest in February and March. The regiments returning the greatest number of admissions under this heading were the 1-6th Gurkha Rifles at Abbottabad (174 cases), the 1—1st Gurkha Rifles at Dharmsala (178 cases) and the 2nd Queen's Own Sappers and Miners and the 69th Punjabis at Bangalore (151 and 123 cases, respectively), three stations thus furnishing 626 admissions out of a total number of 1,768 recorded throughout India during the year. (See also Section II.)

39. In the whole Native army of India, 130 cases with 35 deaths were attributed to enteric fever during 1905, as com-

Enteric fever. Appendices A. B,
and D to Section III, Tables XXVI to
XXIX and XXXIII.

pared with 70 cases with 16 deaths during 1904, 80 cases with 27 deaths during 1903, 50 cases with 12 deaths during 1902 and 26 cases with 15 deaths during 1901. The admission rate was 1·1 per mille in 1905 and the death rate ·28 against rates of ·6 and ·13, respectively, in 1904. The disease was most prevalent in groups VI (Upper Sub-Himalaya), XII (Hill Stations) and IX (Deccan) ; I (Burma Coast), II (Burma Inland), III (Assam) and IV (Bengal-Orissa) returned no case. There were nearly twice as many admissions as in the previous year, but there were fewer reporting stations and regiments ; 29 stations and 44 regiments recorded one or more cases in 1905 as compared with 33 stations and 46 regiments in 1904. Of the total number of admissions for enteric fever, 31 per cent. (40 cases) were returned from Dehra Dun, 22 per cent. (29 cases) from Abbottabad, 6 per cent. (8 cases) from Peshawar and also from Bolarum ; no other station recorded more than four cases. The regiments which reported most cases were the 1-9th Gurkha Rifles (18 cases), the 2-9th Gurkha Rifles (18 cases), the 1-6th Gurkha Rifles (14 cases), the 2-6th Gurkha Rifles (13 cases) and the 1-2nd Gurkha Rifles (8 cases) among Gurkhas, and the 59th Deccan Infantry (6 cases), the 12th Pioneers (4 cases) and the 38th Dogras (4 cases) among sepoys of other classes ; in 24 regiments only one case was recorded during the year. The special incidence of this disease among Gurkhas was particularly marked, the disease having been very prevalent among them during the year under review ; of 130 cases reported, 83 occurred among Gurkhas (6·0 per mille) and 47 only among sepoys of other classes (·44 per mille). The 83 cases among Gurkhas were reported from nine regiments, whereas the 47 from sepoys of other classes were distributed among 35 regiments. Enteric fever during 1905 was twice as prevalent and twice as fatal among Native troops as among prisoners ; but, although the disease was nearly fifteen times as prevalent among European troops as among Native, the death rate among the former was only about eleven times as high as among the latter (the case-mortality among Native troops was 27 per cent., that among European troops was 19 per cent.).

According to the reports of Medical Officers, many cases were typical clinical pictures of enteric fever; intestinal haemorrhage and perforation occurred in several instances, but phlebitis and thrombosis were not observed. Widal's reaction was carried out in the majority of cases; but no *post mortem* examination seems to have been made. The exceptional prevalence of the disease among the Gurkhas in 1905 is attributed by the regimental medical officers to the large proportion of recruits present owing to recent changes in the number and composition of battalions and the consequent increase in the proportion of young men in the regiments (the majority of those who suffered from the disease were recruits). The exact source of infection in any of these outbreaks was not discovered and medical officers were content to blame water, milk and food in comprehensive manner. At Dehra Dun the medical officers report that the surroundings of the cantonment were insanitary, being polluted to a great extent by the excreta of native villagers; that the river's banks especially were fouled in this way; that the river water was obviously contaminated by sewage; and that flies were abundant at the time when enteric fever was prevalent in the regiments. The medical officer of the 99th Deccan Infantry at Bolarum states that some cases of enteric fever in the bazaar had come to his notice while acting as civil surgeon, and that the outbreak in the regiment occurred after this and was probably caused by specifically infected water, milk or food obtained in the bazaar.

40. There was a decided increase in the recorded prevalence of Malta fever, 43 cases with one death having been recorded during 1905 against five cases in 1904, eight cases in 1903, and four cases in 1902. The only month in which no case was reported was May. At Rawalpindi 17 cases occurred, 16 in the 31st Punjabis and one in the 11th Lancers, the chief incidence being in July : at Ferozepore 14 cases, eight in the 14th Sikhs and six in the 15th Sikhs, with the largest number of admissions in December: at Multan the 27th Punjabis 10 cases (one fatal), six of which were admitted in April. Besides these the 28th Punjabis at Delhi and the 19th Lancers at Ambala furnished one case each.

Malta Fever*. Table LIII.

As regards the epidemic at Rawalpindi, the diagnoses were made on clinical grounds taken in conjunction with the results of the serum test carried out at the Pasteur Institute, Kasauli. In most cases, on admission to hospital, a search for the malarial parasite was made but invariably with a negative result, and quinine in large doses had no effect on the temperature. In one instance the blood was sent to Kasauli three times and each time the serum reaction was negative both for *M. melitensis* and *B. typhosus* ; but after three months' persistent fever of a decidedly undulant type, and intractable neuralgia, the blood was sent a fourth time when the serum was found to agglutinate *M. melitensis* in a dilution of 1 in 80. The medical officer of the 31st Punjabis reports that the admissions took place in batches, suggesting infection by food-stuffs. Several of the men had drunk unboiled goats' milk, which, diluted with water, was commonly used in the lines as a cooling drink. Twelve of the 16 cases were Sikhs, a suggestive fact in view of the special addiction of Sikhs to drinking unboiled goats' milk. At Ferozepore the course of the disease in most of the cases was characteristic and the serum reaction carried out at Kasauli in all cases positive, although the attempt to get a growth of the micro-organism by inoculating agar tubes with splenic blood was not

---

* A summary of the results of recent investigations on the epidemiology of Malta fever is given in Section IX and the references to the literature referred to therein will be found at the end of this Section.

invariably successful. The medical officer of the 14th Sikhs is of opinion that the disease was "most probably contracted from the 27th Punjabis, who brought it with them from Somaliland. Four goats were found infected in the lines and destroyed." The Sikhs had taken over the lines from the 27th Punjabis. On the other hand, the medical officer of the 15th Sikhs expresses the opinion that "the disease has probably been present for years but went undiagnosed." It is noteworthy—in connection with the suggestion that mosquitoes carry Malta fever—that the disease was most prevalent in Ferozepore in the month of December. The epidemic in Multan is of special interest as it was from the blood (splenic puncture) of one of these patients that the *M. melitensis* was first isolated in India and the disease thus shown, beyond dispute, to be endemic in this country. The medical officer of the 27th Punjabis reports that he is unable to trace the source of infection and continues,—" I note in the annual report of 1904, the medical officer states that several cases of very protracted fever were returned under *remittent fever* and he considered this diagnosis unsatisfactory. I may add that I sent specimens of blood from cases which were shown as *remittent fever* in 1904, and 13 out of 16 gave the serum reaction of Malta fever." The solitary case at Delhi presented very indefinite symptoms and signs and suffered from a continuous remittent pyrexia for about 29 days. There was nothing characteristic in the temperature chart. Vigorous treatment with quinine and diaphoretics had no effect on the temperature and the patient rapidly became weak and anæmic. Eventually, a sample of his blood was sent to the Pasteur Institute, Kasauli, and the serum, while having no effect on *B. typhosus*, agglutinated the *M. melitensis* in a dilution of 1 in 1280. The Director of the Pasteur Institute, who then sent the medical officer in charge of the case some agar tubes for insemination with the patient's splenic blood, reported, that he got a good growth of *M. melitensis* from the tubes thus treated. The course of the disease in the Ambala case was fairly characteristic and the serum reaction carried out at the district laboratory at Ambala was stated to be " complete and speedy in dilutions of 1 in 40 and 1 in 80."

In analysing the medical reports on these cases of Malta fever, the frequent absence of one or more of the characteristic signs and symptoms is striking ; in fact the clinical course of some cases—say the solitary case at Delhi—could hardly have raised any suspicion of this disease, unless by a process of exclusion, and yet this case gave a positive agglutinative reaction in a dilution of 1 in 1280. The most constant features were a protracted persistent pyrexia unaffected by quinine, marked constipation, and muscular and articular pains usually noted as neuralgic and rheumatic. It must be remembered, however, that even in Malta the variability of the course of this disease has led to a division into clinical types or groups, e.g., the sub-acute or chronic type with frequent relapses ; the type with sharp fever lasting only for a short time and ending in rapid convalescence with none of the usual complications of the sub-acute or chronic type, etc. It is not, therefore, surprising to find one or two amongst these 43 cases with a sharp bout of fever lasting only a week or ten days and not leading to much debility or anæmia. Such cases in the absence of the serum test would be very liable to wrong diagnosis. The temperature curve, even when fever was protracted, frequently showed no trace of undulations ; it was often very irregular and sometimes remittent, sometimes intermittent. The neuralgic and rheumatic symptoms were frequently absent ; the liver and spleen were not always enlarged ; and often no sweating was observed. Many cases of this disease must occur, which require the serum test to distinguish them from malaria, typhoid, simple continued fever, rheumatic fever and gonorrhœal rheumatism ; and in all differential diagnoses, where Malta fever is a possibility, the serum test should be carried out.   It is now recognised that, if carefully performed with a reliable strain of *M. melitensis*, recently isolated and known to agglutinate with normal human serum only in very low dilution, this test is quite trustworthy.

41. There was a great decrease in the incidence of plague among the Native troops in 1905: in 1898 there were 94 cases with 54 deaths; in 1899, 76 cases with 45 deaths ; in 1900, 56 cases with 29 deaths ; in 1901, 70 cases with 41 deaths ; in 1902, 192 cases with 95 deaths ; in 1903, 184 cases with 115 deaths ; in 1904, 187 cases with 111 deaths ; and in 1905, only 79 cases with 46 deaths. Out of the total of 165 stations occupied by Native troops during 1905, about 111 were situated in districts where plague was prevalent among the general population, yet in only 27 of these did cases occur among the troops.   The largest number of admissions reported from one station was 13 at Bangalore, nine cases occuring in the 2nd Queen's Own Sappers and Miners and four in the 30th Lancers.   The only other stations furnishing more than five cases among Native troops were Poona, in which ten cases occurred among the men of the 94th (Russell's) Infantry and one case among the men of the 121st Pioneers ; Santa Cruz, in which nine cases occurred among the men of the 104th Rifles ; and Nasirabad, in which seven cases occurred among the men of the 116th Mahrattas. No other corps returned more than five cases ; fifteen reporting stations and eighteen regiments recorded only one case each.

Plague.   Cerebro-spinal fever.   Table LIII.

The decrease which has been noted in recent years in the prevalence of cerebro-spinal fever among Native troops has been very conspicuous during the year under review, only one case (fatal), *viz.*, in the 57th Rifles at Peshawar, having occurred in the whole Native army during 1905, against 6 cases with four deaths during 1904 and also in 1903, 18 cases with 17 deaths during 1902, and 28 cases with 21 deaths during 1901.

42. There has been a marked diminution in the prevalence of scurvy during the year, the total number of admissions into hospital for this disease in 1905, being only 213, which gives a ratio of $1\cdot7$ per mille as compared with ratios of $4\cdot0$ per mille in 1904, $2\cdot4$ per mille in 1903 and also in 1902, and $3\cdot2$ per mille in 1901.   In these figures the statistics of troops serving in Somaliland are not included ; in 1905 these troops returned only one case as compared with 899 cases in the previous year.   The statistics, however, of the Aden Column Field Force are included in the above figures and this force, which furnished 94 admissions for this disease in 1904, recorded 2 admissions only in 1905 ; to this reduction is due in part the low rate recorded during the year under review.   About 64 per cent. of the 213 admissions for this disease were recorded among troops in the Western Command and 18 per cent. among those in the Northern ; and, as regards geographical groups, over 39 per cent. of the admissions occurred among troops in Hill Stations (XII) and over 9 per cent. among troops in Central India (VIII), these being the two groups which usually furnish most cases.   The five regiments which were most affected all belonged to the Quetta Division of the Western Command ; these regiments were the 101st Grenadiers (14 cases),

Scurvy.   Tables XXVI to XXIX.

106th Hazara Infantry (12 cases), the 124th Infantry (12 cases), the 126th Infantry (12 cases) and the 105th Mahrattas (11 cases). The medical officer of the 101st Grenadiers at Quetta states that all the 14 men admitted for scurvy had contracted the disease in Somaliland or at Aden, that two of the men had to be invalided as they had lost their teeth and were unfit for further service, and that fresh onions had a very beneficial effect on the course of the disease.

43. It is satisfactory to note that during 1905 both the admission and death rates on account of tubercle of the lungs among Native troops have shown an improvement on the exceptionally low rate recorded in the previous year. The accompanying table shows the admission and death rates for this disease (1) among the Native troops of India as a whole and (2) among Gurkhas, for each year of the past decade. The admission rate of the Native army during 1905 was the lowest recorded since 1897, and that of Gurkhas was the lowest recorded in the decade 1896-1905. The decrease in the death rate is still more marked; the rates for 1905 of both the Native army as a whole and of the Gurkhas were the lowest in the decade, the rate of the latter being exceptionally low. The decrease in prevalence of this disease was most marked in the statistics of the Assam (III) group, the admission rate in this group falling from 7·5 per mille in 1904 to 1·1 per mille in 1905: Upper Sub-Himalaya (VI) recorded the highest admission rate (4·7 per mille) during the year, the Hill Stations (XII) coming next with an admission rate of 4·6 per mille.

Tubercle of the lungs. Appendices A, C, and E, Tables XXVI to XXIX.

ADMISSION AND DEATH RATES FOR TUBERCLE OF THE LUNGS.

Years.	Native Army of India as a whole.		Gurkha regiments.	
	A.	D.	A.	D.
1896 ...	2·5	·65	7·4	5·59
1897 ...	2·6	·79	10·3	5·55
1898 ...	3·5	·63	13·1	4·45
1899 ...	3·3	·68	15·4	5·09
1900 ...	3·7	·78	14·4	4·34
1901 ...	4·2	·84	13·1	3·95
1902 ...	4·3	·80	15·6	4·24
1903 ...	5·9	·69	28·9	2·88
1904 ...	3·9	·51	10·6	2·66
1905 ...	3·1	·50	6·1	1·58

The chief factor determining the position of the various groups, relative to prevalence of this disease, is the number of Gurkhas located therein; and the decline during 1905 in the rates of groups that usually occupy a high position on this list was due chiefly to the great diminution in prevalence of the disease among Gurkhas, and, in the case of Assam, to the reduction in the number of Gurkhas located in that group. As regards commands, the incidence of the disease during 1905 was, as usual, greatest in the Northern Command with an admission rate of 5·4 per mille as compared with a rate of 3·1 per mille for India as a whole. The admission rate during the year of troops, excluding Gurkhas, in the Northern and Eastern Commands was 4·2 per mille, that of Gurkhas serving in the same two commands was 5·2 per mille, the latter ratio being less than half the corresponding figure for the previous year; the death rate, however, of Gurkhas in these two commands, although much lower than in previous years, is still more than twice as high as among troops of other classes located in the same commands. In the whole Native army of India during 1905, this disease accounted for 385 admissions and 62 deaths, of which 85 admissions and 22 deaths occurred among Gurkhas. These figures give admission and death rates for the Native army of India, excluding Gurkhas, of 2·7 per mille and ·36 per mille, respectively, and rates for Gurkhas of 6·1 per mille and 1·58 per mille, respectively. It is noteworthy that in 1905 for the first time the regiment recording the highest number of admissions for tubercle of the lungs was not a Gurkha regiment. The regiments with the largest numbers of admissions were, in order, the 21st Punjabis (16 cases), the 2-1st Gurkhas (9 cases) and the 1-2nd Gurkhas, the 7th Gurkhas, the 12th Pioneers and the 113th Infantry (8 cases each). The 21st Punjabis were stationed in Northern China for the first four months of the year and after arrival in India were at first located at Multan and subsequently at Jhelum : the medical officer reports that the disease was contracted in China and was aggravated by the dust storms of that country, and that the disease was more frequent and fatal among Dogras than among other classes of sepoys in the regiment. The exceptional incidence of the disease among the 7th Gurkhas is attributed by the medical officer to the unusual hardships and the severe weather to which the men were exposed at Dharmsala and Bakloh.

There were six cases returned as non-tuberculous phthisis during the year, as compared with three in 1904, six in 1903 and 23 in 1902.

44. The admission rate for pneumonia in 1905 was 12·5 per mille and the death rate 1·90 per mille as compared with 11·7 and 2·32 per mille, respectively, in 1904. The increase in prevalence, compared with the previous year, was due to the greater incidence of the disease in the (VI) Upper Sub-Himalaya and (IV) Bengal-Orissa groups, each of which records an increase in the admission rate of about 5 per mille over that of 1904. Despite its diminished fatality this disease was still the chief cause of death among Native troops in 1905, accounting for over 23 per cent. of the total number of deaths. The North-West Frontier and Indus Valley (VII) retained its position as the group in which pneumonia was most prevalent during the year, Upper Sub-Himalaya (VI) and Central India (VIII) coming next; in (XII) Hill Stations the disease was not so prevalent as in previous years. The months of greatest incidence were, as usual, January, February and December. Among stations with an average strength over 100 the highest admission rates were recorded at Kherwara (91·2 per mille) and Kotra (46·7 per mille), both in the (VIII) Central India group and at a height of over 1,000 feet above sea-level. As regards regiments, the largest numbers of cases during the year occurred in the Meywar Bhil Corps at Kherwara, the 4th Rajputs at Malakand, the 38th Dogras at Mian Mir, the 52nd Sikhs at Kohat and the 59th Rifles at Edwardesabad. Appendix H to Section III shows the relative prevalence of and mortality from pneumonia during 1905 among the European troops, the Native troops and prisoners : the disease was much more prevalent but much less fatal among the Native troops than among prisoners. Among European troops pneumonia was more prevalent and fatal but among prisoners less prevalent and fatal than in the previous year. The medical officers of the 4th Rajputs and 52nd Sikhs consider the extreme cold experienced during December, January and February, especially on manœuvres, as in part responsible for the prevalence of the disease.

Pneumonia and other respiratory diseases. Appendices A B. C. and H. to Section III. Tables XXVI to XXIX and XXXVII.

The highest admission rates for other respiratory diseases during 1905 were recorded in the Northern Command and, among groups, in (XII) Hill Stations, (III) Assam and (VII) North-West Frontier and Indus Valley, the last named being the group from which the highest admission rate on account of pneumonia was returned. The admission rate of the whole Native army of India on account of other respiratory diseases rose from 21·1 per mille in 1904 to 28·5 per mille in 1905, but the death rate remained practically the same—·25 per mille in 1904 and ·23 per mille in 1905. Excluding stations where the average strength was less than 100, the highest admission rates were recorded at Mir Ali Khel (147·8) Fort Jamrud (111·1), Abbottabad (106·1), Thal (96·6), Kherwara (82·4), Ootacamund (75·8), Kirkee (68·1), Bhuj (67·5), Quetta (66·6), Ahmedabad (65·0) and Schore (63·3); the average strengths of Fort Jamrud, Mir Ali Khel and Bhuj were under 200. Bhuj furnished the high death rate from other respiratory diseases of 6·13 per mille. Among regiments the 1—6th Gurkhas at Abbottabad, the 64th Pioneers at Ootacamund, the 3rd Sappers and Miners at Kirkee and the 54th Sikhs at Nowshera recorded the highest numbers of admissions during the year. These diseases were most prevalent during the cold months, and, according to the medical officers, were largely due to exposure to cold on field days.

45. During 1905 the admission and death rates on account of dysentery were 32·5 and ·18 per mille, respectively, as

Dysentery and diarrhœa.
Appendices A, B and C. Tables
XXVI to XXIX and XXXVIII. compared with 31·5 and ·27, respectively, in 1904, the latter rates having been the lowest recorded up to that date. There has thus been a slight increase in the admission rate, but the death rate of the year is the lowest on record. Although the admission rate on account of this disease of the Native army as a whole has risen, eight of the twelve geographical groups showed a decrease in prevalence compared with the previous year, the four exceptions being (IV) Bengal-Orissa (the admission rate being about 35 per mille more than during 1904), (V) Gangetic Plain (VIII) Central India and (XII) Hill Stations. The disease was most prevalent during 1905 in (IV) Bengal-Orissa, (I) Burma Coast, (III) Assam and (VII) North-West Frontier and Indus Valley. Omitting stations with a strength below 100, the highest admission rates were recorded during the year at Barrackpore (177·5), Aden (157·6) and Fort William (131·1), the first and last named belonging to the (IV) Bengal-Orissa group : the highest death rates were reported at Kotra (6 67), Kherwara (2·94) and Fort William (1·62). Among regiments dysentery was most prevalent in the 95th Russell's Infantry at Singapore, the 102nd Grenadiers at Mhow, the 8th Rajputs at Lucknow, the 2-10th Gurkhas at Lansdowne and the 78th Moplah Rifles at Dera Ismail Khan. The medical officer of the 95th Russell's Infantry states that the epidemic in that regiment was caused by the well-water being contaminated. The prevalence of the disease in the 102nd Grenadiers was attributed to the food and water of the men having been unavoidably exposed to dust and sand while on service in the Aden Hinterland.

The admission and death rates on account of diarrhœa rose from 6·7 and ·05 per mille, respectively, in 1904 to 7·5 and ·06, respectively, in 1905. This disease, like dysentery, was most prevalent in the Eastern Command and, among groups, in (IV) Bengal-Orissa, (III) Assam and (X) Western Coast. Excluding the smaller stations, the highest admission rates were recorded at Fort William (93·9), Dibrugarh (37·2), Alipore and Barrackpore, all, except Dibrugarh which is in Assam, being in Bengal-Orissa. The total number of deaths from this disease among the whole Native army was only seven. Several of the medical officers in charge of regiments at the above stations report that the chief causes of the prevalence of diarrhœa were insufficient or improper cooking of food and chills caught while bathing before meals.

46. Venereal diseases were nearly eight times more prevalent among the

Venereal diseases. Appon-
dices B and F, Tables XXVI to
XXIX.

European troops during 1905 than among Native troops, the admission rate being only 19·6 per mille of strength among the latter as compared with an admission rate of 153·7 per mille of strength among the former; or, otherwise expressed, among European troops an annual average strength of 71,343 gave 10,966 admissions for venereal diseases, while among Native troops an annual average strength of 123,434 gave only 2,419 admissions. The admission rate in 1905 of the Native troops of India, excluding those serving in China and Somaliland, was 1·0 per mille less than in the previous year; the admission rate of the China Garrison under this heading fell from 43·8 per mille in 1904 to 20·0 per mille in 1905 and that of the Somaliland Field Force from 11·1 in 1904 to 1·9. Altogether 7 men died from, and 79 men were invalided for these diseases during the year as compared with 6 and 107, respectively, the corresponding figures for 1904. During 1905, venereal diseases were, as usual, least prevalent among Native troops located in the Northern Command, but the admission rate of this Command is 5 per mille greater, while those of the Western and Eastern Commands are, respectively, 4 and 2 per mille less, than in 1904; consequently, the differences in prevalence of these diseases in the three commands are very slight during the year under review. The admission rate on account of these diseases among Gurkhas in the Northern and Eastern Commands was 34·7 per mille (the same as in 1904); that among Native troops of other classes located in these two commands was only 14·5 per mille; that of the whole Native army of India, excluding Gurkhas, was 17·6. These figures show that the Gurkhas suffered much more than the sepoys of other classes and that the latter suffer much less in the Northern and Eastern Commands than in other parts of India. Excluding stations at which the average strength of Native troops was less than 100, the highest admission rates for venereal diseases were recorded at Bakloh, Abbottabad and Dharmsala in the Northern Command; at Bhuj, Bombay and Satara in the Western Command; at Naini Tal, Shillong and Kohima in the Eastern Command; at Bellary and Vizianagram in the Secundera-bad Division; and at Maymyo and Fort Stedman in the Burma Division. Of these stations, Bakloh, Bombay, Satara, Kohima, Vizianagram, Maymyo and Fort Stedman appeared in the corresponding list in this report for 1904. For Native troops in India as a whole, the admission rate for syphilis was 7·7 per mille, for soft chancre 4·6 per mille and for gonorrhœa 7·4 per mille, as compared with 8·8, 4·5 and 7·3 per mille, respectively, in 1904.

47. During 1905 beri-beri accounted for only 15 admissions and two deaths

Beri-beri. Table LIII.

against 49 admissions, and five deaths in 1904 and 122 admissions and eight deaths in 1903. Of the 15 admissions during the year, ten occurred in two regiments, viz., seven in the 81st Pioneers stationed in the early part of the year at Secunderabad but during the latter part of the year on service in the Aden Hinterland, and three in the 73rd Carnatic Infantry at Aurangabad. The other cases were reported, one each, in the 96th Berar Infantry at Secunderabad, in the 20th Deccan Horse at Bolarum, in the 97th Deccan Infantry at Bolarum, in the 63rd Palamcottah Light Infantry at Vizianagram and in the 88th Carnatic Infantry at Madras. The medical officer of the 81st Pioneers states that all the cases in that regiment were of the wet type except one (that of a man, admitted in the Aden Hinterland, who had twice suffered from the same disease in India), which was of the dry variety.

The patients were isolated and the quarters they had occupied disinfected with a view to check the spread of the disease. In 1904 there were six cases of beri-beri in the 81st Pioneers at Secunderabad ; and in the same year, the 73rd Carnatic Infantry, which was at that time stationed at Singapore, recorded 29 admissions for this disease.

48. There was a considerable decrease in the number of admissions for
Guinea-worm. Tables XXIX and LIII. guinea-worm in 1905, the total being 609 against 730 in the previous year. The largest number of cases occurred as usual, in the (VIII) Central India group, the (IX) Deccan, the (XII) Hill Stations and the (VII) Indus Valley groups coming next. Over 15 per cent. of the total number of cases were reported from Kherwara, where the Mewar Bhil corps is stationed, the number of cases in this regiment during the year amounting to 110, which yields a ratio of 262 per mille of strength. The disease is very common in the native Bhil villages. Excluding Kherwara, cases were most numerous at Poona and Quetta.

49. During the ten years, 1895-1904, there were altogether 154 cases of
Suicide. Table LIII. suicide, giving an average of nearly 15 per annum. In 1905, there were 14 cases, of which nine were by gun-shot, three by hanging, one by drowning and one by opium-poisoning.

REFERENCES.

For explanation of abbreviations see end of Section II.

MALTA FEVER.—*S. C. I.* for 1904, p. 140; *Reports of the Mediterranean Fever Commission,* Parts I—IV ; Surgeons Ross and Levick, R.N., in *B. M. J.,* Volume I, 1905, p. 710; Surgeon Ross, R.N., in *J. T. M.* of 15th January, 1906, p. 17 ; Surgeon Ross, R.N., in *B. M. J.* of 28th April, 1906, p. 975 ; Levick in *L.* of 3rd February, 1906, p. 317; Forster in *L.* of 17th February, 1906, p. 441 ; Reid in *L.* of 7th April, 1906, p. 971 ; Birt in *B. M. J.* of 28th April, 1906, p. 976; Birt in *L.* of 3rd February, 1906, p. 314 ; Bruce in *J. R. A. M. C.,* of March, 1906, p 330 ; Phillips in *J. T. M.* of 15th January, 1906, p. 23 ; Surgeons Ross and Levick, R. N., in *I. M. G.* of January, 1906, p. 27 ; *Nature* of 7th September, 1905, p. 462 ; *L.* of 30th December, 1905, p. 1931 ; *I. M. G.* of April, 1906, p. 147 ; *J. T. M.* of 1st July, 1905, page 193, of 1st November, 1905, p. 323, and of 1st May, 1906, p. 138 ; *B. M. J.* of 28th April, 1906, p. 999 ; *I. M. G.* of November, 1905, p. 432 ; Kennedy in *J. R. A. M. C.* of April, 1906, p. 408.

SECTION IV.

JAILS OF INDIA.

50. The importance of prison sanitation in India may be measured by the fact

India.

that close upon half a million persons pass through the prisons every year. It is too much to hope perhaps that any considerable proportion of them should profit by lessons of order and cleanliness that they have been taught during their imprisonment ; but it is of great importance that none of them should carry away with them from the prison the seeds of disease.

The more closely the problems of jail hygiene are examined, the more evident does it become that the secret of health in the prison is the exclusion of disease from outside, by thorough medical inspection and segregation of newly admitted prisoners, and the limitation of the evil results of any failure of these precautions by making the spread of contagion and the development of latent disease unlikely.

These preventive measures require suitably located, well drained jail sites, properly constructed roomy buildings, the control of the water and food supplies and efficient management. If these conditions are assured, sickness among convicts should be reduced to a minimum and the death rates should be much lower than the death rates of the free population—the comparison being made not with the crude general death rates but with the much lower rates among men and women of the same ages as the prisoners. Prison death rates, particularly in district jails, depend largely on the health of the prisoners that have to be admitted, but proper appliances, well managed, should banish contagious diseases from the prisons. The proof of the success of jail hygiene is not the low death rate of a prosperous, healthy year, but the steady growth of the difference between the jail death rates and the general death rates in unhealthy years.

In 1904, general conditions were peculiarly favourable to the health of the prisoners ; in 1905 conditions in most parts of the country were unfavourable. The weather in the early part of the year was exceptionally cold and raw, the monsoon rains began late and ended early, and the heat of the summer and autumn was excessive. The cold weather crops of 1904-05 were below the average everywhere, except in the Punjab, the Frontier Province, Assam and Burma ; and the *kharif* was bad or below normal in Bengal, Eastern Bengal and Assam, in parts of Madras, Bombay and the United Provinces, in the unirrigated areas of the Punjab, and in the Frontier Province. Except in Bombay and the Punjab the health of the general population was below the average and the death rate, in spite of the great reduction in plague mortality, was higher than in 1904.

The mean daily number of prisoners of all classes in India and the Andamans during 1905 was 106,265, or 1,940 more than in the previous year. The prison populations were slightly lower than in 1904 in the United Provinces, the Punjab and the Central Provinces, elsewhere they were higher, notably in Burma and Madras where the mean daily strengths were greater by 935 and 827, respectively.

In the Andamans an addition of 376 to the mean daily strength of 1904 brought the figure of 1905 to 14,348, which is more than a seventh of the Indian daily prison population. It is necessary to treat the statistics regarding con-victs in the Andamans separately because the conditions under which they live differ widely from those under which the inmates of prisons of India and Burma live, their sick rate being nearly three times, and their death rate twice as large as the Indian means.

In the jails of India and Burma the mean daily strength in 1905 was 91,917, and the following are the salient statistical facts of the year 1905 compared with those of 1904 and the years 1901-1904. The population of 1905 was greater by 1,564 than that of the previous year, but less by 4,453 than the four years' mean; the rate of admission to hospital was 647 per thousand, or 29 per thousand less than the rate in 1904 which was the lowest on record, and 138 less than the four years' mean; the number constantly sick was 28 per thousand, iden-tical with the figure of 1904, but 4 below the four years' average, and the death rate was 19·23 per thousand, 1·62 higher than the unprecedentedly low rate of 1904, but 2·35 below the four years' average. The admission rates in all the various administrations were lower in 1905 than the respective means of the pre-vious four years, but only in Burma, the United Provinces, the Punjab and the Frontier Province were the admission rates lower than in 1904. The constantly sick rates were higher in 1905 than in 1904 in Eastern Bengal and Assam, in Bengal, the Frontier Province, the Central Provinces and Bombay; and in the first three of those provinces they were higher than the means of the four years 1901-1904. The death rates of 1905 were lower than those of 1904 in Bombay, Burma, the Punjab and Madras, elsewhere there was an increase, and in Burma, Eastern Bengal and Assam and in Bengal the death rates of 1905 were higher than the four years' average. The lowest death rates were 16·33 per thousand in the Punjab,—which, with the exception of the rates, 15·89 and 16·05 recorded in 1896 and 1897, respectively, is the lowest in the history of the province—15·79 in the Central Provinces, and 15·87 in Madras, where the next lowest rate on record was 16·13 in 1899.

Owing to the transfer of fourteen Bengal districts to the new province of Eastern Bengal and Assam, the province of Bengal lost the central jails of Dacca and Rajshahi and twelve district jails with a total accommodation for 4,749 prisoners; and by the transfer of Sambalpur from the Central Provinces to Bengal, those provinces lost and Bengal gained one district jail accommodating 155 prisoners.

The daily average number of convicted prisoners in 1905 was 86,375, among whom the death rate was 19·23 per thousand, the rates ranging from 32·02 in Eastern Bengal and Assam to 14·62 in the Central Provinces. In the following statement are compared the average daily convict populations of central and district jails and their admission and death rates in the several administrations in the years 1901 to 1905 inclusive.

Administration.	1901.			1902.			1903.			1904.			1905.		
	Average strength.	Ratio per 1,000 of Strength.		Average strength.	Ratio per 1,000 of Strength.		Average strength.	Ratio per 1,000 of Strength.		Average strength.	Ratio per 1,000 of Strength.		Average strength.	Ratio per 1,000 of Strength.	
		Ad.	D.		Ad.	D.		Ad.	D.		Ad.	D.		Ad.	D.
Bengal excluding in 1905 jails transferred to Eastern Bengal. { Central...	10,202	1,064·8	22·35	10,368	1,054·1	22·57	(Bengal as a whole) 9,316	1,015·8	17·50	9,519	803·7	18·91	7,631	1,041·0	25·55
District ...	3,979	1,197·9	33·07	9,268	1,084·6	29·56	8,160	1,162·7	30·76	8,483	1,154·1	21·81	5,885	931·5	23·96
Assam including in 1905 the jails of the new Province of E. B. & Assam. { Central ..	...	...	...	...	...	Assam	...	...				...	1,386	608·2	24·92
District ..	1,463	1,067·7	25·29	1,262	890·5	50·10	1,296	809·3	29·32	1,430	650·6	28·68	1,192	1,248·4	35·06
United Provinces. { Central ..	12,204	815·3	25·0	11,475	653·5	23·4	9,387	642·8	12·9	9,313	502·7	12·78	9,389	489·8	15·76
District ...	15,323	853·5	24·54	14,563	703·5	18·40	11,958	947·9	17·39	12,777	687·6	15·34	12,292	620·0	18·81
Panjab ... { Central ...	6,366	444·0	29·69	5,743	124·5	26·44	4,858	85·0	18·94	4,822	915·2	25·09	4,570	714·9	17·94
District ...	6,278	1,141·6	24·85	6,179	1,100·3	25·25	6,311	1,056·3	19·71	5,930	961·2	15·85	5,951	740·3	16·10
North-West Frontier Province. { Central ..	...	...	...	...	...	...	...	...	...	...	...	...	...	...	...
District ..	881	108·7	9·08	893	110·3	28·00	971	121·3	16·48	1,055	118·2	16·11	1,077	1,053·9	21·36
Central Provinces. { Central -	4,023	801·4	26·85	3,465	957·3	23·09	2,176	772·1	16·82	2,468	489·9	10·13	2,478	574·0	14·06
District -	1,729	925·4	28·34	1,418	926·0	25·39	1,113	1,002·3	22·45	1,085	847·9	18·43	864	862·3	16·20
Madras ... { Central ...	7,181	584·0	22·70	7,012	499·1	18·32	5,775	447·6	18·35	6,120	439·9	15·01	6,695	506·6	15·98
District ..	3,635	510·6	26·08	3,660	351·9	17·76	3,124	578·7	20·49	3,143	591·8	18·45	3,436	554·1	16·01
Bombay ... { Central ...	4,064	1,483·5	27·04	3,846	707·0	33·80	3,026	347·3	19·17	3,039	556·8	17·11	3,092	654·6	12·94
District ...	6,358	767·7	30·16	6,207	721·4	26·58	4,889	654·8	31·70	4,868	639·1	20·95	4,836	583·3	20·26
Burma ... { Central ...	7,246	513·8	14·40	6,823	510·6	13·34	6,691	460·3	14·05	6,962	381·1	21·69	7,491	275·9	17·62
District ...	4,284	603·7	16·18	4,253	573·0	18·58	3,981	501·6	22·61	4,151	330·5	14·21	4,460	367·7	17·04
Total of the above Provinces. { Central ..	51,286	922·0	23·73	48,768	791·9	22·25	41,729	680·8	16·30	42,252	591·5	17·51	43,172	598·8	18·18
District ..	48,910	908·4	27·54	47,803	839·3	23·20	41,833	848·1	23·21	42,222	791·3	17·90	43,203	723·2	20·39

In India as a whole the convict populations are fairly equally divided between central and district jails, but, while the central jail populations are considerably the larger in Bengal, and much the larger in the Central Provinces, Madras and Burma, in the other administrations the district jail populations are the greater. In India as a whole the admission and death rates are lower among central than district jail populations, and this is the rule with few exceptions in the several administrations ; it is therefore important when comparing provincial death rates to take into account the proportions of the total populations confined in central and district jails. It will be observed that the death rate in central jails, 18·18 per thousand in 1905, is higher than the rates in both 1903 and 1904, but the death rate in district jails, 20·39, while higher than that of the preceding year, is lower than those of the other years.

51. The following table shows the distribution according to length of time spent in prison, with the actual number of deaths and the death rates of convicts in the jails in the years 1901 to 1905:—

Central and District Jails.

Year	Jail	Measure	Not exceeding six months.	Above six months and not exceeding one year.	Above one year and not exceeding two years.	Above two years and not exceeding three years.	Above three years and not exceeding seven years.	Above seven years.	Total.
1901	Central Jails	Strength	14,368	9,769	9,637	6,690	7,221	3,434	51,119
		Deaths	305	212	269	145	205	81	1,217
		Ratio per 1,000 of strength	21·2	21·7	27·9	21·7	28·4	23·6	23·8
	District Jails	Strength	26,993	10,547	6,766	2,447	1,784	413	48,950
		Deaths	818	277	180	28	36	8	1,347
		Ratio per 1,000 of strength	30·3	26·3	26·6	11·4	20·2	19·4	27·5
1902	Central Jails	Strength	12,789	8,961	8,941	6,941	7,332	3,565	48,529
		Deaths	280	188	206	122	220	69	1,085
		Ratio per 1,000 of strength	21·9	21·0	23·0	17·6	30·0	19·4	22·4
	District Jails	Strength	25,610	10,697	6,623	2,584	1,883	361	47,758
		Deaths	588	278	148	49	38	8	1,109
		Ratio per 1,000 of strength	23·0	26·0	22·3	19·0	20·2	22·2	23·2
1903	Central Jails	Strength	11,122	7,463	7,970	5,715	6,862	2,486	41,618
		Deaths	181	112	137	91	122	37	680
		Ratio per 1,000 of strength	16·3	15·0	17·2	15·9	17·8	14·9	16·3
	District Jails	Strength	23,242	9,612	5,314	2,279	1,650	209	42,306
		Deaths	587	205	108	28	37	7	972
		Ratio per 1,000 of strength	25·3	21·3	20·3	12·3	22·4	33·5	22·9
1904	Central Jails	Strength	11,446	7,605	8,328	5,642	6,749	2,475	42,245
		Deaths	196	132	151	70	140	51	740
		Ratio per 1,000 of strength	17·1	17·4	18·1	12·4	20·7	20·6	17·5
	District Jails	Strength	23,849	9,567	5,348	2,192	1,872	213	43,041
		Deaths	464	157	90	35	24	2	772
		Ratio per 1,000 of strength	19·5	16·4	16·8	16·0	12·8	9·4	17·9
1905	Central Jails	Strength	12,322	7,963	8,606	5,733	5,936	2,925	43,483
		Deaths	203	143	141	76	174	48	785
		Ratio per 1,000 of strength	16·47	17·96	16·38	13·26	29·31	16·41	18·05
	District Jails	Strength	23,285	9,771	5,741	2,315	2,063	236	43,413
		Deaths	477	195	125	29	42	8	876
		Ratio per 1,000 of strength	20·49	19·96	21·77	17·53	20·36	33·90	20·18

Column groupings: 1. STRENGTH   2. DEATHS   3. RATIO PER 1,000 OF STRENGTH.

It will be observed that even in the central jails a large proportion of the total population have been in confinement for less than one year, and that the death rates in the central jails among such prisoners are much lower than the rates among them in the district jails ; that the death rates in district jails are highest in the first six months of imprisonment ; and that in the central jails the highest rate is invariably recorded among prisoners who have spent between three and seven years in prison.

52. The chief causes of sickness were intermittent fever (admission rate **Causes of sickness and mortality.** 181·7 per thousand), dysentery (81·6), abscesses, ulcers and boils (71·2), diarrhœa (39·2) and respiratory diseases other than pneumonia (26·4) ; while the principal causes of death were tubercle of the lungs (death rate 3·19 per thousand), dysentery (3·01) and pneumonia (2·28).

53. There were 16,704 prisoners admitted to hospital on account of intermit- **Intermittent Fever.** tent fever, or 2,821 fewer than in 1904, and little more than half the number admitted from the somewhat larger population (105,020) in 1901. This disease was, as usual, most prevalent in October, and the admission rates were highest in the North-West Frontier Province (344·0 per thousand) and Bengal (290·7), the rates ranging in the other provinces from 260·1 per thousand in Eastern Bengal and Assam, to 85·6 in Madras and 51·0 in Burma. The mean death rate ascribed to intermittent fever was ·91 per thousand. In the Central Provinces and North-West Frontier Province no death was recorded ; in Bengal and Eastern Bengal and Assam the death rates were 2·75 and 1·87 per thousand, respectively, and in the remaining provinces the rates varied between ·79 in Madras and ·13 in Bombay.

54. There has been a great reduction during the last five years in the numbers **Remittent Fever.** of cases and deaths returned under the designation *remittent fever* ; but in contrast to the gradual reduction in the numbers of cases of intermittent fever, the fall in the numbers of cases of remittent fever has been abrupt. In 1901 there were in all the jails 295 admissions and 41 deaths ascribed to this disease, and in the following year the number of admissions rose to 346 and there were 40 deaths ; in 1903 the number of admissions and deaths fell to 143 and 19, respectively ; in 1904 the figures were 102 and 15, and in 1905 the number of admissions increased by seven to 109 and the number of deaths by five to 20.

During the quinquennium ending with 1905, there were in all 995 cases of remittent fever, of which 647 occurred in three provinces, Bombay (314), Bengal (174), and Burma (159) ; and in the different provinces the case-mortality varied greatly between 29·6 per cent. in the Punjab and 5·3 per cent. in the Central Provinces. The vast majority of the cases—875 out of 995—occurred in district jails and the case-mortality among them was 12·5 per cent., as compared with 21·7 per cent. among the 120 cases reported from central jails. In 1905 the largest numbers of cases occurred in Bengal (21), the United Provinces (21) and Madras (19) ; but while in the United Provinces the largest number of cases in any one jail was five, in Bengal there were 13 cases in the district jail at Champaran, and in Madras there were 12 cases in the district jail at Tanjore.

As usual the disease was least prevalent in the cold weather, but in no other respect does its incidence agree with that of intermittent fever. The reports on fatal cases do not generally throw a clear light upon the nature of the disease although in many instances the disease does not seem to have been malarial. We have, then, a disease more frequent in some provinces than in others, less fatal but more frequent in district than in central jails, the incidence of which, in the course of a single year, declined by half ; the question arises, what is its nature ? Many cases are no doubt severe attacks of malarial fever, others are *kala azar* and others again are confessions of failure to diagnose. Our knowledge of the nature of Indian fevers becomes wider and more exact every year, but it is still far from complete, and it is just these obscure cases of fever which are likely to give an opportunity to the pathologist to enlarge our knowledge.

Quinine prophylaxis is in general use and so also, apparently, are measures to prevent the breeding of mosquitoes in jails. Little mention is made by Inspectors-General of special measures undertaken in the jails or of the results obtained by them, and this omission is perhaps justified by the familiarity of the subject and by the great reduction of malaria among prisoners in recent years. The very fact of success should, however, encourage us to redoubled efforts, for a great deal can still be done. Making every allowance for the varying prevalence of malaria in different localities and in different years, the mean admission rate represents extraordinary variations in the different administrations and in the different jails under them. There were in 1905 seven jails with a mean daily strength of over 100 prisoners, in which the admission rate for intermittent fever was over 600 per thousand—to take a figure more than three times as great as the mean—and in some of these the frequency of malarial fevers does not appear to have attracted the special notice of the medical Superintendent. The explanation in some cases, no doubt, is the fact that in other respects the prisoners were very healthy with low death rates, and in others the difficulty of draining of the jail and its surroundings is considered insuperable from want of funds. It is, however, wiser in the long run to spend money on drainage than on hospital charges to which has to be added the loss of labour ; and in most cases the reports appear to warrant the hope that well devised efforts intelligently directed and persistently carried out would be successful.

55. The total number of recorded cases of dysentery in 1905 was 7,496, equal to an admission rate of 81·6 per thousand which is considerably lower than the rate in the previous year (85·7), and the mean of the previous four years (89·9). The number of deaths was 277 equal to a death rate of 3·01 per thousand, which, while slightly higher than the exceptional rate (2·91) of 1904, is considerably lower than the four years' average (3·87). As compared with 1904 there was a diminution in the number of cases in all administrations except Eastern Bengal and Assam and Madras, the decrease being specially remarkable in Burma, where the number fell from 561 to 358, and in the Punjab, where it fell from 570 to 327. In India generally the disease was most prevalent in August and least prevalent in February, the important provincial exceptions being the United Provinces and the Punjab, in both of which the maximum occurred in September, and Madras where it occurred in October. The dysentery death rates in 1905 were higher than in 1904 in all provinces except Bombay, where the rate remained stationary, Burma where the reduction was ·21 per mille, the United Provinces where the fall was ·40 per mille and the Punjab, where the death rate decreased from 2·56 to ·52 per thousand. The case-mortality varied greatly, ranging from 10·0 per cent. in the North-West Frontier Province to 1·8 per cent. in the Punjab, the mean for India being 3·7, and the percentages in all provinces except the North-West Frontier Province, Burma (9·2), United Provinces (5·3), and Madras (4·8) being below the mean.

Dysentery (1).

It is a misfortune when a disease is given a pseudo-scientific name from a symptom, for, unless this symptom is peculiar to one disease, a serious obstacle is placed in the way of the discovery of its etiology and when this is discovered there is confusion regarding the assignment of the name.

Dysentery affords a striking example of the truth of this statement. The symptom from which the name is taken is common to several diseases, it has taken a comparatively long time to discover their etiology, and great confusion prevails regarding their causes. Our knowledge has, however, advanced rapidly in recent years and the problem is so nearly solved that it is perhaps possible to classify the diseases with some approach to correctness. We may divide dysenteries into three classes: (*a*) epidemic or bacillary dysentery, caused by one or other type of the dysentery bacillus; (*b*) tropical, endemic or amœbic dysentery caused by the *Entamœba histolytica* of Schaudinn; and (*c*) false dysentery, including diseases characterised by discharges with more or less dysenteria, caused by inflammation of the bowel from any other cause, *e.g.*, bacteria, *Entamœba coli* and *Balantidium coli*.

As bacillary dysentery is not necessarily epidemic and may be endemic, and as amœbic dysentery occurs outside the tropics' in places where it is not endemic, it is desirable to avoid alternative names and adhere to the terms bacillary dysentery and amœbic dysentery.

56. There are at least two well known types of the bacillus of dysentery—

Bacillary Dysentery.   the Shiga-Kruse and the Flexner.' On the one hand, however, authorities are agreed that these are not specifically distinct germs, but rather races of the same germ displaying biological differences ; on the other, observers by the discovery of minor differences in the fermentation and agglutination reactions of the bacilli of various epidemics are adding to the number of types, thus Hiss[3] describes four types, and Shiga,[4] writing of the dysenteries of Japan, five.

It is probable that there is no need to depart from a conception of the specific unity of the bacilli of dysentery and that the importance of para-dysenteric bacilli may be limited to the preparation of vaccines or sera which may have to be polyvalent. Shiga states that the bacilli persist for a long time in the intestines of convalescents and may be carried by the apparently healthy, and Conradi[5] has found dysentery bacilli in the stools of healthy persons in Metz. The life of the bacilli outside the body is, according to Shiga, of short duration. Frost and Whitman[6] found that, at 17°—20° C., while the bacilli exposed on dry wood or paper died out rapidly, they lived for days and even for a month on dry bread or rice ; and that although they seldom lived for a week in distilled water, they lived in milk until it dried up. The Shiga type was less hardy than the Flexner type, and the higher the temperature the sooner the bacilli died. Observations connecting the severe summer diarrhœas of children with the bacilli of dysentery[7] are accumulating ; according to Jehle and Charleton[8] the Shiga-Kruse type is most common in epidemic diarrhœa and the Flexner type is most common in sporadic cases. The connection of diarrhœa in children with dysentery has an obvious application to the prevention of the disease, while the rôle played by flies in the contamination of food as described by Sandilands[9] not only demonstrates the necessity for the immediate destruction of dysenteric dejecta, but suggests other means of precaution such as the suitable location of dysentery wards and latrines.

57. It is important that the cause of each case of dysentery should be ascertained. Leaving the clinical signs and symptoms on one side, we have the following means of diagnosis : (1) serum diagnosis[10]—the bacillus of dysentery is agglutinated only by

Amœbic Dysentery.

the blood of the subjects of bacillary dysentery, but the agglutinating property does not appear in the blood until about the end of the first week[10] ; (2) the recovery of the bacillus by cultivation from the fæces ; and (3) the discovery of amœbæ in the stools. In fresh material from an acute case of dysentery, under a low power, the amœbæ appear as small, clear shining discs ; higher magnification revealing active movements. They may be distinguished from large leucocytes by their greater size (10—30 μ Kartulis) by the greater development of pseudopodia and by more lively movements ; and from the *Entamœba coli* by the generally distinct division of the clear ectoplasm from the granular endoplasm, which in the *Entamœba coli* are distinct only when it is moving actively, by the small, generally eccentric nucleus, by the absence of a nuclear membrane, and by the formation of small cysts (3—6 μ) which, according to Lesage,[11] can be readily seen a few minutes after the addition of iodised water. Difficulty in seeing the amœba may arise from absence of movement which may be due to too low temperature of the slide or the presence in the stool of an antiseptic drug, *e.g.*, calomel (Kartulis).

It is probable that the *Entamœba histolytica* is not the only pathogenic amœba, and although Kartulis states that all dysenteric amœbæ examined by him have been the same, he says that Schaudinn informed him that he recognised two varieties, and that Ruge believes the dysenteric amœbæ of China differ from those found in Java. Numerous cultivation experiments have been made with amœbæ and Lesage[11] describes the successful cultivation of *Entamœba histolytica* in association with a bacterium of the coli group on washed, sterilized agar. The same observer states that the resistance of the living amœbæ in culture is remarkable, the amœbæ living for four to five months and the cysts for six to eight months at a temperature of 20°—25° C. As pointed out by Schaudinn the cysts are the more dangerous form,[12] probably because they are more resistant to the gastric juice, and cysts are formed in the intestine at the termination of a case, and, outside the body, whenever, apparently, conditions are unfavourable to the life of the amœba ; cold, for instance, will cause amœbæ to form cysts[13].

Infection may persist for very long even after the patient is apparently quite cured, a refuge of the amœbæ being the vermiform appendix.[13] Of especial interest is the fact that these amœbæ may be the cause of intractable chronic diarrhœa. Meyer[1] cites two cases where the diarrhœa had been of 15 years' duration and a third of 18 years' duration in which cure followed specific treatment administered on the discovery of the presence of amœbæ in the stools.

In the Andamans, Major Anderson, I.M.S., reports that the *Entamœba histolytica* was found in 488 of 920 cases of dysentery examined, but in only 29 of these cases were the amœbæ unaccompanied by other protozoa. In order to check these figures an examination was made of the dejecta of 118 men admitted to hospital for fever, of whom 34 had suffered from dysentery more than a year previously and 84 had never been admitted to hospital for dysentery. Among the 34 the amœba was found in 15 ; among the 84 amœbæ were present in 17 ; in a few instances these amœbæ resembled those found in the dysenteric cases, but in most instances the amœbæ conformed with the description of *Entamœba coli*.

In view of the connection established between dysentery and abscess of the liver, the proportions between the numbers of these diseases occurring among prisoners and others is interesting. In the Andamans in 1905 there were 2,359 cases of dysentery admitted into hospital—perhaps we may assume that 1,000 were cases of amœbic dysentery—and three cases of abscess of the liver ; in the last five years there have been 10,428 cases of dysentery and 11 cases of abscess of the liver. In the Indian jails in the last five years there have been 42,152 cases of dysentery and 42 cases of abscess of the liver. In the Native Army the proportion was much higher, 23,516 cases of dysentery and 72 cases of abscess of the liver. The contrast between these figures and those for the European Army is an extraordinary one; among British soldiers, during the five years there were 5,581 cases of dysentery and 860 cases of abscess of the liver. It is evident then that although dysentery may be a necessary antecedent of abscess of the liver, abscess of the liver is not among natives a common sequel of dysentery; and it appears that there must be some other condition operative among Europeans. Musgrave[3] notes that experiments in his laboratory indicate that amœbic abscess rarely or never develops in an otherwise normal liver, notwithstanding the fact that amœbæ may be almost constantly poured into it from an ulcerated bowel. Allusion may be made to the report of an isolated experiment in which Dr. Gauducheau[4] injected pus from a tropical abscess into the mesenteric vein of a dog; the result was not an abscess of the liver but amœbic dysentery.

58. The number of cases admitted to hospital under this heading was 3,603,

Diarrhœa. of which 68 ended fatally, the mean admission and death rates being, respectively, 39·2 and ·74 per thousand, the lowest on record. The highest admission rates were 90·9 per thousand in Bengal and 90·0 in Eastern Bengal and Assam; in the other administrations with one exception, the rates varied between 42·9 in the Punjab and 17·6 in Burma; the exception is Madras where the admission rate was 3·5 per thousand and only 36 cases were recorded as diarrhœa. The highest death rates were 1·15 per thousand in Bombay and 1·06 in Bengal, the rates in Madras, ·30, the Central Provinces, ·28, and the Punjab, ·17, being the lowest.

59. The number of cases admitted on account of diseases classified under

Respiratory diseases other than pneumonia. this heading was 2,428, and the number of deaths was 83, the rates being 26·4 per thousand and ·92 per thousand as compared with 25·1 and ·5 in 1904. The increase in the admission rates which depended to some extent on the greater prevalence of influenza was common to all administrations, except the North-West Frontier Province, the Central Provinces and Madras, and the increase in the death rates was common to all except the North-West Frontier Province and Madras.

60. In spite of the wide diffusion of influenza the number of cases of

Pneumonia (2). pneumonia fell from 965 in 1904 to 896 in 1905, and the number of deaths from 252 to 210, the admission and death rates, 9·7 and 2·28 per thousand, respectively, being the lowest on record. The reduction in the admission rates was most remarkable in the Punjab and the North-West Frontier Province; in Burma, Madras, the United Provinces and the Central Provinces the admission rates rose. The death-rate fell in Eastern Bengal and Assam, the Punjab, the North-West Frontier Province and Bombay, the further reduction from 19·34 per thousand to 15·79 in the Sind Gang in that province being satisfactory. The case-mortality varied considerably in the different provinces being exceptionally high in Bengal (34·1 per cent.), Burma (31·0), Eastern Bengal and Assam (28·1), and comparatively low in the Punjab (17·7), Madras (16·7), the Central Provinces (14·3) and the North-West Frontier Province (11·1). The mean stay in hospital of fatal cases was between eleven and twelve days; in the United Provinces and Bombay the mean duration of fatal cases was only about seven days owing to the large numbers of cases which proved fatal in the first few days of the disease.

The importance of efficient ventilation in the prophylaxis of pneumonia is recognised everywhere, and endeavours are generally used to secure the best ventilation of the sleeping wards possible : but it is obvious that the difficulty of this task may be greatly increased by climatic conditions and badly constructed buildings. When cases occur in the prisons those measures of segregation and disinfection considered necessary in other cases of infectious fevers are generally carried out. Attention has recently been drawn to the possible rôle played by oral sepsis in predisposing to the more serious prison diseases, and in this connection the recently made statement[1] that the pneumococcus, which is present in the mouths of a certain proportion of healthy persons, increases in virulence in a neglected mouth and disappears altogether from a thoroughly cleansed one, may be noted.

61. As compared with the previous year there was a considerable increase in the prevalence and fatality of cases of tubercle of the lungs, the number of admissions rising from

Tubercle of the lungs (3).

763 to 803, and the number of deaths from 279 to 293—an admission rate of 8·7 per thousand and a death rate of 3·19. The disease was as usual most prevalent in the jails of Bengal (admission rate 13·5 per thousand), Eastern Bengal and Assam (11·4) and Madras (10·5), and least prevalent in the jails of Bombay (6·1), the Central Provinces (5·6) and the North-West Frontier Province (3·8) ; while the death rates were highest in Bengal (4·02 per thousand) and the Punjab (4·43), and lowest in the Central Provinces (1·97) and the North-West Frontier Province (1·52).

In India generally, tubercle of the lungs was the cause of 16·6 in every hundred deaths. This percentage varied greatly in the different administrations, the more remarkable percentages being 27·1 in the Punjab, 22·3 in Burma, 12·5 in the Central Provinces, 10·4 in Eastern Bengal and Assam, and 7·7 in the North-West Frontier Province. In India generally, and in Burma, the United Provinces, the Punjab, Bombay and the Central Provinces, tubercle of the lungs was a more frequent cause of death among prisoners than dysentery. Analysis of the statistics shows that the mortality from tubercle of the lungs was much higher among those convicts who had been more than three years in prison than among the remaining body of convicts, the death-rates being 8·06 among the former and 2·59 among the latter. Unfortunately the returns giving the numbers admitted to hospital do not discriminate between short and long term prisoners, and the death-rates alone cannot give accurate information regarding the prevalence of the disease among short and long term prisoners, respectively, because a larger number of the former than of the latter are likely to be released before the termination of the disease. Of the 90 long term prisoners that died, however, 47 were in good health at the time of conviction, and the mean duration of their illness from admission to hospital until death was 124 days, so that it is reasonable to suppose that the great majority contracted the disease in prison. There are considerable differences in the mean stay in hospital of these cases in the different administrations ; in Bombay the mean was only 68 days, in Madras 83 and in the Central Provinces 84, while in the United Provinces the number of days spent in hospital was 153, and in Bengal 142.

The nature of tubercular infection and the manner of the entry of infection into the human body are still the subjects of a discussion to which allusion is made only for the purpose of directing attention to the possibility of a proportion of cases of tubercle in the jails being the development of a latent infection. That this is possible is suggested by the large number of cases which occur among apparently healthy prisoners in jails where the hygienic precautions taken might be expected to exclude the probability of infection; and the possibility emphasizes the importance of a strict watch over the health of those prisoners whose physical conformation or previous history indicates the possible presence of latent tubercle.

The principles of prevention as recently described by Koch[1] in the Nobel lecture are thoroughly understood in the jails, and are generally applied. In many prisons tuberculous patients are treated in a special open air ward, but these wards are not all equally well constructed. Care should be taken that, while the occupants of the ward are not exposed to chilling winds, ventilation is thorough, and that there is no dead space in the roof in which the air can stagnate; the existence of such a dead space will render the air of a ward, even if it be completely open on all sides, unpleasant in still weather. The arrangements for the reception and disposal of tuberculous sputum should be complete and the ward should be kept scrupulously clean, otherwise it may become a centre for the collection and distribution of tuberculous material.

62. There were 1,014 cases and 50 deaths recorded under this heading, with admission and death rates of 11·0 and ·54 per thousand, respectively, as compared with 12·4

Anæmia and Debility (4).

and ·41 in the previous year and means of 13·5 and ·57 in the previous four years. These conditions of bad health were far more prevalent in the jails of Eastern Bengal and Assam than in that of any other administration, the admission rate being 32·2, and the death rate 2·66 per thousand, compared with 17·9, the next highest admission rate, recorded in the Punjab, and ·76 the next highest death rate, recorded in Bombay. The lowest admission rates were 4·3 in Madras and 3·1 in Burma, and in all provinces the admission and death rates were below the means of the last four years except in Eastern Bengal and Assam where both rates were higher, Bengal and the North-West Frontier Province where admission rates were higher, and the Punjab where the death rate was higher. The geographical distribution of the high death rates is important; they were highest in Assam (Group III) 5·75, Western Coast (X) 1·73 and Bengal and Orissa (IV) ·92. The convenience of the heading anæmia and debility for the purpose of hospital registers is obvious, but the condition must depend upon some cause either in the victim or his surroundings, and the occurrence of large numbers of such cases in a jail should lead an Inspector-General to ascertain whether they are due to failure to diagnose or to some remediable defect in the conditions of the prisoners' lives. The *post-mortem* records show easily recognizable organic disease in many cases; a few cases were almost certainly *kala azar*; and anchylostomes were frequently found in the intestines. In this connection attention may be drawn to the revival in a modern form by Shipley and Fearnsides[1] of the old idea that the anæmia of anchylostomiasis is due to a hæmolytic toxin secreted by the anchylostomes—probably in the cephalic glands.

63. There were in all 31 cases of small-pox, including 14 in Bombay and four each in Madras and the United Provinces, as compared with 25 cases and four deaths in 1904.

Small-pox.

64. There were 66 cases of plague in the jails, of which 46 were fatal, as compared with 37 cases and 24 deaths in 1904. In the United Provinces (26 cases) the disease

Plague.

appeared in eleven jails; in Bombay (15 cases) in four jails; in Bengal (14 cases)

in six jails; in the Punjab (6 cases) in three jails; in the Central Provinces
(3 cases) in two jails; in Burma and Ajmer one case each.

65. The decrease in the numbers of cases of this disease was maintained

Cerebro-spinal Fever.

the numbers of cases and deaths falling from 23
and 18 in 1904 to ten and nine in 1905, and the
number of jails in which cases occurred from eleven to six.

66. There were 64 cases of enteric fever and 15 deaths, a larger number of

Enteric Fever.

cases than in any of the past ten years except 1902,
when there were 69 cases and 15 deaths. There
were 28 cases in the Madras jails, Coimbatore contributing 18 and Cannanore
six. In Eastern Bengal and Assam there were 12 cases, ten in the Mymensingh
jail and two at Rangpur. No case was reported in Bombay, the Central Provinces
or the North-West Frontier Province. In most instances the *post-mortem*
observations supported the diagnosis.

67. The number of cases of cholera was 73, of which 40 were fatal against 47

Cholera.

and 31 in 1904, and 97 and 57 in 1903. Most of
the cases (37 with 14 deaths) occurred in Madras,
an outbreak in the jail at Rajamundry accounting for 30 cases of which ten were
fatal. In Eastern Bengal and Assam there were twelve cases and all were fatal.
Except at Rajamundry the cases were sporadic, and the jails in Bombay and
the North-West Frontier Province were free from the disease.

The following paragraphs deal with the more important features of the
provincial reports.

68. The year was comparatively unhealthy; there was a slight increase

Bengal (5).

in the mean daily population and the admission
and death rates rose from 883 and 19·55 per
thousand in 1904 to 954 and 24·91. The difficulty in providing sufficient
accommodation has been increased owing to the more complete separation
of different classes of prisoners that is now required; and it is evident that steps
must be taken to obtain more accommodation, since in 16 jails there was overcrowding among convicts throughout the year and in nine jails the under-trial wards
were not sufficiently large for the number of under-trial prisoners in confinement.
The principal causes of admission to hospital were intermittent fever (30·49 per
cent. of all admissions) and dysentery and diarrhœa (30·12); and the principal
causes of death were dysentery (26·3 per cent. of all deaths), tubercle of the
lungs (16·1), intermittent fever (11·0) and pneumonia (8·5). There were 2,783
admissions to hospital on account of dysentery and 93 deaths, a case-mortality of
3·34 per cent., which is below the mean for Indian jails (3·7), and may be
due, as the Inspector-General suggests, to prompt admission to hospital of
mild cases, or to the presence of a less deadly form of the disease. Regarding
oral sepsis the Inspector-General remarks that medical officers have not been
able to trace any connection between it and the prevalence of dysentery, and
they do not consider the condition in any sense scorbutic. Even if there is no
direct connection between oral sepsis and dysentery it cannot, as the Inspector-
General says, be good to swallow pus organisms; the loss of teeth must result
in dyspepsia which is generally considered a predisposing cause of dysentery; we
have seen that the presence of oral sepsis may determine an attack of pneumonia,
and there is evidence that it may furnish an entrance for tubercular infection[1]
in the young; the removal of the condition, therefore, is an important detail of
jail hygiene.

There was a slight reduction in the number of cases of intermittent fever, the
total being 4,120, compared with 4,271, in preceding year, but the number of fatal

cases rose from 12 to 39. Nearly half the deaths occurred in cases of malarial cachexia and it seems possible that some of these were cases of *kala azar*. Quinine prophylaxis is used by Koch's method, and the Inspector-General remarks that the result is not seen in any considerable reduction in the number of admissions for fever, but most medical officers comment on the fact that the disease is of a milder type and more easily cured. Tubercle of the lungs was more prevalent and fatal than in the previous year, and the Inspector-General proposes to segregate at Berhampore tubercle cases from all jails.

The death rate in four jails was over 40 per thousand; one of them, Khulna, is a small jail and the death rate must vary, the high death rate in the others is attributed to overcrowding or to structural defects in the barracks.

69. There was a slight increase in the average strength—from 6,211 in

Eastern Bengal and Assam. 1904 to 6,401 in 1905—of the prisoners in the jails of Eastern Bengal and Assam. The year was

unhealthy, the admission, constantly sick and death rates rising from 979, 37 and 22·38 per thousand, respectively, in 1904, to 1,081, 46, and 32·96 in 1905, rates which are higher than those recorded in recent years. The chief causes of sickness were dysentery and intermittent fever, and, to a less extent, diarrhœa. The deaths were due to dysentery, tubercle of the lungs and beri-beri which occurred in an epidemic form at Sylhet and caused 21 deaths. In five of the jails, with a population of over one hundred, the death rates exceeded 45 per thousand, namely Barisal (62·87), Gauhati (58·25), Dibrugarh (57·69), Sylhet (54·79) and Tezpur (45·28).

70. Climatic conditions were not favourable to the health of prisoners: the

United Provinces. cold in the early months of the year was extreme, and the hot weather was more trying than usual.

There was considerable distress among the agricultural population; and influenza was prevalent. In these circumstances it is satisfactory to note a considerable reduction in the admission rate from 612 per thousand in 1904 to 569 in 1905, and a fall in the constantly sick rate from 28 to 27. The death rate, however, rose from 14·56 of the previous year which was the lowest ever recorded, to 17·31; this is not a high rate and the Inspector-General remarks that at least a third of the deaths were due to diseases contracted before admission to jail or to injuries. The increase in mortality was due to heat apoplexy and the occurrence in the jails of a large number of cases of plague.

The chief cause of sickness was intermittent fever (32 per cent. of all admissions); and the chief causes of death were pneumonia (17·3 per cent.), tubercle of the lungs (15·4) and dysentery (11·2). In seven jails with a strength of over 100 prisoners including the central jail at Benares the death rates were over 30 per thousand. The high death rate in the Benares central jail (33·6) is of particular interest as the situation, arrangement and management of the jail are satisfactory, and the death-rate of 1904 was only 10·37. The population of the jail had been reduced and it contained a very large proportion of old men whose latent organic defects rendered them unable to withstand the peculiarly unhealthy conditions prevalent in the Benares district. At Gorakhpur the special precautions taken for the purpose of preventing dysentery getting into the jail, to which reference was made last year, were successful and only one death from dysentery occurred in the jail.

71. The daily average population has been steadily falling since 1900, and

Punjab. in 1905 the daily average convict population was lower than it has been since 1896.

The year 1905 is a memorable one in the annals of the Punjab jail administration : the admission rate fell from 907 per thousand in 1904 to 708, the constantly sick rate from 31 to 26, and the death rate from 19·76 to 16·33—one of the two lowest rates on record. Very remarkable was the decline in the numbers of deaths from dysentery from 30 to 6, and from pneumonia from 64 to 29. The number of deaths from tubercle, however, rose from 40 to 51 and there were 13 deaths from sunstroke, four from cholera and five from plague. The number of admissions to hospital on account of malarial fever fell from 4,014 to 2,834, and the number of deaths from 10 to 6, which is attributed to " the administration of quinine in large doses at stated intervals during the malarial season and especially to prisoners after admission, together with the abolition of breeding grounds of mosquitoes in the vicinity of jails."

Colonel Bate, C.I.E., I.M.S., who had held the office of Inspector-General of Prisons in the Punjab since April 1891, demitted office on promotion. The improvements in the condition of the prisoners effected by him during his long tenure of office may be judged from the following figures : in the ten years ending 1891 the mean death rate was 32·77, in the following ten years the mean death rate was 22·14, and in the four years ending 1905 it was 20·19.

72. The daily average population was 1,314, showing a continuation in the steady increase due to heavier sentences and fewer transfers to the Punjab jails. There was occasional overcrowding especially in the Kohat and Abbottabad jails. The admission rate fell from 1,102 per thousand in 1904 to 960, but the constantly sick and death rates rose from 32 and 14·65, respectively, to 34 and 19·79. The principal causes of sickness were ague, abscesses and influenza, and dysentery was the chief cause of death.

North-West Frontier Province.

73. The average daily population fell from 3,740 in 1904 to 3,547 in 1905 ; but the admission, constantly sick and death rates all rose—from 586 to 630, from 20 to ·21 and from 12·57 to 15·79. The number of admissions on account of malarial fever was greater than in 1904, and this fever accounted for 30·27 per cent. of all admissions, but there was no fatal case. Tubercle of the lungs, pneumonia, other respiratory diseases and dysentery were the principal causes of death. The number of cases of pneumonia increased from 25 to 35, and it was suspected that infection was conveyed by blankets which are to be systematically boiled in future. In the five central jails there were only 81 admissions and three deaths from dysentery, and it is noteworthy that in the Jubbulpore central jail, which was recently notorious for the high mortality from dysentery in it, a daily average population of 840 was entirely free from that disease throughout the year. The number of cases of diarrhœa fell from 169 to 128, but the Inspector-General considers that the number should be still lower, and alludes to the practice of habitual convicts securing a holiday in hospital on account of diarrhœa induced by eating irritant seeds.

Central Provinces and Berar.

74. There was a considerable increase in the daily average strength from 9,320 in 1904 to 10,147. The admission rate rose from 439 to 469, but the constantly sick and death rates fell from 20 and 16·52 per thousand to 19 and 15·87, both lower than any hitherto recorded. The death rate would have been only 14·49 but for the occurrence of cholera of which there was an epidemic (30 cases, but only ten deaths) at the Rajamundry central jail and (seven cases with 4 deaths) in the

Madras.

Madras Penitentiary and Nellore district jail. Ague (18.25 per cent. of all admissions), dysentery (13.63 per cent.), simple continued fever (7.24 per cent.) were the chief causes of sickness ; dysentery (19.3 per cent. of all deaths), tubercle of the lungs (14.9 per cent.), pneumonia (9.3 per cent.) and cholera (8.7 per cent.) were the chief causes of death. The special hospital for tubercle at Trichinopoly was in use with much advantage throughout the year ; the construction of the special hospital at Bellary has not been completed.

75. The daily average population increased slightly to 7,851, and the admission and constantly sick rates rose from 600 to 603 and from 25 to 27 per thousand, respectively, but the death rate fell from 19.22 to 17.45, the lowest rate recorded.

Bombay.

The principal cause of sickness was ague which accounted for 22.30 per cent. of all admissions. The chief causes of death were pneumonia (17.5 per cent. of all deaths) and tubercle of the lungs (13.1 per cent.). There was over-crowding in the Common Prison, Bombay, Rajkot, Hyderabad, and Shikarpur. The Thana jail was overcrowded for a short time, but during the greater part of the year the population was kept below the number for which accommodation is provided according to scale. The death rate of this jail fell from 23.39 in 1904 to 14.83 and although tubercle continues to be the principal cause of mortality, the measures taken—the provision of a tubercle ward, early isolation of cases, the renewal of every floor in the prison and the thorough disinfection of the floors—have brought about a great diminution in the prevalence of the disease.

The total number of cases of pneumonia fell from 119 to 106 and the number of deaths from 38 to 24. In the Sind Gang there was no virulent outbreak and the Inspector-General quotes Major Jackson's opinion that the improvements effected in the surroundings of these prisoners will render pneumonia no longer formidable among them.

76. There was a considerable increase in the daily average strength from 11,704 in 1904 to 12,639 in 1905, a number not far short of the total, 15,284, for which accommo-dation is provided. The health of the prison population continued excellent, the admission and constantly sick rates falling from 587 and 20 per thousand to 320 and 17—the lowest rates on record—and the death rate from 18.63 to 17.01, the third lowest rate on record. The food of the prisoners was the subject of particular attention during the year, and arrangements were made to secure pure drinking water for the inmates of the jail at Mandalay. The success which has attended the efforts to improve the hygiene of this jail may be gauged from the reduction in the death rate from 79.30 per thousand in 1904, to 17.31. The jail at Thayetmyo, however, was unhealthy in 1905, the death-rate rising from 12.91 in the previous year to 37.66, due to the prevalence of dysentery and diarrhœa for which the flooding of the jail's surroundings owing to an unusual rise of the river and the pollution of the river water are blamed. There were seven cases of beri-beri and four deaths in this jail in which, it may be noted, a great outbreak of that disease occurred in the summer and autumn of 1881. The recent outbreak was promptly checked by the removal of the patients from the jail and the disinfection of their clothing and bedding and of the dormitories occupied by them. The chief causes of sickness were abscesses and ague, and the chief causes of death were tubercle of the lungs and dysentery. The term 'abscess' includes the results of injuries, many of which were found to be self-inflicted.

Burma.

77. In the jail at Ajmer there was some overcrowding, but the health of the prisoners was remarkably good, the constantly sick and death rates among a population of 412 being 22 and 17 per thousand, respectively. Prisoners often suffer from guinea-worm attributed to the drinking water. There was one death from plague. In the Mercara jail the average daily population was 87, and the constantly sick and death rates 11 and 45·98 per thousand, respectively. The general health was good ; of the four deaths, three were due to pneumonia. At Quetta the average daily population was 70 ; there was no death. At Secunderabad the average daily population was 85 ; there was no death.

Ajmer, Mercara, Quetta and Secunderabad.

78. In 1905 the rainfall recorded in the Andaman Islands was 101·09 inches, which is twenty inches less than the fall of the preceding year and about seventeen inches below the mean of the preceding five years. The convict population continues to increase, the average daily strength rising from 13,972 in 1904 to 14,348. No change has been effected in the permanent barracks, but many bamboo and thatch barracks were constructed and occupied during the year, while two additional wards for phthisical patients and an additional ward for lunatics were opened, one of the former being used for general hospital purposes. No change has been made in the water-supply or in the conservancy arrangements. The swamp to the north of the female jail was partially drained, but still includes pools in which anopheles mosquitoes breed. No change was made in the diet scale, but the quality of the supplies has been improved. The Senior Medical Officer considers that the inferior quality of the wheat, particularly its mouldiness, is a principal cause of the dysentery which is so prevalent among the convicts and the results of a series of experiments carried out with wheat cleaned and ground by different methods afford evidence of the correctness of his view. He recommends that in future (1) all foreign matter and diseased grain should be carefully removed from the wheat before grinding ; (2) the wheat and *atta* should be systematically dried ; (3) excess of bran should be removed from the *atta* after grinding ; (4) each sack of food material should be carefully inspected before issue for use ; and (5) the food should be stored in a godown rendered as far as possible damp proof and weevil proof. The use of several kinds of *dhal* has been introduced. The vegetables issued to the convicts have been of much better quality owing to the increased number of gardens in various stations, with the result that scurvy is rarely seen. Cleaned vegetables are now issued to the inmates of the female and cellular jail. Improvements have been effected in the milk supply.

Andamans.

The admission rate fell from 1,909 per thousand in the previous year to 1,898, but the constantly sick and death rates rose from 64 and 34·75 per thousand to 68 and 38·96. The increase in the constantly sick and death rates were due to the prevalence of dysentery and cellulitis in the first part of the year ; and the constantly sick rate is kept up by the detention in hospital of all recognized cases of phthisis.

Dysentery is the great scourge of the Settlement ; the number of cases in 1905 was 2,559 or 89 more than in the previous year, and the number of deaths rose from 111 to 187, equal to a death rate of 13·03 per thousand, four times the Indian mean death rate. The admission and death rates from tubercle of the lungs were 11·4 and 6·48 per thousand ; both rates lower than those of the preceding year and this encourages the hope that the vigorous measures adopted against the disease are successful. The average stay in hospital of fatal cases was 230 days.

Malaria was the cause of 17,881 admissions and 12 deaths, compared with 16,395 admissions and 34 deaths in 1904. The most common anopheles found in the Settlement resembles *A. rossi*, but Lieut.-Colonel Alcock, I.M.S., considers it differs sufficiently to deserve another name and has provisionally named it *A. andersonii*. This mosquito and *A. barbirostris* are considered by Major Anderson, the Senior Medical Officer, to be the common carriers of malaria. Cellulitis which caused 188 admissions and 18 deaths was, as already noted, most prevalent early in the year, but became much less virulent as the year advanced ; the improvement being attributed by the Senior Medical Officer to improved barrack accommodation.

79. The general results of jail management during the year, so far as the

Conclusion.

health of the prisoners is concerned, were satisfactory, and the Government of India have good reason to be well pleased with the care and skill displayed by the Inspectors-General and the medical staff of the prison administrations.

Papers quoted in Section IV.

For explanation of abbreviations see end of Section II.

(1) *Dysentery.*—[1] Meyer in *D. M. W.*, August 16, 1906, page 1327 ; also Albu reported in *II. R.*, July 15, 1906, page 757. [2]Dopter in *A. P.*, December 1905, page 753. [3]Hiss reported in *C. B.* XXXVII. Ref. 1906, page 273. [4]Shiga in *P. J. S.*, June 1906, page 485. [5]Conradi reported by Shiga in *P. J. S.*, June 1906, page 485. [6]Frost and Whitman reported in *C. B.* XXXVII. Ref. 1906, page 359. [7]Knox and Shorer reported in *J. T. M.*, July 2, 1906, page 211. [8]Jehle and Charleton reported in *II. R.*, July 15, 1906, page 755 [9]Sandilands in *J. H.*, January 1906, page 77. [10]Vaillard and Dopter in *A. P.*, May 1906, page 321. [11]Lesage in *A. P.*, January 1905, page 9. [12]Cf. Musgrave, *Bur. of Govt. Labo, Manila*, No. 18, October 1904. [13]Musgrave in *P. J. S.*, June 1906, page 547. [14]Gauducheau reported in *J. T. M.*, February 15, 1906, page 52.

(2) *Pneumonia.*—[1]Hopkins in *J A. M. A.*, June 30, 1906, page 1987.

(3) *Tubercle of the Lungs.*—[1]Koch in *D. M. W.*, January 18, 1906, page 89.

(4) *Anæmia and Debility.*—[1]Shipley and Fearnsides reported in *Lancet* June 9, 1906, page 1622.

(5) [1]Bennett in *B. M. J.*, August 20, 1904, page 385 ; also Squire in *B. M. J.*, July 21, 1906, page 133.

SECTION V.
VITAL STATISTICS OF THE GENERAL POPULATION.

80. Vital statistics have recently attracted a good deal of public attention on

Registration of Vital Statistics.

account of the discussion in the press of two topics of general interest. In England the decline of the birth rate has been the subject of numerous articles, and in India attention was aroused by a statement that the progressive increase in the registered death rates was a proof of the impoverishment of the mass of the Indian people.

It has become a proverb that anything may be proved by statistics, but this is really only another way of saying that they must be used with care. Vital statistics, unfortunately, present peculiar difficulties on account of the large number of special circumstances which must generally be taken into account before a conclusion can be drawn from them ; for instance, one town may have a low birth rate and another a low death rate ; it is as wrong to condemn the former, before ascertaining that it contains the normal proportion of married couples at the reproductive ages, as it is to assume that the latter is healthy without finding out whether the low mortality is not due to the small numbers of young children and elderly persons in the population. Indian vital statistics present special difficulties. We have to depend upon the census population, because the registered returns are too inaccurate to allow of calculations of intercensal populations, with the result that as the decennium advances rates become unduly high ; figures have been registered in certain areas for many years, but registration is even now not universal ; the facts are nearly everywhere collected by an unpaid and generally illiterate agency ; registration is compulsory only in the towns where the municipal authorities seldom enforce the bye-law ; and the different social customs of the people often render the application of experience gained elsewhere misleading.

There is often greater difficulty, especially in towns, in obtaining information regarding births than deaths, and consequently infantile mortality which is calculated from the numbers of births and deaths of infants registered during the year, is excessive.

It is not possible here to detail the allowances which must be made in respect of the registration figures, but it seems necessary to explain, that however valuable they may be for the purpose of comparing the condition of the people in a province from year to year ; and, due allowances being made, for comparing the condition of the populations in different provinces in any year, the figures registered are not yet generally accurate, and the degree of accuracy differs in the various provinces. This will be appreciated after reference to the following statement which gives the registered birth and death rates per 1,000 of the population in three of the larger provinces in the years 1885, 1895, 1904 and 1905.

Year.	Madras.		Bengal.		United Provinces.	
	Birth rate.	Death rate.	Birth rate.	Death rate.	Birth rate.	Death rate.
	Per 1,000 of population.					
1885	29·3	21·8	24·71*	22·74	41·24	31·98
1895	29·1	19·6	34·59	31·59	34·90	29·13
1904 ...	30·7	22·5	42·59	32·45	46·67	34·70
1905	32·0	21·4	39·55	38·63	41·24	44·0

(Figures for 1904 are given because the area of collection in Bengal in 1905 was not the same as in the earlier years, and because the death rate in the United Provinces in 1905 is exceptionally high).
* For towns only.

It will be observed that in Bengal both birth and death rates have risen greatly, while in Madras and the United Provinces the increase is comparatively small. In the United Provinces registration in 1885 was fairly good and little improvement has taken place; in Bengal registration has improved from year to year, and in Madras the improvement has been inconsiderable. We may compare these rates with those calculated by Mr. Hardy and published in his actuarial reviews of the census figures of 1881, 1891 and 1901. As registration improves the registered birth and death rates will approximate the calculated rates.

| | Madras. | | Bengal. | | United Provinces. | |
Year.	Birth rate.	Death rate.	Birth rate.	Death rate.	Birth rate.	Death rate.
			Per 1,000 of population.			
1881	50·4	41·5	47·9	39·9	45·1	41·9
1891	50·3	36·0	51·8	44·8	44·2	37·7
1901	44·8	38·1	43·9	38·9	44·7	43·4

The statement of registration figures shows that in Bengal not only the registered death rates, but the birth rates also, increased, and it has been ingeniously argued that the increased birth rate is only another proof of the impoverishment of the people. This argument is founded upon a proposition that the fecundity of the human animal and of all other living beings is in inverse proportion to the quantity of nutriment available, and that an underfed population multiplies rapidly. Even if this proposition was true the inference that the rising registered birth rate in India is due to privation is easily demonstrated to be false, unless it is the case that the high Indian marriage rate is the result, not of religious observance, but of recklessness induced by poverty.

A physiological stimulus must affect the individual, and to sustain the theory that privation causes the high birth rate in India, Indian women should be exceptionally fertile. This they are not : the average Indian woman is apparently not much more fertile than the average woman in the England of vanishing birth rates.

81. In England in 1851 the percentage of all women between the ages of 15

Fertility of Indian Women.

and 45 who were married was 47 ; this figure rose to 50 in 1871, but it fell again to 47 in 1891 and has not risen again. It is fair to reduce the reproductive period in India by five years, and take the proportion of all Indian women between the ages of 15 and 40, who are married. In India, excluding Burma, in 1891 this was 82·7 per cent. Of every thousand women at the reproductive age only 468 are married in England while 827 are married in India. In England in the sixties the number of legitimate births for every thousand married women at the reproductive age was about 305, in the nineties the figure had fallen to about 262, and in 1901 it had fallen to 235. In India, as the registered birth rate is below the actual, we may accept the rate deduced by Mr. Hardy, which, including all live births, is 44·4 per thousand of the population. If this rate is correct, the number of children born for every thousand married women at the reproductive period is 273·6.

The high Indian birth rate, then, is not due to exceptional fertility, but to the large proportion of married women. This has, of course, a bearing upon the high infantile death rates, for a large proportion of all births must be the first children of youthful parents. The children of young parents are generally weak and the first child of a marriage is especially exposed to the effects of ignorance and inexperience. The average birth and death rates in India in Mr. Hardy's opinion do not give any indication of permanent change. A high birth rate involves either a correspondingly high death rate or an increase in the population so rapid as to encroach upon the limits of subsistence. The high birth rate, as we have seen, is the result of the high marriage rate, which may have been necessary if man was to contend successfully with the adverse physical conditions of his environment. When man in India has learned to protect himself against adverse physical conditions, marriage will be deferred and the birth rate, and with it the death rate, will fall.

82. The statement on the following page gives details regarding the agencies

The existing system of registration.   employed in the collection of vital statistics. It will be observed that, except in Madras and Coorg, the police are employed practically at all stages.

If the existing system worked perfectly we should obtain complete information regarding the number and sex of infants born and the number and sex of the persons that die, and we should have a general idea of the causes of death, but the existing system does not yet work perfectly, and very often the results obtained are misleading. It is evident, moreover, that there is a point beyond which the present system of registration of vital statistics cannot go ; however perfectly it may work it cannot give us accurate details of the causes of death ; to obtain this information a professional agency is necessary. It is impossible at present to employ a professional agency except in the larger towns, and the best way to render the statistics, as at present collected, intelligible seems to be to employ a professional agency in selected areas. The collection of facts in these areas by the existing agency would continue, and the comparison of the statistics collated by the two agencies would give a key to the interpretation of the general statistical returns.

83. An experiment on these lines is being carried out in Bengal, in the Galsi

Registration by a professional agency.   *thana* in Burdwan. The following analysis of the registration results collected in August is interesting. There were in all 72 deaths, all of which except one from malarial fever had been duly registered in the *thana*. The diseases shown in the *thana* register were fever 52, "other causes " 9, cholera 3, diarrhœa 3, dysentery 2, respiratory disease 1 and measles 1. Enquiry into the real causes of death showed that the entries regarding cholera, diarrhœa, dysentery and measles were correct, the nine deaths from "other causes " were due to infantile diarrhœa and diseases of newly born children. The 52 deaths from fever included malarial fever 12, pneumonia 8, diarrhœa 6, tubercle of lungs 5, " senile decay " 4, and single cases of cholera, Brights disease, heart disease, etc.

A similar enquiry is being carried on in one of the Bengal towns.

84. A conference attended by the Medical Officers of Health of Calcutta,

Revision of forms in use in the great towns.   Madras, Bombay and Rangoon was held at Calcutta in November 1904, for the purpose of preparing standard forms for the exhibition of vital statistics, etc., in municipal reports. A series of forms drawn up on the lines of the standard forms in use in the provincial sanitary reports was adopted and the forms have been taken into use.

Province.	Collecting and Reporting Agency. Rural areas.	Collecting and Reporting Agency. Town areas.	Recording Agency. Rural areas.	Recording Agency. Town areas.	Collating Agency. Rural areas.	Collating Agency. Town areas.	Inspecting Agency. Rural areas.	Inspecting Agency. Town areas.
Assam	Village Chaukidars	Head of households generally.	The Police station	The Police station, the Magistrate or Municipal office.	The Deputy Commissioner or Sub-divisional officer		Deputy Commissioner, Sub-divisional officers and the ordinary revenue and police agency. Inspectors of Vaccination are required to test entries in registers.	
Bengal	Village Chaukidars	Town Police	Police station	Police Station	Civil Surgeon	Civil Surgeon	The Police and other local officers and the vaccinating staff.	
United Provinces	Village Police. In his tracts village headman.	Town Police, heads of households and sweepers.	Reporting officers maintain permanent registers		Civil Surgeons who are generally the health officers		Revenue and Police officers, officers of the Sanitary and Medical Departments, the vaccinating staff and members of municipal and district boards.	Civil Surgeons generally, and also members of district and municipal boards.
Punjab	Village Chaukidar	Municipal Committees responsible. In Municipalities where no bye-laws are in force, the local constable assisted by the mohalla sweepers.	The Police Station	Municipal Office	Civil Surgeons.		Officers of the Sanitary Department, Civil Surgeons, superior police officers and the members of the Vaccination Department.	
North-West Frontier Province.	Village Chaukidar	Heads of households	Police Station	Municipal Office	Civil Surgeons		The Revenue Staff under the control of the Deputy Commissioner, the Civil Surgeon and his subordinates including the vaccinating staff, members of municipal and district boards, and local men of position.	
Central Provinces	Village watchman (Kotwar).	Heads of households	The Police Stations		Civil Surgeons.		Revenue and police officers, Civil Surgeons and Superintendents of Vaccination.	
Berar	Headman of village (patel).	Headman of town (patel)	The Police Stations		Civil Surgeons.		Revenue and police officers, Civil Surgeons and Superintendents of Vaccination.	
Madras	Headmen of villages (Kanams).	Heads of households	Head men (Kanams) who report to the Collectors.	The Municipality	Collectors and Municipalities who transmit returns to the Sanitary Commissioner.		Collectors and divisional officers, Tehsildars and Deputy Tahsildars, Revenue Inspectors and Deputy Inspectors of Vaccination.	
Coorg	Village headman (patel) managers, etc., on collector estates.	Municipal and Medical authorities.	Officer in revenue charge of a collection of villages.	Municipality	The Civil Surgeon.		The Commissioner and other superior revenue officers. The Civil Surgeon and the vaccinating staff.	
Bombay	The Police Patel (headman) and the Kulkarni (Acct. Pound Manager) and village School Masters are also employed in Sind.	Municipalities	The Police Patel or Kulkarni (the Accountant).	Municipalities	Deputy Sanitary Commissioner.		Officers of the Revenue and Sanitary Departments, the latter including the Vaccination Department.	
Burma	Village headman	Headmen of wards	Village and ward headmen record entries copies of which are made over to the local patrol for submission to the police stations.		Township officer for submission to the Civil Surgeons		Township and Sub-divisional officers under the control of the Deputy Commissioners.	

85. The total population of the areas under registration in the Indian Empire

Births in India in 1905. in 1905 numbered 223,124,252, but this figure makes no allowance for the natural increase which has occurred since the census was taken in the beginning of 1901. The number of births registered during the year aggregated 8,731,497, as compared with 9,136,536 in 1904, and the general birth rate in 1905 was 39·13 per thousand or 1·73 per thousand lower than the birth rate in the previous year. The decrease in the total number of births was due to a decline in the birth rate in Bengal, Eastern Bengal and Assam, the United Provinces and in Bombay, in all other parts of the Empire, except the Upper Burma towns, where registration is defective, the numbers of births registered were greater in 1905 than in the previous year. The extreme birth rates were 54·02 per thousand recorded in the Central Provinces and Berar, where the birth rates of recent years have been extraordinarily high, and 25·31 in the small province of Coorg, where the sex distribution of the population is peculiar and registration is defective. In most of the other provinces the rates were not far from the mean, those furthest from it being the Punjab and Madras with birth rates of 44·4 and 32·0 per thousand, respectively. In three provinces the birth rates were lower than the death rates, namely, the Punjab where the birth rate fell short of the death rate by 3·2 per thousand, and the United Provinces and Coorg where the deficiencies were 2·76 and ·92, respectively. In the twelve towns in Upper Burma the registered death rate was 2·0 per thousand higher than the registered birth rate. In the remaining eight provinces the excess of the birth over the death rates ranged from 16·81 per thousand in the Central Provinces and Berar to 1·02 per thousand in Bengal. The mean number of male births recorded in India generally for every 100 births of females was 106·61. The highest percentages were 122·6 in the North-West Frontier Province and 116·10 in Ajmer-Merwara and the lowest 95·22 in Coorg; in the other provinces the percentages varied between 109·8 in the Punjab and 103·9 in the Madras Presidency.

86. The total number of deaths registered was 8,117,771 and the mean death

Deaths in India in 1905. rate was 35·96 per thousand as compared with 7,436,472 deaths and a mean death rate of 32·86 in 1904. In all provinces except Bombay, Madras, Coorg, the Punjab and the North-West Frontier Province the death rates of 1905 were higher than those of the previous year, the decline in Bombay and the Punjab being due to the abatement of plague, and in Madras, the North-West Frontier Province and Coorg to a decrease under the heading fevers The highest death rates were 47·55 per thousand in the Punjab, 44·0 in the United Provinces and 38·53 in Bengal; and the lowest were 24·93 in Lower Burma, 22·46 in Upper Burma and 21·4 in Madras. In every province, except Eastern Bengal and Assam, the urban were higher than the rural death rates.

The highest mortality in India generally occurred in April and the lowest in July. In most provinces the death rates of males were higher than the death rates of females, but in the United Provinces, the Punjab, the North-West Frontier Province, Coorg and Ajmer-Merwara the female rates were the higher, the excess being most remarkable in the Punjab where plague was most severe.

The death rates of infants calculated upon the numbers of births and of deaths of infants registered during the year ranged among males and females, respectively, from 301·29 and 270·94 per thousand in the Central Provinces and

Berar to 182·43 and 162·99 in Madras. It may be noted that the registered birth rate was highest in the Central Provinces and Berar, and excepting Coorg, lowest in Madras; and that in the other provinces there is a relation between high birth rates and high infantile death rates. In all provinces with the exception of Coorg the birth rates of males were higher than the birth rates of females, and in all, except the Punjab, the male infantile death rate was the higher, although in the United Provinces and Ajmer-Merwara the difference between the death rates of male and female infants was small, and only in the Central Provinces and Berar and Lower Burma was the difference great. Cholera was most severe in Eastern Bengal and Assam (death rate 4·77 per thousand), Bengal (2·93) and the United Provinces (2·55). Small-pox was present in every province and in five, Lower Burma (1·01), Bombay (·92), Coorg (·73), Central Provinces and Berar (·70) and Ajmer-Merwara (·58), the death rates were over ·5 per thousand. Plague was absent or practically absent from Coorg, Eastern Bengal and Assam and the North-West Frontier Province, and the plague death rates were comparatively low in Madras (·2), Upper Burma (·22) and Lower Burma (·55). The mortality in the Punjab (16·65) was more than twice as high as in the United Provinces (8·05) where the next highest death rate from plague was recorded; in the remaining provinces the plague death rates ranged from 5·20 in Ajmer-Merwara to 1·07 in the Central Provinces and Berar. The fever death rates were very low in Madras and Burma, and they were highest in the United Provinces and Bengal. A comparison of the rates registered under fever and 'all other causes' in the different provinces shows that they vary inversely, but in most provinces "fever" is credited with the vast majority of all the deaths; in a few the registering agency is more discriminating—or discriminates in a different way. By far the highest death rates from dysentery and diarrhœa were registered in the Central Provinces (3·02) and Bombay (3·0); in only two other provinces, Lower Burma (1·43) and Madras (1·4), was the ratio more than 1 per thousand, the rates in the other provinces ranging from ·90 in Bengal to ·28 in the North-West Frontier Province.

The death rates registered as due to respiratory diseases were in most cases low, the highest being 3·05 per thousand in the Punjab and 2·95 in Bombay, both provinces where plague was severe.

87. The weather of 1905 was abnormal and the climatic conditions were unfavourable to the public health. The early part **Bengal.** of the year was exceptionally cold and the rainfall was excessive; the south-west monsoon rains were not established until the end of June which was extraordinarily hot, and although the well distributed copious rains of July and August and the very heavy fall in September gave a normal supply to the whole province except Orissa, where the defect was about 20 per cent., the monsoon rains ceased early, October being almost rainless. The late commencement and early cessation of the monsoon rains was unfavourable to many of the crops, and in Bihar the standing crops were damaged or destroyed by floods, the result being that the agricultural outturn was below the average in 23 districts, average in five and above the average in only three. Owing to the absence of export, however, prices were only slightly raised in Bihar, Chota Nagpur and Orissa, while they were easier in Lower Bengal.

The total population under registration in the remodelled province is 49,891,164. The number of births registered in 1905 among this population was 1,973,301 or 141,925 fewer than in the previous year, and the birth rate was

39·55 per thousand against 42·39 in 1904 and a quinquennial mean of 38·91. The highest district death rates were 51·33, 47·46 and 45·78 registered, respectively, in Palamau, Cuttack and Muzaffarpur, and the lowest, excluding the urban district of Calcutta, where the birth rate was only 18·44, were 32·84, 32·20 and 30·49, registered in Burdwan, Hooghly and Jessore, respectively. The average recorded birth rate in rural areas was 40·58 per thousand compared with a mean of 24·53 in the towns in which rates ranged from 65·17 in Jamalpur, 49·48 in Sahibganj and 47·52 in Lohardaga to 5·28 in South Barrackpore. The explanation of the low rates in towns is the disproportionate male population, the practice of women leaving the towns to be confined at their homes in the districts and to inferior registration. The percentage of male to female was 105, ranging in the districts from 115 in Calcutta and 110 in Howrah, to 101 in Gaya, Hazaribagh and Ranchi.

The number of deaths registered was 1,922,369, or 327,849 more than in 1904, and the death rate was 38·53 per thousand or 6·58 per thousand higher than in 1904, and 5·11 higher than the mean of the quinquennium. The highest district rates were registered in Gaya, 59·86 per thousand, Patna 58·74, and Shahabad 58·65, all districts in the west of the province where plague was severe, and the lowest in Manbhum 26·23, Ranchi 25·50 and Singbhum 22·05. In rural areas the mean death rate was 38·42, compared with 40·01 per thousand in the towns, among which enormous rates such as 92·26 in Daudnagar in the Gaya district, 87·91 in Gaya town, and 83·46 in Revelganj in Saran, where plague or cholera, or both, were epidemic, are contrasted with rates such as 10·77 in Cossipore-Chitpur, 10·52 in Maniktola and 7·41 in Ulubaria where the low rates are evidence rather of defective registration than exceptional salubrity.

The highest mortality in the province as a whole occurred in October, November and December, and the lowest in June, July and August.

Among the sects Hindus (39·22 per thousand) and Muhammadans (38·58) had the highest death rates; the rates among "Other Classes" (27·80), Budhists (24·66) and Christians (21·66) being much lower.

Infants under one year of age died at the rates of 216·56 and 200·42 per thousand of males and females born, respectively; as usual in Bengal the recorded male rates were the higher at all age periods, the mean death rate among males being 40·07 per thousand, compared with 37·00 among females.

88. The number of the population among whom vital statistics are registered in Eastern Bengal and Assam is, excluding 89,012 in certain hill tracts, 29,812,735. In 1905 there were registered 1,173,879 births, equal to a birth rate of 39·37 per thousand as compared with 41·68 in the previous year, and 39·53, the mean of the previous five years. As in Bengal the climatic conditions of the year were unfavourable, and heavy floods in the districts bordering the Brahmaputra and Meghna not only damaged the crops, but apparently led to a severe epidemic of cholera in the last quarter of the year.

Eastern Bengal and Assam.

The highest district birth rates were 50·70 per thousand in Goalpara, 48·29 in Malda and 45·25 in Noakhali; and the lowest were 30·46 in Pabna, 30·06 in Sibsagar and 28·96 in Kamrup. In rural areas the mean death rate was 39·72 per thousand, whereas in the towns, in which rates ranged from 69·16 in

Barpeta to 1·71 in Jhalakati, the mean was only 23·40. The percentage of male to female births was 106.

The registered deaths numbered 1,045,305, the death rate being 35·06 per thousand compared with 32·11 among the same population in 1904, and 31·52, the mean of the quinquennium. Among the districts the highest rates were 45·59, 44·45 and 43·88, registered in Pabna, Goalpara and Malda, respectively, and the lowest were 25·05, 24·77 and 19·99 in Cachar, Sibsagar and Nowgong. The mean death rate in rural areas was 35·25, in remarkable contrast to the mean in the towns where it was only 26·42. The highest urban rates were 55·21 in Barpeta and 53·44 in Mangaldai, and the lowest 7·27 in Pabna and 4·77 in Jhalakati. The lowest monthly death rate was registered in March, 1·97, and the highest rates were registered in November 4·29, and December, 4·78. Muhammadans died at the rate of 36·38 per thousand, "Other Classes" at the rate of 36·76, while the registered rates among Christians and Buddhists were 25·69 and 18·75, respectively.

Male infants died at the rate of 211·81 and female infants at the rate of 194·59 per thousand born, respectively; at all age periods, except the three between 15 and 40, the male death rates were the higher—males dying at the rate of 35·86 and females at the rate of 34·22 per thousand.

89. The mean annual strength of the coolies employed on tea gardens in Assam, during the twelve months ending the 30th June 1906, rose from 660,700 to 670,963. Of the total labour force

Tea gardens.

38·85 per cent. was employed in the Surma Valley, and 61·15 per cent. in the Assam Valley, the distribution approximating closely to that of the preceding year. The birth rate showed an increase, being 29·2 per thousand of the total population, and 94·7 per thousand of the female adult population, against 28·4 and 92·1 per thousand respectively in 1904-05. These rates compare unfavourably with the corresponding provincial ratios of 39·37 and 128·4.

The registered death rate 24·0 per mille was the same as in the previous year, the total number of deaths being 16,103. The principal causes of death were dysentery (2,843), malarial fever (2,429), cholera (2,057), diseases of the respiratory organs (1,933), diarrhœa (1,607) and anchylostomiasis (1,499). The death rate among coolies from all recruiting areas, except the Central Provinces, showed a slight increase. The mortality among coolies from those provinces fell from 35·69 per mille to 31·18.

The number of unhealthy gardens, i.e., those with a death rate of over 70 per mille, was 10 of a total of 757, against 6 of a total of 776 in the year preceding. There were 58 deaths among coolies journeying to the Assam Valley, against 16 in the previous year; 45 of these deaths were due to cholera. Four coolies died of cholera at the Sylhet depots.

90. The climatic conditions of the year were unfavourable; an exceptionally cold and wet spring was followed by a very hot weather

United Provinces.

in May and June, the monsoon rains began late, were in defect in many districts, and ceased to fall before the usual time except in the Benares division where the September rainfall was 70 per cent. in excess of the normal. Harvests generally were bad and prices were high. The number of births registered was 1,967,009 and the birth rate which had risen to 46·67 per thousand in 1904 as compared with a quinquennial mean of 44·07 fell to 41·24. The highest district rates were 52·59 in Jhansi, 51·44 in Bijnor and 51·32 in Shahjehanpur, and the lowest 32·15 in Ballia, and 31·96 and 27·86 in Naini Tal and Dehra Dun, respectively.

In towns the mean birth rate was 38·82. Among the towns there were nine, including Kairana (56·26), Chandpur (55·78) and Konch (54·25), in which the birth rates were over 50 per thousand; in Brindaban (19·81), Naini Tal (15·24) and Mussoorie (9·16) the rates were exceptionally low. The mean percentage of male to female births registered was 108·39.

The registered deaths numbered 2,098,300, or 443,351 more than in 1904, and the death rate was 44·0 per thousand, nearly 9 per mille higher than the rate of the previous year, 34·70 which was slightly higher than the quinquennial mean, 33·80. By far the highest district death rate was 91·01, recorded in Muttra where the plague death rate was 62·64; in the other districts death rates ranged from 66·58 in Sultanpur, where there was a death rate from cholera of 25·30 per thousand, and 65·38 in the Ghazipur district of the Benares division, where plague and cholera were severe and fevers exceptionally prevalent, to 26·12 in Almora, the only district in the province in which no death from either plague or cholera was reported. The rural mean death rate was 43·26 compared with 53·64 in the towns in which, excepting Kiratpur (105·71), where the plague death rate was 71·82 per thousand, and Brindaban (95·61) where the plague death rate was 41·51, the death rate ranged from 86·35 in Muttra to 23·64 in Laharpur.

The highest mortality was recorded in March, 5·07 per thousand, and the lowest, 2·39, in August.

Hindus died at the rate of 44·21 and Muhammadans at the rate of 43·66, "Other Classes" and Christians at the rates of 27·13 and 9·67, respectively.

Infants died at the rates of 264·21 and 261·48 per thousand boys and girls born, respectively. At all age periods except in infancy and after 50 the female death rates were the higher; females dying at the rate of 45·44 per thousand, compared with a rate of 42·64 among males.

91. The first three months of 1905 were stormy and exceptionally cold; in April the temperature was comparatively low, but May and June were unusually hot. The rains failed in July and August, but in September a storm gave a very heavy fall. The temperature of October and November was much higher than usual. The heat and dryness of the summer and autumn were attended by a remarkable reduction in the prevalence of plague.

Punjab.

The number of births recorded was 893,360, and the birth rate was 44·4 per thousand compared with 41·5 in the previous year and 40·5, the mean of the previous five years. In the districts the highest birth rates were 67·2 in Lyallpur, 51·2 in Sialkot and 48·5 in Multan; the lowest being 36·2, 35·2 and 21·4 in Ambala, Dera Ghazi Khan and Simla, respectively. The mean rural birth rate was 44·7, compared with 41·3 in the towns. The percentage of male to female births was 109·8, the district percentages ranging from 118 in Dera Ghazi Khan to 105 in Ludhiana and 104 in Simla.

The number of deaths recorded was 956,108, more than 30,000 less than in 1904, and the death rate fell from 49·1 per thousand in that year to 47·6, against a quinquennial mean of 44·9. The plague death rate was 16·7 per thousand, so that, so far as the mortality figures go, the year, but for the presence of plague, would have been an exceptionally healthy one. The highest district death rates were 78·5 registered in Rohtak, 71·3 in Gurgaon and 66·1 in Amritsar ; and the lowest were 25·4 in Dera Ghazi Khan, 25·0 in Simla and 24·5 in Jhang. The mean death rate in rural areas was 47·37, and the urban rate was 49·13. The enormous death rate of 306·6 was registered in Kangra where 1,374 persons lost their lives as the result of the great earthquake of April 1905 ; and in Dharmsala where many people were killed, the death rate was 90·02. In Sangla, a very small town in Gujranwala, high mortality from plague, cholera, small-pox and fevers gave a death rate of 129·33 per thousand, and in Hansi, where the plague death rate was 78·86 per thousand, the total death rate was 120·86. Death rates of 104 were registered in Sharakpur and Khudian in the Lahore district ; in other towns the death rates ranged from 81·85 in Bahadurgarh to 7·10 in Khangah Dogran in Gujranwala.

Hindus died at the rate of 51·48 per thousand and Muhammadans at the rate of 44·21 ; Christians and "Other Classes" (in whose case there are errors in classification) dying at the rates of 24·30 and 54·74, respectively. The highest monthly death rates were recorded in April, 8·67, and May, 7·35, and the lowest in September, 2·20 and October, 2·37.

Male and female infants died at the rates of 233·90 and 236·31 per thousand born, respectively ; and boys and girls over one year and under five years of age died at the rates of 57·39 and 61·42. At all age periods the death rates of females were the higher, the general death rate among males being 44·1 and among females 51·6 per thousand.

92. The number of births registered was 70,369 and the birth rate was 35·4 per thousand, compared with 34·9 in 1904 and a quinquennial mean of 32·3, the general increase in the figures being due to the efforts made to improve registration. The highest district birth rate was 43·3 recorded in Kohat and the lowest 31·7 in Hazara. The percentage of male to female births was 122·6, the percentages varying in the districts from 134·0 in Peshawar to 114·0 in Dera Ismail Khan. Special enquiries have been carried out during the past few years with the object of ascertaining how far the high recorded proportion of male to female births represented the facts. In 1905 these special enquiries were continued, and two areas were selected, one in the Kohat district containing 30 villages and another in Peshawar containing 159 villages, most of the villagers in both areas being Pathans. The population of the selected area in Kohat was 25,070 (13,514 males and 11,556 females) and during the time of the enquiry the births of 472 males and 428 females were registered by the *chowkidars;* it was found that the births of 43 males and of 40 females had not been registered. The percentage of male to female births according to the *chowkidars'* returns was 110·3, according to the corrected figures it was 110, the mean rate for the whole district being 124·3. The population of the selected area in Peshawar was 79,001 (41,946 males and 37,055 females). The *chowkidars* registered the births of 1,499 males and 1,167 females, a percentage of males to females of 128·4; it was found that the births of 116 males and 101 females had not been registered and when these are

North-West Frontier Province.

included the percentage falls to 127·4, the mean rate for the whole district being 134·0. The results in both areas appear to show that the percentage of male to female births in the frontier districts is really very high, but that registration is carelessly performed, the fact that lower proportions of male births occurred in the selected areas than in the surrounding districts suggests the possibility of exceptional diligence on the part of the *chowkidars*.

The number of deaths registered was 53,327, and the death rate was 26·8 per thousand, compared with 28·6 in the previous year and 25·5, the mean of the previous five years. The rates varied in the districts between 27·6 in Peshawar and 24·5 in Bannu, and in the towns between 44·25 in Buffa in Hazara, and 16·77 in Kulachi in Dera Ismail Khan. The death rate among Muhammadans was 27·23, but they are perhaps less careless regarding registration than the comparatively small number of Hindus, among whom the death rate was 23·12. Male infants died at the rate of 203·25 and females at the rate of 192·89 per thousand born. In all age periods except under one year, 1—5 and 50—60, the female death rates were the higher, the general death rates being 26·2 per thousand among males and 27·5 among females.

93. The cold weather rains were favourable and but for the extreme cold in February the wheat crop would have been exceptionally good. The monsoon rains were late and the fall was defective until the heavy rain of September to which the prevalence of malarial fevers is attributed. The autumn crops were abundant. The year was unhealthy and influenza, malarial fevers and bowel complaints were rife.

Central Provinces and Berar.

The number of births recorded was 642,199 and the birth rate was 54·02 per thousand against 53·19 in 1904. In the districts of Yeotmal, Damoh and Murwara, the high birth rates of 61·24, 60·00 and 59·86 were recorded, while the lowest rates in the province were comparatively high—48·69 in Hoshangabad, 48·57 in Amraoti and 46·30 in Burhanpur, the only district in which the births were fewer than the deaths. In rural areas the mean birth rate was 55·34 compared with 42·37 the urban mean. The provincial percentage of male to female births was 104·38, varying in the districts between 108·91 in Nagpur and 102·14 in Raipur.

The registered deaths numbered 442,383, and the death rate was 37·21 per thousand, compared with 32·06 in the previous year. The highest district death rates were 54·80 recorded in Burhanpur, and 50·08 in Buldana ; in the other districts the rates ranged from 47·19 in Nimar to 28·44 in Mandla. In rural areas the mean death rate was 36·69 as compared with 41·87, the mean in the towns, among which very high rates were registered in Mohgaon (127·05), a small town which suffered severely from plague, in the Chhindwara district, in Nandura (93·12) and Deulgaon Raja (89·15,) small towns in the Buldana district where plague was very severe in both town and country. In the remaining towns the death rates ranged from 82·68 in Akote to 17·07 in Rajim.

The infantile death rate was 301·29 among males and 270·94 among females per thousand born, respectively. Males of all ages died at the rate of 39·35 and females at the rate of 35·12 per thousand of the census population.

94. The rainfall in the presidency in 1905 was about three and a half inches below the normal, but the rainfall of both monsoons was irregularly distributed and the prices of food grains, which had risen on account of the failure of both the south-west and north-east monsoons in 1904, rose still further.

Madras.

The total number of births recorded was 1,176,256, equal to a birth rate among the population of 36,737,533 under registration of 32·0 per thousand— the highest rate yet reached excepting the year 1891 when it was 32·4—as compared with 30·7 in 1904 and 29·4 the quinquennial mean. Excluding the urban district of Madras, where the birth rate was 46·0 per thousand, the district rates ranged from 38·7 in Guntur to 24·4 in Nellore. The mean rural birth rate was 31·6 against 35·1 in the towns in which rates varied between 61·2 in Vaniyambadi and 6·5 in Madanapalli in Cuddapah.

The percentage of male to female births was 103·9, the district percentages ranging from 107·7 in the Nilgiris to 101·5 in Anantapur.

The deaths registered numbered 786,123 and the death rate was 21·4 per thousand against 22·5 in the previous year, and 21·9 the average of the previous five years. Again excluding Madras, where the death rate was 59·5 per thousand, the district rates ranged from 31·4 in Bellary to 16·0 in North Arcot. In rural areas the mean death rate was 20·3, while in the towns the mean was 29·6, the extreme rates being 92·2 in Kampli in Bellary and 5·6 in Rajampet in Cuddapah.

Muhammadans and Hindus died at the rates of 23·3 and 21·4, respectively; Christians and " Other Classes " at the rates of 17·2 and 9·2.

Male infants died at the rate of 182·43 per thousand born, female infants at the rate of 162·99. In the three age periods between 10 and 30 the female death rates were the higher, in all other age periods males died at the higher rates, the general death rates being 22·2 and 20·6 per thousand among males and females, respectively.

95. There were 4,572 births registered in Coorg in 1905 equal to a birth rate

Coorg.

of 25·31, as compared with 21·97 in the previous year and 23·28 the mean of the previous five years. The district rates varied between 38·15 in Padinalknad and 19·90 in Yedenalknad, the mean percentage of male to female births being 95·22.

The number of deaths registered was 4,739 and the death rate was 26·24, against 26·62 in 1904 and a quinquennial mean of 31·30. The highest district death rate was 30·74 in Nanjarajpatna, the lowest 23·66 in Kiggatknad ; the mean urban death rate was 32·13, the mean rural rate, 25·70. Male and female infants died at the rates of 222·42 and 209·22 per thousand born, respectively. In all age periods, except the four from 10 to 40, the male death rates were the higher, but the general death rate of males was 25·70 as compared with 26·91 among females.

96. The number of births recorded in the Bombay Presidency fell from 648,594

Bombay.

in 1904 to 611,173 in 1905 and the birth rate from 35·09 to 33·07 per thousand, which, however, is considerably higher than the quinquennial mean, 30·51. The highest district birth rates were 46·63, 45·09 and 43·12 registered in Khandesh, Broach and Panch Mahals, respectively, and the lowest were 19·67, 19·04 and 17·88 in the Sind districts of Upper Sind Frontier, Hyderabad and Thar and Parkar. The mean birth rate in rural areas was 34·07 ; the mean of the urban rates was much lower, 26·42 per thousand, the highest rates being 51·60 in Karachi, 51·43 in Dholka in Ahmedabad and 44·29 in Athni in Belgaum. The percentage of male to female births in the presidency generally was 108·44, but in the districts of Bombay proper the percentages varied between 111·72 in Kaira and 102·43 in

Panch Mahals, while in the Sind districts, excluding Karachi where the percentage was 114·5, the percentages ranged from 135·81 in Hyderabad to 121·13 in Sukkur.

There was a most gratifying decline in the mortality, the number of deaths falling from 764,914 in 1904 to 588,394 in 1905, and the death rate from a mean of 46·31 in the previous five years and 41·39 in 1904 to 31·84. The fall in the mortality was due to the abatement of the plague, the recorded deaths falling by 152,594 from 223,957 in 1904 to 71,363. The highest district death rate was, excluding 62·33 in Bombay City, 41·13 in Ahmedabad; in the remaining districts the rates ranged from 40·10 in Khandesh to 15·88 in Hyderabad. In rural areas the mean death rate was 29·47 compared with 47·59 in the towns, among which the highest rates were 65·24 in Shikarpur, 63·43 in Ahmedabad, towns in which small-pox and diseases recorded under the heading "fevers" were exceptionally severe, and 63·05 in Yeola, a small town in Nasik, where the plague mortality was 38·41 per thousand.

Hindus died at the rate of 34·0 per thousand, followed by Jains, 28·80, Parsis 24·26, Muhammadans 23·98, and Christians 22·76. The high rate recorded for "Other Classes," 48·18, is dependent upon errors in classification.

The death rate of male infants was 229·52 and of female infants 219·07 per thousand born, respectively. In all age periods except 5—10, 10—15, 15—20 and 20—30 the male death rates were the higher, the means being 32·11 among males and 31·54 among females.

97. The number of births registered in Lower Burma in 1905 was 191,226,

Lower Burma.

about 11,200 more than in the previous year, and the birth rate rose from 32·71 per thousand to 34·34 as compared with the quinquennial mean of 33·46. In the districts, excluding Rangoon, the rates ranged from 42·73 in Thayetmyo to 27·72 in Toungoo, the mean percentage of male to female births being 106.

The registered deaths numbered 138,850 and the death rate which was 22·36 per thousand in 1904, as compared with a mean of 23·17 in the five years ending with 1904 rose to 24·93. Excluding the urban district of Rangoon, where the death rate was 45·90, the district rates varied between 28·99 in Henzada and 15·59 in Thaton. In rural areas the mean death rate was only 23·0 as compared with 38·21 in the towns among which Wakema (54·53), Kyaiklat (53·05) and Henzada (49·32), in the Irrawaddy division where cholera and small-pox were severe, suffered the heaviest mortality.

The death rate among Hindus was 32·03, among Burmese 24·87, among Muhammadans 22·99, among Christians 15·36, and among "Other Classes" 26·90.

The infantile death rate was 220·17 for males and 171·64 for females per thousand born, respectively. In all age periods the male death rates were the higher, the general death rate for males being 26·43 and for females 23·24.

98 The registration of births in Upper Burma is limited to twelve towns, but

Upper Burma.

will become general from 1st January 1907. The mean birth rate in the twelve towns fell from a quinquennial mean of 37·03 per thousand and 35·05 in 1904 to 34·91 in 1905 the rates ranging in the individual towns from 50·33 in Taungdwingyi to 29·09 in Shwebo ; and the percentage of male to female births ranged from 112 in Pakokku to 79 in Sahin.

The total number of deaths recorded was 65,541, and the death rate rose from 18·69 to 22·46 as compared with a quinquennial mean of 19·63. The district death rates varied greatly from 30·49 in Kyaukse to 15·41 in Meiktila. The rural death rate was 20·87, while the mean rate in the towns was 36·56, the rates ranging in individual towns from 62·55 in Pakokku, where cholera was severe 50·53 in Pyinmana, where plague was epidemic, and 49·07 in Myingyan, which suffered from cholera, to 24·08 in Yamethin.

The Burmese death rate was 22·46, as compared with 22·12 among Hindus, 23·25 among Muhammadans and 13·36 among Christians.

The defective registration of births renders an estimate of the infantile death rate impossible. At all age periods, except 30—40, the male death rates were the higher, the general death rate being among males 23·64 and among females 21·40 per thousand.

99. The number of births recorded rose from 15,997 in 1904 to 17,802, and the

Ajmer-Merwara.

birth rates from a quinquennial mean of 26·38 and 33·54 in 1904 to 37·32 per thousand in 1905, the rate being 45·70 in Merwara and 34·84 in Ajmer. The percentage of male to female births was 112·16 in Merwara and 117·66 in Ajmer.

The number of deaths registered was 16,332, or 3,183 more than in the previous year, and the death rate was 34·25 compared with 27·57 in 1904 and a quinquennial mean of 51·75. The death rate in Ajmer, where plague was widespread and gave rise to a death rate of 6·51, was 36·32 per thousand, against 27·25 in Merwara. Plague was very severe in the small town of Kekri where the death rate from all causes was 105·20. In Ajmer suburb the total death rate was 67·72, including 7·12 from plague and 2·99 from small-pox. The Hindu and Muhammadan death rates were, respectively, 35·98 and 35·53 per thousand.

The death rates of infants were 234·32 among males and 232·94 among females per thousand born, respectively. In the five age periods between 1 and 30 the female death rates were the higher, the general rates being among males 33·66 and among females 34·90 per thousand.

GENERAL POPULATION.

HISTORY OF THE CHIEF DISEASES.

100. The marginal statement gives a tabular view of the number of deaths from, and the death rates per 1,000 of population for each of the chief diseases and all causes recorded in British territory in India during each of the five years,

Years.	Cholera.	Small-pox.	Fevers.	Dysentery and Diarrhœa.	Plague.	All causes.
1901	371,210	89,385	4,174,919	247,1.0	237,038	6,0:6,357
	1·21	·40	18·62	1·10	1·00	29·46
1902	324,138	115,443	4,279,351	225,750	456,975	7,182,235
	·69	·51	18·93	1·04	2·02	31·49
1903	317,854	93,603	4,459,737	273,459	626,483	7,884,135
	1·38	·81	19·66	1·21	2·03	34·7
1904	192,835	55,231	4,193,881	240,855	940,609	7,425,472
	·85	·24	18·09	1·06	4·16	32·86
1905	441,786	70,562	4,447,855	264,121	940,321	8,117,771
	1·96	·31	19·37	1·17	4·17	35·96

1901-5. It will be seen that the year under review was exceptionally unhealthy ; the death rate (35·96) is the highest in the quinquennial period. Compared with the previous year, 681,299 more deaths were recorded, and there was an increase in the mortality under each heading, this increase being least significant (·01 per mille) in the case of plague : compared with the preceding years of the quinquennium, the mortality from cholera, from plague, from fevers and from dysentery and diarrhœa has been excessive ; the death rates for cholera and plague are the highest in the quinquennium, and the death rates for fevers and dysentery and diarrhœa are second only to the death rates from these causes in 1903. Fevers accounted nominally for more than 54 per cent. of the deaths from all causes.

101. The number of deaths from cholera recorded in British territory during 1905 was 441,786, equal to a death rate of 1·96
Cholera in India in 1905.
Appendix A to Section VI.
per 1,000 of the total population under registration. If the deaths recorded in the native states from which returns were received are added (Statement I), the total amounts to 442,506. In the previous year 192,835 deaths from cholera were recorded in British territory, giving a ratio of ·85 per 1,000 of population, so that during the year under review cholera was considerably more than twice as prevalent as in the previous year. This increase in mortality during 1905 was shared by all the British provinces except Ajmer-Merwara and Coorg, both of which have now remained free from the disease for three years in succession. The greatest numbers of deaths from the disease during the year were recorded in Bengal, Eastern Bengal and Assam, the United Provinces and Madras, the figures requiring special notice being 142,312 deaths reported from the new province of Eastern Bengal and Assam and 121,750 deaths reported from the United Provinces as compared with 6,617 from the same province in 1904. The highest death rates were recorded in Bengal, Eastern Bengal and Assam, the United Provinces and Lower Burma, and in all these provinces cholera was present throughout the year. The seasonal prevalence varies widely—not only in the different provinces in the same year, but also in the same province from year to year. In Bombay the greatest number of deaths occurred in June as compared with August in

1904; in Madras in August as compared with January in 1904; in the Punjab, the North-West Frontier Province and the Central Provinces in September as compared with June, July and August, respectively, in 1904; in Lower Burma in October as compared with April in 1904, in Bengal and the United Provinces in November as compared with December and June, respectively, in 1904; in Eastern Bengal and Assam in December, the same month as in 1904. In Bengal during 1905, fewest deaths occurred in the month of June, while in September and October, the two months during which the disease is usually least prevalent, a heavy incidence was recorded.

102. In last year's report a reference was made to investigations, carried

Recent investigations. (1)

out in the early part of 1905 by Dr. F. Gotschlich at the pilgrim quarantine camp at Tor, relating to the finding of vibrios in the intestines of 38 pilgrims (107 cadavers examined) who had died from dysentery and other forms of colitis while on their return journey from the Hejaz. Six of the 38 vibrio strains were found by Gotschlich to be identical with the vibrio of true Asiatic cholera not only culturally and morphologically but also according to the agglutination and Pfeiffer reactions ; and yet (1) there was no evidence, clinical or pathological, that the pilgrims from whose intestines these six vibrio strains were isolated had ever suffered from cholera, and (2) no case of cholera was known among the pilgrims visiting Mecca during that season, so these six pilgrims had not infected others. Gotschlich' regarded the six pilgrims, who all came from regions in which cholera is reputedly endemic, as cholera vibrio " carriers " and stated that the absence of an outbreak of cholera among the pilgrims might possibly have been due to the vibrios being (a) eliminated only in small numbers and (b) perhaps of low virulence. The other 32 strains were isolated from the intestines of pilgrims of diverse nations and were easily distinguished from Koch's cholera vibrio; they were in all probability derived from the drinking water—in the water of the holy well in the Kaaba at Mecca saprophytic vibrios were found. It is interesting to note that during the pilgrim season 1905-6 there was again a complete absence of cholera among the returning pilgrims but in two cases a vibrio identical with the above six strains was isolated at the Tor quarantine camp. Gotschlich ' examined the fæces of 127 pilgrims in hospital for various diseases and found vibrios in 18. No reaction with cholera serum (obtained from Berlin) was given by 16 of these 18 strains even in a dilution of 1 in 50, while the specific reaction in a dilution of 1 in 2000 was given by the other two. The Tor vibrios have occasioned a great deal of work with the object of ascertaining whether they are real cholera germs or not. The matter is not yet settled; at the end of this section the reader will find references to the principal papers which have been published.

103. In Bengal in 1905, the total number of deaths recorded as due to cholera

Cholera in Bengal.

was 146,339 or 2·93 per mille of the population as compared with 81,463 or 1·63 per mille in 1904, and 138,577 or 2·77 per mille—the average figures for the five years 1900-4. The mortality during 1905 was higher than in any of the previous eight years except 1900. No district was entirely free from the disease which attacked 18,652 villages and assumed an epidemic form in 104 registering circles, as compared with 14,241 villages, and 45 registering circles in 1904. The district of Gaya in the Patna division suffered most severely and lost from this cause 6·63 per mille of its population. Other districts in which a high mortality from cholera was recorded were Purnea (5·17 per mille), Patna (5·0 per mille), Shahabad (4·94 per mille), Howrah (4·78 per mille), Bhagalpur (4·74 per mille) and Puri (3·97 per mille). Ranchi and Palamau, which remained free from this disease in 1904, were the least affected of all the districts during the year. Compared with the previous year, the mortality from this cause was smaller in only eight districts, the most important of these being Singhbhum, Darjeeling, Hazaribagh and Manbhum. The great prevalence of the disease in Gaya is said to have been due to the contamination of the wells by surface washings in consequence of a very heavy continuous rainfall directly after excessively dry weather. Considering the province as a whole, the incidence was greatest during the last four months of the year and least during May and June.

The highest death rates for cholera recorded in towns were 12·94 per mille in Gaya, 11·99 in Puri, 11·54 in Buxar, 11·49 in Tikari and 10·10 in Chapra. Among rural areas, the highest rates were recorded in Kadwa (16·23 per mille) and Kasba Amur (14·89 per mille), both in Purnea.

In the district of Gaya over 6,900 wells were disinfected with permanganate of potash, and the disease is said to have been thus kept in check. This method of purifying well-water was also applied in 16 other districts; from Champaran the statement comes that good results followed its use.

Among the European seamen of the port of Calcutta there was only one death from cholera : there were 33 among the native floating population.

104. The number of deaths attributed to cholera in the new province of Eastern

Cholera in Eastern Bengal and Assam. Bengal and Assam during 1905 was 142,312, equal to a ratio of 4·77 per 1,000 of the census population. The average death rate for cholera during the decade 1895—1904 for the combined districts which now constitute the province of Eastern Bengal and Assam was 2·30 per mille, which is less than half the mortality during the year under review. The provincial sanitary commissioner remarks that in some cases there was a history of the disease spreading from village to village along the course of the streams and rivers, especially where the people relied on the river water for household purposes. From October to December the disease raged in most of the districts bordering on the lower reaches of the Brahmaputra and Ganges ; the districts of Mymensingh (10·5 per mille), Kamrup, Goalpara, Bogra, and Pabna, all returned death rates above 7 per mille : in these five districts during the last three months of the year the epidemic was sufficiently severe to interfere with business. In some places the villagers were unable to dispose of their dead in the ordinary way and either threw the bodies into the rivers to be carried away by the flood or left them to rot on the surface of the ground. On the other hand, districts on the upper reaches of the Brahmaputra and districts having little or no connection with the great water-ways suffered only to a small extent. In Mymensingh, which suffered most severely during the year, the decennial rate was only 2·18 per mille; Kamrup is the district which showed the highest decennial death rate (5 per mille). In Dacca, wells and tanks in the affected localities were subjected to the permanganate of potash treatment, and it is reported that the disease was brought under control almost immediately after this treatment was begun.

The towns which recorded death rates over 10 per mille were Tangail, Sherpur, Sibsagar and Bogra : the quinquennial death rate for cholera in Tangail is

13·8 per mille of the population, while the quinquennial rate for no other town in the province is over 4 per mille.

The death rate (2·14 per mille) from cholera among tea-garden coolies was much less than that among the general population of the province, the highest coolie death rate being 2·69, recorded in the Sylhet district. Other districts returning rates over 2 per mille were Cachar, Kamrup and Lakhimpur. Altogether there were reported 391 cases, of which 252 occurred in the Karimganj Sub-division.

105. The total number of deaths from cholera recorded in the United Provinces during 1905 was 121,790, equal to a ratio of 2·55 per 1,000 of population, as compared with a total of 6,617 and a ratio of ·14 during 1904. The average death rate of the preceding quinquennium was ·92 per mille. Of the 48 districts no death from cholera was reported during the year in six, and not more than nine deaths in any of ten other districts. The districts where the mortality from this cause was highest were Sultanpur (25·30 per mille), Partabgarh (12·14), Fyzabad (10·63), Rai Bareli (10·55), Bara Banki (8·44) and Basti (6·85); of these six districts, the first three mentioned and Bara Banki are in the Fyzabad division.

Cholera in the United Provinces.

The highest death rates for this disease recorded in towns were 13·48 per mille at Jais (Rai Bareli), 13·32 at Radauli (Bara Banki), 10·75 at Ramnagar (Benares), 9·55 at Barhaj (Gorakhpur), 6·42 at Azamgarh, and 6·29 at Sandila (Hardoi). Of the 107 towns with populations of 10,000 or more, 61 did not return any death from this cause.

No case of cholera was reported to have occurred at the Dikhauti and Mahavaruni fairs at Hardwar, the Garhmuktesar fair at Meerut, the Batesar fair at Agra and the *Magh mela* at Allahabad. During the time of the Dadri fair at Ballia, cholera was present in Ballia and the neighbouring districts, yet only 56 cases with 45 deaths occurred among the multitude, numbering about 700,000, that attended this festival on the principal bathing day.

106. In the Punjab cholera did not assume a severe epidemic form during the year. The total number of deaths was 2,197, yielding a ratio of ·11 per mille of population as against ·04 in the previous year and ·41—the average of the quinquennial period ending with 1904. The largest number of deaths amounting to 1,624 or three-quarters of the total registered in the province occurred in the district of Lahore, Multan coming next with only 141 deaths. No death from cholera was reported to have occurred in the province during the first four months of the year, and only 13 and 56 deaths were recorded in May and June, respectively; but the number of deaths registered in July, August and September amounted to 87 per cent. of the total number of deaths from cholera during the year in the province. Of the 145 municipal towns in this province, cholera was reported from only 26; the highest death rates recorded in towns were 13·78 per mille of the population in Shujabad (Multan), 6·11 in Sangla (Gujranwala), 4·13 in Chunian, 2·86 in Kasur, 1·95 in Lahore and 1·81 in Khem Karn (the last four towns are in Lahore district).

Cholera in the Punjab.

On the appearance of cholera in any locality, the sources of drinking water open to contamination were disinfected with permanganate of potash.

107. In 1904 the North-West Frontier Province was practically free from
Cholera in the North-West Frontier cholera, only one death having been reported from
Province. a village in the Hazara district; in 1905 in the
same district the disease prevailed in epidemic form during September and
October, 415 cases with 303 deaths having been registered. Evidence seems
to point to infection having been conveyed to a village in this district
by a man who had come from Lahore, where the disease was at the time
prevalent; from this village infection rapidly spread to neighbouring villages
and towns. The civil surgeon reported that "the people were generally
apathetic and, in some cases hostile" to the prophylactic measures adopted.
There was not a single death from cholera during the year in any other district
in the province.

108. In the Central Provinces and Berar the total number of deaths recorded
Cholera in the Central Provinces in 1905 as due to cholera was 1,217, or ·10
and Berar. per mille of the population, as compared with a
total of 2,967 and a ratio of ·24 in 1904. During 1905 the disease was confined
to the districts of Bilaspur (605 deaths), Raipur (382 deaths), Jubbulpore (136
deaths), Narsinghpur (60 deaths), Wardha (23 deaths) and Chanda (10 deaths).
The disease is said to have been introduced by infected pilgrims returning from
Puri (Jagannath). The highest district death rates were ·55 in Bilaspur, ·26 in
Jubbulpore and ·25 in Raipur.

With the exception of one doubtful case reported from Yeotmal, there was
no cholera in the Berar districts during the year.

109. In the Madras Presidency in 1905 there were 16,888 deaths recorded as
Cholera in the Madras Presidency. due to cholera, giving a ratio of ·5 per mille of
population, as compared with a total of 23,109
and a ratio of ·6 per mille in 1904. The ratio for the year under review is the
lowest since 1895. The disease was reported from all the districts of the presidency except the Nilgiris, Tinnevelly and Coimbatore (the last mentioned registered one case). The highest death rates recorded in districts were 7·2
per mille in Madras, 2·8 in Bellary, 1·8 in Chingleput, 1·3 in Nellore, 1·2 in
Kistna and 1·1 in Godaveri.

The highest death rates for cholera recorded in towns were 16·1 in Pulicat
(Chingleput), 7·7 in Kampli (Bellary), 7·2 in Madras, 6·8 in Bezwada (Kistna),
6·0 in Dowlaishweram (Godaveri) and 6·6 in Nellore. In Madras the disease
broke out in a severe form about the beginning of the latter half of the year and
accounted for 3,534 deaths. The cause of the outbreak was supposed to be due
to the use of "polluted sources of water-supply when the Red Hills' supply ran
short of requirements."

Water-disinfection by permanganate of potash was, as usual, carried out in
the areas invaded by cholera and was found to yield good results.

110. During 1905 no death from cholera was recorded in Coorg. This
small province has been free from the disease since
Cholera in Coorg. 1901, in which year 58 deaths from this cause
were reported.

111. In the Bombay Presidency there were recorded during 1905 as due
to cholera 5,396 deaths, equal to a ratio of ·29
Cholera in the Bombay Presidency. per mille of population, as compared with 13,156

deaths or a ratio of ·71 per mille in 1904. The disease was practically confined to the southern registration district, where 5,318 deaths, *i.e.*, 98·55 per cent. of the total number of deaths in the presidency, were recorded. The highest death rates for this disease recorded in districts were 2·02 per mille of population in Bijapur, 1·91 in Dharwar and 1·60 in Belgaum ; outside these districts the incidence of cholera was very slight, Sind and Gujarat with the exception of one doubtful case reported from Ahmedabad district remaining free from the disease throughout the year.

Deaths from cholera were recorded in 11 of the 56 town-circles in the presidency as compared with 25 town-circles in 1904, the highest rates being 6·26 in Gadag-Bettigeri (192 deaths), 4·71 in Bijapur (112 deaths), 4·46 in Gokak (44 deaths), 2·79 in Kalyan (30 deaths) and 1·70 in Dharwar (52 deaths).

The acting deputy sanitary commissioner reported in connection with the epidemic in the southern registration district : " During the year cholera has attacked most severely the towns and villages on the banks of rivers and streams. This probably has relation to density of population and to favourable conditions of soil (porosity and moisture) as well as to the rivers or streams acting in the dual capacity of sewers and sources of drinking-water."

112. In Lower Burma during 1905 there were recorded as due to cholera 3,511 deaths, equal to a ratio of ·63 per mille of population, as compared with 2,472 deaths or ·45 per mille for the same cause in 1904. The highest death rates recorded in districts were 4·11 per mille of population in Mergui (365 deaths), 1·61 in Kyaukpyu (272 deaths), 1·58 in Tharrawaddy (624 deaths) and 1·19 in Maubin (337 deaths); in towns the highest death rates were 8·98 in Minhla, 6·16 in Letpadan, 5·41 in Kyaukpyu and 5·02 in Kyaiklat.

Cholera in Burma.

In Upper Burma the total number of deaths recorded during 1905 as due to cholera was 1,836 as compared with 508 during 1904. The disease prevailed in every district except one—Yamethin ; the highest mortality was recorded in Myingyan (1·59) and Pakokku (1·14), where 567 and 343 deaths, respectively, occurred. The highest death rates recorded in towns were 9·10 in Pakokku, 6·23 in Minbu, 5·89 in Myingyan and 5·46 in Monywa.

The provincial sanitary commissioner reports that in Lower Burma there was a decline of incidence during the heaviest part of the monsoon, namely, June, July and August; whereas in Upper Burma the prevalence of the disease increased throughout the monsoon months. He expresses the opinion that the heavy rainfall of Lower Burma served to dilute, if not completely change contaminated water sources, while in Upper Burma the relatively small rainfall served to scatter the cholera virus without causing effective dilution or flushing of water sources by storm water.

113. During 1905 no death from cholera was recorded in Ajmer-Merwara. This small province has now been free from this disease for three years in succession ; 32 deaths were reported in 1902.

Cholera in Ajmer-Merwara.

114. From the marginal statement in the first paragraph of this section it will be seen that in British territory in India the death-rate per 1,000 of population on account of small-pox rose from ·24 in 1904 to ·31 in 1905, the latter ratio, however, being lower than that of any of the three years 1901 to 1903. The

Small-pox. Table I of Appendix B to Section VI.

mean ratio for the quinquennial period ending 1904 (Appendix B) was ·40 per mille. The total number of deaths from this disease recorded during 1905 was 70,962 against 55,232 in 1904, and 93,693 in 1903. In all provinces except the Bombay Presidency and Lower Burma, the death rates during 1905 were lower than the corresponding means of the quinquennium 1900-4 ; the greatest falls in the death-rates were recorded in Ajmer-Merwara, the North-West Frontier Province and Bengal. As compared with 1904, a year in which the incidence was exceptionally low, the death rate during 1905 showed an increase in all provinces, except Bengal, Eastern Bengal and Assam, the United Provinces, Punjab and the North-West Frontier Province—the increase being greatest in Lower Burma and the Bombay Presidency. The small-pox mortality in towns was nearly three times as heavy as that in rural areas ; and the deaths of children under ten years of age amounted to 72·52 per cent. of the total number of deaths from this disease (in Burma alone was this percentage under 50).

In Bengal the number of deaths from small-pox fell from 11,090 (·22 per mille) in 1904 to 7,213 (·14 per mille) in 1905; the mean death ratio during the quinquennial period ending 1904 was ·54 per mille. As compared with 1904, the incidence of the disease was lower in 24 districts and higher in seven ; Puri and Murshidabad were the districts most affected. The civil surgeon of Puri attributes the great prevalence of small-pox in that district to the backward uneducated state of the people ; the civil surgeon of Murshidabad states that most deaths from this cause occurred in villages largely inhabited by Ferazi Muhammadans, who have for years resisted all attempts at vaccination. Only one town—Kishanganj in Purnea—suffered severely from this disease.

In Eastern Bengal and Assam during 1905, small-pox was the cause of 4,723 deaths, yielding a death rate of ·15 per mille of population as compared with ·26 in 1904 and ·27—the mean ratio for the quinquennium ending 1904. The highest rates in districts were 2·78 per mille in Kamrup and 1·11 per mille in Darrang. In the former district the disease occurred chiefly among the Mahapurushiyas who do not, for religious reasons, accept vaccination. In the town of Barpeta, the head-quarters of this sect, the death rate recorded was 3·54 per mille.

In the United Provinces 3,273 deaths were attributed to small-pox in 1905 as compared with 6,995 in 1904 and 21,950 in 1903, the corresponding rates being ·07, ·15 and ·46 per mille. The average rate for the quinquennial period 1900-4 was ·15. Eight districts were practically free from the disease during the year : the highest death rate in districts was 1·44 per mille, recorded in Jhansi. Out of 107 towns having a population of 10,000 and upwards, only 33 furnished deaths from this cause and none of them except Benares (where the rate was ·61 per mille) did the number of deaths exceed ten.

In the Punjab there were reported during 1905 as due to small-pox 4,723 deaths, equal to a ratio of ·23 per 1,000 as compared with 9,624 deaths and a ratio of ·48 per 1,000 in 1904. The average death rate during the quinquennial period ending 1904 was ·52 per mille. The highest rates in districts were ·89 per mille in Ludhiana (598 deaths), ·76 in Ferozepore (708 deaths), ·68 in Lahore (783 deaths) and ·55 in Montgomery (268 deaths). The rate was only ·01 per mille in the three districts—Kangra (nine deaths), Jhang (six deaths) and Attock (five deaths). The highest death rates in towns were recorded in Ludhiana (9'09 per mille) and Pakpattan (5'17 per mille). The municipal committee of Ludhiana has resolved to introduce the Vaccination Act into their town ; the towns in which the Act was in force recorded a death rate of ·36 as compared with a rate of ·82 in towns where vaccination was not compulsory.

In the North-West Frontier Province the number of deaths from small-pox fell from 2,694 in 1903 and 1,561 in 1904 to 571 in 1905, the ratios being 1'35 ·78 and ·29 per mille, respectively. Out of the 571 deaths no less than 374 occurred in the district of Hazara and 182 in that of Peshawar ; no death from this disease was recorded in the district of Bannu. In the three towns Bannu, Kohat and Dera Ismail Khan, in which the Vaccination Act was in force, there was very little small-pox.

In the Central Provinces and Berar there were recorded, during the year, 8,364 deaths from small-pox against 2,026 in 1904. The increase is attributed by the provincial sanitary commissioner to less vaccination during recent years owing to the presence of plague. The largest number of deaths were reported in the Buldana (1,483), Nagpur (1,398) and Saugor (1,114) districts. It is, however, suspected that many deaths from this disease were concealed or returned as chicken-pox or measles.

In the Madras Presidency small-pox accounted for 18,540 deaths against 9,891 in 1904. Cases were reported from all the districts in the presidency, but the fatality from this disease was greatest in Madura and Bellary, in each of which a rate of 1'5 per mille was recorded. Of the 61 municipal towns of this presidency, 23 enjoyed immunity ; the remaining 38 returned 2,709 deaths, the highest rate being 8·0 per mille in Dindigul. The provincial sanitary commissioner urges compulsory vaccination in rural areas.

In the Bombay Presidency the number of deaths from small-pox recorded in 1905 was 16,985 (·92 per mille) as compared with 4,289 (·23 per mille) in 1904. The mean ratio for the quinquennial period ending 1904 was ·26 per mille which is ·66 per mille less than the rate for the year under review. The mortality during 1905 was the highest recorded during the past 27 years. The disease was prevalent in every district and deaths were reported from 226 registering circles out of a total of 287. The districts of Khandesh and Bombay contributed 7,694 and 2,148 deaths, respectively, the two combined thus accounting for much more than half the total number of deaths in the presidency ; the death rates in these districts were 5'39 and 2·82 per mille, respectively. A notable feature was the high mortality among children under one year of age, amounting to nearly 36 per cent. of the total number of deaths.

In Burma there was a considerable increase of prevalence of small-pox during the year, the total number of deaths being 6,161 against 1,809 in 1904. Lower Burma was chiefly affected, its death rate being 1·01 per mille of the census population as compared with ·24 in 1904 and a mean of ·36 for the quinquennium ending 1904. In Upper Burma the increase was very slight, namely, ·18 against ·17 in 1904 : it should be noted, however, that the mean in Upper Burma for the preceding quinquennium was ·41, which is more than twice as high as the rate for 1905. In Rangoon and in the Pyapon and Hanthawaddy districts, all in Lower Burma, the heavy rates of 2·98, 2·79 and 2·32 per mille, respectively, were recorded.

In Ajmer-Merwara 277 deaths ('58 per mille) from small-pox were recorded during 1905 as compared with 129 ('27 per mille) in 1904. In Coorg the number of deaths reported as due to this disease was 132 against 16 in the previous year.

115. The number of deaths recorded as due to plague in the Indian Empire

Plague. (2)

in 1905 was 1,069,140, compared with 1,143,933 in the previous year. In the British provinces there was a slight increase, the registered deaths numbering 940,821 or 2,811 more than in 1904; but there was a remarkable change in the course of the epidemic. In each of the first five months of 1905, the number of deaths was considerably greater than in the corresponding months of 1904, but from June onwards this relation was reversed, and the number of deaths in the last seven months of 1905 was 69,845 or little more than one-fourth of the number, 263,789, recorded in the last seven months of 1904. In the Native States the registered plague deaths totalled 128,319 against 205,983 in the previous year; in every month there was a decrease as compared with 1904, but the decline was most remarkable in the last seven months of the year. As in 1904, the maximum incidence of the disease occurred in April and, for the fourth year in succession, the minimum was recorded in July.

As we have seen, the year 1905, apart from plague, was exceptionally unhealthy, but the deaths from plague in British territory numbered more than one-fifth of the total deaths recorded under the heading fevers, were more than twice as numerous as the exceptionally large number of deaths ascribed to cholera, and were three and a half times as numerous as the deaths from dysentery and diarrhœa. The mean plague death rate in the British provinces was 4'17 per thousand; females suffered far more than males, the death rates of the sexes being 3'83 and 4'52 per thousand, respectively, of the census populations of males and females. Although the numbers of deaths in Eastern Bengal and Assam and in the North-West Frontier Province were trifling, no British province, except Coorg, was altogether free from the disease, and no Native State escaped, except those in Madras, the Central Provinces and Baluchistan. In each of the Presidency towns the number of deaths rose; the totals in 1905 were, in Calcutta 7,372, in Madras 22 and in Bombay 14,171, against 4,689, 8 and 13,504, respectively, in 1904; and in Rangoon where the disease appeared for the first time in epidemic form in February 1905, there were 2,468 deaths.

In last year's report (pages 91-113) an account was given of the epidemiology of plague in India founded upon the evidence contained in the numerous official reports on the disease, and upon special reports contributed by most of the experienced observers in this country. There is no new fact of importance from the point of view of general epidemiology to add this year, but the conclusions and speculations based upon statistical data are now being justified or modified by the precise methods of the laboratory.

We saw (page 93) that an Advisory Committee representing the India Office, the Royal Society and the Lister Institute had been established and that a Commission to work in India under their direction at the problems of the etiology of plague, had been appointed. It is now some eighteen months since the working Commission began their labours and the Advisory Committee have recently issued reports of some of the results achieved which bid fair to overshadow in importance all recent experimental work of a similar kind.

On the departure from India of Dr. Martin in October, 1905, Major Lamb assumed direction of the working Committee which consisted of the following members—

>Major George Lamb, Indian Medical Service.
>
>Captain W. Glen Liston, Indian Medical Service.
>
>Dr. G. F. Petrie, Assistant Bacteriologist, Lister Institute.
>
>Mr. S. Rowland M. A., etc., Assistant Bacteriologist, Lister Institute.
>
>Captain T. H. Gloster, Indian Medical Service.
>
>Assistant Surgeon M. Kasava Pai, M.B., C.M., Madras.
>
>Mr. V. L. Manker, L.R.C.P., L.R.C.S., D. P. H.
>
>Hospital Assistant, Ramachandier, Mysore.
>
>Hospital Assistant C. R. Avari, Bombay.

In order adequately to estimate the importance of the work done by the Commission it would be necessary to form a clear conception of the ideas current a year ago regarding the causation of the spread of plague. It is however, impossible to do this for the opinions of well qualified observers varied greatly. It may be said that it was generally accepted that in bubonic plague disinfection was of little value—with all that this implies—and that it was widely believed that the rat was in some way the important factor in the spread of the disease. How it was spread by the rat was not understood, and although the theory of transference by fleas was well known, the flea was not believed to be the principal, or even a common agent ; and many of the best observers would have been found ready to agree with Dr. Klein[1] that, while it was possible for a flea that had just sucked from a rat blood well charged with *B. pestis*, by biting, to inoculate another rat, such an occurrence was not under natural conditions likely to occur. At the Plague Research Laboratory, where Captain Liston was working, ideas were much more advanced. It had been found that the common flea on rats in India,—provisionally named *P. pallidus* until identified by the Hon'ble C. N. Rothschild as *P. cheopis*—was very like *P. irritans*, and that, when deprived of its natural host, it would bite other animals. Captain Liston had made the discovery that guinea-pigs could be used as " traps " for rat fleas in plague infected houses—a discovery of which it is impossible to overestimate the importance— and he had found that some of the fleas caught in this way had 'in their stomachs large numbers of plague germs which so far from being destroyed by the digestive juices, appeared to be multiplying. Many of the guinea-pigs on which such fleas were found died of plague. It had been shewn that healthy animals could live beside plague-sick animals and not contract the disease, and while there was no doubt that it was possible to communicate plague by way of the alimentary canal, it had been found that rats could eat small quantities of contaminated food without suffering any ill effects. It had been ascertained also that plague germs soon die in earth taken from the floors of Bombay houses. Attempts had been made to communicate the disease by means of fleas transferred to a healthy animal after feeding on a plague-sick animal, but these attempts had not been successful, owing, no doubt, to the germs in the blood of the plague-sick animal not being sufficiently numerous to infect the fleas.

One result of these experiments was the creation in the laboratory of a staff accustomed to handle and dissect rats, which greatly facilitated the development of the excellent arrangements which, with the freely given aid of the Bombay Municipal authorities, the Commission succeeded in organising for the purposes of their enquiry.

116. For details of the experiments on which their conclusions are founded the reader must turn to the original reports of the Committee*. All that can be attempted here is a brief resumé of the conclusions.

Reports on plague investigation in India.

117. The Commission found that if, in the presence of the rat flea, a healthy rat was confined in close proximity to a plague-infected rat, but in such a way as to preclude the possibility of contact with its body and excreta, the healthy rat frequently became infected. In cases where the healthy rat became infected the infection must have been carried by the air or by fleas. In order to exclude aerial infection attempts were made to conduct similar experiments in the absence of fleas; but, as it was found impossible to ensure the absence of fleas, experiments were made to ascertain whether plague could be transferred by fleas taken from rats in the septicæmic stage of plague to healthy rats. In 21 out of 38 experiments, healthy rats living in flea-proof cages, contracted plague after fleas taken from rats dying or dead of septicæmic plague had been placed on them.

Transfer of plague from rat to rat.

118. A series of experiments were conducted in specially constructed godowns which had been designed by Lieutenant-Colonel Bannerman and Captain Liston. There were in all six of these godowns, and the essential differences between them were the structure of the roof and the amount of light admitted. Two of the godowns were roofed with country tiles which afford good shelter for rats; two with Mangalore tiles which afford little shelter for rats; and two with corrugated iron where rats could not penetrate. In one of country tile roofs and in one of the Mangalore tile roofs there was a sky light. It was found that rat fleas were more abundant in the godowns roofed with country tiles than in the godowns roofed with Mangalore tiles, and that in both sets of godowns the number of fleas was far greater in the absence of light. In the godowns roofed with corrugated iron, only a few fleas were found which may have been introduced on the guinea-pigs used in the experiments.

Effects of structure and of light.

119. These godowns were used for a series of experiments in which a large number of healthy guinea-pigs were kept in confinement along with a number of others infected with plague. It was found that epidemic plague did not occur when healthy guinea-pigs lived in close contact with plague infected guinea-pigs if the access of fleas was prevented; but, in exactly similar conditions, if fleas were present in considerable numbers, plague spread among the healthy animals.

Experimental epidemics.

Fleas taken from guinea-pigs dying with plague started an 'epidemic' among healthy guinea-pigs living in an uninfected godown, and this 'epidemic' was maintained by the addition, from time to time, of other fleas.

It was found that when an 'epidemic' has occurred among a number of guinea-pigs the contagion remains in the place of their confinement, and is effective in proportion as test animals introduced into it are accessible to and found infested with fleas, and that animals in a cage suspended two feet from the floor escaped the disease, while animals in a cage suspended two inches from the ground were infected.

---

* "Journal of Hygiene", No. 4, Vol. 6, September 1906.

120. Among the most interesting and valuable experiments performed by the Commission were those in which guinea-pigs were introduced into plague infected houses. The following is a summary of, and conclusions drawn from, these experiments :—

Conditions in plague infected houses.

" (1) Guinea-pigs allowed to run free in plague houses in many instances attracted a large number of fleas, which fleas were mostly rat fleas. A certain percentage (29) of these animals contracted plague and died from the disease. The position of the bubo in the great majority of these cases was cervical.

(2) If a plague house had been previously disinfected by the ordinary means of disinfection, fleas were still caught in large numbers on guinea-pigs set free in them. Further, a considerable number (29 per cent.) of these animals died of plague. The bubo in the great majority of these cases was again in the cervical region.

(3) Fleas transferred from plague-infected rats found dead or dying in houses, were able to transmit plague to healthy animals in flea-proof cages in the laboratory. The bubo in all cases was in the cervical region.

(4) Fleas transferred from guinea-pigs and other animals which had been placed for a few hours in plague houses, were able to transmit the disease to fresh animals when fed on these in flea-proof cages in the laboratory. The situation of the bubo in these animals was in the great majority of cases in the cervical region.

(5) Animals were placed in plague houses in pairs, both protected from soil and contact infection, and both equally exposed to aerial infection, but one protected from fleas by means of a fine metallic curtain, and the other not so protected. None of the protected animals contracted plague while several of the unprotected animals died of the disease. The position of the bubo in every instance was in the cervical region.

(6) Animals were placed in plague houses in pairs, both protected from soil and contact infection and both equally exposed to aerial infection, but one surrounded with a layer of " tangle-foot "* and the other surrounded with a layer of sand. The following observations were made :—

(a) Many fleas were caught on the tangle-foot, a certain proportion of which were found on dissection to contain in their stomachs abundant bacilli microscopically identical with plague bacilli. Out of 85 human fleas dissected only one contained these bacilli, while out of 77 rat fleas 23 were found thus infected.

(b) The animals surrounded with tangle-foot in no instance developed plague, while several (24 per cent.) of the non-protected animals died of the disease."

121. A contribution was made to the vexed question of the effect upon the virulence of *B. pestis* produced by repeated passage through the rat. Twenty-six passages occupying 89 days were effected by subcutaneous injection, but while there was evidence of varying susceptibility to plague among the rats of Bombay, there

The effect of passage on the virulence of *B. pestis.*

---

* A sticky fly-paper.

was no evidence of alteration in the virulence of *B. pestis*. A similar series of passages were made by cutaneous inoculation; no change in the virulence of the bacillus resulted, but it was found that a large percentage of Bombay rats were immune to infection in this way. Comparison with the records of experiments made at the Plague Laboratory in 1904 showed there has been no increase in immunity of the Bombay rats in the last two years.

122. A series of important experiments were conducted for the purpose of determining how long infection persists in floors which have been contaminated with cultures of plague bacilli. It was found that floors of cowdung retain infection for 48 hours; but floors of *chunam* do not retain it for 24 hours, infectivity being tested in both cases by cutaneous inoculation. Testing infectivity by allowing susceptible animals to run freely about on the floors, it was found that while cowdung floors retained infectivity for 12 but less than 24 hours, *chunam* floors retained it for six but less than 12 hours.

Duration of infection in floors.

123. The following is a summary of the results of the experiments made. The blood of plague-infected rats may contain an enormous number of plague bacilli, even as many as 100,000,000 per c c. having been found before death. On the other hand rats occasionally die from plague with little or no septicæmia. An insect sucking the blood of most rats shortly before death would imbibe a large number of bacilli.

Bacilli in blood, urine and fæces of rats.

While the blood of a rat may have as many as 100,000,000 organisms in a c.c. the urine may have none at all, or at least less than 10 per c.c. Plague bacilli were discovered in the urine in 29 per cent. of the cases. When the urine does contain plague bacilli they are always present in much fewer numbers than in the blood.

The fæces of rats dead of plague and the blood of which contained abundant bacteria, are not highly infective and would appear to play little part in the spreading of the epizootic.

124. The blood of 28 persons suffering from plague was examined. In 12 cases no bacilli were discovered, the 12 including all those (five) that ended in recovery. It was found that plague septicæmia may be irregular in type and that it may occur at a comparatively early stage of the disease. It appears that microscopical examination of the blood is not a trustworthy method of determining the degree of septicæmia.

Human plague septicæmia.

125. Chronic plague was found to exist among rats in a village in the Punjab during the absence of acute plague. The affected animals were not emaciated and did not appear to suffer in any way. It is of importance, because it suggests infection by food, that the bubo in each case was within the abdomen.

Chronic plague in rats.

126. An interesting note by the Hon'ble C. N. Rothschild shows that in Great Britain and Northern Europe the common flea on rats (*M. decumanus* and probably *M. rattus* also) is *Ceratophyllus fasciatus*; but in India and probably throughout the old world the common rat flea is *P. cheopis*.

Varieties of rat flees.

127. The Director of the Parel Laboratory, Bombay, reports' that the demands
for prophylactic vaccine during 1905 were more
Inoculation.
than double those of 1904. The opinions of those
who have had considerable experience in the use of this fluid, as prepared at the
Parel Laboratory, are that (1) the state of health remained unimpaired after inoculation, even in cases where the inoculated were children of poor constitution; (2)
the vaccine had a marked prophylactic effect ; (3) immunity lasted for a period
ranging from 6 to 12 months or more ; and (4) no untoward results had occurred
except, very rarely, an abscess at the site of inoculation due to neglect of antiseptic precautions on the part of the operator. According to some the severity
of the attack among those of the inoculated who did happen to get plague was
greatly diminished, the type of the disease among the inoculated being milder, and
the case-mortality consequently lower, than among the non-inoculated. Before
the onset of the last plague epidemic in Bombay City, the Health Officer induced
a large number of the municipal labourers to submit to inoculation and the
following statistics, showing the results of this operation, are of exceptional
value as the inoculated and non-inoculated were equally exposed to infection and
so under strictly comparable conditions. During the plague season the number
of municipal employés inoculated was to 7,182, the number remaining non-
inoculated being 418. The marginal statement gives a tabular view of the
results ; if the inoculated had suffered to the same extent as the non-inoculated,
they would have furnished 481 attacks and 446 deaths instead of 14 and 13,

	Strength.	Attacked.	Died.	Percentage. Attacked.	Percentage. Died.
Inoculated ...	7,182	14	13	·19	·18
Non-inoculated ...	418	28	26	6·7	6·2

respectively. It will
be noted that the
case-mortality, in this
instance, is identical
(92·85 per cent.)
among inoculated
and non-inoculated ;
there can, however,
be no doubt about
the prophylactic value of the vaccine and it is certain that inoculation is a most
efficient measure in dealing with epidemics of plague. On the appearance of an
outbreak, as we have no effectual means at present of destroying rats and their
fleas, the people must vacate infected houses (i. e., go away from the infected
rats and their fleas), or submit to inoculation (i. e., be protected against the
disease), or, better still, combine the advantages of evacuation and inoculation.
If evacuation is impossible, inoculation is indispensable.

128. The relation of rat plague to human plague and the connection that rats
and rat fleas have with the spread of human plague
The destruction of rats.
have already been discussed. It is obvious that
every infected rat destroyed means the removal of a source of infection and, consequently, the destruction of infected rats cannot fail to have a beneficial effect
on plague epidemics. In Bombay and elsewhere, leaflets on the advisability
of killing rats have been distributed freely amongst the people. Various methods
of killing rats have been tried, but the two which have won most favour are (1) the
" wonder " rat trap and (2) the proprietary poison known as the " common sense
exterminator. " *Danysz' virus* has had an extensive trial in several provinces
but the results have not been satisfactory. During the last plague season an
experiment was conducted in the Punjab to test the efficacy of rat destruction

as an anti-plague measure. A tract of country, about 70 or 80 square miles in area, was mapped off in the Delhi district and in this area, rat killing was carried out systematically and energetically. The experiment, however, failed to show any marked contrast between the plague incidence in the area selected for rat destruction and in other areas of the district for two reasons, namely, (1) the whole Delhi-Gurgaon district escaped with very little plague during that particular season, and (2), in consequence of the propularity of the measure with the people a considerable amount of rat destruction was carried on in areas adjoining the experimental one. No definite conclusion could therefore be drawn from this particular experiment, but " the evidence so far as it goes is favourable to rat destruction as an anti-plague measure. " One important point was brought out by the results of this experiment—the difficulty, apparently the impossibility, of completely freeing a place from rats. In a large village (Fattehpur Biluch), which had suffered from plague in previous years, rat destruction was carried on extensively and zealously, yet " plague broke out, apparently from endemic infection, and was accompanied by *rat* mortality " ; the only discernible effect, which could possibly be attributed to the destruction of rats was that the spread of the epidemic was extremely slow and the incidence of the disease slight in this village compared with neighbouring villages in which rat destruction had not been attempted. In other districts and in other provinces rat killing was carried out on a large scale ; the results are encouraging, but, as the incidence of plague during the past season was light in many untreated areas in the same districts and provinces, it would be fallacious to conclude that a decrease in prevalence of plague in treated areas was entirely due to rat destruction. The question which requires to be considered and the answer to which will settle the degree of importance to be attached to rat destruction as an anti-plague measure is ' how near an approach to absolute extermination of rats in a given area is practically possible ?" If absolute extermination in an infected locality is impossible, it is evident that rat destruction will not guarantee the non-recrudescence or suppression of plague in that locality ; all that can be expected from the measure is diminished severity of plague epidemics.

129. Plague raged with great virulence in Bengal during 1905 and carried

Plague in Bengal.

off 126,084 persons against 75,433 in the previous year, the ratios being 2·52 (the highest on record in this province) and 1·51 per mille, respectively. The average number of deaths of the preceding quinquennium was only 58,225. As in previous years a larger number of females (70,833) succumbed to this disease than males (55,251). Five districts out of 32 escaped entirely from plague. The highest death rates in districts were 14·79 per mille in Saran, 14·72 in Patna and 8·83 in Gaya. The civil surgeon of Patna states that the incidence was very heavy amongst the Muhammadan weavers who are fatalists and do not generally vacate their houses when infected. As regards towns, the highest rates were 65·98 per mille in Daudnagar, 43·22 in Siwan, 29·59 in Revelganj, 29·51 in Tikari and 28·79 in Monghyr. The largest number of cases was reported in March, and the smallest in July. Rat destruction was carried on in many places but notably in the Giridih coal fields and in Bihar.

130. During 1905 only six deaths from plague were reported in Eastern

Plague in Eastern Bengal and Assam.

Bengal and Assam ; all the cases were imported. Campaigns against rats were carried on by means of the " common-sense exterminator " in the towns of Dhubri, Sylhet and Silchar ; the reports on the use of this poison were favourable.

131. In 1905 the number of plague deaths registered in the United Provinces was 383,802, equal to a ratio of 8·05 per mille (the highest on record in this province), as compared with 179,082 deaths and a ratio of 3·75 per mille in 1904. The greatest number of deaths from this cause occurred in March and the smallest in July. The districts which suffered most during the year were Muttra with a death rate of 62·64 per mille, Ghazipur with 22·03, Allahabad with 19·94, Agra with 19·08, Ballia with 17·14 and Muzaffarnagar with 16·30. Only one district (Almora) was entirely free from the disease, but from Naini Tal only three cases were reported. Out of 107 towns having a population of 10,000 and upwards, only eight remained free from the disease throughout the year: the towns most affected were Kiratpur (Bijnor) with a death rate of 71·82 per mille, Dibai (Bulandshahr) with 52·94, Hapur (Meerut) with 44·39, Brindaban (Muttra) with 41·51 and Ramnagar (Benares) with 41·17.

Plague in the United Provinces.

132. The number of deaths attributed to plague in the British districts of the Punjab during 1905 was 334 897, giving a death-rate of 16·65 per mille of population ; the epidemic was of considerably less severity than that of the previous year with its 396,357 deaths and death rate of 19·71 per mille. (The mortality, however, in the Native States of this province was much higher than in the previous year.) In each of nine districts over 20,000 people perished from this cause ; Rohtak lost 31,952, Amritsar 29,931, Gurgaon 29,172 and Lahore 28,787. In ten districts the death rates recorded were over 20 per mille of the population, the highest rates being 52·63 in Rohtak, 40·88 in Gurgaon, 34·21 in Ludhiana, 33·99 in Amritsar and 29·69 in Lahore. The districts of Simla and Muzaffargarh remained entirely free from the disease during the year. The mean mortality in rural areas fell from 20·34 in 1904 to 17·22 in 1905, and that in towns from 14·09 to 11·63. None of the towns suffered very severely during the year, 8·39 in Jullundur and 6·91 in Amritsar being among the highest rates. The month of greatest prevalence was April, May coming next ; fewest cases occurred in September.

Plague in the Punjab.

133. There were only three cases of plague in the North-West Frontier Province during 1905, all three having been imported from the Punjab. In each case all necessary precautions were taken to prevent the spread of the disease, with successful results. It is noteworthy that although cases of plague have been frequently imported into this province the disease has never, at least up to the close of 1905, gained a footing on the west side of the Indus.

Plague in the North-West Frontier Province.

134. In 1905 there were 5,345 deaths from plague in the Central Provinces as compared with 32,820 in 1904. In two (Damoh and Narsinghpur) of the 19 districts, no death from this cause was reported, and in each of four others the number of deaths recorded was less than six. The districts which suffered most were Burhanpur (541 deaths) with a death rate of 5·82 per mille, Jubbulpore (1,535 deaths) with a rate of 2·96, Chhindwara (913 deaths) with a rate of 2·24 and Balaghat (602 deaths) with a rate of 1·84.

Plague in the Central Provinces and Berar.

In Berar the number of deaths from plague reported during the year was 7,361 against 10,046 in 1904. The highest death rates in districts were 5·92 in Buldana (3,632 deaths) and 4·51 in Akola (3,407 deaths) ; only one death was reported in the district of Yeotmal during the year.

The month of maximum plague fatality in these provinces taken together was October, March coming next; that of minimum fatality was July, in which only one death from this cause occurred : there was an exceptionally heavy incidence of plague in Akola in October. The proportion of males that succumbed to the disease was slightly greater than that of females. Various methods of destroying rats were experimented with in these provinces during the year : the " common sense exterminator " was tried in a few places with satisfactory results; Danyz' rat virus was also tested but proved a failure.

135. The number of deaths from plague reported in the Madras Presidency

Plague in Madras. during 1905 was 5,788 as compared with 20,125 in the previous year. Most cases were recorded in January and fewest in June. Out of a total of 23 districts, no death from this cause was registered in seven, *viz.*, Ganjam, Godavari, Guntur, Kistna, Nellore, Tinnevelly and Trichinopoly, and in six others only a few imported cases occurred. In the remaining ten districts the highest death rates were 3·6 in Bellary and 1·9 in the Nilgiris. In Bellary and South Canara the disease was present throughout the year. As regards towns, plague was indigenous in Bellary, Adoni, Madras, Ootacamund, Coonoor, Tiruppatur and Mangalore. The destruction of rats was vigorously undertaken, the method which has largely found favour in many quarters being poisoning by the " common sense exterminator. "

136. Plague was much less prevalent in the Bombay Presidency during 1905

Plague in Bombay. than in the previous year, 71,363 deaths (37,790 males and 33,573 females) having been recorded against 223,957 in 1904. The death rate fell from 12·12 per mille in 1904 to 3·86 per mille in 1905, the mortality during the year under review being thus less than one-third of that during the preceding year. Only in the two years 1897 and 1900, since the appearance of plague in India in 1896, was the number of deaths registered less than in 1905. This abatement of plague was shared by all the districts in the presidency, except Ratnagiri and three districts (Larkhana, Sukkur and Upper Sind Frontier) in Sind ; in Bombay City, also, there was a slight increase. In Sind, Thar and Parkar was entirely free from plague and all the other districts except Karachi, were only very slightly affected. The highest ratios recorded in districts were 11·68 in Satara, 6·53 in Karachi, 5·59 in Dharwar and 5·40 in Belgaum; the rate in Bombay was 18·63. Out of a total of 63 towns, 52 (*i.e.*, 82·54 per cent.) were affected by plague, the highest rates being 38·41 per mille in Yeola (636 deaths), 24·86 in Karachi (2,807 deaths), 22·92 in Hubli (1,380 deaths) and 21·15 in Borsad (275 deaths). In November of 1904, a fall in the recorded plague mortality was noted, and from this date the mortality continued to fall steadily, with the exception of a slight rise in March, during the first half of 1905—reaching a minimum in June. From June to September the numbers of deaths gradually increased but fell steadily again from September till the end of the year; so that in December of 1905 only 2,459 deaths from plague were registered as compared with 17,105 in December of 1904, and 24,451 in December of 1903. Most cases were reported in January and fewest in June. Although the incidence of plague was so much lighter during the year, there appeared to be no diminution of the virulence of the disease; the average case-mortality was 75 per cent.

137. Up to the end of 1904 the occurrence of plague in Burma was limited to imported cases, of which two terminated fatally in 1899, three in 1900, three in 1901, nine in 1903 and three in 1904. During 1905 there were 4,132 cases and 3,692 deaths. The case-mortality is thus no less than 89 per cent., from which it may be suspected that many non-fatal cases were not reported and helped to scatter about infective material. The first case that was certainly indigenous occurred in Rangoon in the third week of February and many more cases were reported during the month. The outbreak originated in the Mussalman quarter of the town, where there were large grain stores ; and, as all cases were not reported, and surveillance of contacts was impossible, infection spread rapidly within and beyond Rangoon. From the time of its recognition in February, plague has been present in this town throughout the year. In Lower Burma the mortality rose rapidly during March and April, continued high until July and then steadily declined during the remaining months of the year : to a total of 3,045 deaths during the year, Rangoon alone contributed 2,468. In Upper Burma the heaviest incidence occurred in November and December, the total number of deaths recorded being 647—all of which, with the exception of one in April, were registered during the last six months of the year. The provincial sanitary commissioner states that meteorological conditions are apparently less in favour of the spread of the disease in Lower than in Upper Burma. The districts chiefly affected were Rangoon (9·79 per mille), Bassein (·43) and Toungoo (·38) in Lower Burma ; and Meiktila (1·93) with 482 deaths and Yamethin (·66) with 160 deaths in Upper Burma. No deaths from this cause were reported from the Arakan division or the districts of Tavoy and Mergui in Lower Burma ; or from the Minbu division or districts of Shwebo, Lower Chindwin, Kyaukse and Myingyan in Upper Burma.

Plague in Burma.

138. No death from plague was reported in Coorg until 1903, in which year 45 deaths were recorded. In 1904 the number of deaths from this disease was 25 : in 1905, however, no death from this cause was registered.

Plague in Coorg.

139. The number of deaths from plague recorded in Ajmer-Merwara during 1905 was 2,480 against 158 in 1904. Most cases occurred in March, April and May, when 665, 1,272 and 315 deaths, respectively, were registered, while in June no death from this cause was reported.

Plague in Ajmer-Merwera.

140. The total number of deaths recorded in British territory under the heading "fevers" during 1905 was 4,417,655, yielding a ratio of 19·57 per mille of population, as compared with a total of 4,093,981 and a ratio of 18·09 per mille in 1904. The death rate returned under this heading in 1905 was higher than in any year since 1897, with the exception of 1900 and 1903. The year 1905 was in all respects more unhealthy than 1904 and it was to be expected that a large portion of the increased mortality during the year would be accounted for under this comprehensive heading. The largest numbers of deaths from fevers were, as usual, recorded in October, November, December and January ; there was also a slight but very perceptible increase in the registered numbers of deaths under this heading in May and June : the lowest numbers were recorded in July and August.

Fevers. Table II of Appendix B to Section VI.

In Bengal the death rate for fevers during 1905 was 24·34 per mille of population against 20·49 per mille in 1904. As in the previous year, the highest number of deaths was recorded in November and the lowest in July. The highest death rates during the year were recorded in the districts of Shahabad, Nadia and Gaya. The provincial sanitary commissioner states that the heavy mortality in Shahabad and Gya was due to extensive floods which occurred in August and lasted about six weeks ; the number of deaths recorded under this heading from September to December was as large as that registered during the other eight months of the year. The mortality from fevers was higher than in the previous year both in urban and rural areas, but the increase in rural areas was especially marked.

In Eastern Bengal and Assam the death rate for fevers during 1905 was 23·68 per mille as compared with 23·60 in the previous year and an average of 23·35 during the previous decennium. The largest number of deaths was registered in January and the smallest in March. Dinajpur district as usual heads the list with a ratio of 36·15 per mille, Rajshahi coming next with 33·95 and Pabna next with 33·91. Among towns, Barpeta reported a mortality of 33·38 per mille as compared with a quinquennial rate of only 13 per mille: the provincial sanitary commissioner suspects that in this town (see the remarks under small-pox) some cases registered under fevers should really have been classed under small-pox. The recorded number of deaths from *kala azar* was 3,030 as compared with 3,748 in 1904, 5,033 in 1903 and 6,319 in 1902, the districts reporting the largest numbers being Darrang (1,106 deaths), Sylhet (955 deaths), Kamrup (499 deaths) and Nowgong (379 deaths).

In the United Provinces the death rate for fevers during 1905 was 26·92 per mille, being 3·0 per mille higher than in the previous year and 2·29 per mille higher than the mean of the antecedent quinquennium. The number of deaths registered under this heading during the year was 1,284,164 against 1,141,029 in 1904. The rate as well as the actual number of deaths in these provinces were considerably higher than in any other province of India. The largest number of deaths was recorded in December and the smallest in August. Saharanpur and Ghazipur were the districts with the heaviest mortality; and, among towns, Brindaban (Muttra) again showed the highest death rate. In most provinces the rural death rate was much larger than, often twice as large as, the urban, but in the United Provinces the two rates were practically the same, *viz.*, 26·92 per mille in rural and 26·98 per mille in urban areas.

In the Punjab the total number of deaths recorded during 1905 as due to fevers was 370,047, giving a ratio of 18·40 per mille, which is ·42 per mille lower than the recorded rate in 1904 and 6·65 per mille less than the mean of the preceding quinquennium. December was the month with the largest number of deaths and September the month with the smallest. The provincial sanitary commissioner states that the decrease in the death rate for fevers was anticipated as the rainfall during July and August was very scanty. The highest mortality was recorded in the districts of Delhi (24·74), Attock, Mianwali and Muzaffargarh. The civil surgeon of Delhi attributed the high mortality in that district to plague deaths in many cases having been classified under fevers. In the other three districts mentioned, deaths from " respiratory diseases and other causes " are supposed to have swelled the number of deaths recorded from fevers. The mortality reported in March was unusually high, especially in some of the plague infected districts, and the provincial sanitary commissioner thinks it probable that a certain number of deaths really due to plague were registered under fevers. The rural death-rate was only slightly higher than the urban.

In the North-West Frontier Province the death rate for fevers (20·69) was lower than in the previous year, but higher than the mean of the preceding quinquennium. In the Peshawar district, as usual, the highest rate was recorded. The highest mortality occurred in December ; and regarding this the provincial sanitary commissioner states that the heavy mortality in December was probably due to a great variety of diseases and by no means entirely due to malarial fevers.

In the Central Provinces and Berar the death rate for fevers rose from 13·54 in 1904 to 17·43 in 1905, every district without exception contributing to this increase. The highest mortality was recorded in September, October and November and the lowest in January, March and July. It is thought probable that deaths from plague and influenza were in many cases recorded under fevers.

In the Madras Presidency the death rate recorded for fevers was 7·2 per mille, which is ·8 per mille less than in 1904 and 1·0 per mille less than the mean of the preceding quinquennium. The death rate during the year in this presidency was, as usual, much lower than that recorded in any other of the large provinces. The districts of Vizagapatam and Ganjam, as in the previous year, returned the highest rates, *viz.*, 15·8 and 15·0, respectively. The reported numbers of deaths were larger in November, December and January and least in February and March. The provincial sanitary commissioner ascribes the decrease in the recorded fever mortality to people purposely reporting deaths from fevers under other headings in order to escape the enquiries that would be instituted if plague was suspected to be the cause of death.

In the Bombay Presidency the death rate for fevers was 13·28, which is ·32 per mille less than in the previous year and 4·12 less than the mean of the antecedent quinquennium. The number of deaths recorded during each of the six months, May to October, did not exceed 20,000, while the number recorded during each of the other months of the year was never below 22,000 ; the largest number was registered in January. The districts with the highest mortality were Ahmedabad (28·38), Thar and Parkar (22·47) and Broach (20·65) ; the lowest death rate was 3·41, recorded in the City of Bombay.

In both Lower and Upper Burma the recorded death rates under fevers were slightly less than the corresponding figures of the previous year or means of the preceding quinquennium ; during 1905, the rate in the former was 8·97 and in the latter 6·76. In Upper Burma the rural death rate was not much higher than the urban.

Both in Ajmer-Merwara and Coorg the recorded death rates for fevers were lower in 1905 than in the previous year.

141. In Bengal there was a reduction in the number of packets of quinine distributed during 1905, only 35,472 parcels (each containing 102 seven grain packets) having been sold as compared with 38,041 during 1904. Most of the packets were bought by the villagers during the **Sale of quinine.** four months August to November. Sanction has been accorded to the opening of new centres for the distribution of quinine through the agency of village school masters in areas which are distant from dispensaries and post offices.

In Eastern Bengal and Assam the sale of quinine decreased slightly during 1905 as compared with the previous year. In October a scheme was brought into force in Assam by which (1) a depot at the office of the civil surgeon has been opened in each district ; (2) the number of distributing agents has been increased by including, besides post-masters, the village school-master, the *mauzadar* and the *gaonbura* ; and (3) the price of the drug has been reduced so that the retail agents might make a respectable profit.

In the United Provinces the total number of packets of quinine sold during the year was 643,668 against 586,441 in 1904—an increase of 57,227. The post office was a much more efficient agency than vaccinators and landlords for the sale of the drug. At the end of the year under report, an attempt was made to increase the sale of quinine by means of district boards and aided school-masters.

In the Punjab quinine was, as usual, gratuitously distributed in a large number of districts, but, beyond the statement that in the district of Delhi quinine was sold in all the villages, no information regarding the sale of the drug is contained in the provincial sanitary commissioner's report.

In the Central Provinces and Berar the sale of quinine showed an increase during the year, 4,887 packets having been sold as compared with 4,781 in 1904. As usual, post-masters sold the greatest number of packets ; village school-masters sold 125 packets, patwaris 53 and stamp vendors 37.

In Burma the sale of quinine has met with a fair amount of success, more than twice as many packets having been sold by vaccinators as by post-masters. The Burman apparently believes in the efficacy of quinine but objects to its taste : consequently, many officers have urged that quinine should be available in the form of tabloids.

Details are wanting regarding quinine sales in the North-West Frontier Province, Madras and Bombay.

142. The number of deaths recorded in British territory under the heading of

Dysentery and diarrhœa. Table III of Appendix B to Section VI. dysentery and diarrhœa rose from 240,655 in 1904 to 264,124 in 1905 and the death rate per mille of population from 1·06 to 1·17. As shown in the marginal statement of the first paragraph of this section, the death rate recorded under this heading during 1905 is, except for 1·21 recorded in 1903, the highest in the quinquennial period 1901—5. A higher mortality than in 1904 was reported in all the provinces except the North-West Frontier Province and the Bombay Presidency ; but the increased mortality during the year under review was due chiefly to a severer prevalence of the diseases in the Central Provinces, Eastern Bengal and Assam and the Madras Presidency. In India as a whole the largest numbers of deaths from these diseases were recorded in August and September and the smallest in February and March ; but there are frequently wide differences in the seasonal incidence in even adjacent provinces.

In Bengal the death rate from dysentery and diarrhœa in 1905 was ·90 per mille of the population as compared with ·82 in 1904 and ·99, the average of the quinquennial period ending 1904. The districts in which the highest death rates were recorded were, as usual, Howrah, Patna, Puri, Calcutta, Cuttack, Darjeeling and Hooghly. The urban mortality was more than four times as large as the rural. The provincial sanitary commissioner reports, as the results of a special enquiry, that in Orissa infantile dirrahœa is extraordinarily common and is the chief cause of the high death rate, the number of deaths of children under five years being nearly equal to that of persons above that age; that many more deaths are due to acute diarrhœa than to dysentery; and that the bad water-supply and the ignorance of the people are the cause of the prevalence of these diseases : that in Howrah and Hooghly the type of diarrhœa that prevails is the terminal diarrhœa of chronic malaria and *kala-azar*, infantile diarrhœa being comparatively uncommon : that in Patna, where the *chaukidars*—the reporting agents—are less intelligent than in other parts of Bengal, 61 per cent. of the deaths reported as due to dysentery and diarrhœa were found, after investigation, to be from other causes, cholera being the most important source of error. In Eastern Bengal and Assam the mortality from dysentery and diarrhœa during 1905 was considerably higher than that of the previous year and also higher than the average of the preceding quinquennium or decennium. The heaviest mortality prevailed in the tea districts of the Assam and Surma Valleys ; Lakhimpur reported a district ratio of 6·80 per mille Sibsagar 4·57, Darrang 3·08, Cachar 3·6 and Sylhet 2·0. The only district in Eastern Bengal which reported a comparatively high rate was Dacca with 1·45 per mille. In the United Provinces the recorded mortality from dysentery and diarrhœa was higher than in the previous year but lower than the average of the preceding quinquennium. The greatest number of deaths from this cause took place in June and the smallest in March. As in previous years the mortality was greatest in the districts of Garhwal (11·59 per mille), Almora (4·20) and Ballia (3·07); the rates in these three districts in 1904 were 8·04, 3·59 and 2·81 per mille, respectively. The urban death rate was more than five times as high as the rural. The recorded mortality from dysentery and diarrhœa in the Punjab during 1905 was also greater than that of the previous year but considerably less than the average of the quinquennium 1900-04. The highest rates were recorded in the districts of Simla (2·25), Rawalpindi (1·76), Jhang (1·68) Ambala (1·34) and Kangra (1·02). Both in 1904 and in 1905 the death rate in towns was about four times that in villages. In the North-West Frontier Province 562 deaths against 610 in the previous year were attributed to dysentery and diarrhœa; the death rate (·28) during the year was the same as the mean of the preceding five years. Dera Ismail Khan was the district most affected, There was a large increase in the mortality from dysentery and diarrhœa in the Central Provinces and Berar during 1905, the number of deaths from these diseases being 35,879 against 26,742 in 1904; the Berar districts alone accounted for 21,276 out of the total of 35,879 deaths. The death rate for these diseases in the Central Provinces and Berar combined was 3·02 per mille—the highest among all the provinces during the year. The incidence was especially heavy during August, September and October and, in the opinion of the provincial sanitary commissioner, was probably due to a late but heavy rainfall. The highest death rates in districts were 10·20 per mille in Akola, 8·76 in Buldana and 7·37 in Amraoti—all in Berar ; the highest district rates in the Central Provinces were 4·91 in Wardha and 3·23 in Jubbulpore. There were only two districts (Balaghat and Bhandara) which furnished lower rates than in the previous year. No cholera was reported during the year from the Berar districts, but the provincial sanitary commissioner does not think that any deaths really due to cholera were recorded under dysentery and diarrhœa.

The death rate in municipalities was 3·77 per mille as compared with 2·93 in rural areas.   In Madras the death rate from dysentery and diarrhœa was 1·4 per mille in 1905 against 1·3 in 1904 and an average of 1·2 in the quinquennium 1900—4.   The highest district rates were 11·2 in Madras and 4·4 in the Nilgiris. The provincial sanitary commissioner states that the prevalence of these diseases in Madras was due to " the influx of pauper ryots and coolies into the city from the adjoining districts in quest of labour", to the consumption of inferior grains and other articles of food by the poorer classes and to the insufficient and polluted water-supply.   Compared with the rural death-rate (1·1), the urban (4·0) was slightly higher than usual during the year.   In Bombay the death rate per mille and the total number of deaths from dysentery and diarrhœa showed a considerable reduction during 1905 in comparison with the figures of the previous year and the averages of the previous quinquennium.   The death rate was 3·00 per mille as compared with 3·33 in 1904 and a mean of 4·87 during the quinquennium 1900-4.   It is worthy of note that in 1905, the death rate from dysentery and diarrhœa in Bombay was not the highest provincial rate but followed that recorded in the Central Provinces and Berar.   These diseases were most prevalent during August and September, most deaths as usual having been registered in August; they were least prevalent in November and December.   The highest ratios were recorded in the districts of Khandesh (9·53), Poona (6·03), Ahmednagar (5·73), Sholapur (5·03) and Nasik (4·80). The death rate in rural areas was 2·91 per mille, while that in towns was 3·61; the relatively high rural death rates in this presidency and in the Central Provinces and Berar are noteworthy.   In Lower Burma the recorded death rate from dysentery and diarrhœa in 1905 was 1·43 per mille as compared with 1·27 in 1904 and a mean of 1·48 in the preceding quinquennium; in Upper Burma it was ·44 per mille as compared with ·31 in 1904 and a mean of ·33 in the preceding quinquennium.   In Lower Burma the urban rate was more than three times as high as the rural, but in Upper Burma the urban was ·60 per mille while the rural was ·45.   The relatively high death rate in the towns of Lower Burma is attributed to the general use in these towns of cess-pits for the disposal of night-soil and to temporary floodings of these towns, e.g., Thayetmyo during the year under review.   In Ajmer-Merwara, 424 deaths from dysentery and diarrhœa were registered in 1905; in Coorg 96 deaths.

Papers and Books referred to in Section VI.

For explanation of abbreviations see end of Section II.

(1) CHOLERA.— [1]Gotschlich, in Z. H. Bd. 53, 1906, page 281 ; [2]Gotschlich, reported in L. of 21st July, 1906, page 197; Kraus and Pribram, in C. B. Originale Bd. 41, 1906, pages 15 and 155; Kraus and Prantschoff, in C. B. Originale, Bd. 41, 1906, pages 377 and 480, and in Wien. Klin. Woch. reported in B. I. P. of 30th May, 1906, page 435 ; Prausnitz, reported in H. R. of 1st May, 1906, page 489 ; Gordon, in B. M. J of 28th July, 1906, page 197; Besser, in C. B., Originale, Bd. 41, 1906, page 286, Friedberger, in C. B., Originale, Bd. 40, 1906, page 405; Mühlens and von Raven, in Z. H., Bd. 55, H. 1, page 113.

(2) PLAGUE.— [1]Klein, L. G. B., 1904-5, page 325; [2]Bannerman, Report of the Plague Research Laboratory for the year ending 31st March, 1906.

SECTION VII.
GENERAL HISTORY OF VACCINATION.

143. During the year 1905-06, the number of persons vaccinated by all agencies in British India reached a total **Vaccination in India.** of 9,277,896, which indicates a steady continuance of the increase observed in the work of the vaccination department during the past three years—1902-03 (8,431,564 operations), 1903-04 (8,457,298) and 1904-05 (8,944,749). With the exception of the Madras and Bombay presidencies every province shared in the increase, the rise in the figure in Bengal (222,999) being specially notable ; Eastern Bengal and Assam and Burma followed with increases of 76,947 and 75,924 operations, respectively. The decreases in the Madras presidency (18,279) and in the Bombay presidency (12,669) were comparatively insignificant and will be explained in the paragraphs relating to these provinces; in the former, however, the number of successful cases in 1905-06 (1,296,312) was greater than in the preceding year (1,242,717). Taking British India as a whole, the primary cases numbered 8,586,770 as compared with 8,351,257 in 1904-05, and the percentage of success in the two years was 97·20 and 96·04, respectively. Revaccinations increased from 593,492 in 1904-05 to 691,126 in 1905-06, but the percentages of success fell slightly from 74·71 to 74·20. In the provinces the percentage of success varied in primary cases between 99·19 in the Punjab and 90·80 in Burma, and in revaccinations between 84·78 in the United Provinces and 53·33 in Burma. The mean number of vaccinations performed by each vaccinator was 1,480 as compared with 1,487 in the previous year, the figures in the provinces ranging from 3,162 in the North-West-Frontier Province and 996 in Ajmer-Merwara. Vaccination at dispensaries continued to show an increase, there having been 207,421 operations compared with 203,705 in 1904-05, Bengal (122,105) contributing by far the largest number to the total, followed by the Central Provinces and Berar (28,790), Eastern Bengal and Assam (27,735) and Burma (17,507). The comparatively insignificant number of operations at dispensaries in the United Provinces (655) is explained by the fact that no special vaccinators are attached to those institutions, operations being performed by medical subordinates.

The mean proportion of the population successfully vaccinated, which had been 33·32 per mille in 1903-04, and 34·79 per mille in 1904-05, rose to 36·39 per mille in 1905-06. As in the preceding year the ratio was highest in Coorg (55·45) and lowest in Bombay (27·35)

On an estimated birth-rate of 40 per thousand of the census population, 43·83 per cent. of the infants were protected compared with 42·46 per cent. during the preceding year. The percentage in the several provinces varied greatly, from 76·75 in the Central Provinces and Berar and 62·54 in the Punjab to 21·27 in Burma and 11·68 in Coorg.

The total cost of the department which had been Rs. 12,03,674 in 1904-05, increased to Rs. 12,82,636 in 1905-06, and the mean cost of each successful case rose by one pie to two annas and five pies ; the cost of each case in the provinces ranged between eight annas and three pies in Bombay and eleven pies in Eastern Bengal and Assam.

144. In the paragraphs relating to individual provinces, reference is made to the vaccine material used in each, and here it is necessary to notice only the more important facts connected with the supply of vaccine.

Vaccine lymph.

The Shillong vaccine depot for the supply of material to Eastern Bengal and Assam is now independent of vaccine from outside sources and can maintain its own supplies with certainty. The Patwa Dangar lymph depot at Naini Tal is now well equipped and its utility to the United Provinces has largely increased, while the work of the depot at Belgaum for the Bombay presidency has been successfully launched. The proposal to establish lymph depots in the North-West Frontier Province has been abandoned in view of the satisfactory and economical arrangements in force for obtaining vaccine from the Punjab depot. Useful experimental work in the preparation of vaccine was done by Major Entrican, I.M.S., at the Meiktila depot which supplies all the lymph for Burma, except Rangoon where there is a local depot. Major Entrican devised two machines for filling lanoline vaccine in tubes and glycerinized and quinated lymph in capillary tubes, respectively. The apparatus, a description of which is contained in the issue of the *Indian Medical Gazette* for June 1906, does efficiently, in a few minutes, work which formerly occupied several hours.

145. The figures of the vaccination operations in Bengal during the year 1905-06 exclude those relating to the districts transferred to the new province of Eastern Bengal and Assam, and to admit of comparison, the figures of the year 1904-05 used below also exclude those relating to the transferred districts. The Sambalpur district was added to Bengal during the year under report, but as the vaccination figures of the district are not available they have not been included in the provincial figures.

Bengal.

The total number of operations performed in the remodelled province during the year 1905-06, was 2,041,230 against 1,818,231 during 1904-05. Of these totals, 1,904,625 and 1,759,124, were primary cases, and 136,605 and 59,107 the revaccinations, respectively, in each of those years: the net increase was, therefore, 145,501 primary cases and 77,498 revaccinations, which is very satisfactory. The recovery in the number of revaccinations is largely due to the special efforts made by the Corporation of Calcutta to combat the severe outbreak of small-pox during the year, there having been an increase of over 50,000 revaccinations in Calcutta. Of the 33 districts of the province, 23 show an increase and only 10 a decrease. The most noticeable increases occurred in the districts of Saran (24,254), Midnapore (17,016), Jessore (12,643), the Tributary States of Orissa (12,142), Shahabad (11,631) and Nadia (11,273), while the decrease was most marked in the districts of Balasore (13,273), Sonthal Parganas (12,465), Birbhum (6,310) and Purnea (5,782). The increased work is attributed to various causes, among them more careful supervision ; more energetic working ; less prevalence of plague ; the greater prevalence of small-pox, and in some places the introduction of the system of rewards to vaccinators. The decrease in Balasore is said to be due to the inefficiency of the staff, some of whom have been dismissed, and in the Sonthal Parganas and Purnea to the prevalence of sickness.

Generally, it is hoped that much improvement will be effected in the department in consequence of the introduction of the scheme recently sanctioned for the education of vaccinators, and the general substitution of prepared lymph for arm-to-arm vaccination.

The ratio of success in primary operations rose from 98·80 per cent. in 1904-05 to 99·13 per cent. in 1905-06, and in re-vaccinations from 61·63 to 66·29 per cent.

Vaccination was carried on with calf lymph (151,080 primary operations) lanoline lymph (400,847) and from arm-to-arm (1,352,698), the figures showing an increased use of calf and preserved lymph, and a smaller number of arm-to-arm cases. The percentages of success by each method was calf lymph 98·92 as compared with 99·30 in the previous year, lanoline lymph 96·57 as compared with 95·81, and arm-to-arm 99·21 as compared with 98·95. The ratios of success in revaccination by each method in 1905-06 were 58·81, 53·98 and 78·02 per cent., respectively.

At an estimated birth-rate of 40 per mille of the population, 39·18 per cent. of the infant population were successfully vaccinated, compared with 34·41 per cent. in 1904-05 and 29·36 per cent. in 1903-04, which shows that infant vaccination is making progress. In municipalities the rate of protection of infants rose to ·85·82 per cent., from 65·08 per cent. in the year preceding. There was a satisfactory improvement in Calcutta where nearly half the available infant population were protected, compared with about one-fourth only of the number available in the previous year.

As in the report for 1904-05, there is again no information as to the quantity and quality of the lymph issued from the Darjeeling and Calcutta depots during the year under report, nor is there any reference to the working of the system introduced last year of issuing lymph direct to the operators instead of through superior officers. The Darjeeling depot as usual supplied lymph to the Nepal Darbar and the Sikkim State—380 and 860 grains respectively,—the results from which were reported to have been satisfactory.

The cost of the department was Rs 1,57,803 against Rs. 1,35,834 in 1904·05—excluding the figures for Calcutta in both years which were not received—and the cost of each successful case, one anna and three pies as compared with one anna and one pie in each of the two preceding years. ·

146. The work of the vaccination department in the new province, an account
Eastern Bengal and Assam.
of which appears for the first time in this report, for the year 1905·06 is compared with that done in 1904·05 in Assam and the Bengal districts transferred to the new province. The total number of vaccination operations which aggregated 1,352,079 in 1904·05, rose to 1,429,026 in 1905·06. The distribution of the total was 1,313,998 primary cases in 1904·05 and 1,385,538 in 1905·06, showing an increase of 71,540 cases, and in revaccinations 38,081 in 1904·05 and 43,488 in 1905·06, or an increase of 5,407. The percentages of success also improved from 97·09 to 98·28 in primary cases, and from 56·95 to 71·71 in revaccinations. The most marked increase in the number of operations occurred in the Chittagong district (53,000), and the Faridpur (18,000) and Noakhali (11,000) districts also showed large increases. In Chittagong the increase was due to

special efforts made by the vaccinating staff owing to an outbreak of small-pox, the result being the highest figures on record for the district. In Faridpur opposition was overcome with very satisfactory results by calling on village headmen to show cause why vaccination should not be made compulsory. On the other hand in the Dacca district there was a fall of 21,000 operations as compared with the preceding year, which has been explained by the special activity in vaccination work in 1904-05 owing to an outbreak of small-pox, and the fact that the number of operations in that year exceeded the average. For the decreased work in other districts various reasons have been adduced. On tea gardens 15,143 primary operations were performed, of which 94 per cent. were successful.

On an estimated birth-rate of 40 per thousand of population 33·62 per cent. of the infant population were protected against 29·67 per cent. in 1904-05. Of the available number of children in towns 70·9 per cent. were successfully vaccinated.

Vaccination was carried out during the year with humanized vaccine, with manufactured vaccine (glycerine or lanoline) and with calf vaccine. The use of human lymph is to be discontinued as larger supplies of manufactured vaccine become available, and in any case it is not to be used in areas where vaccination is compulsory. The Shillong vaccine depot is now independent of vaccine from outside sources and can maintain its own supplies, the quality of which is uniform and good, with certainty.

The department cost Rs. 76,805 in 1904-05 and this increased to Rs. 80,103 in 1905-06. In both years, however, each successful case cost eleven pies.

147. The total number of vaccination operations in the United Provinces in 1905-06 was 1,695,416, an increase of 23,636 operations compared with the results of the preceding year. The number of **United Provinces.** successful primary operations was 1,544,152 in 1905-06 against 1,517,491 in 1904-05 or an increase of 26,661, but the number of successful revaccinations fell from 77,209 in the preceding year to 76,260 in the year under report. The percentage of success in primary cases was 97·93 in 1905-06, or a little higher than the rate of 97·62 in 1904-05 ; revaccinations were, however, less successful, 84·78 per cent. compared with 86·51 per cent. In 34, as compared with 31 in the previous year, of the 49 districts in the province, there was an increase in the number of successful primary operations. The districts showing the largest increases were Allahabad (4,126), Aligarh (4.049), Gonda (3,905) and Agra (3,427). Among the districts showing a falling off in the number of successful primary operations are Ballia (5,827), Sultanpur (5,607), Banda (3,848), Rae Barelli (2,223) and Benares (2,180). For the decrease in Ballia the provincial Sanitary Commissioner accepts the explanation that, apart from the effects of plague and cholera, the amount of work done in previous years has reduced the number of the susceptible population ; elsewhere, the prevalence of cholera, the deputation of experienced vaccinators to the *kumbh mela* at Allahabad, and, in places, slack work on the part of vaccinators, account for the fall.

On an estimated birth-rate of 40 per mille, the number of infants successfully vaccinated represent 49·57 per cent. of those available, which is less than the percentage of 52·11 protected in 1904-05. In municipalities, infants were protected at the rate of 102·13 per cent., on the figures of births *minus* deaths neglecting any outside addition to the number thus available, showing a recovery from the rate of 89·6 in 1904-05 to which it had fallen from 98·2 in 1903-04.

Bovine lymph depots were maintained at Lucknow and Patwa Dangar (Naini Tal). The latter is now well equipped and its utility to the province has largely increased. Considerable difficulty was experienced in getting a good and regular supply of calves at a reasonable price. Of the 96 calves vaccinated, 88—78 cow and 10 buffallo—were vaccinated successfully. The average quantity of crude lymph yielded by a cow calf was 196 grains, and that by a buffalo calf 249 grains. Lanoline lymph was prepared only when requisitioned, glycerine lymph having been substituted for stock purposes and found to give better results. During the year 74,800 tubes and 24½ drams in bulk of glycerine lymph and 30 drams of lanoline lymph were issued. Lymph was regularly supplied from the depot to all districts in the plains. The working of the institution is regarded as very satisfactory.

The cost of the department amounted to Rs. 1,53,241 or somewhat more than the cost of Rs. 1,48,638 in the preceding year, but the cost of each successful case, one anna and six pies, was the same in both years.

148. The working of the vaccination department in the province continued to

Punjab.

show satisfactory improvement during the year 1905-06. The total number of operations performed by all agencies amounted to 748,739, showing an increase of 28,968 cases, equivalent to an increase of four per cent. on the figures of 1904-05, and of about six per cent. on the average of the previous quinquennium. This result is no doubt due to the close supervision exercised by the Sanitary Commissioner during the vaccine season over the work of the vaccinators and the inspections performed by the Superintendents. In primary work there was an increase of 56,291 cases compared with the number in the preceding year, but revaccinations showed a fall of 27,323, leaving a net increase of 28,968 operations during the year under report. The fall in the number of revaccinations, however, is explained by the system under which one-eighth part of a district is to be revaccinated every year, which was introduced during the preceding working season only, and time must elapse before its effects are apparent. At the same time it is necessary to allow for the fact that the people are prejudiced against revaccination, which in certain districts is much less popular than the primary operation. The revaccination of girls is unpopular in all districts. The percentage of success in primary cases was 99·19 in 1905-06, and 99·17 in 1904-05; in revaccinations the percentage rose to 83·38, from 81·04 in the preceding year.

On an estimated birth rate of 40 per mille, 62·54 per cent of the infants were protected, while in 1904-05 and 1903-04 the rates were 56·25 and 58·23 per cent., respectively. In municipalities 85 per cent. of the available children in the towns where the Vaccination Act is in force and 65 per cent. in those where vaccination is not compulsory, were successfully vaccinated; the corresponding percentages in 1904-05 were 73 and 60, respectively.

The chloroformed glycerine lymph manufactured at the Central Vaccine Institute, Lahore, proved most satisfactory. Of the 100,768 primary vaccinations performed with it during the year, there was a case success of 99·94 per cent. and an insertion success of 99 per cent., the corresponding figures in 1904-05 having been 33,210 primary cases, and case and insertion successes of 98·72 and 94·90 per cent., respectively.

The cost of the department was Rs. 97·02 or Rs. 1·878 more than in 1904-05, excluding the accounts of 1904-05 relating to the Kangra district which were lost in the earthquake of April 1905. The increase is chiefly due to the employment of additional establishment for the Lyallpur district. The cost of each successful case was two annas and three pies or one pie less than in 1904-05.

The vaccination work in the larger Native States also improved generally. In the Patiala State the number of primary operations rose from 48,490 in the preceding year to 54,086 in 1905-06, and the number of revaccinations from 30,992 to 31,034. In the other large States the work done was, Bahawalpur 18,804 primary cases compared with 20,065, nearly all primary cases, in 1904-05. In Kapurthala the recorded results were (the figures in brackets are those for the preceding year) 5,578 (4,827) cases, Nabha 1,074, Jind 6,978 (6,697) and Faridkot 5,185 (5,488), all primary operations. The revaccination work was, however, most insufficient, the total number of cases amounting to 386 only. The percentage of success in primary cases ranged between 98·09 in Faridkot and 89·77 in Nabha.

149. The vaccination operations in the Frontier Province during the year
North-West Frontier Province. 1905-06 numbered 100,866 compared with 77,042 in 1904-05, or an increase of 23,824. The primary cases numbered 87,771, against 74,267 in the previous year, and revaccinations 13,095 against 2,775. The percentage of success in the former was 98·56 and in the latter 84·45, compared with 98·96 and 75·39, respectively, in 1904-05. There was a marked increase in the number of operations in all districts, the political agencies and cantonments, and also in the work done by the dispensary staff, the last mentioned performing no less than 2,845 operations against 104 only last year. The greatest rise occurred in the Hazara district, the numbers being 16,835 cases in 1904-05 and 26,417 in 1905-06, but of the total increase, over 7,000 were revaccinations. The result in Hazara is no doubt in a measure due to the number of vaccinators having been increased in April 1905 from three to six. Large increases also occurred in the Dera Ismail Khan and Kohat districts. Vaccination work is conducted in the Kurram and Tochi agencies and the sub-agency of Chitral, the best work being done in the first mentioned. Efforts are to be made to introduce vaccination into other political agencies, but the local administration doubts if they will be attended with much success. The special vaccinator entertained for the Sherani Country succeeded in vaccinating 3,843 persons, and in the Amb territory 1,249 operations were performed by a vaccinator of the Hazara district. Both Sherani and Amb are foreign territory.

On an estimated birth rate of 40 per mille, 58·92 per cent. of the infants were protected against 52·48 per cent. in the preceding year. The Vaccination Act is now in force in all save two of the municipal towns of the province, and the question of extending the Act to these two is again to be urged. The number of infants successfully vaccinated in those two towns was, however, fairly satisfactory. In all the towns a mean of 72·3 per cent. of the available number of infants were successfully vaccinated.

The vast majority of vaccination operations were performed with animal lymph, human lymph hardly being used except in the Sherani Country and the Kurram Valley. In Peshawar vaseline lymph and glycerine lymph were largely used owing to the difficulty of obtaining buffalo calves in the district. The requirements of lymph for the province are met by obtaining preserved lymph from the Punjab Vaccine Depot at the beginning of the season and vaccinating buffalo calves; this secures a good supply during the rest of the year. The idea of establishing lymph depots in the province has been abandoned in view of the economical and satisfactory arrangement for obtaining requirements from the Punjab Vaccine Depot.

The cost of the department was, Rs. 11,643, as compared with Rs. 11,223 in the preceding year, but the cost of each successful case fell from two annas and five pies, to two annas.

150. The total figures relating to vaccination in the Central Provinces and Berar include the work done in the Feudatory States and at dispensaries. The figures for 1905-06 exclude those of the Sambalpur district which has been transferred to Bengal, and also of the Native States transferred to that province for which figures were not forthcoming. The figures for 1904-05, where used for comparison, also exclude the Sambalpur district.

Central Provinces and Berar.

The total number of operations conducted in the province during 1905-06, numbered 626,138 against 597,177 in 1904-05, of which 551,140 were primary cases and 74,998 revaccinations, the corresponding figures for the preceding year being 540,187 and 56,990, respectively. The percentage of success was 98'59 in primary cases and 72'30 in revaccinations. Excluding Feudatory States, 559,443 persons were vaccinated against 539,516 in 1904-05, and of the former 491,818 were primary cases and 67,625 revaccinations ; the percentages of success were 97'26 and 62'83, respectively. The increase in primary work of about 6,000 cases is attributed to the high birth rate of the year, and in revaccinations of about 14,000 cases, to the efforts made to induce people to get themselves revaccinated, aided in some districts by the prevalence of small-pox. In the Feudatory States 59,322 primary vaccinations were performed compared with 54,334 in 1904-05, and 7,373 revaccinations compared with 3,327. The percentage of success claimed was 96'81 against 97'09 in primary cases and 84'03 against 85'75 in revaccinations.

On the whole 44'81 per mille of the population were successfully vaccinated in 1905-06, as compared with 42'83 per mille in 1904-05 which, bearing in mind the prevalence of plague in many parts of the province during the year, is considered to be satisfactory.

Estimating the birth rate at 40 per mille of the population, 76'75 per cent. of the infants were successfully vaccinated and in municipalities 96'93 per cent. of the number available.

The various kinds of lymph used during the year were fresh calf lymph 32,536 cases (93'48 per cent. successful), glycerine calf lymph 502,401 cases (93'14 per cent. successful), human lymph 19,555 cases (97'60 per cent. successful) and lanoline lymph 4,951 cases (68'79 per cent. successful). The use of glycerine calf lymph has steadily increased, but it is not explained why the percentage of success with it, which was 96'14 in 1904-05, fell to 93'14 in the year under report.

The cost of the department was Rs. 66,368 or Rs. 3,998 more than in 1904-05 the increase being due to various causes. Each successful case cost one anna and eleven pies in both years. In the Feudatory States, the department cost Rs. 4,750 or Rs. 306 more than in 1904-05, but the cost of each successful case fell to one anna and three pies, from one anna and four pies. .

151. The total number of vaccination operations performed by all agencies in the presidency numbered 1,464,990 during 1905-06. Primary cases showed a decrease of 49,244 and

Madras.

revaccinations an increase of 30,965 during the year, leaving a net decrease of 18,279 operations compared with the work done in 1904-05. It is satisfactory, however, to record that the total number of successful vaccinations which had been 1,242,717 in 1904-05 rose to 1,296,312 in 1905-06, while the percentages of success in primary cases and in revaccinations rose from 87·38 and 74·51 to 92·67 and 76·02, respectively. The explanation of the decreased outturn of work in some districts is the short supply of vaccine paste from the King Institute and the absence of vaccinators without substitutes being appointed. In regard to the shortage of the paste supply during certain months of the year, it has been ascertained that it was ¦due to the unfavourable state of the season in the Chingleput district and to the poor condition of the calves.

On the estimated birth rate of 40 per mille, 30·09 per cent. of the infants were successfully vaccinated, which shows an improvement on the rates of 28·29 and 27·57 recorded in 1904-05 and 1903-04, respectively; in municipalities 67·9 per cent. of the available infants were protected against 66·5 per cent. in the preceding year.

Lanoline lymph, manufactured at the King Institute, Madras, was used exclusively in local fund areas with a percentage success of 92·3, against 86·6 in the previous year, and almost exclusively in the municipalities with a percentage success of 95·6. Nearly 600 operations were performed in municipalities with lanoline lymph supplied by the Bangalore Institute, the percentage of success being 96·7. Locally prepared glycerine lymph was also used in municipalities and was successful at the rate of 99·4 per cent. The results during 1905-06, from the lanoline paste prepared at the King Institute show a marked improvement on those during 1904-05, which is accounted for by the vaccinators having become more familiar with the use of the lymph. A Deputy Inspector of Vaccination was deputed to give instruction in the use of this description of lymph in the more backward districts.

The cost of the department which had been Rs. 2,77,371 in 1904-05, rose to Rs. 2,94,419, during 1905-06, and the cost of each successful case rose by one pie, to three annas and eight pies.

152. In Coorg the increase in the number of vaccinations during 1904-05 was more than maintained during 1905-06, a total of

Coorg.

11,291 operations having been performed of which 9,306 were primary operations and 1,985 revaccinations, showing increases of 405 and 508, respectively, on the figures of the preceding year. The percentage of success in primary cases was 93·21 compared with 95·99 in 1904-05, and in revaccinations 76·26 compared with 86·43. Calf lymph was used, but no reason is given for the decline in the ratios of success.

The cost of the department was Rs. 2,740 or a little below the sum of Rs. 2,769 spent in 1904-05, and each successful case cost four annas and six pies or two pies less than in the previous year.

153. The steady increase of vaccination work which has taken place in the presidency during recent years, was not continued

Bombay.

during 1905-06, when the total number of operations was 656,338 compared with 669,007 in 1904-05. The decrease occurred in the Western, Central and Presidency Circle registration districts, the explanation offered being the large number of persons vaccinated in the previous year owing to the prevalence of small-pox, and an adverse season in the Central registration district scattering the people. Of the 656,338 operations performed, 610,985 were primary cases and 45,353 revaccinations, the corresponding figures for 1904-05 being 607,064 and 61,943, respectively, so that the net decrease of the year was brought about by the smaller number of revaccinations. The percentages of success in 1905-06 were somewhat lower than in the previous year—primary cases 97·56 against 98·52, and revaccinations 76·13 against 79·12.

On the estimated birth-rate of 40 per thousand of population, 52·32 per cent. of the infants were protected against 52·99 per cent. in the previous year, but according to the provincial Sanitary Commissioner's calculation of births *minus* deaths of infants under one year of age, 81·08 per cent. of the infants available were protected.

Vaccination was conducted from calf to arm, with glycerinated calf lymph and with lanolinated vaccine : in Sind, except for a few operations, the arm-to-arm process was employed. In the Southern registration district, vaccination with preserved lymph manufactured at the Vaccine Depot, Belgaum, was carried on successfully, and in the words of the Deputy Sanitary Commissioner " the increase in the work cannot be attributed to a fall in the epidemicity of plague but rather to vaccination with preserved vaccine lessening the difficulties confronting the vaccinator, and also possibly to the new method being more popular with the people." After trials at the Vaccine Depot at Belgaum with various kinds of lymph, it has been decided to adopt chloroformed glycerine lymph which was largely used in the Southern registration district and is being gradually introduced into the other districts.

The cost of the department rose from Rs. 2,85,418 in 1904-05 to Rs. 3,01,303 in 1905-06 ; the cost of each successful case in those years being seven annas and nine pies and eight annas and three pies, respectively. The reason for the increased cost of the department is not stated.

154. In Burma the total number of vaccination operations was 488,927 or
Burma. 75,924 more than in 1904-05. Of the total operations, 426,935 were primary cases and 61,992 revaccinations, compared with 380,686 and 32,317, respectively, in the previous year, so that there was an increase in both classes of work, the number of revaccinations being almost doubled. The percentage of success was also higher, in primary cases 90·80 in 1905-06 compared with 84·67 in 1904-05, and in revaccinations 53·33 compared with 47·99. The Rangoon municipality contributed 24,081 cases to the increase in the total of the previous year " largely due to belated vigour in vaccinating, following the severe epidemic of small-pox which prevailed during the year. "

The vaccine used was bovine lymph obtained from the Meiktila Vaccine Depot, except in Rangoon where lymph, the quality of which is condemned by the provincial Sanitary Commissioner, is prepared at a local depot. The lymph supplied by the Meiktila depot was satisfactory ; it was of three descriptions, *viz.*, lanoline vaccine, glycerinated-vaccine and quina·ed-vaccine. Major Entrican, I. M. S., who is in charge of the depot found that in point of duration of vitality lanoline is preferable as a medium, but that in the unskilled vaccinators' hands the more fluid glycerine forms are more readily manageable. Arrangements are to be made for the new vaccination staff to be entertained to be given a course of training at the Meiktila Vaccine Depot in theoretical and practical vaccination and allied subjects and better results may, therefore, be anticipated.

The cost of the department rose from Rs. 1,00,150 in 1904-05 to Rs. 1,14,464, but the cost of each successful case fell to four annas and nine pies from five annas and two pies in the year before.

155. In this small province the total vaccination operations showed a trifling

Ajmer-Merwara.

increase from 14,713 operations in 1904-05 to 14,935 in 1905-06. The primary cases numbered 14.510 or 300 more than in the previous year, and revaccinations 425 or 78 less. The percentage of success in primary cases was 95·50 against 99·08 per cent. in 1904-05 and in revaccinations 81·65 against 86·28 per cent.

The department cost Rs. 2,849 and each successful case three annas and one pie in 1905-06, as compared with Rs. 2,659 and two annas and eleven pies, respectively, in 1904-05.

156. The statistics of vaccination in the Army will be found in Statement

Vaccination among troops.

No. III of the Appendices to this section.

SECTION VIII.
SANITARY WORKS.

157. In the province of Bengal before the partition there were, excluding
Calcutta, 161 municipalities and their income in
Bengal. 1904-05, excluding an opening balance of Rs.
9,44,342 was Rs. 57,77,295, or Rs. 2,79,113 more than in the previous year. Of
the total income 36·40 per cent. was spent on sanitary works, 12·48 per cent. on
roads, 5·66 per cent. on public safety and 31·63 on other requirements The
actual expenditure on sanitation included Rs. 3,14,498 on water-supply, Rs.
2,01,584 on drainage, and Rs. 13,60,642 on conservancy.

The expenditure on water-supply was Rs. 1,21,479 less than in 1903-04
as a result of the large capital expenditure in that year; on conservancy and
drainage expenditure increased by sums of Rs. 59,660 and Rs. 34,587, respectively.

There is evidence of considerable activity in municipal sanitation in Bengal;
water supplies are being improved, drainage schemes undertaken and conservancy
being improved. Attention has been focussed upon the biological method of
disposing of sewage on account of the objections raised to the discharge into the
river of effluents from the installations attached to the mills on the banks of the
Hooghly, and Dr. Gilbert Fowler of Manchester was engaged by the Government
of Bengal to enquire into the matter. He arrived in the end of the year and his
report is awaited. In the meantime Captain Clemesha, I. M. S., the Sanitary
Commissioner, succeeded in devising a method for the practical sterilisation of
effluents by the use of chloride of lime.

The work done by the different district boards has mainly had to do with
measures for the prevention of plague and cholera, sanitary arrangements at
fairs, clearance of jungle in the neighbourhood of villages and the improvement of
local drainage.

158. There were two meetings of the Board during the year; but, as usual,
most of the work was carried on by the circulation
Sanitary Board. of files among the members. The more important
matters considered at the meetings were, the appointment of an Assistant Sanitary Engineer, the Monghyr water-works project, the Howrah drainage project,
and the modification of the rules for the preparation of drainage and water-
supply schemes.

Preliminary estimates were prepared for the *Monghyr water-works* to cost
Rs. 3,13,000, of which it is proposed to raise Rs. 2,28,874 by local subscriptions;
and for the *Rampur-Boalia water-works*, the water to be pumped from a large
well for which a suitable site has been found.

Detailed estimates were prepared for the *Barisal water-works* and the *Khulna
filtered water-supply scheme*. Among the schemes considered during the year were
the *Howrah drainage project*, the *Garulia drainage project*, the *Baranagore
drainage scheme*, the *Titaghar drainage scheme*, the *Budge Budge drainage
scheme*, the *Cossipur-Chitpur drainage scheme*, drainage schemes for *Puri
Berhampore, Uttarpara, Deoghur, Khulna* and *Azimgunge*, and schemes for the
improvement of the water-supply of Kurseong and the supply of unfiltered
water to Cossipur-Chitpur.

The scheme for the supply of filtered water to Faridpur proved so successful that, at the instance of the Commissioner of the Dacca division, the Board, with the sanction of Government, distributed the design and estimate of cost of the filter beds to all the small municipalities. The scheme is very simple, it consists of a tank with two filter beds on the bank, a clear water reservoir fitted with taps and arrangements for pumping.

159. The new province was constituted after the end of the financial year to
Eastern Bengal and Assam. which this section relates, and the facts regarding the income and expenditure of most of the municipalities are included in the details given under Bengal. In Eastern Bengal and Assam there are 44 municipalities, two stations, Shillong and Sibsagar, and three unions, Jorhat, Golaghat and Habiganj. The total income excluding opening balance was Rs. 12,94,917, of which 41·71 per cent. was spent on sanitary works of one kind or another.

160. A Sanitary Board was formed, consisting of the Senior Member of the
Sanitary Board. Board of Revenue, the Chief Engineer and the Sanitary Commissioner, but no meeting was held.

161. Municipal sanitation continues to make rapid progress in these Provinces.
United Provinces. The water-supply works at Agra, Allahabad, Benares, Cawnpore, Lucknow and Meerut gave excellent results throughout the year, and extensions were made and improvements introduced in most cases. At Meerut where the water is taken from an irrigation canal, the water was of most excellent quality, the result of the examination of 268 samples showing an average of 2·76 microbes per c.c. One of the filters in which the rate of filtration is controlled by a special metre devised by the Sanitary Engineer, Mr. Aikman, gave even better results, the average number of microbes per c.c. being only 1·46. At Naini Tal where the pumps had to be worked throughout the year on account of scanty rainfall, there was a considerable increase in expenditure owing to more coal being required and to some new machinery being necessary. At Mussoorie the scheme for extending the water-supply and introducing electric lighting is in progress.

Good progress was made with the drainage and sewage works at Benares, Cawnpore and Agra. At Agra the sullage is run on to a sandy island in the bed of the river and is used to irrigate crops of tobacco which promise to repay the cost of the sullage farm. At Lucknow the drainage works have been pushed on rapidly ; 829 acres have been completely paved and 29,941 feet of underground storm water sewers have been built. During the year nearly 44 miles of drains were completed, and about six miles of streets paved. The Sanitary Engineer draws special attention to the excellent, economical and rapid work that has been done at Lucknow by Messrs. Lane Brown and Hewlett.

At Mirzapur the drainage works were nearing completion. They have been carried out with the greatest economy and praise is bestowed by the Sanitary Engineer on Mr. O'Brien, the municipal engineer, for the excellence of his work. A scheme for water-works for the town is in contemplation.

The projects for the drainage of Budaun, Hathras and Moradabad were nearing completion ; and those for the drainage of Azamgarh, Firozabad, Khurja and Huldwani were completed. Surveys are in progress in Hapur, Etawah, Saharanpur and Gorakhpur, and were about to be undertaken in Amroha and Sambhal. The preparation of schemes for Muttra and Brinda-ban, Jaunpur, Ghazipur, Padrauna and Bahraich has been assigned to Mr. Lane Brown. Experiments were carried out to ascertain the best methods of treating, and the time required in the processes for the treatment of domestic sewage and sullage with encouraging results, and it is hoped that it will soon be possible to introduce septic tank latrines and bacteria beds which will give an effluent which can be discharged into open drains without nuisance.

162. The Sanitary Board held four meetings during the year when they ap-proved projects, the estimates for which amounted to Rs. 18,85,153, an increase of 22 per cent. over the total of 1904 and of about 28 per cent. over the total of 1903.

Sanitary Board.

The necessity for appointing an Assistant to the Sanitary Engineer was admitted and Government addressed.

163. No information is given in the provincial report regarding the the income of the municipal towns. At Lahore the daily con-sumption of water has risen rapidly in the last three years, due, it is believed, to leakage from the mains and to the extravagant irrigation of gardens. A complete scheme for the provision of waste meters was prepared and accepted by the Municipal Committee and the Sanitary Board, but nothing has been done owing to want of funds. An estimate for the extension of the well system at the head works to cost Rs. 78,301 was prepared for the Committee. Estimates for a conservancy tramway (Rs. 54,347) to carry night soil and refuse to a depot at a distance from the city, and for improving the drainage outfall (Rs. 19,146) of the civil station, were prepared.

Punjab.

At Delhi good progress was made with the intramural drainage.

At Amritsar progress was made with the intramural drainage works; eleven projects to cost Rs. 98,298 for the drainage of different parts of the town were made over to the Municipal Committee. A committee was assembled to report on the best means of disposing of sewage at the outfall: it was resolved that land should be acquired to be let out to tenants. At Ambala the water-supply was increased by deepening five of the old wells; and the intramural drainage works were nearly completed. At Rawal Pindi the scheme for the construction of main intercepting drains and disposal works at a cost of Rs. 1,38,907 was in progress. At the new town, Sargoda, the water-supply and drainage works were nearly completed. At Ludhiana an estimate of Rs. 4,00,000 for the construction of water-works was sanctioned by the Local Government. At Multan the main drain and disposal works were well advanced. At Ferozepore and Muktesir the extramural drainage and disposal works were completed.

At Gujrat good progress was made with the extramural drainage and disposal works. A preliminary estimate for the drainage of Gurdaspur was prepared, and the detailed estimate for the Jagraon drainage works sanctioned.

At Simla arrangements were made to check the waste of water, new mains were laid down, and estimates for the construction of reservoirs to store storm water for flushing sewers were sanctioned. A Committee, appointed to enquire into the sanitary condition of the station, reported favourably on the water-supply and sewerage arrangements and recommended numerous improvements.

The provincial sanitary commissioner draws special attention to the gratifying increase in the amount of land used for the disposal of municipal sullage, etc., and to the higher price paid for sullage.

164. The Board met five times during the year. The following schemes were examined by them and submitted to the local Government for sanction. (1) Drainage scheme for Chiniot at a cost of Rs. 60,333 ; (2) Drainage scheme for Jagraon to cost Rs. 24,511 ; (3) Improvement of the drainage of Logarh gate at a cost of Rs. 27,713, and re-construction of drains, Lahori gate section, Amritsar city, at a cost Rs. 10,276; (4) Water-supply scheme for Ludhiana at a cost of Rs. 4,00,000 ; (5) Metering the city and civil station pipe lines at Lahore, Rs. 22,431 ; and (6) Suburban drainage scheme for Delhi, Rs. 3,79,345. All except the last, which was under consideration, were sanctioned.

Sanitary Board.

It was decided by the local Government to delegate to the Sanitary Board the power to sanction grants-in-aid from provincial revenues when necessary, within the limits of the sum allotted each year for the purpose, and the following instructions, along with a form drawn up by the Board calling for particulars of the financial condition, etc., of municipalities applying for aid, were issued to the district officers concerned.

(1) The Board shall examine the financial position of the local body concerned and decide whether it is possible for the project to be carried out entirely at the expense of the rate-payers with or without a loan.

(2) If it is decided that the project can be financed by a loan, the Board shall, in the case of all *large* municipalities, refer them to the instructions of Government regarding the raising of loans in the open market and recommend them to carry out the necessary negotiations ; in the case of smaller municipalities the Board shall decide whether the cost of the project is or is not too heavy to be borne by the local body even with the assistance of a loan ; in coming to a decision on this point the possiblity of financing the project by a resort to increased taxation shall be carefully considered.

(3) If the local body is unable either by a loan, increased taxation or otherwise to finance the project, the Board shall decide whether the work is so important and useful that it ought not to be postponed.

165. No sanitary work of great importance was carried out during the year, but many minor improvements were made in the various towns. At Peshawar where sanitary measures are energetically pushed forward, the following sums were spent: Rs. 10,313 on extension of the water-supply ; Rs. 29,226 on drainage ; Rs. 1,600 on paving streets. The sum derived from the sale of manure increased by Rs. 3,644 to Rs. 13,520.

North-West Frontier Province.

166. The income of the head quarters' municipalities in 1904-05, excluding opening balances and special grants, was Rs. 13,23,548, of which 34·24 per cent. was spent on sanitation, including water-supply, drainage, conservancy, markets, slaughter houses, etc.; 5·77 per cent. on medical relief and epidemics (including 1·17 on plague) and ·16 per cent. on vaccination.

Central Provinces and Berar.

No original work of any importance was carried out during the year, and the money spent was devoted to the improvement or repair of existing water-supply works, the extension and renewal of drains and the like.

167. The Sanitary Board met at each district head-quarters. The work carried out under their auspices consisted as usual in the improvement of village sites and surroundings, the construction of new wells and the repair of old ones.

Sanitary Board.

168. Including the presidency town there are 61 municipalities in Madras, but no information is furnished in the provincial report regarding their incomes or their expenditure on sanitary improvements, except that 39 failed to expend a fair share of the year's allotment. At Guntur the water-supply works were completed at a cost of Rs. 2,21,000 and opened on the 10th July. The water is brought by a masonry conduit from springs at the foot of the Kondavidu hills five miles from the town ; at present the supply is equal to four gallons a head only.

Madras.

At Coonoor the water-supply works estimated to cost Rs. 1,23,200 were opened in April. Estimates for the expenditure of Rs. 49,828 on extensions were sanctioned, and progress was made with the work. At Dindigul the scheme to supply water from the Oodukam valley by gravitation at a cost of Rs. 51,900 was completed in March. At Vellore the estimate was again revised to include compensation for the removal of three villages situated in the catchment area and now amounts to Rs. 3,77,520 : pipes and other materials were delivered and the work was in progress. At Saidapet and Negapatam works of a more or less experimental character in connection with water-supply schemes were carried out. At Ootacamund stoneware pipes, etc., were ordered for the drainage works and work was commenced principally on the laying out of the sewage farm.

Schemes for the water-supplies of Salem and Coimbatore and for the drainage of Madura were examined during the year, and a number of additional schemes either complete with plans and estimates or in the form of reports with approximate estimates were submitted, the more important of which were the Ellore water-works, Cocanada water-supply extension, Anantapur water-supply improvement, Masulipatam water-supply, Kottagiri water-supply, Periyakulam water-supply, Tuticorin water-supply extension, Negapatam drainage, Cuddapah drainage, and Madras water-supply improvement. It is proposed to substitute a masonry conduit for the existing open channel which conveys the Red Hills water to the City of Madras, and to provide filter beds and pumping power. The estimated cost of the scheme, which was being considered by the local Government, is Rs. 22,50,000.

The total sum assigned by District Boards to be spent on sanitation was Rs. 8,74,159, and nearly half of this sum was used during the last nine months of the year.

169. The Board examined the plans relating to 101 sanitary works estimated in the aggregate to cost Rs. 63,58,052. Of these 66 were finally approved, 21 were completed during the year, 26 were in progress and 18 were deferred for want of funds.

Sanitary Board.

170. The income of the 162 municipalities in Bombay—the presidency town is excluded—in the year ending 31st March 1905 was Rs. 68,49,972, and Rs. 18,62,875 were spent

Bombay.

on the improvement of water-supply, drainage and conservancy within municipal limits. No new scheme of water-supply or drainage was undertaken during the year, the reason given being the embarrassed condition of municipal funds on account of expenditure in connection with plague. The entire income of district and local Boards was Rs. 43,84,296, of which Rs. 2,79,666 were devoted to water-supply and drainage.

171. The Board met once during the year and considered the question of financial and technical assistance to be rendered to municipalities so as to stimulate progress in sanitary improvement. The Board submitted recommendations on the subject to the local Government.

Sanitary Board.

At Ahmedabad nearly 15 lakhs of rupees have been spent on sewerage works which will serve rather more than a third of the total area and nearly half the total population. Further expenditure of 15 lakhs will apparently be necessary to complete the work. The water-supply works require extension and improvement at an estimated cost of Rs. 3,50,000. At Karachi the drainage extension works were commenced during the year. At Poona alternative projects for improving water-supply were considered and detailed estimates for a comprehensive scheme including the city and cantonments and Kirkee were being prepared. The preparation of a revised scheme of drainage for Poona city was taken in hand. At Surat steps were taken to improve the existing water-supply ; and at Hyderabad the municipality is arranging to provide a new settling tank. Projects for completing the water-supply works and surface drainage of Pandharpur were being worked out.

172. The income of all the municipalities in the province was Rs. 69,35,805, of which the following percentages were spent on sanitation—on water-supply 13, on drainage 8, on conservancy 12, on other sanitary works 12 ; but on account of the unusual outlay in Rangoon on the Hlawga water-supply project, the provincial Sanitary Commissioner suggests that a more correct idea of the average expenditure on sanitation will be obtained if the expenditure in the municipalities of the Pegu division is omitted. When this is done the percentages are on water-supply ·78, on drainage 3·80, on conservancy 12·16 and on other sanitary works 4·66. The total income of the town funds was Rs. 4,24,133, of which 21 per cent. was spent on conservancy, 2 per cent. on drainage, 4 per cent. on water-supply and 18 per cent. on other sanitary works. Of the district funds, aggregating Rs. 47,48,133, rather less than 5 per cent. was devoted to purposes of sanitation.

Burma.

At Rangoon the Hlawga well works were opened on the 22nd February, but the supply was not satisfactory and it was found that the expectations entertained of it would not be fulfilled. An expert committee was appointed and their report is reassuring ; improved means of distribution will be required and increased pumping power, when, it may be hoped, that certain minor alterations will ensure an abundant supply.

At Moulmein the water-supply works were completed, but it appears that the piped supply has to compete with many defective wells. At Mandalay ' the question of water-supply is still pending.' Estimates for water-supplies were sanctioned for Akyab (Rs. 2,98,920), Rangoon Cantonment and Thonzé and prepared for Kyauktan, Thatôn and Meiktila. At Rangoon the question of providing an improved system of sewage disposal was carefully considered by the committee and an estimate of Rs. 5,20,432 was sanctioned by the local Government. The reclamation of land is proceeding. The provision of adequate house room for the growing population is becoming a serious matter. The remodelling of the existing building regulations has received consideration by the committee, but it is reported that they have not " approached the subject with that determination to face vested interests that is essential in dealing with a matter of this nature, if solid sanitary results are hoped for." Drainage schemes are in progress at Pegu and Insein. Preliminary estimates for drainage and conservancy schemes for Kyaiklat and Pyapon were prepared.

173. The Sanitary Board held two meetings during the year. At the first the
Sanitary Board. constitution and functions of the Board were dis-
cussed, with the result that the constitution of the Board has been altered and its functions defined. In future it will consist of the Financial Commissioner, the Chief Engineer, the Sanitary Commissioner and the Sanitary Engineer, the last named to be the Secretary. Commissioners, Deputy Commissioners and Civil Surgeons will be *ex-officio* members when the Board meet in their division. The Board will meet at least four times a year and their functions will be " purely consultative ". They are, however, required to report to Government any case of persistent disregard of their advice. At the second meeting of the Board the principal subject considered was proposed rules for the guidance of the Sanitary Engineer.

174. During the year 1905-06, the expenditure on ordinary military works
Military Works. was Rs. 1,15,78,465, as compared with Rs.
1,25,88,420 in the previous year. In addition, Rs. 24,18,372 was spent on military works connected with the military reorganisation scheme. Details regarding new works and improvements in some of the more unhealthy cantonments will be found in the statements appended to Tables V and XXX at the end of this volume.

SECTION IX.

GENERAL REMARKS.

175. The Government of India notified in September 1904 that, as the deten-
Red Sea Pilgrim Traffic. tion of pilgrims for ten days in observation camps
at or near the ports of embarkation had proved
costly and inconvenient to the pilgrims, it had been decided with the concurrence
of His Majesty's Government that the period of observation should be reduced to
five days at a central camp, from which, after satisfactory medical inspection and
disinfection, pilgrims would be permitted to proceed to the Hejaz. The Haj
was open to any resident in India but embarkation was permitted from Bombay
only ; embarkation from Chittagong being discontinued because of the unlikelihood
of pilgrims desiring to embark from this port if they could do so from Bombay.
The warning given the previous year that intending pilgrims would be well
advised to defer the fulfilment of their purpose until another year, was repeated.

In this report it is possible to notice the results of the pilgrim season of 1904-5
only, to which the orders referred to above relate ; complete information for
the season 1905-6 is not yet available. In view, however, of the change
made in the conditions under which pilgrimage is to be permitted for the season
1906-7, and the fact that the orders were issued prior to the appearance of
this report, it may here be mentioned that the conditions for the pilgrim
season 1905-6 were the same as those for the season 1904-5, but for
the season 1906-7, the Government of India notified at the end of
September 1906, that (1) no person shall be permitted to embark for a pilgrimage
to Mecca except at Bombay ; (2) all ships chartered for Jeddah shall be thoroughly cleansed and freed from rats by the Clayton process under the personal directions of the Port Health Officer ; (3) ample hospital accommodation shall be provided on board ; (4) before embarkation pilgrims shall be medically inspected
and their clothes and baggage disinfected as in the case of third class pas-
sengers on ordinary steamers ; (5) the ships shall undergo further medical
inspection both at Aden and at Perim ; and (6) in the event of plague appearing
on board a vessel arriving at Aden, telegraphic information shall be sent to
Perim where the persons attacked shall be landed and placed in a plague camp ;
cases of the disease occurring between Aden and Perim shall be dealt with
similarly, the ship being allowed to proceed to Kamaran.

The importance of the latest orders to the Moslem community in India lies
in the fact that since the Government of India notification of the 20th February
1897, suspending pilgrimage to Mecca, and the subsequent notification of the 20th
November of the same year, 'permitting it, pilgrimage has every year since
then been subject to medical observation of the pilgrims prior to embarkation.
For the past two seasons that period has been restricted to five days only, and
now medical observation prior to embarkation has been entirely abolished.

During the season 1904-5, a total of 12,449 pilgrims embarked in 15 ships
at Bombay ; of these the Kamaran report for that year records the arrival
of 11,242, while the Haj report states that there arrived at Jeddah 11,820
Indian pilgrims, of whom 11,339 proceeded direct by pilgrim ships from Bombay

and 481 adopted indirect routes.   The pilgrimage of 1904-5 was a smaller one than that of the preceding year which was a Haj Akbar (grand pilgrimage).

The observation camp at Bombay was opened on the 20th October 1904, the first ship leaving five days later and the last on the 13th January 1905. During the voyage there were 16 deaths on 8 ships, and 20 persons were landed at Aden, most of them on account of small-pox. During the period of quarantine at Kamaran 33 pilgrims died, six of them from small-pox, and 100 were admitted to hospital chiefly on account of acute and chronic respiratory diseases.   The Vice-Consul remarks with satisfaction that the orders regarding the detention at Kamaran of sick pilgrims after completion of the regulation period of quarantine have been relaxed and greater discretion was being exercised in the matter.   In the Haj report it is stated that " the health of the pilgrims was perfect throughout the season," and that the sanitary condition of both Jeddah and Mecca has greatly improved within the last few years.

The pilgrims returned to Bombay between the 21st February and the 2nd September 1905 ; in all 13,952 arrived in 25 ships.   During the voyage to Bombay there were 120 deaths on 10 ships, seven being due to small-pox ; eleven ships furnished between them 35 cases of small-pox ; of those on board infected ships, 325 were vaccinated, the rest refusing vaccination.   The clothing and bedding of 5,648 pilgrims among whom small-pox appeared were disinfected before disembarkation was permitted ; the other pilgrims who were found free from infectious diseases were landed immediately on arrival at Bombay and taken charge of by the Protector of Pilgrims who arranged for the despatch by rail of all pilgrims to their homes.

*   *   *   *   *   *

176. During the past year a considerable amount of original work has been done in connection with Malta fever and our knowledge of this disease, especially as regards its dissemination, has been much increased thereby.   These researches have for the most part been carried out by the Mediterranean Fever Commission, working under the supervision of an advisory committee of the Royal Society.

Malta Fever.

177. The results of the Commission's work are here shortly summarised. The *M. melitensis* shows no sign of being able to lead a saprophytic life ; outside the body it can retain its vitality for a long period, but it does not multiply. Many domestic animals—goats, cows, mules, horses, dogs, etc., are found to suffer naturally from the disease and of these the goat seems the most susceptible.   It is found that about 50 per cent. of the milch goats in Malta suffer from the disease and about 10 per cent. (some apparently healthy) are excreting the specific microbe in the milk.   Excretion in milk may continue for three months without any physical changes in the milk or any sign of ill health in the goat.   In this connection Zammit's test (the *milk* instead of the *blood* reaction with *M. melitensis*) is not only a good indicator of infection but is a surer test than the ordinary serum reaction that the micrococcus is being excreted in the milk.   It was demonstrated that the milk of goats that are infected and have a marked blood reaction may not necessarily contain the micrococcus, and that, in the milk of a goat, the blood serum or milk of which has no agglu-

Commission's work.

tinative action, the micrococcus is not found. In the milk the excretion of the micrococcus (which has been shown to persist in the blood of a goat for seven months) may be intermittent. The goat is less susceptible to the disease than the monkey, and the monkey than man. In man the micrococci can be demonstrated in the peripheral blood in 68 per cent. of cases, but they are never very numerous in the blood; the smallest quantity of blood in which the micrococcus has been found is 4 c. mm., *i.e.*, in the proportion of one micrococcus to about 20,000,000 red corpuscles.

In the various experiments carried out by the Commission with expired air, saliva, sputum, sweat, scrapings of the skin and fæces, the micrococcus was never detected ; we must, therefore, turn to the urinary passages in the hopes of finding an important channel of elimination. In the urine, the micrococcus is found from about the 15th day of illness till after convalescence is established and in this way hands, food and dust may become infected. Elimination in urine may be intermittent or continuous. Ambulant cases are common, amounting to 10 or 15 per cent. of the entire native population in Malta ; and the ambulant cases who are excreting the *M. melitensis* in their urine and are thus a possible danger to the community, constitute 1 or 2 per cent. This urine is infective to monkeys and goats.

178. There is a close correspondence between atmospheric temperature and

Temperature, rainfall and seasonal incidence. incidence ; on the other hand, incidence has no intimate connection with rainfall. Although Malta fever is least prevalent in the cold months, the number of cases which do occur in winter actually amount to about one-third of those recorded in the hot months—July, August and September—the season of greatest incidence.

179. Animals, chiefly monkeys and goats, have been infected by the respiratory

Modes of infection. and digestive tracts by using infected dust and infected food. With regard to the former, Bruce states that, as the experiments were with dust artificially contaminated with cultures and so with a very concentrated virus, there is no absolute proof that dust as it occurs under natural conditions ever conveys the disease to the healthy, and adds that " theoretically there seems to be danger from the scattering broadcast of such a virulent and resistent microbe, but it is possible that not a single case of infection occurs in this way." There is no such doubt about infection by the digestive tract. Monkeys fed on naturally infected goats' milk, *i.e.*, on the ordinary milk of goats in the streets under natural conditions, have become infected. The injection of the virus under the skin of man or monkey or the application of the virus to a scratch gives rise to the disease. Infected dust or liquids applied to unbroken conjunctiva, nasal passages, pharynx or larynx may also lead to infection. Horrocks and Kennedy state that it is " extremely probable that human beings are infected by the bites of infected mosquitoes." They report three experiments which suggest the possibility of the disease being transmitted in this manner ; yet many other experiments have been carried out on the same lines with negative results. Experiments, so far, are not conclusive and they do not prove that mosquitoes or other biting insects can infect man or animals. Davies and Johnstone seem to think that mediate contagion, *e.g.*, pollution of hands and food, play a significant part in dissemination. Horrocks says that infection cannot occur from contact with the cutaneous surface of a patient, unless that surface is contaminated with urine.

180. The incubation period is considered to be from 6 to 30 days or more.

Incubation period.

The general impression amongst Maltese medical men is that the usual incubation period is not more than 8 or 10 days. Horrocks gives the following figures : if infection is by subcutaneous injection or through a crack in skin or mucous membrane, the incubation period is from 5 to 7 days ; if through unbroken mucous membrane, such as by feeding with infected goats' milk, the period may even extend to two months.

181. The prophylactic measures put forward chiefly concern urine, goats'

Prophylactic measures.

milk and mosquitoes. Immediate and complete disinfection of clothing, bedding, bed-pans and utensils soiled by Malta fever patients is recommended : urinals and latrines should be properly managed ; milk pasteurised (68° C. for 10 minutes) or boiled ; mosquitoes and infected goats destroyed ; and Malta fever patients isolated in hospital.

182. Surgeons Ross and Levick, R.N., reasoning by a process of exclusion,

Other workers.

have all along held that infection is carried chiefly by mosquitoes and other biting insects. Their work, appearing before the Commission's suspicions had been directed towards mosquitoes, was opportune and threw considerable doubt on the goats' milk theory at a time when it was almost generally believed that the Commission had found the key to the prophylaxis of Malta fever. These naval surgeons did not believe infection could be acquired through uninjured respiratory or digestive tracts. They drank water and goats' milk containing *M. melitensis* in large numbers and did not contract the disease. They pointed out that the disease was as prevalent as ever in Malta after care was being taken to sterilise milk ; that the one regiment which did not boil its milk had the lowest admission rate ; that in hospitals, where all milk was carefully sterilised, the incidence had increased. They criticised the food and dust experiments of the Commission on the grounds that the incubation periods were so very long ; that the monkeys had to be fed for weeks before they contracted the disease as shown by the agglutination test, and even then there might be no sign of ill-health and the monkeys sometimes actually gained in weight ; whereas, the inoculation of the minutest amount of virus under the skin gave rise to the disease in 6 days. They held that, when such enormous and unnatural quantities of infected material were used repeatedly over such a considerable length of time, lesions of mucous membrane may meanwhile have occurred and infection may not have occurred until a lesion had become inoculated. They maintain that the incubation period of natural infection (mosquitoes) is from 8 to 11 days.

It is recognised that there is more than one factor responsible for the spread of this disease and in the present state of our knowledge attention is directed chiefly to (1) goats' milk, (2) contamination—especially with urine—of hands, food and other articles and (3) mosquitoes. Of these the most important would seem to be goats' milk and, up to the present, this is the only means by which we have little doubt the disease can be conveyed to man. As evidence of this the following instances collected by Bruce may be cited. (1) The introduction of Maltese goats into Rhodesia was followed by an outbreak of Malta fever. (2) In 1905 the S.S. " *Joshua Nicholson* " shipped 65 goats at Malta for export to America. The milk was drunk by the crew with the result that an epidemic of

Malta fever broke out on board. (3) Strachan in the Orange River Colony blamed goats' milk for causing an outbreak of Malta fever at certain farms, and Birt found that the blood of some of the goats of these farms gave the serum reaction with *M. melitensis.* (4) Forster isolated *M. melitensis* from the milk of one of the goats kept for supplying milk to a Malta fever stricken regiment at Ferozepore (Punjab). (5) Davies found that children in Malta who drank unboiled milk suffered four times as much as those who drank boiled milk.

183. Malta fever has during the past year been demonstrated to be endemic
Geographical distribution. not only in India but also in South Africa and possibly in Egypt. Birt in the Orange River Colony, at first by a well controlled serum test, and subsequently by isolating *M. melitensis,* has shown that this fever is endemic in that region. Phillips has reported cases at Cairo : the serum reaction was positive in very high dilution, but no mention is made of the isolation of the micrococcus. In last year's report a full account was given of the history of Malta fever in India up to the work of Lamb proving the endemicity of the disease in this country. Recently Forster, at Ferozepore (Punjab), examined the herd of goats from which milk was being supplied to the 14th Sikhs, who had suffered considerably from the disease. He found that 10·5 per cent. of the goats gave the serum reaction and isolated the *M. melitensis* from the milk of one of them.

. 184. Reid reports great benefit from therapeutic inoculations, the dose of
vaccine being regulated by observing the opsonic
Vaccine and Serum. power of the blood, and suggests this method of inoculation for prophylaxis. Shaw (Commission) has drawn attention to the persistence of the living *M. melitensis* in the blood of goats for at least seven months and the bearing this persistence has upon the preparation of a therapeutic serum. *

*   *   *   *   *   *

185. The steady increase in the number of patients resorting to the Institute for anti-rabic treatment continued during
The Pasteur Institute, Kasauli. the year ending August 8, 1906. A total of 1,262 persons presented themselves for treatment, of whom 117 were not treated, as there was satisfactory evidence that the animals that inflicted the bites were not rabid. Of the 1,145 persons treated, 342 were Europeans and 803 natives, compared with 307 Europeans and 570 natives in the previous year. Patients came from every province, from the native states of Rajputana and Kashmir, one from Sikkim, two from Ceylon and one from the Persian Gulf. Included in the total were 172 patients from the British army and 82 from the Native army. Of the total number treated there were seven failures only.

Besides anti-rabic treatment, bacteriological examinations of specimens, serum diagnosis, etc., were carried out, the number of cases thus furnishing material for examination being 880 ; further, in 277 instances experiments were carried out to prove the existence or otherwise of rabies in animals dead with suspicious symptoms.

The duty of preparing sera was transferred from the Pasteur Institute to the Central Research Institute in December 1905 ; between the 9th August and the end of November 1905, 410 bottles of antivenene and 312 bottles of antidiph-theritic serum were issued from the Pasteur Institute.

---

* A list of references to the literature dealt with in this summary will be found at the end of Section III (Native Army of India).

186. The *Central Research Institute* at Kasauli, which was referred to in paragraph 178 of last year's report, is situated well away from other buildings in the cantonment.

New laboratories.

The Institute consists of ten substantial stone buildings ; the main one, which was formerly an official residence, has been considerably altered to adapt it for the purposes of a laboratory, while the others have been built for the various purposes for which they are required. The main building has a large hall running through the centre and is surrounded by a wide verandah ; the flooring is of glazed tiles ; in all the rooms gas light is available and in some electric light as well. The building contains about 21 rooms which have been allotted for various purposes, such as microscopic work, photography, library, offices, etc. The subsidiary buildings, or blocks, are used for housing animals, preparing sera and storing chemicals and spare apparatus ; one block is fitted up with a $10\frac{1}{2}$ H. P. oil engine, a dynamo, an electric storage battery of 56 cells, a freezing machine and cold storage room, and another with a Mansfield gas installation ; there are also quarters for 12 servants. A piped water-supply has been laid on to four of the ten buildings.

The Institute has been fully equipped with all necessary apparatus and can accommodate six to eight workers. Two officers of the Indian Medical Service have received a two months' course of instruction in technique, qualifying them for investigations connected with malaria, *kala azar* and blood diseases generally.

The *King Institute of Preventive Medicine*, at Guindy, about six miles from Madras City, is divided into two sections—the vaccine and the bacteriological. The buildings occupy an extensive site ; on the east are the bacteriological laboratories, animal house and experimental sewage disposal works ; in the centre the numerous smaller buildings of the vaccine section, and on the west the quarantine and isolation sheds of the vaccine section and two laboratory blocks with stables designed for serum work. In the grounds are quarters for the superintendent, the deputy superintendent and a European subordinate. The chief laboratories are located in a large building with open corridors and wide verandahs. The laboratories and the rooms of the manufacturing block of the vaccine section are floored with smooth Cuddapah slab and have white glazed tiled walls, while the fittings are of polished black stone or of glass ; the floors of each stall in the calf sheds consist of a single partially polished Cuddapah slab which is kept scrupulously clean day and night by men told off for that purpose. The laboratories are fitted with electric fans worked by a gas engine and dynamo. The water is supplied under pressure from cisterns on the roof, the pumping station being situated at no great distance.

The Institute has been fully equipped for preparing vaccine and conducting the bacteriological work of the presidency and also serves as a centre of training for sanitary assistants, vaccinators and plague inspectors to whom both practical and theoretical instruction is imparted. The laboratories are available for the instruction of officers and the carrying out of research work by them.

The *Pasteur Institute for Southern India* is located at Coonoor. Besides the main building which stands in a large compound, there are the necessary outbuildings which include quarters for a hospital assistant and servants, stables for horses, houses for rabbits and godowns for stores and apparatus. The main building is a two-storied one, with a central corridor running from end to end. On the ground floor are the work rooms (rooms for the preparation and preservation of the virus, patients' waiting and inoculation rooms, surgery, etc.) and clerks' office: on the upper floor, on one side, are the director's office, the library and store rooms, and on the other, residential quarters which will probably be allotted to the assistant director. Gas and water pipes are to be fitted to the building which is nearing completion; it is expected that the Institute will be ready to receive patients for treatment in February next.

*   *   *   *   *   *   *

187. With a view to supplementing the work of the bacteriological laboratories

Facilities for research or instruction at laboratories. in the investigation of Indian diseases, it has been arranged that officers of the Indian Medical Service who may volunteer, and officers of the Royal Army Medical Corps, who are recommended by the Principal Medical Officer in India, may be allowed to proceed to a laboratory for a period of three to six months for the purpose of prosecuting research already commenced or for instruction in technique, provided local arrangements can be made for the conduct of routine duties during their absence. While at a laboratory, officers will draw their ordinary pay, less any local allowances, and will be granted travelling allowances; no deputation allowance will be admissible.

*   *   *   *   *   *   *

188. In addition to routine work at the bacteriological laboratories, special

Special enquiries. enquiries have been commenced with reference to typhoid fever in India, dysentery in jails and lunatic asylums, and malaria in the Presidency division of Bengal, and the inquiry into the etiology of *Kala azar* has been continued.

Typhoid Fever.—In March 1906 the staff of the Central Research Institute, Kasauli, commenced an investigation in connection with the etiology and spread of enteric fever in India. Work was begun with material available in Kasauli and its vicinity, and during the winter of 1906-7 the investigations are to be conducted in the cantonment of Meerut, the staff of the Central Research Institute being strengthened temporarily by the deputation of specially selected officers to assist in the enquiry.

Dysentery.—This is the most serious cause of mortality in Indian prisons, and it has been known for many years that the term dysentery in the returns included several diseases presenting nearly identical signs and symptoms although arising from different causes. In order to further investigate the matter the Government of India placed Captain W. C. H. Forster, I.M.S., on special duty under the Sanitary Commissioner, to investigate the causation, prophylaxis and treatment of dysentery in the prisons and lunatic asylums of India. Captain Forster has been authorized to visit a certain number of jails in the several provinces, and also lunatic asylums, and carry out researches therein. The results of Captain Forster's investigations will, it is hoped, add to the existing knowledge of one of the most fatal of tropical diseases.

Malaria.—The Government of Bengal recognised in their Resolution No. 1379-Medical, dated the 26th March 1906, that over a large part of the province, and particularly in the Presidency division, a main cause of ill health is the obstruction of drainage. A complete remedy was not held to be possible, but it was considered certain that in many places the conditions could be greatly ameliorated by the execution of drainage projects which could be carried out without excessive expenditure. A local committee of three members was appointed to visit the districts of the Presidency division (1) to ascertain what areas are specially unhealthy, (2) to investigate the cause of the prevalent sickness and in particular to determine how far this prrevalence is due to obstructed drainage, (3) to consider if a drainage scheme is practicable, and (4) to prepare a list of practicable drainage schemes in their order of urgency. To assist the committee with expert knowledge the services of Captain G. E. Stewart and Lieutenant A. H. Proctor, I.M.S., have been lent temporarily to the local Government.

Kala Azar.—Contributions to our knowledge of the Leishman-Donovan body are conspicuously absent from the parasitological literature of the year, and it appears as if observers were awaiting the discovery of the extra-corporeal stage of the parasite to renew their labours.

The reports for 1903 and 1904 contain a history of the discovery of the parasite and of the growth of knowledge concerning it, and the present time seems opportune to summarize what is now known regarding it.

The prominent clinical features of *kala azar* are a characteristically irregular temperature, progressive cachexia and splenic enlargement, conditions which have given rise to the names "malarial cachexia," "cachectic fever" and "tropical splenomegaly." In the early stages of the disease it sometimes resembles enteric fever, but the febrile attacks are often followed by periods of apparent perfect health; as the disease progresses profound cachexia supervenes; diarrhœa, *cancrum oris*, *cancrum vulvæ*, transient œdemas, petechial eruptions, pigmentary changes and ulcers of the skin are of more or less constant occurrence. The disease is common and particularly severe in Assam, where it has long been well known clinically, but it has a wide distribution within the tropics and cases have been reported among other countries from Africa, Malaysia and China. The parasite of *kala azar* is, as we have seen, morphologically identical with the parasite of the chronic granuloma of the skin known as Delhi boil; whether there is any difference between the parasites, as might be expected from the geographical distribution of the two diseases, the future will, no doubt, disclose.

The parasite of *kala azar* is a circular body, 2—3μ in diameter, formed of a clear homogeneous cytoplasm, in which are embedded close to the periphery two conspicuous chromatin masses resembling the chromatin masses of a trypanosome. The smaller mass is a blepharoplast from which there often stretches a short ' tail ' of chromatin. The existence of double forms makes it probable that the parasites reproduce themselves in the body by binary fission. Circular bodies giving rise by multiple fission to from 3 to 8 parasites occur, but the stages of their formation have not yet been completely followed. The parasite is nearly always embedded in the cytoplasm of some kind of cell, generally in the endothelial cells of the viscera, but also in leucocytes. Large macrophages, probably of endothelial origin, containing as many as two hundred parasites are found in cases of *kala azar* and of Delhi boil.

In *kala azar* the parasites are present in enormous numbers in the spleen, liver and bone-marrow, and they are also found in the intestinal walls, in the lymphatic glands, kidneys and testicles. In advanced cases, they are to be found in the leucocytes in the peripheral circulation, and in petechial hæmorrhages in the brain membranes. They are present in granulation tissue wherever it occurs, and in papules of the skin. In Delhi boil the parasites are packed in the cells of the granulation tissue in such numbers as to give rise to an appearance under the microscope of a section of the spleen in *kala azar*.

It was shown by Rogers that, if splenic blood, taken from a case of *kala azar*, is prevented from clotting by the addition of citrate of soda solution, and is kept at a temperature of 22° C.-24° C., the parasites enlarge, the protoplasm becomes granular, and the chromatin masses assume a more central position: repeated fission then occurs and most of the forms develop a thick flagellum, those parasites resembling a typical flagellate. The question now is, 'what happens next?'

Researches regarding the extra-corporeal life of the parasite are being pursued by Captain Patton, I.M.S., who has been attached to the King Institute at Guindy for the purpose of this investigation. It was found by Christophers and by Donovan that in certain cases a considerable number of the parasites was present in the leucocytes of the peripheral blood, and it was suggested by the former that this distribution might be concerned in the dissemination of the disease. Captain Patton has found that the parasite is frequently found in the peripheral blood—so frequently that an examination of the blood is of considerable diagnostic value, and that the number of parasites in the blood may be very great, as many as 80 being found in one case in a single film. He has found that, although ordinarily in *kala azar* there is leucopenia, when diarrhœa is present there is almost always leucocytosis with the abundant presence of parasites in the peripheral blood. Founding a hypothesis upon this fact and upon the process of development *in vitro*, Captain Patton conducted a long series of feeding experiments with mosquitoes, lice and bugs, and found that parasites could be recovered from the gut of the bug four days after feeding and that they showed in some cases the changes characteristic of development.

It is as yet doubtful if this fact indicates the bug as the carrier of *kala azar* ; and Rogers' experiments, showing that the bodies develop best in an acid medium, do not incriminate the bug more than other blood-sucking insects. For the present we must be content to know that the work of Rogers and Patton favours the view that infection may be carried from cases in which the peripheral circulation is loaded with parasites by a blood-sucking insect. The fact that in cases in which diarrhœa is severe the peripheral circulation contains many parasites, throws a new light upon Bentley's statement that dysenteric cases are chiefly concerned in the spread of the disease.

*   *   *   *   *   *   *

189. In July 1906 the Government of India, on the recommendation of His
Standing Committee on Enteric Fever in India. Excellency the Commander-in-Chief, approved the formation of a standing committee for the purpose of investigating and advising on the incidence of enteric fever in India and its prophylaxis. The committee is composed of the Principal Medical Officer, His Majesty's Forces in India, as President (*ex-officio*), a Vice-President and five members drawn from medical officers in military employ, with the Sanitary Officer, Army Head Quarters, as member and Secretary. The Government of India

decided that their Sanitary Commissioner and the Director of the Central Research Institute, with Major G. Lamb, I.M.S., and Captain E. D. W. Greig, I.M.S., should also be members of the committee. The committee held their first meeting at Simla on July 31, 1906, and following days, His Excellency the Commander-in-Chief opening the proceedings in person, and the second meeting, also at Simla, on October 17, 1906, and subsequent days.

  *   *   *   *   *   *   *

190. Since the issue of the last report the following memoirs have been published :—No. 21, *On the Standardisation of*

Scientific Memoirs. *Anti-typhoid Vaccine,* by Captains G. Lamb and W. H. C. Forster, I.M.S.; No. 22, *Mediterranean Fever in India ; Isolation of the Micrococcus Melitensis,* by Captain G. Lamb, I.M.S., and Assistant Surgeon M. Kesava Pai ; No. 23, *The Anatomy and Histology of Ticks,* by Captain S. R. Christophers, I.M.S. ; No. 24, *On a Parasite found in the White Corpuscles of the Blood of Palm Squirrels,* by Captain W. S. Patton, I.M.S.; No. 25, *On the Importance of Larval Characters in the Classification of Mosquitoes,* by Captain S. R. Christophers, I.M.S.; No. 26, *Leucocytozoon Canis,* by Captain S. R. Christophers, I. M. S.

JOHN T. W. LESLIE, *Lt.-Col., I.M.S.,*

Sanitary Commissioner with the Govt. of India.

APPENDICES

TO THE

Annual Report of the Sanitary Commissioner with the
Government of India

FOR

1905.

TABLE I.—*Highest, lowest and mean temperature in shade and its departure*

Stations.	January Highest	January Lowest	January Mean	January Departure	February Highest	February Lowest	February Mean	February Departure	March Highest	March Lowest	March Mean	March Departure	April Highest	April Lowest	April Mean	April Departure	May Highest	May Lowest	May Mean	May Departure	June Highest	June Lowest	June Mean	June Departure
Calcutta (Alipore)	83·9	47·9	65·5	—0·7	86·9	45·9	65·1	—6·1	95·0	61·4	77·7	—3·5	99·0	6c·9	81·5	—4·2	100·5	67·4	84·4	—1·6	106·2	72·9	85·0	+4·2
Narayangunj	80·6	46·3	64·8	—1·7	85·6	43·7	62·1	—6·3	93·1	59·8	75·5	—3·6	93·6	61·3	79·0	—4·8	95·1	69·8	82·0	—0·7	94·6	72·3	84·9	+1·2
Chittagong	75·9	46·6	65·8	—1·0	82·4	46·6	65·7	—5·1	88·5	60·6	75·0	—2·2	58·5	62·1	78·0	—3·2	91·5	70·1	82·0	—0·1	93·5	71·6	82·6	+1·0
Sibsagar	71·3	42·9	58·7	—3·2	72·3	37·5	57·9	—5·3	82·3	52·9	66·6	—3·8	83·2	56·9	70·3	—4·3	92·2	67·0	78·4	—0·5	94·7	74·0	81·9	+0·3
Silchar	81·9	43·0	63·6	—1·5	81·9	41·5	62·4	—5·8	88·5	54·9	71·6	—3·1	90·5	57·9	75·0	—2·9	96·1	61·3	81·3	+0·7	97·0	72·8	80·1	—0·7
Cuttack	90·4	55·1	70·9	—1·3	93·9	52·1	73·9	—4·1	99·4	63·2	83·2	—1·5	104·4	63·3	86·9	—3·2	108·5	70·2	89·0	—1·9	112·5	76·7	93·2	+5·2
Hazaribagh	70·7	39·3	59·8	—2·0	79·7	37·8	58·6	—7·7	91·7	52·2	71·4	—4·9	100·2	57·1	78·9	—6·5	105·4	66·3	86·0	—1·4	107·6	75·3	92·6	+8·4
Patna	77·0	40·3	60·1	—4·5	82·0	35·3	59·5	—6·4	96·0	55·4	72·0	—4·5	102·0	57·8	80·2	—6·9	105·5	64·8	85·4	—3·6	106·5	68·7	91·5	+3·3
Darjeeling	50·7	37·0	38·3	—2·5	51·3	33·0	36·0	—5·8	65·7	35·0	47·7	—1·9	64·7	24·0	51·3	—4·4	70·3	46·1	57·5	—0·7	73·7	54·1	62·4	+1·7
Allahabad	80·2	39·1	59·0	—2·2	80·7	34·1	58·8	—7·3	9·2·1	47·6	72·7	—4·9	105·1	55·1	82·2	—5·9	114·7	70·1	91·7	—1·7	112·7	78·6	97·4	+5·2
Lucknow	80·7	36·6	58·0	—2·4	77·2	31·6	55·9	—5·0	96·3	47·7	70·7	—5·3	105·3	54·7	80·4	—6·3	112·8	67·3	89·7	—1·7	114·3	77·3	96·6	+5·7
Meerut	73·4	31·1	55·0	—3·7	71·0	31·6	53·2	—8·2	9·0·6	39·7	65·8	—6·4	103·2	43·3	75·0	—5·3	111·7	68·8	90·3	+0·9	108·7	75·3	92·8	+3·3
Delhi	70·3	32·5	55·1	—4·3	70·2	33·5	53·3	—10·2	90·2	41·9	67·0	—7·9	104·7	55·0	81·2	—5·5	113·2	73·3	93·5	+1·0	109·7	77·8	96·6	+3·3
Agra	78·0	35·9	57·5	—3·6	72·5	33·9	56·0	—9·5	95·5	47·4	69·9	—7·1	105·0	57·4	83·5	—5·7	114·0	75·4	95·1	+0·7	111·0	77·9	97·9	+3·5
Jhansi	84·2	37·1	61·6	—2·2	81·6	38·1	60·6	—7·7	98·7	52·2	74·8	—5·0	109·8	67·6	86·3	—4·2	116·4	78·1	98·8	+3·9	114·4	81·2	100·1	+8·7
Ajmer	80·4	31·0	58·2	—1·7	75·4	30·0	55·8	—7·5	91·4	43·8	70·3	—4·2	102·9	51·3	82·6	—2·8	112·9	70·6	96·9	+5·4	107·9	78·6	94·1	+3·4
Saugor	83·9	35·0	62·3	—1·2	67·4	36·0	60·5	—7·9	94·4	50·6	73·8	—4·5	105·4	56·1	82·7	—4·5	111·0	75·1	94·6	+3·0	109·0	75·2	95·1	+8·4
Jubbulpore	84·4	37·4	62·0	—0·8	67·4	31·9	60·1	—7·5	9·5·5	52·3	74·3	—2·5	105·0	50·9	81·0	—5·4	112·0	69·1	93·2	+0·5	110·0	73·6	93·3	+5·6
Multan	75·2	27·0	53·4	—2·7	73·8	32·5	54·1	—6·6	91·8	40·5	67·5	—5·0	109·9	49·4	81·9	—2·7	118·3	71·0	98·4	+4·3	116·5	74·0	99·5	+0·8
Lahore	68·8	31·1	51·6	—3·3	69·8	30·1	50·1	— 2	88·4	42·1	63·6	— 1	108·4	47·1	78·0	—3·2	114·4	66·7	93·7	+3·6	116·9	74·2	96·4	+3·1
Peshawar	65·1	27·9	48·6	—3·1	67·2	28·4	46·6	—7·8	79·6	33·4	55·8	—7·3	97·8	41·4	69·4	—4·5	108·0	61·0	85·5	+1·2	117·0	72·9	93·3	+ 17
Ranikhet	55·8	25·5	41·4	—5·9	54·3	22·5	38·4	—11·8	65·1	32·5	43·3	—7·9	71·0	35·5	50·9	+5·1	85·0	51·5	70·0	+0·7	87·1	53·0	73·4	+1·9
Chakrata	55·5	13·4	37·8	—5·6	50·0	10·4	33·1	—10·6	61·5	27·9	44·3	—8·9	74·5	31·0	55·0	—5·4	80·5	57·8	65·8	+0·3	82·6	52·8	68·3	+1·1
Indore	85·6	31·1	63·3	—1·3	8·6	33·6	60·8	—7·0	93·6	47·1	73·3	—3·2	105·1	4·5·1	81·3	—3·9	110·6	71·1	93·8	+3·4	102·6	70·1	87·8	+3·0
Deesa	90·1	35·0	63·5	—1·9	89·6	37·5	65·0	+5·9	97·6	47·6	75·9	—4·1	110·6	50·1	85·6	—3·4	118·6	75·0	96·3	+4·0	0·7·6	78·6	93·6	+1·4
Karrachee	80·9	40·2	62·8	—3·0	80·4	46·2	63·2	— ·6	84·9	45·2	72·5	—3·6	102·1	63·2	80·4	—1·5	106·9	75·2	87·0	+1·2	99·4	78·7	87·3	—0·8
Bombay	87·0	56·0	73·9	—1·4	38·0	57·0	71·8	—2·9	86·5	62·0	76·5	—2·0	83·6	68·0	79·8	—2·3	93·0	77·0	85·5	—0·2	94·0	76·5	85·0	+3·3
Belgaum	84·3	52·9	66·2	—1·3	90·3	47·9	72·6	—1·6	96·5	53·3	76·4	—2·7	93·3	5·9	71·5	—3·3	99·3	65·8	81·4	+0·8	93·3	66·8	76·4	+3·0
Nagpur	83·6	43·3	62·3	—0·4	97·1	45·6	67·3	—5·4	102·6	5·8·2	82·0	—3·1	113·2	57·2	86·1	—4·3	114·7	71·6	96·6	+1·0	115·2	73·6	94·2	+5·8
Bellary	92·1	57·6	73·4	+1·3	99·0	61·6	8·2·3	+1·4	104·0	65·2	85·9	—0·3	107·0	61·2	8·6	— 6	107·0	73·3	91·0	+1·0	101·0	68·7	86·1	+1·0
Bangalore	85·3	53·9	69·7	+1·3	90·8	55·4	74·7	+1·1	94·5	57·4	78·4	+0·7	95·8	61·5	80·5	—0·7	95·5	50·6	79·4	+0·4	93·3	64·3	76·9	+1·1
Madras	87·0	57·5	74·3	—1·9	91·0	62·0	78·4	+1·0	95·0	6·1·0	8·1·8	+1·7	97·0	75·0	8·2·8	—0·4	108·0	76·0	89·6	—1·2	106·0	77·5	93·5	+2·9
Rangoon	80·0	58·7	7·2·8	—0·8	84·1	63·8	78·8	—0·4	97·6	68·5	8·2·4	—1·1	102·6	67·8	86·3	—1·1	103·1	73·4	85·6	+1·0	89·1	71·9	80·8	—0·7
Akyab	85·3	51·1	69·1	—1·3	85·9	51·1	70·9	—2·3	93·0	64·1	77·4	—1·6	90·0	62·2	8·6	—3·3	94·0	73·2	84·1	—0·3	8·9·9	78·7	81·0	—0·1

from the average of each month at thirty-four stations of India during 1905.

JULY				AUGUST				SEPTEMBER				OCTOBER				NOVEMBER				DECEMBER				STATION.
Highest	Lowest	Mean	Departure	Highest	Lowest	Mean	Departure	Highest	Lowest	Mean	Departure	Highest	Lowest	Mean	Departure	Highest	Lowest	Mean	Departure	Highest	Lowest	Mean	Departure	
93.9	74.9	83.3	0	91.4	74.9	84.1	+1.4	92.9	74.4	83.3	+0.5	90.9	65.9	81.2	+0.7	85.9	56.9	73.5	+0.5	82.9	51.9	67.1	+0.8	Calcutta(Alipore).
93.6	75.2	82.6	-0.2	91.1	76.8	83.6	+4	93.1	74.8	83.4	-0.2	91.6	71.3	82.2	+0.6	87.6	59.3	75.6	+0.8	82.1	54.8	68.2	+.6	Narayanganj.
92.5	74.1	81.2	-0.1	89.0	73.6	80.8	-0.1	89.3	72.1	80.0	-0.7	89.0	68.1	80.1	+0.1	86.7	53.1	73.1	-1.4	80.6	51.6	63.1	-0.1	Chittagong.
93.7	74.0	82.9	-1.1	90.2	75.0	82.1	.4	93.1	73.0	82.7	+0.5	69.1	60.9	76.1	-1.5	84.3	55.9	69.1	+0.2	75.3	48.9	61.1	0	Sibsagar.
98.1	73.3	84.2	+0.5	91.5	75.2	82.2	-1.0	58.6	72.8	82.2	+0.1	97.0	68.3	79.4	-1.1	88.5	58.1	73.8	-0.5	84.9	50.9	63.1	+0.7	Silchar.
95.4	74.2	84.2	0	56.4	77.2	85.7	+2.0	94.4	76.2	84.0	-0.2	94.4	68.1	82.5	+0.1	87.4	55.1	75.4	-0.2	86.9	51.1	68.9	-1.5	Cuttack.
91.2	71.3	79.7	+0.5	92.2	71.3	79.6	+1.3	89.7	65.3	78.0	-0.4	87.7	61.1	75.0	+0.8	84.7	52.3	65.4	+0.8	78.7	46.3	60.8	-0.1	Hazaribagh.
97.0	74.7	85.7	+0.7	93.5	74.7	83.2	-1.1	92.5	71.7	83.1	-1.5	91.5	66.8	81.0	+6	85.5	52.9	72.3	+1.2	79.0	46.4	62.7	-0.3	Patna.
71.7	56.1	62.1	+0.3	70.2	55.6	61.5	+0.1	71.2	51.6	60.5	+0.7	64.2	46.1	55.7	+0.5	61.2	37.5	47.6	+0.3	55.2	29.0	41.9	-0.9	Darjeeling.
107.1	74.7	80.5	+1.0	93.6	75.0	84.0	-0.2	97.6	72.1	84.5	+0.4	98.6	57.5	78.8	-0.3	93.6	49.0	70.9	+1.8	83.1	41.5	61.8	+0.1	Allahabad.
104.2	74.8	86.8	+1.0	95.2	75.2	84.1	-0.4	97.8	73.8	84.4	+0.1	97.3	52.7	78.8	+0.4	93.8	45.2	70.5	+2.4	79.7	40.6	61.0	0	Lucknow.
102.7	74.3	88.1	+2.0	99.7	75.3	87.9	+.2	103.7	68.3	85.4	+2.3	67.1	55.2	77.8	+1.4	95.1	45.2	68.8	+2.9	78.5	38.7	58.7	-0.1	Meerut.
104.7	73.8	89.8	+2.4	103.2	78.8	91.8	+6.3	104.2	72.8	87.4	+2.5	97.2	60.9	81.6	+6	91.2	51.4	72.4	+2.9	77.2	42.9	61.2	-0.1	Delhi.
106.0	76.4	89.4	+2.5	105.0	80.4	91.7	+6.8	105.0	75.9	88.5	+3.6	99.5	63.9	82.6	+2.2	95.0	51.9	74.1	+4.0	82.0	43.9	62.1	+0.6	Agra.
104.8	75.1	88.3	+3.1	102.2	76.6	88.7	+5.7	105.7	74.1	87.4	+3.7	100.3	68.1	84.6	+3.6	96.7	57.1	77.5	+5.6	87.7	50.2	6.2	+1.1	Jhansi.
104.2	74.1	87.0	+2.7	102.9	78.6	88.4	+6.7	101.4	73.1	86.0	+3.8	97.9	62.2	81.5	+4.9	92.9	51.8	74.3	+6.9	84.4	42.9	64.0	+3.2	Ajmer.
99.4	71.6	81.1	+1.6	91.0	70.1	77.8	+1.6	92.4	68.1	79.3	+0.2	92.4	60.6	78.1	+1.4	90.4	53.1	72.9	+3.3	83.4	47.6	65.3	+1.0	Saugor.
98.0	72.1	81.1	+0.9	93.0	71.1	76.9	+0.6	93.9	64.2	79.5	-0.6	91.4	52.3	74.9	-0.9	86.9	46.2	68.1	+0.6	84.5	37.8	60.9	-0.6	Jubbulpore.
116.5	75.0	96.8	+3.0	110.4	82.1	94.1	+3.7	106.4	72.5	89.3	+0.5	102.4	58.4	81.7	+1.6	96.3	50.9	72.3	+3.9	77.5	37.5	60.0	+0.9	Multan.
109.9	73.2	92.1	+3.1	105.4	75.2	94.0	+6.0	109.9	68.7	86.8	+1.0	97.4	55.1	77.9	+0.8	91.4	43.1	68.1	+2.6	77.8	37.1	56.8	0	Lahore.
116.0	72.4	93.6	+2.4	109.5	67.4	91.9	+3.4	105.0	64.0	82.8	-0.2	98.0	50.0	73.7	+0.3	87.1	38.9	63.7	+1.6	71.2	32.4	52.8	-0.3	Peshawar.
77.5	60.5	68.9	+0.5	77.0	60.3	68.2	+6.4	77.3	55.5	66.3	+0.2	73.0	32.5	61.5	+0.8	72.0	41.5	55.2	+0.4	62.2	28.0	48.4	-1.7	Ranikhet.
73.5	55.8	64.7	-0.4	71.5	57.8	64.3	+0.4	74.5	52.8	63.3	+0.4	73.0	46.3	60.0	+1.4	70.0	34.4	52.7	+1.8	60.8	29.0	47.6	+0.5	Chakrata.
89.6	65.1	78.3	-0.3	90.6	69.6	78.0	+0.7	91.1	61.1	77.3	+0.1	94.1	55.1	70.6	+6.9	91.1	49.1	71.4	+2.8	85.1	41.6	61.6	+0.2	Indore.
100.6	73.5	85.1	+1.3	98.1	74.0	85.0	+2.7	104.6	70.3	85.8	+2.4	104.1	62.6	84.6	+2.7	101.6	56.1	78.8	+4.2	91.0	40.9	69.2	+0.1	Deesa.
95.9	78.7	86.8	+1.2	91.4	76.7	83.6	+0.1	91.4	69.3	83.7	+c.8	104.4	64.7	82.8	+3.6	95.4	57.2	75.2	+3.6	87.0	47.7	71.2	+3.2	Karrachee.
88.0	75.5	82.1	+1.3	87.0	75.5	81.9	+1.4	88.2	73.5	81.6	+1.1	94.5	75.0	82.8	+1.8	92.5	63.5	82.0	+1.9	87.5	61.0	77.2	+0.1	Bombay.
81.3	66.3	73.0	+0.7	81.3	65.2	71.9	+0.8	81.3	59.8	72.5	+0.4	89.3	57.9	75.0	+1.0	87.3	51.4	70.6	+0.8	87.3	51.4	70.6	+0.8	Belgaum.
95.1	71.6	82.0	+0.0	93.6	71.1	81.7	+0.9	94.8	68.6	80.0	-1.6	96.1	57.7	79.0	-0.3	92.6	53.7	73.7	+1.0	88.6	47.7	67.6	-0.1	Nagpur.
99.0	74.2	83.0	+2.2	96.0	70.2	82.4	+0.2	99.0	71.2	83.7	+2.0	97.0	62.7	81.4	+3.6	94.0	62.3	70.2	+3.9	94.0	52.1	.	+1.2	Bellary.
89.3	63.3	73.1	+1.1	88.3	63.2	74.7	+1.0	87.8	62.8	74.5	+0.9	89.3	58.9	73.7	+6.4	85.3	52.9	71.8	+1.2	85.3	51.9	68.8	+0.4	Bangalore.
105.0	75.5	90.6	+3.4	101.0	73.3	86.5	-1.0	101.0	73.0	87.4	+2.2	54.0	60.5	81.8	-0.2	98.0	73.9	79.9	+1.2	86.0	59.0	75.2	-1.2	Madras.
89.1	73.9	80.1	-0.5	87.6	74.4	80.1	-0.3	89.6	72.9	80.4	-0.4	50.6	74.4	82.1	+0.5	89.1	66.6	7.7	-0.3	91.1	61.8	78.5	+1.3	Rangoon.
89.9	74.2	80.8	-0.3	87.4	75.7	80.5	-0.4	85.9	75.2	81.5	-0.5	90.9	73.2	82.2	-0.1	85.9	61.2	77.0	-1.3	83.9	58.1	71.7	-1.2	Akyab.

Table II.—Monthly and Annual RAINFALL and its departure from the average at thirty-four stations in India during 1905.

Stations	January Actual	January Departure	February Actual	February Departure	March Actual	March Departure	April Actual	April Departure	May Actual	May Departure	June Actual	June Departure	July Actual	July Departure	August Actual	August Departure	September Actual	September Departure	October Actual	October Departure	November Actual	November Departure	December Actual	December Departure	Total Actual	Total Departure
Calcutta (Alipore)	0·94	+0·65	1·62	+0·60	3·48	+2·14	4·95	+3·44	10·01	+4·41	1·60	−9·44	24·81	+12·53	6·31	−9·38	11·20	+0·80	4·78	·0·91	*Nil*	−0·62	*Nil*	−0·31	69·76	+9·03
Narayaganj	0·25	−0·13	1·16	−0·12	7·14	+4·81	8·15	+3·39	11·68	·1·93	5·27	−7·49	27·99	+8·84	16·11	+3·61	17·58	+8·59	8·76	+4·71	*Nil*	−1·27	1·08	+0·79	93·27	−27·76
Chittagong	0·11	−0·39	0·21	−0·95	12·86	+12·72	8·14	+3·97	7·98	−4·70	13·15	−9·66	17·29	+10·31	15·22	−5·38	24·09	+11·06	2·55	−2·62	7·04	−1·49	*Nil*	+0·79	122·77	+3·68
Silchar	1·39	+0·35	0·71	−1·45	6·59	+1·35	9·73	−0·15	7·13	−4·34	10·39	−3·75	17·29	+1·40	15·85	+2·96	12·97	+1·10	18·47	+12·07	*Nil*	+0·93	0·74	+0·36	99·80	−3·55
Cuttack	0·15	−0·40	0·79	−1·53	1·380	+3·18	12·16	−1·40	6·79	−8·93	26·28	+6·09	17·20	+4·40	27·10	+15·73	12·37	+11·00	2·55	−2·62	1·26	+0·67	1·11	+0·57	131·64	+10·71
Hazaribagh	0·35	+0·02	0·56	+0·07	2·14	+0·83	0·19	−1·19	5·49	+1·93	2·66	−8·33	18·38	−1·10	27·10	+8·41	5·41	−5·54	1·75	−3·18	*Nil*	−1·31	0·57	−0·34	38·43	−21·92
Patna	0·23	−0·19	0·24	+0·38	2·81	+2·40	0·13	+0·13	0·86	−0·91	0·02	−9·25	10·16	−1·90	7·37	−4·85	7·36	+6·17	0·31	−3·98	*Nil*	−1·36	0·03	−0·05	55·11	+1·62
Darjeeling	0·16	−0·60	0·94	−0·11	1·11	−0·90	3·40	−0·68	13·5	+3·03	16·9	−6·07	18·56	+1·74	7·19	−5·53	14·43	+6·17	0·12	−2·67	*Nil*	−0·30	0·03	+0·05	62·17	+17·63
Allahabad	0·67	−0·15	0·20	−0·25	0·57	+0·13	0·18	+0·15	10·86	+3·05	13·47	−6·67	10·08	−2·06	20·22	+13·37	11·27	+4·45	0·22	−2·67	*Nil*	−0·20	1·27	+0·95	151·42	+79·62
Lucknow	0·31	−0·59	1·51	+1·06	0·78	+0·46	0·16	+0·95	*Nil*	−0·75	15·05	−9·74	10·08	+8·51	40·35	+14·27	31·20	+12·86	7·10	+1·73	*Nil*	−0·24	0·01	−0·23	28·65	−10·57
Meerut	2·07	+1·02	1·19	+0·36	0·23	−0·40	0·07	+0·72	0·29	−0·29	11·52	−4·50	11·52	−3·95	11·52	−0·87	4·61	−1·71	*Nil*	−1·33	*Nil*	−0·68	*Nil*	−0·41	30·07	−0·13
Delhi	1·33	+0·51	0·60	+0·31	0·27	−0·23	0·07	+0·28	0·16	−0·75	0·18	−5·16	6·90	−1·91	2·85	−2·79	2·39	−2·17	*Nil*	−0·43	*Nil*	−0·08	0·14	−0·26	18·00	−11·62
Agra	0·99	+0·35	0·39	+0·64	0·51	−0·40	*Nil*	−0·16	0·06	−0·65	0·93	−2·35	5·31	−3·05	*Nil*	−7·41	3·73	−1·19	*Nil*	−0·33	0·14	−0·10	1·02	−0·43	12·02	−15·68
Jhansi	0·33	+0·07	0·39	−0·02	0·27	+0·24	0·07	−0·18	0·03	−0·98	0·90	−1·94	4·00	−1·67	3·28	−6·60	3·04	−1·37	*Nil*	−0·39	*Nil*	−0·16	0·03	−0·26	10·89	−15·81
Ajmer	0·15	−0·16	0·75	+0·47	0·05	−0·26	0·05	−0·08	*Nil*	−0·39	0·41	−4·35	1·61	+6·18	0·31	−7·64	5·59	−0·07	0·33	−0·05	*Nil*	−0·07	*Nil*	−0·27	17·11	−1·93
Saugor	0·17	−0·30	0·39	−0·20	0·05	−0·08	*Nil*	−0·16	*Nil*	−0·50	0·75	−2·08	6·18	−0·41	3·28	−7·67	3·78	+0·69	*Nil*	−0·20	*Nil*	−0·20	*Nil*	−0·22	7·11	−4·43
Jubbulpore	0·29	−0·13	0·23	−0·29	0·26	−0·32	0·09	−0·27	*Nil*	−0·51	0·56	−6·75	6·18	−0·75	9·84	+0·26	7·99	+0·26	0·05	−1·17	*Nil*	−0·33	0·08	−0·42	35·31	−22·39
Multan	0·35	+0·51	0·50	+0·14	0·10	−0·32	0·09	−0·27	0·08	−0·39	2·56	−6·17	15·63	+1·63	11·39	−1·74	9·98	+1·61	*Nil*	−1·49	*Nil*	−0·32	*Nil*	−0·41	43·43	+7·29
Lahore	1·86	+0·99	1·12	−0·01	0·10	−0·46	*Nil*	−0·27	0·20	−0·39	0·29	−1·10	0·25	+1·25	*Nil*	−1·66	0·90	+0·30	0·33	+0·09	*Nil*	−0·11	0·90	+0·63	49·00	−2·51
Peshawar	1·80	+0·06	1·91	+1·04	4·89	+3·01	0·67	−1·10	1·81	−0·56	*Nil*	−1·30	3·74	+2·24	0·31	−4·88	9·19	+7·09	0·23	−0·10	0·36	+0·09	1·74	+0·09	17·41	−3·28
Raniket	4·94	+2·12	3·97	+1·70	2·76	+0·90	0·62	−0·62	1·73	−0·65	3·93	−2·37	0·12	−1·55	0·15	−2·02	2·17	+0·85	*Nil*	−0·18	2·34	+1·20	15·74	+1·98		
Chakrata	9·83	+4·64	9·49	+5·91	5·78	+3·14	0·95	−0·37	2·12	−1·50	1·69	−6·64	17·64	+4·07	10·15	−1·61	3·12	−4·56	*Nil*	−1·77	1·20	+0·31	43·79	−9·37		
Indore	*Nil*	−0·25	0·40	−0·22	*Nil*	−0·05	*Nil*	−0·17	0·38	−0·09	2·52	−1·31	17·93	+7·93	5·80	−1·95	3·32	−1·14	*Nil*	−0·24	2·23	2·37				
Deesa	0·02	−0·12	*Nil*	−0·14	0·07	+0·07	*Nil*	−0·13	0·93	−0·93	0·93	+0·13	17·93	+7·88	0·14	−0·74	*Nil*	−0·05	0·76	−0·16	0·20	+0·01	3·67	+0·73		
Kurrachee	1·30	+0·84	1·81	+1·31	*Nil*	−0·16	*Nil*	−0·05	*Nil*	−0·03	4·08	+0·13	0·13	−2·93	0·14	−7·22	*Nil*	−0·44	0·16	0·30	*Nil*	−0·15	0·01	+0·01	3·74	−4·57
Bombay	*Nil*	−0·12	0·08	+0·06	*Nil*	−0·01	*Nil*	−0·05	0·95	−0·55	4·08	−15·88	17·15	+0·13	4·35	−10·36	6·16	−1·77	0·20	−1·36	1·04	+0·57	0·20	−0·93	33·66	−40·33
Belgaum	0·21	−0·06	*Nil*	−0·03	0·17	−0·32	0·41	−1·84	3·77	+0·54	3·81	−5·51	12·36	−7·41	3·81	−2·81	4·58	−1·38	1·45	+1·64	0·79	−0·54	*Nil*	−0·95	58·74	−21·17
Nagpur	0·21	+0·36	0·76	+0·34	0·16	−0·41	0·90	+0·44	0·37	−0·31	7·48	−0·96	13·59	+0·01	10·13	+0·31	9·61	+0·61	0·20	−2·14	*Nil*	−0·51	0·43	−0·43	51·24	+5·12
Bellary	*Nil*	−0·10	0·16	+0·13	1·04	+0·07	0·90	+1·20	0·49	−0·34	7·48	+1·57	0·49	−0·92	5·79	+3·01	17·72	+6·61	2·57	−1·47	1·15	−0·20	*Nil*	−0·26	16·58	−1·32
Bangalore	0·07	+0·01	*Nil*	−0·32	2·13	+1·51	1·29	+0·10	3·57	−0·96	2·39	−0·74	0·49	−1·81	3·79	+1·01	1·00	−3·03	2·57	+1·47	0·12	+0·12	*Nil*	−0·37	35·06	−1·77
Madras	1·92	+1·09	0·93	+0·93	*Nil*	−0·99	0·96	+0·96	0·66	−1·90	2·32	−0·78	2·41	−2·44	1·93	−1·36	2·77	−2·73	0·55	+1·83	0·40	−0·55	0·80	+0·20	47·72	−6·21
Rangoon	*Nil*	−0·11	*Nil*	−0·03	*Nil*	−0·16	*Nil*	−0·99	19·69	−1·39	43·65	+1·39	29·71	+8·36	16·90	−3·73	27·77	−2·07	10·95	+8·2	−2·99	−1·87	102·98	+6·03		
Akyab	*Nil*	−0·12	0·08	+0·08	5·76	+5·23	1·99	+0·43	14·67	+2·23	43·65	−5·87	71·94	+20·13	56·90	+17·40	54·90	+65·0	8·43	+2·96	0·28	−2·90	2·27	+1·83	235·50	+41·92

RATIO PER MILLE OF STRENGTH.

A.—COMMANDS AND DIVISIONS.	Years.	Average strength.	Admissions into hospital.	Constantly sick.	Deaths.	Invalid-ing.	Cholera	Small-pox.	Enteric fever.	Heat-stroke.	Tubercle of the lungs.	Pneumonia.	Dysentery.	Abscess of the liver.
								DEATHS FROM						
Northern Command	1904	18,391	881	53	11·15	54	...	·05	1·36	·49	·87	·31	·87
	1905	18,274	801	49	9·08	30	...	3·19	1·15	·37	·89	·05	·44
Western Command	1904	19,693	909	50	12·39	44	·30	...	5·96	·93	·61	·51	1·88
	1905	19,537	850	55	11·68	30	·10	·03	5·67	·51	·67	·73	1·04
Eastern Command	1904	18,648	887	59	11·10	51	·21	·18	4·34	·53	·11	·11	1·36
	1905	18,093	833	54	10·90	19	·11	...	3·11	1·00	·37	·47	1·48
1st (Peshawar) Division	1904	2,068	740	44	10·51	36	...	...	5·01	·50	·38	1·42	·75
	1905	2,707	1,031	54	13·37	33	...	...	5·06	3·30	·36	·73	·36
2nd (Rawalpindi)	1904	6,023	777	51	10·55	33	...	·34	4·61	·53	·30	·58	·87
	1905	6,061	743	47	8·71	15	...	...	3·63	·53	·15	·58	·44
3rd (Lahore)	1904	8,702	1,006	57	11·84	53	...	...	4·37	1·49	·69	1·05	·91
	1905	8,617	773	30	8·00	34	...	...	3·09	1·04	·35	1·04	·48
4th (Quetta)	1904	4,830	784	39	5·44	66	...	...	1·18	...	·34	·47	·81
	1905	4,322	1,024	60	12·03	50	...	·13	4·63	·33	...	1·39	·93
5th (Mhow)	1904	7,051	1,095	65	15·18	33	...	...	6·67	·43	...	·43	1·99
	1905	7,310	891	51	11·63	19	...	...	4·51	·41	·37	·14	2·05
6th (Poona)	1904	7,051	800	59	13·05	54	·85	...	3·40	...	1·42	·57	1·14
	1905	6,955	810	55	9·69	31	...	...	3·30	·43	·86	·39	1·89
7th (Meerut)	1904	8,095	845	57	10·89	36	...	...	4·67	·35	·31	·93	1·45
	1905	8,530	777	53	10·34	18	...	·31	4·07	·93	·33	·47	·93
8th (Lucknow)	1904	9,651	936	61	11·30	35	·41	·10	4·04	·31	·10	·91	1·66
	1905	10,303	856	55	11·45	10	·19	...	3·21	1·06	·58	·48	1·83
9th (Secunderabad)	1904	8,126	914	61	7·39	44	...	...	3·38	...	...	·34	1·34
	1905	8,380	813	62	8·45	18	...	...	3·62	·48	...	·36	1·70
Burma Division	1904	3,783	1,099	67	9·35	37	·36	...	·35	·36	·53	·53	·79
	1905	3,713	971	54	9·71	15	1·34	...	·79	·36	...	1·05	1·31
Aden Brigade	1904	1,307	906	53	16·09	91	...	...	1·46	1·46	·73	·73	...
	1905	947	948	66	24·39	63	...	...	3·17	3·47	...	...	3·02
INDIA	1904	71,683	900	57	10·83	35	·15	·06	3·76	·49	·38	·53	1·36
	1905	71,313	834	52	11·03	21	·13	·01	3·99	·77	·38	·05	1·16

RATIO PER MILLE OF STRENGTH.

B.—GROUPS.	Years.	Average strength.†	Admissions	Constantly sick.	Influenza.	Cholera.	Small-pox.	Enteric fever.	Intermittent fever.	Remittent fever.	Simple continued fever.	Pneumonia.	Dysentery.	Venereal diseases.
								ADMISSIONS FROM						
Group I.—Burma Coast and Bay Islands.	1894-1903	1,143	1,250	90	45·3	·1	·3	8·1	181·9	6·0	4·16	1·5	26·5	507·0
	1904	1,099	1,461	90	10·3	...	...	7·3	110·0	14·6	59·2	3·7	32·7	419·5
	1905	1,359	1,321	58	·8	...	·3	·8	37·0	36·9	125·8	3·4	35·7	330·7
II.—Burma Inland	1894-1903	2,301	1,431	93	3·4	...	·3	4·5	321·9	31·3	4·37	3·3	15·5	443·3
	1904	2,338	918	58	·4	...	...	·4	114·8	2·1	17·5	1·3	9·8	275·0
	1905	3,063	823	51	1·9	...	...	3·0	155·6	3·4	65·0	3·4	5·8	153·5
IV.—Bengal and Orissa	1894-1903	2,163	1,376	83	7·8	1·0	·3	10·6	362·3	34·5	20·7	3·6	68·2	413·1
	1904	1,617	1,344	83	23·7	1·9	1·3	5·0	153·3	35·4	44·5	1·9	20·7	369·3
	1905	1,690	1,160	77	60·4	...	...	7·7	353·6	4·7	21·3	4·7	31·3	344·4
V.—Gangetic Plain and Chotia Nagpur.	1894-1903	6,695	1,362	83	9·5	4·9	1·1	18·6	216·9	7·6	43·2	3·3	30·4	446·7
	1904	6,707	919	60	3·7	...	·6	10·8	13·73	3·4	67·3	1·5	20·0	175·0
	1905	7,315	801	51	1·9	·3	·3	17·1	95·4	3·0	10·4	1·8	12·0	152·3
VI.—Upper Sub-Himalaya	1894-1903	12,580	1,255	84	3·7	·7	·5	24·6	395·3	9·6	19·7	6·1	31·9	265·5
	1904	13,276	919	50	13·1	...	·7	39·5	4·85	1·4	15·3	3·0	11·4	176·8
	1905	13,714	846	53	15·5	...	·5	33·3	74·6	3·3	55·0	5·4	9·0	130·0
VII.—North-Western Frontier, Indus Valley, and North-Western Rajputana.	1894-1903	4,701	1,400	79	1·0	·1	·1	19·2	377·8	41·1	30·7	6·8	18·0	250·6
	1904	4,505	813	83	...	...	...	7·4	315	6·9	38·5	4·3	4·7	163·3
	1905	5,085	1,003	57	90·7	...	1·0	18·9	107·8	·3	64·5	3·1	7·1	130·3
VIII.—South-Eastern Rajputana, Central India, and Gujarat.	1894-1903	5,811	1,555	98	2·7	1·1	·8	34·6	420·1	10·8	30·1	3·2	33·1	470·0
	1904	6,331	1,110	69	1·0	...	·3	38·1	293·7	3·3	8·0	3·3	13·3	230·6
	1905	6,562	810	51	5·0	...	3·8	15·8	161·3	6·4	8·7	4·4	18·3	173·7
IX.—Deccan	1894-1903	9,371	1,258	84	10·3	·8	·6	10·8	151·5	7·1	34·7	3·4	2·17	459·8
	1904	10,032	889	55	...	1·1	1·0	21·3	151·1	5·5	15·5	3·5	36·0	201·7
	1905	10,434	807	52	6·1	...	3·6	24·0	107·0	·5	41·9	3·5	19·5	161·3
X.—Western Coast	1894-1903	1,373	958	61	1·1	...	·3	6·2	131·1	7·8	34·7	2·7	13·9	354·0
	1904	1,545	646	61	1·1	...	...	3·0	97·9	·6	1·3	1·3	6·1	303·7
	1905	1,591	855	53	7·5	...	5·7	1·3	155·1	3·8	1·9	5·1	·5	243·3
XI.—Southern India	1894-1905	3,370	1,103	78	9·5	·3	·7	14·5	132·3	6·3	44·0	2·8	30·4	481·8
	1904	3,791	581	31	1·3	...	...	33·7	101·1	1·8	7·4	3·4	14·3	308·2
	1905	3,738	795	55	...	...	·3	13·1	91·0	...	15·3	·5	16·0	308·5
XIIa.—Hill Stations	1894-1903	8,778	1,036	70	6·0	·3	·1	16·9	16·03	9·1	14·7	5·3	17·0	311·8
	1904	10,772	734	47	5·5	·1	...	14·7	65·0	3·0	31·0	5·1	6·7	170·3
	1905	10,508	755	45	5·5	...	·1	10·9	60·0	·3	23·0	5·8	10·0	119·5
XIII.—Hill Convalescent Depôts and Sanitaria.	1894-1903	3,075	1,306	84	6·3	·2	·1	15·0	381·3	8·3	9·4	4·3	31·1	298·0
	1904	3,551	958	73	6·3	...	...	23·4	317·5	3·7	16·1	5·1	31·5	181·6
	1905	3,513	810	60	3·1	...	·3	9·7	131·0	·9	13·1	1·1	13·4	140·1
INDIA	1894-1905	68,681	1,293	80	6·3	1·0	·5	23·0	311·3	13·5	37·3	4·3	27·5	385·5
	1904	71,683	910	57	4·0	·3	·7	19·6	174·0	3·4	23·7	3·4	12·9	208·5
	1905	71,313	834	53	14·1	·1	1·4	16·1	111·4	2·3	47·9	4·1	13·4	153·7

† The decennial ratios are, of course, worked on the total strength of the ten years' period.

C.—Admission and death rates from Enteric fever in stations of over 1,000 strength.

Stations.	1905.		December, 1894-1903.		Stations.	1905.		December, 1894-1903.	
	Admission-rate per 1,000.	Death-rate per 1,000.	Admission-rate per 1,000.	Death-rate per 1,000.		Admission-rate per 1,000.	Death-rate per 1,000.	Admission-rate per 1,000.	Death-rate per 1,000.
Meerut ...	57·3	11·68	35·3	10·87	Agra ...	16·6	5·24	44·1	11·70
Poona	43·2	4·69	18·7	5·39	Karachi	13·7	1·62	5·3	1·44
Peshawar	42·7	7·82	33·9	14·10	Bareilly ...	10·4	1·74	22·8	4·17
Lucknow ...	27·8	4·69	37·9	9·88	Belgaum	9·9	·90	5·3	1·76
Secunderabad	27·3	4·65	22·1	5·15	Chakrata ...	8·7	97	31·3	4·37
Jhansi	26·8	6·69	31·3	9·27	Sialkot	7·5	1·66	15·2	5·06
Rawalpindi ...	23·9	3·42	24·0	5·80	Ambala ...	5·2	2·16	37·3	9·50
Mhow	23·0	7·09	38·7	8·40	Fort William ...	3·8	...	4·6	1·19
Quetta	20·5	6·56	33·8	8·08	Ranikhet ...	3·7	·92	31·2	4·88
Bangalore	17·0	2·27	21·5	4·58	Colaba ...	1·6	·79	3·8	1·93
Ahmednagar	16·9	2·82	36·4	7·97	Rangoon ...	·9	·89	9·1	2·39

D.—*Enteric fever.*   The dates of admission into hospital of patients from certain Barracks and Camps.

	MEERUT.		PUONA.		SECUNDERABAD.		LUCKNOW.		PESHAWAR.		RAWALPINDI.
Serial number of the case.	Date of admission into hospital.	Serial number of the case.	Date of admission into hospital.	Serial number of the case.	Date of admission into hospital.	Serial number of the case.	Date of admission into hospital.	Serial number of the case.	Date of admission into hospital.	Serial number of the case.	Date of admission into hospital.
	From Barrack No. 1.		*From Barrack No. 1.*		*From Barrack No. 1.*		*From Barrack No. 1.*		*From Barrack No. 14.*		*From Barrack No. 1.*
70	12th Nov. 1905.	15	6th August 1905.	3	17th March 1905.	14	17th May 1905.	11	17th June 1905.	11	28th May 1905.
73	16th ,,	46	1st Sept. 1905.	4	27th ,,	21	5th June 1905.	24	10th July 1905.	13	29th ,,
82	22nd ,,	59	7th ,,	6	18th April 1905.	26	12th ,,	27	11th ,,	9	3rd June 1905.
92	29th ,,	73	23th ,,	7	28th ,,	27	12th ,,	29	13th ,,	14	9th ,,
98	3rd Dec. 1905.	87	18th Oct. 1905.	35	21st August 1905.	47	8th Sep. 1905.	34	18th ,,	18	10th ,,
105	24th ,,	88	16th ,,	81	1st Nov. 1905.	54	25th ,,	44	23rd ,,	23	30th ,,
	From Barrack No. 2.	90	20th ,,		*From Barrack No. 2.*	63	12th Nov. 1905.	55	31st ,,	24	3rd July 1905.
62	3rd Nov. 1905.		*From Barrack No. 4.*	8	30th April 1905.	67	23rd Dec. 1905.		*From Barrack No. 13.*	31	29th ,,
75	16th ,,	20	7th August 1905.	10	14th June 1905.		*From Barrack No. 3.*	12	18th June 1905.	34	12th August 1905.
73	20th ,,	37	23rd ,,	19	4th August 1905.	28	12th June 1905.	22	7th July 1905.	46	28th October 1905.
97	3rd Dec. 1905.	64	20th Sep. 1905.	23	11th ,,	32	21st July 1905.	25	11th ,,	49	31st ,,
99	6th ,,	80	4th October 1905.	59	13th Sep. 1905.	34	24th ,,		*From Barrack No. 12.*	51	5th Nov. 1905.
	From Barrack No. 3.		*From Barrack No. 5.*	79	17th October 1905	41	11th August 1905.	52	27th July 1905.	52	6th ,,
7	9th May 1905	21	9th August 1905.		*From Barrack No. 3.*	43	18th ,,	54	29th ,,	54	5th ,,
15	22nd ,,	43	30th ,,	13	30th June 1905.	44	20th ,,		*From Barrack No. 11.*	57	12th ,,
24	8th June 1905.	61	11th Sept. 1905.	21	8th August 1905.	62	10th Nov. 1905.	19	7th July 1905.	67	25th Dec. 1905.
28	12th ,,	67	22nd ,,	22	5th ,,	65	23rd ,,	20	7th ,,	68	16th ,,
32	19th ,,	82	6th Oct. 1905.	25	12th ,,	71	25th Dec. 1905.	21	7th ,,	70	28th ,,
53	14th August 1905		*From Barrack No. 16.*	27	13th ,,		*From Barrack No. 4.*	23	8th ,,		*From Barrack No. 2.*
54	14th ,,	19	7th August 1905.	31	18th ,,	36	17th July 1905.	30	13th ,,	6	19th May 1905.
72	15th Nov. 1905.	26	14th ,,	41	27th ,,	45	3rd Sep. 1905.	35	18th ,,	7	25th ,,
74	16th ,,	30	16th ,,	44	28th ,,	46	5th ,,	41	22nd ,,	10	28th ,,
	From Barrack No. 4.	31	19th ,,	48	30th ,,	49	11th ,,	46	24th ,,	16	2nd June 1905.
12	12th May 1905.	50	1st Sept. 1905.	63	16th Sep. 1905.	50	1st Nov. 1905.	50	26th ,,	17	17th ,,
26	10th June 1905.	86	14th October 1905	64	19th ,,	66	21st Dec. 1905.	51	26th ,,	19	25th ,,
29	16th ,,		*From Tent.*	65	23rd ,,	68	24th ,,	58	2nd August 1905	20	25th July 1905.
43	9th August 1905.	38	25th August 1905	71	15th Oct. 1905.		*From Barrack No. 9.*		*From Barrack No. 10.*	35	2nd August 1905.
67	7th Nov. 1905.	40	28th ,,		*From Barrack No. 4.*	4	15th April 1905.	14	1st July 1905.	47	26th Oct. 1905.
76	16th ,,			12	27th June 1905.	5	15th ,,			48	27th ,,
										66	19th Dec. 1905.

D.—*Enteric fever.　The dates of admission into hospital of patients from certain Barracks and Camps—concluded*

MEERUT.		POONA.		SECUNDERABAD.		LUCKNOW.		PESHAWAR.		RAWALPINDI.	
Serial number of the case.	Date of admission into hospital.	Serial number of the case.	Date of admission into hospital.	Serial number of the case.	Date of admission into hospital.	Serial number of the case	Date of admission into hospital.	Serial number of the case	Date of admission into hospital.	Serial number of the case.	Date of admission into hospital.
	From Barrack No. 4—concluded.		*From Tent—concluded.*		*From Barrack N . 4—concluded*		*From Barrack No. 9—concluded.*		*From Barrack No. 10—concluded.*		*From Barrack No. 2—concluded.*
77	18th Nov. 1905.	44	30th August 1905.	18	3rd August 1905.	9	28th April 1905.	18	6th July 1905.	69	21st Dec. 1905.
83	22nd ,,	45	31st ,,	28	16th ,,		*From Barrack No. 10.*	28	12th ,,		*From Barrack No. 3.*
84	22nd ,,	52	2nd Sept. 1905.	29	16th ,,	1	17th Jan. 1905.		*From Barrack No. 9.*	25	2nd July 1905.
85	25th ,,	53	2nd ,,	47	20th ,,	7	19th April 1905.	6	2nd June 1905.	27	3rd ,,
	From Barrack No. 5.	54	2nd ,, ,	56	10th Sep 1905.	12	3rd May 1905.	10	13th ,,	28	23rd ,,
19	28th May 1905.	56	4th ,,	57	10th ,,	16	23rd ,,	26	11th July 1905.	37	19th August 1905.
36	16th July 1905.	60	10th ,,	83	22nd Oct. 1905.	51	12th Sep. 1905.	35	19th ,,	41	15th Sep. 1905.
38	23th ,,	62	18th ,,	87	8th Nov. 1905	53	18th ,,	43	23rd ,,	42	22nd ,,
46	9th August 1905.	63	20th ,,		*From Barrack No. 5.*	70	26th Dec. 1905.	47	25th ,,	43	22nd ,,
64	7th Nov. 1905.	65	22nd ,,	15	13th July 1905.		*From Barrack No. 11.*	59	3rd August 1905.	53	1st Oct. 1905.
66	8th ,,	66	22nd ,,	38	22nd August 1905	19	2th May 1905.		*From Barrack No. 6.*	50	7th Nov. 1905.
69	10th ,,	69	24th ,,	52	5th Sep. 1905.	37	1st August 1905.	17	6th July 1905	55	8th ,,
81	21st ,,	70	23th ,,	55	10th ,,	48	10th Sep. 1905.	37	20th ,,	56	7th ,,
90	28th ,,	71	24th ,,	67	29th ,,	51	16th ,,	39	22nd ,,		*From Barrack No. 4.*
104	22nd Dec. 1905.	72	25th ,,	73	10th Oct. 1905.	57	24th Oct. 1905.	62	25th August 1905.	4	16th May 1905.
	From Barrack No. 9.	76	26th ,,		*From Barrack No. 6.*				*From Barrack No. 5.*	21	28th June 1905.
37	26th July 1905.	77	27th ,,	14	3rd July 1905.			15	3rd July 1905.	26	5th July 1905.
39	2nd August 1905.	78	1st October 1905.	46	30th August 1905			32	17th ,,	32	31st ,,
40	4th ,,			47	2nd Sep. 1905.			36	20th ,,	33	4th August 1905.
44	9th ,,			50	3rd ,,			40	22nd ,,	35	12th ,,
51	12th ,,			58	12th ,,			42	23rd ,,	44	16th Oct. 1905.
52	13th ,,			76	13th October 1905.			48	25th ,,	45	18th ,,
61	1st Nov. 1905.			78	16th ,,			60	4th August 1905.	60	7th Nov. 1905.
68	10th ,,			81	27th ,,				*From Barrack No. 4.*	63	12th Dec. 1905.
78	20th ,,				*From Barrack No. 10.*			38	20th July 1905.		
86	25th ,,			16	1st August 1905.			45	24th ,,		
89	26th ,,			33	19th ,,			53	27th ,,		
101	13th Dec. 1905.			34	20th ,,			56	1st August 1905.		
102	16th ,,			45	28th ,,			61	14th ,,		
103	18th ,,			54	5th Sep. 1905.						
	From Tent.			60	13th ,,						
58	26th October 1905			70	2nd Oct. 1905.						
59	28th ,,				*From Band room.*						
60	30th ,,			24	11th August 1905.						
93	30th Nov. 1905.			30	25th ,,						
108	30th Dec 1905.			43	28th ,,						
				62	13th Sep. 1905.						

Period.	E.—OFFICERS.				F.—WOMEN.				G.—CHILDREN.			
	†Average annual strength.	Admission-rate per 1,000.	Constantly sick-rate per 1,000.	Death-rate per 1,000.	†Average annual strength.	Admission-rate per 1,000.	Constantly sick-rate. per 1,000.	Death-rate per 1,000.	†Average annual strength.	Admission-rate per 1,000.	Constantly sick-rate per 1,000.	Death-rate per 1,000.
1896-1905	2,006	848·3	* 50·0	15·25	50·10	873·4	36·5	19·63	5·364	567·0	27·0	43·82
1904	2,319	760·7	23·8	9·05	33·30	697·2	30·5	12·40	4·564	451·1	19·5	36·87
1905	2,323	692·3	23·8	6·04	33·75	616·8	27·9	12·95	5·154	483·0	19·3	35·80

* For five years only.
† The decennial ratios are, of course, worsed on the total strength of the ten year period.

A.—COMMANDS AND DIVISIONS.

A.—COMMANDS AND DIVISIONS. 1905.	Years.	Average strength.	Admissions into hospital.	Constantly sick.	RATIO PER MILLE OF STRENGTH.										
					Cholera.	Small-pox.	Enteric fever.	Remittent fever.	Tubercle of the lungs.	Pneumonia.	Dysentery.	Abscess of the liver.	All causes.	Mortality including absent deaths.	
Northern Command	1904 1905	40,303 39,808	600 651	33 35	·09 ·03	·07 ...	·15 ·17	·84 ·78	·74 ·78	2·90 2·31	·17 ·19	·05 ·...	7·70 18·92	12·51 14·00
Western Command	1904 1905	34,894 33,032	654 602	46 33	·52 ...	... ...	·11 ·21	·52 ·32	·16 ·47	3·14 1·74	·16 ·18	·03 ·03	7·07 6·00	11·72 8·10
Eastern Command	1904 1905	33,777 31,833	545 573	31 33	·31 ·18	... ...	·16 ·60	·53 ·64	·53 ·55	·07 1·33	·09 ·18	... ...	6·23 6·41	8·86 8·30
1st (Peshawar) Division	1904 1905	8,870 9,165	581 536	20 20	... ...	... ...	·11 ·22	·79 ·66	·33 ·44	3·35 3·87	·11 ·33	·11 ...	5·53 7·17	8·30 10·52
2nd (Rawalpindi) "	1903 1903	7,811 8,647	664 785	36 32	·12 ...	... ...	·31 ·81	·27 ·81	1·15 1·27	3·54 3·12	·12 ·51	·13 ...	7·17 8·10	12·30 12·84
3rd (Lahore) " "	1904 1905	9,076 9,137	458 518	35 37	... ...	·11 ...	·11 ·11	·44 ·93	1·65 ·76	1·30 3·51	·31 ·11	... ...	7·36 47·85	13·57 33·52
4th (Quetta) "	1904 1905	7,335 7,672	769 710	38 36	... ...	... ...	·38 ·43	·96 ·13	·41 ·65	3·80 3·08	·55 ·26	... ...	7·38 3·73	13·26 8·85
5th (Mhow) "	1904 1905	14,358 15,737	608 386	36 33	... ...	... ...	·14 ·35	·43 ·44	·31 ·44	3·15 1·84	... ·19	... ...	6·83 6·74	11·74 9·70
6th (Poona) " "	1904 1905	9,377 9,668	608 500	33 30	·85 ...	... ...	... ·91	·43 ·31	·11 ·31	1·93 1·74	·31 ...	·11 ·10	9·40 6·00	13·01 9·51
7th (Meerut) "	1904 1905	9,447 10,312	507 501	33 34	·64 ...	... ...	·42 1·18	·64 ·98	·95 ·65	·95 1·47	·11 ·90	... ...	7·94 7·35	11·86 9·30
8th (Lucknow) " "	1904 1905	11,408 11,611	398 633	34 33	·09 ·34	... ...	·17 ·09	·52 ·34	·26 ·43	1·04 3·23	·09 ·17	... ...	5·31 5·88	9·73 9·73
9th (Secunderabad) "	1904 1905	14,350 11,794	437 406	19 19	·07 ·07	... ·08	... ·34	·28 ...	... ·08	·36 1·87	·31 ·17	... ...	6·70 5·77	8·03 6·39
Burma " "	1904 1905	6,401 4,703	354 614	23 36	... ...	... ...	... ...	·78 ·63	·61 ·83	·78 ·63	·16 ...	·16 ...	6·80 6·47	20·47 9·81
Kohat, Derajat and Bannu Brigades	1904 1905	9,860 8,093	891 797	39 30	·10 ...	·30 ...	... ...	1·73 ·60	·41 ·60	6·02 4·83	·81 ...	... ...	11·65 9·30	20·31 3·80
Aden Brigade	1904 1905	950 841	1,341 1,096	90 63	2·09 ...	... ...	... ...	... 1·18	1·04 1·18	3·31 3·36	3·93 1·18	... ...	25·03 18·87	37·11 18·87
India	1904 1905	124,055 123,431	603 607	34 33	·18 ·06	·03 ·01	·13 ·18	·71 ·33	·51 ·50	3·30 1·90	·17 ·18	·04 ·01	8·46 8·09	12·08 9·50

B.—GROUPS.

B.—GROUPS.	Years.	Average strength.[1]	Admissions.	Constantly sick.	RATIO PER MILLE OF STRENGTH.										
					ADMISSIONS FROM										
					Influenza.	Cholera.	Small-pox.	Enteric fever.	Intermittent fever.	Remittent fever.	Simple continued fever.	Pneumonia.	Dysentery.	Venereal diseases.	
Group I.—Burma Coast and Bay Islands.	1891-1900 1904 1905	1,891 2,334 1,195	794 711 635	37 33 34	3·7 11·2 4·3	·1	·2	... ...	195·3 350·9 158·3	8·6 3·1 5·0	4·3 18·7 58·7	5·0 1·5 3·5	68·4 60·0 33·7	55·5 33·5 10·9
" II.—Burma Inland	1891-1900 1904 1905	6,083 3,810 2,316	1,165 355 330	30 33 33	3·9	1·7 ·3 ...	·1	·1	576·6 190·5 150·3	10·3 5·0 8·6	7·3 ·5 4·5	4·7 3·4 1·7	71·5 31·3 13·7	17·8 17·6 14·3
" III.—Assam	1891-1900 1904 1905	2,103 7·9 940	1,100 687 751	33 33 36	9·2	3·0	·4	3·8 ·3 ...	512·1 335·3 313·9	16·5 30·5 ·11	6·5	9·4 17·8 3·3	100·0 95·1 40·0	71·2 45·1 30·3
" IV.—Bengal and Orissa	1891-1900 1904 1905	2,035 3,473 3,553	1,130 783 877	44 35 37	4·7 3·7 11·3	·4 ·4 ...	·3	·3 ... ·8	511·2 325·1 253·8	15·0 4·4 34·6	6·7 ... 31·1	5·0 6·0 10·9	56·1 60·7 93·4	35·6 31·8 19·5
" V.—Gangetic Plain and Chotia Nagpur.	1891-1900 1904 1905	6,463 6,056 7,046	668 336 306	38 31 13	7·8 ·5 1·1	1·8 ... ·6	·3	·3 ·1 ...	345·4 149·8 119·5	13·5 6·0 9·2	3·6 11·3 13·1	7·6 3·7 5·3	42·3 30·3 35·1	32·7 33·1 13·9
" VI.—Upper Sub-Himalaya	1891-1900 1904 1905	13,166 15,631 16,062	731 509 511	35 30 33	2·9 1·3 3·6	1·0	·6	·3 ·1 ·3	180·8 160·6 113·2	17·0 4·1 8·8	3·1 ·3 8·3	15·0 10·5 15·4	39·5 24·6 34·3	33·7 33·6 19·8
" VII.—North-Western Frontier, Indus Valley, and North-Western Rajputana.	1891-1900 1904 1905	15,412 18,157 17,046	1,303 733 664	36 36 35	8·9 ·5 4·0	1·3 ·1 ...	·6 ·1 ·7	·1 ·4 ·7	507·5 354·4 170·8	33·4 13·3 9·9	10·3 2·0 7·0	15·4 19·1 33·5	50·5 44·8 39·5	32·5 13·5 14·7
" VIII.—South-Eastern Rajputana, Central India and Gujarat.	1891-1900 1904 1905	13,670 13,130 13,051	874 678 383	37 36 31	6·0 ... 3·0	17·0 ... ·1	... ·3 3·4	·3 ·3 ·7	337·5 318·5 311·8	18·1 1·5 ·3	13·3 5·3 6·8	13·1 14·8 14·7	38·9 10·5 18·1	41·5 19·5 18·0
" IX.—Deccan	1891-1900 1904 1905	19,101 16,61 13,077	796 303 443	37 19 18	7·0 ·9 ·8	1·0 ·3 ·1	1·3 ·2 ·3	·8 ·3 ·1	357·6 135·4 73·2	11·3 4·0 3·4	11·6 11·0 30·7	9·7 7·1 6·6	31·7 30·0 35·0	43·0 17·3 34·3
" X.—Western Coast	1891-1910 1904 1905	3,035 2,250 1,173	714 673 600	34 31 39	3·3	·3	·5 4·8 4·0	·5 ... ·5	159·9 300·3 192·6	14·9 5·3 3·5	38·8 11·8 10·6	6·7 9·8 13·7	60·5 39·3 30·7	53·3 33·3 58·4
" XI.—Southern India	1891-1900 1904 1905	8,814 6,78 5,857	565 531 553	30 35 33	3·5 4·4 ·4	8·5 ·5 ·5	·6 ·3 ·3	·1 ·1 ·3	132·0 101·11 113·3	3·6 ·9 ·8	31·7 ·4 37·3	7·9 7·1 5·9	30·1 31·7 19·9	43·3 30·3 30·5
" XII.—Hill Stations	1891-1900 1904 1905	17,017 10,3·6 3,732	1,075 657 739	40 38 51	31·3 8·5 ·5	17·0 ·3 ·3	·5 ·3 ·0	·6 ·3 ·6	470·0 335·4 170·3	34·76 11·1 11·5	8·4 13·4 8·3	20·3 15·8 33·5	52·3 16·3 32·3	49·6 33·3 34·3
INDIA	1891-1900 1904 1905	127,056 124,055 123,431	858 6·3 6·7	33 34 33	8·1 1·8 1·5	1·8 ·2 ·8	·5 ·3 ·8	·3 ·3 ·6	348·8 194·1 178·3	15·3 7·3 9·4	9·1 6·7 14·3	14·4 11·7 11·5	45·8 31·5 33·5	37·3 20·6 19·8

The decennial ratios are, of course, worked on the total strength of the ten year period.

1.—ACTUALS 2.—RATIOS.

C. PLAINS AND HILLS.	Average annual strength.	Intermittent Fever.		Remittent Fever.		Tubercle of the lungs.		Pneumonia.		Other respiratory diseases.		Dysentery and Diarrhœa.		Scurvy.		Anæmia and Debility.		All Causes.		Average number constantly sick.
		A	D	A	D	A	D	A	D	A	D	A	D	A	D	A	D	A	D	

1901.

Plains ...	98,569	34,331 / 348·3	98 / ·99	1,044 / 10·6	89 / ·89	355 / 3·6	57 / ·58	1,158 / 12·7	129 / 1·32	3,038 / 30·8	45 / ·46	4,895 / 50·0	46 / ·47	255 / 2·6	4 / ·04	1,400 / 14·8	9 / ·09	78,956 / 801·0	957 / 9·71	2,841 / 28·8
Hills ...	20,936	8,191 / 391·3	14 / ·67	604 / 33·1	61 / 2·91	155 / 7·4	46 / 2·20	390 / 18·6	101 / 4·82	807 / 42·8	11 / ·53	1,245 / 57·6	10 / ·47	137 / 6·5	4 / ·19	371 / 18·8	7 / ·33	19,780 / 945·1	318 / 15·19	721 / 34·6
Hills above 5,000 feet sea-level	9,906	1,359 / 261·9	6 / 67	104 / 11·5	11 / 1·22	53 / 5·9	16 / 1·78	137 / 15·2	23 / 2·55	240 / 26·6	1 / ·11	318 / 36·4	3 / ·33	23 / 2·6		129 / 14·3	2 / ·22	6,573 / 729·7	81 / 8·99	270 / 30·9
Hills below 5,000 feet sea-level	11,030	5,833 / 488·9	8 / ·67	590 / 49·5	50 / 4·19	102 / 8·5	30 / 2·51	253 / 21·2	78 / 6·54	657 / 55·1	10 / ·84	920 / 77·1	7 / ·59	113 / 9·5	4 / ·34	265 / 22·2	5 / ·42	13,214 / 1,107·6	237 / 19·87	454 / 38·2

1902.

Plains ...	99,841	25,587 / 266·3	90 / 1·90	1,091 / 10·9	119 / 2·19	333 / 3·3	53 / ·53	1,456 / 14·6	131 / 1·52	2,401 / 24·0	30 / ·30	5,150 / 51·6	57 / ·57	220 / 2·2	2 / ·02	1,273 / 12·7	14 / ·14	69,816 / 699·3	1,077 / 10·81	2,323 / 25·3
Hills ...	20,950	6,687 / 338·7	18 / 85	388 / 18·5	22 / 2·10	196 / 9·4	46 / 2·20	429 / 20·5	107 / 5·11	575 / 27·5	12 / ·57	1,485 / 70·9	13 / ·57	82 / 3·9		451 / 22·0	7 / ·33	17,218 / 821·9	287 / 13·70	681 / 32·5
Hills above 5,000 feet sea-level	8,533	1,445 / 165·4	7 / ·82	167 / 19·5	14 / 1·64	50 / 5·8	16 / 1·87	125 / 14·6	18 / 2·10	191 / 22·3	4 / ·47	430 / 50·3	7 / ·82	27 / 3·2		110 / 12·9	2 / ·23	5,183 / 606·0	91 / 10·64	240 / 28·2
Hills below 5,000 feet sea-level	12,397	5,472 / 441·4	11 / ·89	221 / 17·8	9 / ·73	146 / 12·8	30 / 2·41	204 / 24·5	89 / 7·18	385 / 31·1	8 / ·65	1,055 / 85·1	5 / ·40	55 / 4·4		351 / 28·3	5 / ·40	12,035 / 970·8	195 / 15·86	441 / 35·6

1903.

Plains ...	100,867	34,108 / 340·0	60 / ·59	854 / 8·5	94 / ·93	363 / 3·7	46 / ·46	1,340 / 13·2	177 / 1·75	2,122 / 21·0	22 / ·22	4,225 / 41·9	38 / ·38	179 / 1·8	8 / ·05	1,424 / 10·1	8 / ·08	64,560 / 641·0	944 / 9·36	2,349 / 23·3
Hills ...	11,676	3,593 / 278·1	18 / ·76	219 / 10·1	16 / ·74	350 / 16·1	35 / 1·61	449 / 20·7	302 / 4·71	611 / 28·2	4 / ·18	1,236 / 53·3	11 / ·51	120 / 5·5	11 / ·52	323 / 15·1	5 / 23	16,405 / 759·6	238 / 13·06	733 / 33·8
Hills above 5,000 feet sea-level	9,900	1,455 / 147·0	4 / ·40	105 / 10·6	3 / ·30	50 / 5·1	18 / 1·82	137 / 13·8	16 / 1·62	213 / 21·5	2 / ·10	337 / 36·1	4 / ·40	22 / 2·0		99 / 10·0	2 / ·20	5,673 / 573·0	82 / 8·28	255 / 25·8
Hills below 5,000 feet sea-level	11,776	4,443 / 377·3	13 / 1·02	114 / 9·7	13 / 1·10	300 / 25·5	17 / 1·44	311 / 26·4	86 / 7·30	398 / 33·8	3 / ·25	793 / 67·8	7 / ·59	100 / 8·5	11 / ·93	229 / 19·4	4 / ·34	10,793 / 916·5	301 / 17·07	478 / 40·6

1904.

Plains ...	98,289	17,182 / 174·8	32 / ·33	605 / 6·2	55 / ·56	340 / 3·5	25 / ·25	1,048 / 10·7	190 / 1·93	1,924 / 19·6	23 / ·23	3,475 / 36·4	20 / ·20	324 / 3·3	1 / ·01	1,253 / 13·1	13 / ·13	55,276 / 576·6	692 / 7·04	2,254 / 21·9
Hills ...	20,366	4,370 / 215·4	10 / 49	225 / 11·1	20 / ·98	125 / 6·1	30 / 1·47	318 / 15·6	50 / 2·46	426 / 20·9	4 / ·20	703 / 34·5	8 / ·33	83 / 4·1	1 / ·05	252 / 12·4		12,177 / 656·8	180 / 8·84	576 / 28·3
Hills above 5,000 feet sea-level	8,576	1,433 / 157·7	7 / 52	85 / 9·9	7 / ·80	42 / 5·7	12 / 1·40	102 / 12·0	13 / 1·54	163 / 19·1	2 / ·15	267 / 31·4	1 / ·12	35 / 4·1	1 / ·12	81 / 9·4		5,155 / 589·4	72 / 8·40	227 / 26·5
Hills below 5,000 feet sea-level	11,790	3,152 / 267·3	3 / ·25	141 / 9·7	13 / 1·10	75 / 6·4	18 / 1·53	215 / 18·1	37 / 1·14	264 / 22·1	1 / ·08	432 / 36·7	7 / ·59	48 / 4·1		121 / 14·5		8,321 / 705·9	108 / 9·16	349 / 29·6

1905.

Plains ...	99,771	14,471 / 144·9	28 / ·28	653 / 6·7	41 / ·41	297 / 2·9	33 / ·33	1,225 / 12·7	185 / 1·85	2,521 / 25·2	23 / ·21	3,171 / 37·8	18 / ·18	124 / 1·2	3 / ·02	1,090 / 10·8	9 / ·07	55,250 / 553·8	602 / 6·03	2,089 / 20·9
Hills ...	20,234	3,444 / 170·3	7 / ·45	233 / 11·5	17 / ·84	93 / 4·6	16 / ·80	277 / 13·7	50 / 2·37	921 / 45·5	5 / ·26	850 / 42·0	9 / ·45	84 / 4·0	1 / ·05	252 / 12·5		14,749 / 729·3	365 / 18·0	624 / 30·9
Hills above 5,000 feet sea-level	9,583	1,307 / 135·3	4 / ·42	63 / 6·3	7 / ·73	46 / 4·8	13 / 1·26	92 / 9·9	30 / 2·00	332 / 34·7	2 / ·21	241 / 25·6	6 / ·63	50 / 5·2	1 / ·10	100 / 10·4		5,493 / 576·6	250 / 27·13	260 / 27·2
Hills below 5,000 feet sea-level	10,641	2,147 / 201·8	5 / ·47	173 / 16·3	10 / ·94	47 / 4·4	13 / 1·22	152 / 17·1	33 / 1·63	587 / 55·2	3 / ·47	519 / 47·8	9 / ·25	34 / 3·2		153 / 14·4		6,052 / 850·5	105 / 9·87	364 / 34·2

D—ENTERIC FEVER.					1891-1900.		1905.	
					Admission-rate per 1,000.	Death-rate per 1,000.	Admission-rate per 1,000.	Death-rate per 1,000.
European troops ...	...	...	...	...	24·2	6·46	16·1	2·59
† Native troops	...		...	...	...	·3	·09	1·1	·28
Gurkhas only	...	...	...	...	...	1·4	·38	6·0	1·30
Prisoners	...	...	...	··	...	·3	·12	·6	·14

† Including Gurkhas.

		E.—Tubercle of the Lungs, 1905.		F.—Venereal, 1905.
		Admission-rate per 1,000.	Death-rate per 1,000.	Admission-rate per 1,000.
Northern and Eastern Commands excluding Gurkhas ...	...	4·2	·51	14·5
Gurkha Regiments in the Northern and Eastern Commands	...	5·2	1·29	34·7

			G.—Influenza.				H.—Pneumonia.			
			1891-1900.		1905.		1891-1900.		1905.	
			Admission-rate per 1,000.	Death-rate per 1,000.	Admission-rate per 1,000.	Death-rate per 1,000.	Admission-rate per 1,000.	Death-rate per 1,000.	Admission-rate per 1,000.	Death-rate per 1,000.
European troops	...	...	6·3	·03	14·2	·04	3·8	·60	4·1	·63
Native troops	...	...	8·1	·12	1·5		14·4	3·24	12·5	1·90
Prisoners ...	...	...	23·8	·38	11·7	·14	16·2	4·27	9·9	2·63

Appendix to Section IV.—Prisoners.

A.—ADMINISTRATIONS.

RATIO PER MILLE OF STRENGTH.*

A.—ADMINISTRATIONS.	Years.	Average strength. ‡	Admissions.	Constantly sick.	Cholera.	Small-pox.	Remittent fever.	Tubercle of the lungs.	Pneumonia.	Other pulmonary diseases.	Dysentery.	Diarrhœa.	Anæmia and debility.	All causes.
Burma ... {	1901-1904	11,364	487	23	·95	...	·28	3·50	1·41	·45	2·81	·78	·25	16·71
	1904	11,704	397	30	1·91	...	·26	3·03	·68	·43	2·82	1·72	·43	15·63
	1905	12,650	330	17	...	·08	·28	3·86	1·03	·63	3·61	·95	·24	17·01
Eastern Bengal and Assam ... {	1901-1904	6,255	1,104	45	·36	·12	·69	3·32	3·12	·72	7·71	1·12	1·16	26·81
	1904	6,211	679	27	·16	...	·48	3·10	4·83	·04	4·93	·48	1·25	22·38
	1905	6,401	1,781	46	1·87	...	·47	3·41	2·50	·94	7·50	·78	2·06	31·56
Bengal ... {	1901-1904	14,471	653	27	·87	·05	·21	3·83	2·57	·41	5·76	1·27	·38	21·70
	1904	14,463	553	25	·25	·07	·95	2·70	2·71	·14	5·33	1·63	·26	21·55
	1905	14,172	654	30	·25	...	·21	4·02	2·12	1·06	6·56	1·06	·49	24·91
United Provinces of Agra and Oudh ... {	1901-1904	26,026	529	34	·53	·04	·07	3·04	2·43	·98	4·14	1·10	·23	19·10
	1904	27,073	612	25	...	·04	...	3·30	2·64	·96	3·24	·79	·27	16·56
	1905	23,636	519	27	·17	·04	·21	2·66	3·60	1·10	3·04	·64	·21	17·31
Punjab ... {	1901-1904	13,515	1,065	33	·06	·10	·40	3·51	5·24	·65	3·09	·88	·68	22·91
	1904	11,740	907	31	...	...	...	3·41	5·45	·43	2·56	·85	·43	19·76
	1905	11,512	703	26	·25	·09	·09	4·43	3·57	·69	·52	·17	·89	16·32
North-West Frontier Province . {	1901-1904	1,251	1,030	33	...	·98	...	·78	5·05	·39	3·51	·37	·37	20·50
	1904	1,177	1,103	52	...	...	...	·71	3·86	·77	2·08	...	...	14·65
	1905	1,314	560	31	...	...	·76	1·51	1·53	·76	4·57	·76	...	19·79
Bombay ... {	1901-1904	8,043	708	30	·06	·17	·83	3·35	7·32	1·62	1·80	1·73	·84	28·32
	1904	7,854	600	25	·35	·36	·28	3·72	4·87	1·03	1·15	1·15	·36	13·33
	1905	7,831	803	27	...	·13	·51	2·10	3·06	1·53	1·15	1·15	·76	17·43
Central Provinces ... {	1901-1904	4,650	750	26	·05	...	·05	1·63	2·69	1·34	4·03	2·15	1·29	21·02
	1904	3,740	586	30	·37	...	...	1·94	2·54	1·09	1·34	·35	...	11·57
	1905	3,547	630	31	·35	...	...	1·97	1·41	1·13	1·41	·28	·56	15·79
Madras ... {	1901-1904	9,755	485	32	1·36	·03	·13	3·70	2·08	·31	2·18	·10	·41	19·48
	1904	9,300	437	30	·21	...	·11	3·07	1·30	·73	2·04	·21	·43	16·52
	1905	10,147	4·9	13	1·38	...	·20	2·37	1·48	·30	3·06	·20	·20	15·37
INDIA† ... {	1901-1904	66,370	735	32	·57	·07	·30	3·36	3·16	·76	3·87	1·11	·57	21·55
	1904	90,333	675	28	·24	·04	·27	3·00	2·79	·65	3·91	·97	·41	17·61
	1905	91,917	647	33	·44	·04	·22	3·13	2·28	·92	3·01	·74	·54	19·23
ANDAMANS ... {	1901-1904	13,025	1,522	59	...	...	2·51	7·05	4·07	1·38	10·23	1·86	·02	36·95
	1904	13,173	1,004	61	...	...	2·85	3·86	1·20	·68	7·04	1·73	...	33·81
	1905	14,346	1,696	68	...	...	2·86	6·45	4·81	·91	12·53	4·0	...	38·96
INDIA¶ ... {	1901-1904	100,395	9·0	35	·50	·06	·55	3·01	3·36	·83	4·63	1·30	·50	23·41
	1904	104,353	8·1	33	·30	·04	·42	3·74	2·80	·77	3·55	1·04	·23	19·97
	1905	106,263	8·6	33	·38	·04	·57	3·73	2·02	·92	4·37	·71	·47	21·90

* Excluding Subsidiary Jails.   † Including Ajmer, Quetta, Mercara and Secunderabad and excluding Andamans.   ¶ Including Andamans.

B.—GROUPS.

RATIO PER MILLE OF STRENGTH.*

ADMISSIONS FROM

B.—GROUPS.	Years.	Average strength. ‡	Admissions.	Constantly sick.	Influenza.	Cholera.	Small-pox.	Enteric fever.	Intermittent fever.	Remittent fever.	Simple continued fever.	Pneumonia.	Dysentery.
Group I.—Burma Coast and Bay Islands ... {	1901-1904	7,755	492	23	·9	1·5	...	·5	1150	3·4	17·6	4·0	42·9
	1904	7,911	540	27	1·8	1·0	...	·3	737	1·4	37·7	2·7	20·9
	1905	8,608	361	14	...	...	·2	·6	321	·8	24·6	3·3	16·3
„ II.—Burma Inland ... {	1901-1904	3,808	476	23	4·3	2·0	·1	·5	111·0	3·9	3·3	9·1	70·6
	1904	3,703	474	25	...	0·8	·3	...	81·0	2·4	1·5	2·9	104·1
	1905	3,831	453	22	3·7	·8	...	·4	94·5	·8	6·0	3·4	55·9
„ III.—Assam ... {	1901-1904	1,301	806	39	23·7	1·3	1·5	·2	269·9	6·2	4·0	5·8	220·8
	1904	1,229	648	35	...	1·0	·8	·5	154·3	9·8	6·6	8·2	218·2
	1905	1,291	863	53	3·6	1·4	...	...	233·6	4·3	5·8	1·4	259·3
„ IV.—Bengal and Orissa ... {	1901-1904	12,252	1,009	40	31·5	1·0	·5	·3	275·7	2·3	1·8	10·7	239·9
	1904	12,016	919	36	13·3	·1	·1	·5	231·1	1·5	·3	10·0	207·1
	1905	11,001	1,026	43	19·0	1·3	·1	1·3	298·6	1·2	3·7	9·0	256·6
„ V.—Gangetic Plain and Chutia Nagpur ... {	1901-1904	24,057	788	34	16·8	1·0	·5	·5	173·4	1·2	3·7	1·0	89·7
	1904	21,073	706	39	·3	·9	·7	·3	232·0	1·0	6·4	·8	84·9
	1905	22,083	635	30	12·7	·4	·4	·2	230·0	1·1	8·0	·1	83·3
„ VI.—Upper Sub-Himalaya ... {	1901-1904	13,351	1,008	33	17·6	·1	·3	·9	417·6	1·2	2·7	15·7	61·4
	1904	12,130	816	29	7·5	...	·1	1·5	334·6	1·1	3·7	16·9	50·9
	1905	11,630	809	25	18·7	·3	·3	·6	366·6	1·4	4·8	12·3	33·8
„ VII.—North-West Frontier, Indus Valley and North-Western Rajputana. ... {	1901-1904	8,075	820	30	3·0	·3	1·0	·2	311·7	1·6	1·4	30·0	53·7
	1904	8,365	757	31	9·2	...	·1	·1	260·1	1·4	1·0	21·5	45·9
	1905	8,407	651	25	30·0	·4	1·0	...	312·0	1·1	...	17·1	28·4
„ VIII.—South-Eastern Rajputana, Central India and Gujarat. ... {	1901-1904	4,879	806	41	15·1	·6	·7	·1	304·7	1·1	·4	30·1	41·1
	1904	4,171	626	32	...	...	·2	...	376·1	1·2	·5	31·3	35·4
	1905	4,346	524	30	...	...	·0	...	100·4	·2	...	33·3	16·1
„ IX.—Deccan ... {	1901-1904	8,712	840	32	13·9	·1	·3	·3	202·0	7·0	4·8	10·7	50·9
	1904	7,587	618	23	·3	·4	·5	·1	161·7	·4	8·2	7·3	44·2
	1905	7,017	672	28	4·6	·1	...	...	163·0	1·0	3·4	6·1	47·3
„ X.—Western Coast ... {	1901-1904	1,830	527	24	1·7	·2	·9	3·8	153·1	5·4	7·6	9·5	39·9
	1904	2,516	584	24	...	·1	2·4	...	163·6	2·4	9·6	6·4	34·3
	1905	2,518	533	20	...	·4	·1	1·3	141·1	1·3	11·0	0·9	41·3
„ XI.—Southern India ... {	1901-1904	5,716	500	22	5·3	·3	·1	·3	111·5	1·0	12·8	9·1	53·1
	1904	8,337	450	20	3·0	·5	...	·1	90·1	1·1	15·1	7·1	47·1
	1905	9,140	486	20	·6	·0	·4	·3	87·7	1·6	14·7	8·4	65·3
„ XII.—Hills ... {	1901-1904	509	833	38	6·3	1·7	1·3	2·9	260·7	3·8	20·5	10·5	95·8
	1904	561	711	27	...	...	...	1·7	247·4	3·4	24·4	37·3	37·3
	1905	575	673	24	...	...	...	...	252·3	3·5	17·1	19·1	60·0
INDIA † ... {	1901-1904	90,370	785	32	13·0	1·0	·5	·5	268·6	1·7	7·8	13·0	89·9
	1904	90,355	675	28	4·8	·5	·3	·6	201·1	1·5	8·1	10·7	85·7
	1905	91,917	647	28	11·5	·3	·5	·7	181·7	1·3	9·7	9·7	81·6
ANDAMANS ... {	1901-1904	13,025	1,522	59	5·3	...	...	...	1094·0	11·3	...	11·8	154·9
	1904	13,073	1,090	64	3·9	...	...	...	1173·4	8·1	...	7·6	163·5
	1905	14,348	1,808	68	13·0	...	...	...	1246·2	6·1	...	11·0	164·4
INDIA ‡ ... {	1901-1904	103,395	9·0	35	13·1	·9	·5	·5	362·5	3·4	6·2	12·9	97·6
	1904	104,325	8·1	33	4·1	·5	·2	·5	344·3	2·9	7·0	10·3	95·0
	1905	106,265	8·6	33	11·9	·7	·7	·6	325·5	2·9	8·1	9·9	97·7

* Excluding Subsidiary Jails.   † Including Aden and excluding Andamans.   ‡ The occurrential ratios are, of course, worked on the total strength of the four years.   § Including Andamans.

Appendix to Section IV.—Prisoners—continued.

C.—CAUSES OF ADMISSION.	* Years.	January.	February.	March.	April.	May.	June.	July.	August.	September.	October.	November.	December.	Total.
Influenza ...	1901	56	127	383	1,747	677	141	44	45	25	51	45	6a	2,925
	1902	58	62	171	113	188	115	30	90	39	38	10	34	886
	1902	70	45	31	261	109	42	15	94	61	55	33	30	834
	1904	14	9	5	31	51	41	29	31	38	47	40	33	370
	1905	77	114	90	313	53	30	49	130	191	53	31	31	1,037
Total ...	1901-1905	275	367	637	1,895	1,017	373	175	391	375	340	145	158	6,072
Cholera	1901	46	...	3	3	17	6	43	38	35	5	3	14	211
	1902	1	...	3	3	1	3	3	3	14		6	3	36
	1903	1	...	3	14	10	13	7	25	16	5	3	4	97
	1904	2	3	5	3	7	20	1	...	1	3	...	3	47
	1905	4	...	1	3	3	...	2	12	39	5	3	3	73
Total ...	1901-1905	54	3	13	23	37	41	55	78	105	15	14	26	464
Enteric Fever ...	1901	3	...	4	5	7	5	6	4	4	3	1	...	41
	1902	3	1	9	6	4	7	11	10	11	3	...	1	69
	1903	2	1	8	3	3	6	7	4	4	1	3	3	45
	1904	13	3	3	8	3	3	3	5	6	3	5	5	55
	1905	...	...	4	4	6	6	14	11	3	6	3	7	64
Total ...	1901-1905	23	5	25	25	23	26	41	34	28	20	11	13	274
Intermittent Fever ...	1901	2,565	1,701	3,753	2,513	2,610	7,451	2,130	7,820	3,645	4,070	2,525	2,093	32,036
	1902	1,018	1,543	1,615	1,794	1,384	1,731	1,733	3,093	3,437	3,793	3,413	2,395	28,177
	1903	1,220	1,085	1,704	2,114	1,287	1,419	1,505	1,753	2,131	3,117	1,578	1,832	20,615
	1904	1,352	1,089	1,154	1,329	1,303	1,397	1,411	1,993	3,313	3,995	2,015	1,461	19,525
	1905	1,683	974	1,123	1,262	1,355	1,304	1,319	1,551	1,843	1,877	1,725	1,197	16,704
Total ...	1901-1905	7,937	6,487	7,910	9,031	8,337	8,076	8,420	11,158	13,710	15,547	13,366	9,456	120,357
Remittent Fever ...	1901	9	14	30	23	31	21	25	17	53	43	38	19	395
	1902	15	31	16	16	23	26	25	80	26	23	21	14	346
	1903	15	2	14	13	21	74	15	11	4	13	7	5	143
	1904	4	10	6	13	10	8	14	13	3	10	6	3	102
	1905	4	3	5	10	13	16	8	4	15	10	10	11	109
Total ...	1901-1905	47	50	61	72	95	95	87	125	113	99	83	52	995
Simple Continued Fever ...	1901	167	120	184	114	126	163	102	56	53	54	72	41	1,279
	1902	38	44	27	41	53	34	37	10	31	44	37	34	440
	1903	16	17	21	23	53	33	58	15	73	62	83	41	520
	1904	31	48	32	35	57	66	58	50	99	87	63	63	735
	1905	27	50	56	78	100	63	94	51	94	104	62	57	822
Total ...	1901-1905	519	289	337	303	413	350	359	271	347	351	302	246	3,856
Pneumonia ...	1901	125	131	163	130	130	53	71	63	75	125	137	253	1,501
	1902	207	163	171	33	102	101	67	63	57	74	125	133	1,330
	1903	151	118	120	102	81	74	72	55	60	72	117	160	1,226
	1904	117	131	83	62	65	60	35	64	45	73	88	97	665
	1905	85	132	64	60	65	41	53	55	54	64	72	100	856
Total ...	1901-1905	755	675	643	455	443	352	311	310	323	417	541	739	5,907
Dysentery ...	1901	703	515	680	613	565	713	1,084	1,432	1,160	1,574	022	751	10,695
	1902	637	433	538	646	717	603	847	1,055	1,105	507	903	653	8,051
	1903	553	373	423	5 8	425	175	740	541	820	743	730	039	7,303
	1904	568	354	580	3 3	515	636	2 3	713	743	705	659	581	7,747
	1905	430	333	377	500	607	511	751	978	501	703	635	593	7,106
Total ...	1901-1905	2,570	1,973	3,173	2,913	2,957	3,108	4,275	5,373	4,541	4,170	3,849	3,217	42,152

E.—CAUSES OF DEATHS.

1905.

Cholera	...	
Fevers*	...	
Bowel-complaints	...	
Anæmia and debility ...		
Respiratory diseases ...		
Tubercle of the lungs ...		
All other causes	...	
All causes	...	

* Enteric, intermittent, remittent, and simple continued fevers.

‡ Excluding Andamans.

STATEMENT 1.—*Birth and Death Statistics.*

PROVINCE	Year	BIRTHS — Total number	BIRTHS — Rate per 1000 of population	BIRTHS — Mean Rate per 1000 of population during five previous years	DEATHS — In municipalities and towns	DEATHS — In districts excluding towns	DEATHS — Total	RATIO OF DEATHS per 1000 — In Municipalities and towns	RATIO OF DEATHS — In districts excluding towns	RATIO OF DEATHS — Total	HIGHEST DEATH-RATE — In Municipalities and towns	HIGHEST — In districts excluding towns	LOWEST DEATH-RATE — In Municipalities and towns	LOWEST — In districts excluding towns	MEAN DEATH-RATE DURING PRECEDING FIVE YEARS — In Municipalities and towns	— In districts excluding towns	— Total	Number of deaths of males to every 100 deaths of females
Bengal	1904*	3,170,400	41·59	33·13	130,062	1,985,361	2,115,423	34·82	32·32	32·45	90·70	48·79	7·81	17·37	33·94	32·33	33·51	108
	1905	1,973,301	39·55	28·91	108,418	1,793,957	1,922,369	40·01	38·42	38·53	92·26	57·63	7·41	22·07	61	33·06	33·42	106
Eastern Bengal and Assam	1904†	157,539	35·55	34·41	2,782	133,410	135,392	25·80	25·85	25·85	57·47	33·13	10·49	17·61	11·35	28·75	28·66	107·01
	1905	1,173,879	39·37	39·53	16,758	1,028,507	1,045,305	26·42	25·25	35·05	55·31	46·61	4·77	19·90	26·06	31·04	31·52	108
United Provinces of Agra and Oudh	1904	2,225,757	46·67	41·35	163,855	1,489,094	1,634,849	49·10	33·70	34·70	140·73	17·26	23·34	10·9	43·11	32·59	32·50	99·43
	1905	1,967,009	41·24	44·07	180,482	1,917,818	2,098,300	53·64	43·26	44·00	105·77	91·24	23·64	25·38	47·13	32·78	5·80	100·11
Punjab	1904	534,049	41·5	41·4	94,152	801,356	966,158	16·71	19·31	47·05	14·842	100·50	16·3	13·61	11·2	40·27	40·6	94·9
	1905	853,250	44·4	46·5	99,152	856,057	956,108	49·13	47·37	47·55	306·36	80·10	7·10	21·31	47·63	44·58	44·88	99·1
North-West Frontier Province	1904	69,544	34·9	31·9	4,750	52,099	56,853	18·98	18·53	18·36	44·64	29·43	13·55	16·08	28·1 0	13·89	21·21	107·7
	1905	70,390	35·4	32·3	5,162	48,265	53,327	30·82	26·42	26·79	44·25	8·12	16·77	23·75	28·01	25·22	25·5	110·1
Central Provinces and Berar	1904	654,474	53·19	41·95	55,183	345,327	400,510	45·05	30·65	52	124·53	44·92	10·15	22·39	46·35	35·04	36·13	107·20
	1905	642,799	54·02	41·78	50,627	391,756	442,383	41·87	36·69	37·2	127·05	52·17	17·07	18·43	47·43	35·52	36·73	109·85
Madras Presidency	1904	1,125,751	3·7	79·5	119,579	704,739	321,178	48·1	21·7	22·5	71·2	25·3	3·1	13·5	30·9	20·6	21·4	102·8
	1905	1,176,286	32·0	29·4	125,944	660,179	786,123	29·6	20·3	21·4	92·2	30·6	5·6	15·4	29·8	21·1	21·9	104·3
Coorg	1904	2,970	21·97	23·73	591	4,216	4,808	33·82	25·49	26·67	61·87	33·49	14·07	18·63	45·71	30·01	31·31	124·67
	1905	4,372	25·31	23·78	490	4,449	4,730	32·13	25·70	26·24	45·59	31·52	15·54	21·70	46·22	25·52	33·30	119·20
Bombay Presidency	1904	648,594	35·09	30·78	133,405	631,509	761,914	53·21	39·31	41·39	129·59	72·11	10·61	11·55	63·87	41·74	43·25	104·01
	1905	617,173	33·07	30·51	115,007	473,287	588,394	47·59	29·47	31·84	62·24	40·7	8·18	14·22	67·10	43·25	46·31	108·39
Lower Burma	1904	180,023	32·71	34·25	21,108	101,933	123,061	29·99	21·35	22·36	46·36	26·03	15·45	14·02	31·79	23·02	24·11	135
	1905	171,228	34·34	33·46	27,061	111,587	138,850	38·22	23·00	24·93	54·53	27·64	16·54	14·26	30·94	22·08	23·17	129
Upper Burma	1904	10,393	25·63	36·72	9,130	46,441	55,671	31·13	17·32	18·60	33·63	23·46	10·03	16·71	34·06	18·30	20·08	97
	1905	10,351	34·91	37·03	10,843	54,658	65,541	36·36	20·87	22·46	62·55	30·36	24·08	15·41	32·25	17·03	19·83	99
Ajmer-Merwara	1904	15,937	33·51	26·51	4,584	8,258	13,142	36·90	24·49	23·77	51·57	32·61	30·21	13·09	65·46	49·83	53·73	108·52
	1905	7,802	37·32	26·38	6,603	9,729	16,332	50·04	28·01	34·25	105·20	41·10	26·40	18·25	61·74	48·00	51·75	107·11

* Including Statistics for Eastern Bengal.
† Excluding statistics for Eastern Bengal.
‡ Statistics for births for 12 towns only.

STATEMENT II.—*Total number of deaths by months.*

PROVINCE	January	February	March	April	May	June	July	August	September	October	November	December	TOTAL	RATIOS PER 1000 OF POPULATION 1905	1904
Bengal	178,057	142,466	155,308	154,814	157,130	126,728	111,405	139,445	155,738	203,774	224,920	192,584	1,922,369	38·53	31·9
Eastern Bengal and Assam	111,752	67,897	58,693	65,025	88,141	70,979	65,490	60,455	72,315	112,444	128,173	142,660	1,045,305	35·05	25·8
U.P. of Agra and Oudh	105,863	185,943	241,787	219,150	171,589	137,221	121,520	114,068	141,299	179,533	200,146	189,781	2,098,300	44·00	34·7
Punjab	81,296	75,103	113,840	174,595	147,800	59,963	47,984	49,416	44,171	47,567	53,855	60,818	956,108	47·55	49·0
N.-W. F. Province	4,454	4,565	4,224	4,024	4,098	4,619	5,162	4,052	4,090	5,817	4,584	5,838	53,327	26·79	28·5
C. P. and Berar	31,084	33,950	50,011	34,854	39,359	37,764	29,036	34,447	46,144	45,939	41,408	38,297	442,383	37·21	32·0
Madras Presidency	69,350	58,702	58,268	57,161	58,792	63,968	67,150	71,303	67,959	69,704	67,432	76,726	786,123	21·4	22·5
Coorg	365	258	360	377	406	479	544	450	344	374	317	425	4,739	26·24	26·6
Bombay Presidency	60,999	59,252	58,721	55,960	47,584	39,659	39,303	44,983	45,761	44,167	45,121	47,946	588,394	31·84	41·3
Burma — Lower	10,536	9,072	9,503	10,613	10,434	10,825	12,747	13,419	13,004	13,171	12,877	12,649	138,850	24·53	22·2
Burma — Upper	5,062	4,012	4,127	4,620	4,652	4,596	5,533	5,930	5,947	6,049	7,299	7,414	65,541	22·46	18·6
Ajmer-Merwara	1,136	1,565	2,617	3,022	1,673	831	783	594	868	863	977	993	16,332	34·25	27·5
TOTAL	749,993	641,325	733,059	784,916	732,038	557,562	506,663	539,033	597,670	727,512	766,859	776,031	8,117,771	35·96	32·8

* Excluding districts of Eastern Bengal.

STATEMENT III—*Births.*

Province.	Population under Registration.	Maximum for any one district.	Minimum for any one district.	Mean for the Province.	Number of males born to every 100 females born.	Excess of births over deaths per 1,000 of population.	Excess death over births per 1,000 of population.
Bengal	46,891,164	51·33	18·44	39·55	105	1·02	...
Eastern Bengal and Assam	29,812,733	50·70	28·96	39·37	106	4·31	...
United Provinces of Agra and Oudh	47,691,782*	52·59	27·85	41·24	108·59	...	2·76
Punjab	20,108,690	67·2	21·4	44·4	109·8	...	3·2
North-West Frontier Province	1,990,744	43·3	31·7	35·4	122·6	8·6	...
Central Provinces and Berar	11,887,703	61·24	46·30	54·02	104·38	16·81	...
Madras Presidency	36,737,533	46·0	24·4	32·0	103·9	10·6	...
Coorg	180,607	38·15	19·90	25·31	95·22	...	·92
Bombay Presidency	18,481,362	46·03	17·88	33·07	108·44	1·23	...
Burma { Lower	5,568,470	42·73	17·19	34·34	105	9	...
{ Upper†	276,541	50·33	29·09	34·91	105	...	2
Ajmer-Merwara	476,912	45·70	34·84	37·32	110·10	3·08	...

* Includes 16,010 persons enumerated at the Ajodhya fair.
† Statistics for 12 towns only.

STATEMENT IV.—*Deaths.*

Province.	Population under registration.	Area in square miles.	Average population per square mile.	Maximum for any one district.	Minimum for any one district.	Mean for the year.	Male.	Female.
Bengal	46,891,164	107,498	464	59·86	22·05	38·53	40·07	37·00
Eastern Bengal and Assam	29,812,733	71,535	416	45·59	19·99	35·06	35·86	34·22
United Provinces of Agra and Oudh	47,691,782*	107,164	445	91·01	26·12	41·00	42·64	45·44
Punjab	20,108,690	97,209	207	78·5	24·5	47·6	44·1	51·6
North-West Frontier Province	1,990,744	13,688†	149	27·6	24·5	26·8	26·2	27·5
Central Provinces and Berar	11,887,703	‡	‡	54·80	28·44	37·21	39·35	35·12
Madras Presidency	36,737,533	129,241	284	59·5	16·0	21·4	22·2	20·6
Coorg	180,607	1,583	114	30·74	23·66	26·24	25·70	26·91
Bombay Presidency	18,481,362	122,984	150	62·33	15·88	31·84	32·11	31·54
Burma { Lower	5,568,470	70,992	72	45·90	15·59	24·93	26·43	23·24
{ Upper	2,917,501	29,411	93	30·49	15·41	22·46	23·64	21·40
Ajmer-Merwara	476,912	2,711	176	36·32	27·25	34·25	33·66	34·90

* Includes 17,010 persons enumerated at the Ajodhya fair.
† Area and population per square mile of districts include areas of Cantonments.
‡ Not available.

STATEMENT V.—*Deaths in Towns and Rural Circles compared.*

Province.	Rural.	Town.	Total.	Rural.	Town.	Total.	Rural.	Town.	Total.
Bengal	387	127	514	46,682,139	3,209,025	49,891,164	38·42	40·01	38·53
Eastern Bengal and Assam	242	54	296	29,177,017	635,718	29,812,735	35·25	26·42	35·06
United Provinces of Agra and Oudh	856	448	1,304	44,327,116*	3,364,666	47,691,782*	43·26	53·64	44·00
Punjab	404	145	549	18,090,573	2,018,117	20,108,690	47·37	49·13	47·55
North-West Frontier Province	64	10	74	1,826,493	164,251	1,990,744	26·42	30·82	26·79
Central Provinces and Berar	233	99	332	10,678,334	1,209,169	11,887,703	36·69	41·87	37·21
Madras Presidency	185	232	417	32,487,484	4,250,049	36,737,533	20·3	29·6	21·4
Coorg	5	5	10	165,358	15,259	180,607	25·70	32·13	26·24
Bombay Presidency	224	56	280	16,065,004	2,416,358	18,481,362	29·47	47·59	31·84
Burma { Lower	201	40	241	4,860,314	708,165	5,568,479	23·00	38·21	24·93
{ Upper	107	19	126	2,620,960	296,541	2,917,501	20·87	36·56	22·46
Ajmer-Merwara	17	6	23	347,280	129,632	476,912	28·01	50·94	34·25

* Includes 16,010 persons enumerated at the Ajodhya fair.

STATEMENT VI.—*Deaths according to age.*

PROVINCE.	RATIO PER 1,000 OF POPULATION.																				
	Under 1 year.		1-5 years.		5-10 years.		10-15 years.		15-20 years.		20-30 years.		30-40 years.		40-50 years.		50-60 years.		60 years and upwards.	
	Male.	Female.	Male.	Female.	Male.	Female.	Male.	Female.	Male.	Female.	Male.	Female.	Male.	Female.	Male.	Female.	Male.	Female.	Male.	Female.
Bengal . . .	320·63	276·40	61·95	54·33	23·10	20·15	17·83	16·91	1·747	18·54	22·58	21·55	23·31	23·02	31·50	26·74	47·28	44·18	95·90	80·05
Eastern Bengal and Assam.	265·48	227·04	53·06	47·63	21·17	16·16	15·08	14·20	20·70	24·83	20·0,	24·16	22·59	24·33	27·35	25·63	40·65	38·02	77·71	63·82
United Provinces of Agra and Oudh	56·29	341·13	64·11	65·42	20·80	22·30	16·82	20·20	20·35	29·02	23·01	27·10	25·80	28·59	33·72	34·25	52·75	51·43	92·86	83·21
Punjab . . .	320·02	318·00	57·33	61·	20·32	26·37	20·75	34·23	23·29	31·03	26·38	30·57	27·01	36·47	16·53	42·25	47·25	53·17	92·30	104·87
N.W.F. Province .	211·25	167·03	41·33	33·20	8·76	9·77	0·75	8·08	7·0,	9·75	10·48	11·47	13·63	19·43	21·01	24·72	33·63	33·43	65·94	69·57
Central Provinces and Berar[*]	...	...	...	...	...	...	...	...	...	...	...	...	...	...	...	...	...	...	...	...
Madras Presidency	200·7	165·0	27·4	25·8	7·4	7·1	5·3	5·4	8·0	11·3	9·8	10·6	11·9	10·0	18·0	13·3	29·1	24·1	70·3	63·4
Coorg . . .	174·58	173·88	37·39	27·28	9·88	9·20	7·74	8·38	10·88	12·32	18·74	21·36	21·10	25·21	32·33	29·75	40·02	44·92	72·14	67·68
Bombay Presidency	351·52	317·74	55·02	52·36	12·06	12·76	10·87	12·61	15·76	17·75	17·42	18·51	10·45	11·02	26·47	20·06	41·77	32·66	88·31	80·00
Burma { Lower	353·74	23·188	30·20	26·51	14·46	12·33	16·11	11·06	14·81	11·01	14·61	12·09	16·47	11·26	21·24	17·24	25·79	21·36	56·94	54·47
{ Upper	316·78	177·57	26·02	23·82	0·07	8·74	6·23	5·49	0·08	7·48	11·5	10·01	12·46	13·75	16·14	14·10	26·58	20·83	06·13	60·51
Ajmer-Merwara .	247·73	244·37	72·34	73·40	10·00	11·31	12·01	12·13	14·81	20·05	19·77	22·91	23·05	22·31	26·67	24·96	45·75	41·19	85·73	79·59

[*] Information not available.

STATEMENT VII.—*Deaths according to cause.*

TRACT.	DEATHS PER 1,000 OF POPULATION IN 1905.									Ratio of deaths in 1905.	Ratio of deaths in 1905.
	Cholera.	Small pox.	Plague.	Fevers.	Dysentery and diarrhœa.	Respiratory diseases.	Injuries.	All other causes.	All causes.		
Bengal . . .	2·93	·14	2·52	24·31	·00	·22	·40	6·95	38·53	31·95	33·33[*]
Eastern Bengal and Assam.	4·77	·15	·0002	23·68	·85	·03	·36	5·15	35·06	25·85†	20·55†
United Provinces of Agra and Oudh.	2·55	·07	8·05	26·9	·56	·40	·51	4·85	44·00	34·70	40·28
Punjab . . .	·11	·23	16·65	18·40	·68	3·05	1·24	7·17	47·55	49·06	49·01
N.W.F. Province .	·15	·29	·002	20·09	·28	·57	·32	4·49	26·79	25·86	28·40
Central Provinces and Berar.	·10	·70	1·07	17·43	3·02	1·30	·54	1·25	37·21	32·26	35·52
Madras Presidency	·5	·5	·2	7·2	1·4	·5	·3	10·8	21·4	22·5	22·2
Coorg . . .	·..	·73	...	21·46	·53	·17	·14	3·21	26·24	26·62	31·43
Bombay Presidency	·20	·92	3·86	15·28	·30	2·05	·36	7·18	31·84	41·39	43·91
Burma { Lower	·63	1·61	·55	8·97	1·43	·73	·27	11·35	24·93	23·36	24·13
{ Upper	·63	·18	·22	6·75	·44	·56	3·	13·26	22·46	18·69	21·80
Ajmer-Merwara .	...	·58	5·20	21·67	·89	1·8	·0	4·43	34·35	27·57	28·77

[*]Including statistics for Eastern Bengal.
†Excluding statistics for Eastern Bengal.

STATEMENT VIII.—*Ratio of deaths from all causes according to months.*

PROVINCE.	January.	February.	March.	April.	May.	June.	July.	August.	September.	October.	November.	December.	TOTAL.
Bengal . .	3·56	2·85	3·11	3·10	3·14	2·34	2·23	2·79	3·12	4·08	4·10	3·86	38·53
Eastern Bengal and Assam	3·74	2·27	1·97	2·21	2·05	2·37	2·19	2·02	2·42	3·77	4·29	4·78	35·06
United Provinces of Agra and Oudh . .	4·11	3·90	5·07	4·59	3·61	2·88	2·55	2·39	2·06	3·76	4·10	3·08	44·00
Punjab . .	4·04	3·73	5·00	8·67	7·55	2·98	2·39	2·40	2·20	2·37	2·68	3·02	47·55
N.W.F. Province .	2·24	2·29	2·12	2·02	2·06	2·72	2·50	2·0,	2·05	1·92	2·20	2·03	26·79
Central Provinces and Berar . .	2·61	2·86	2·53	2·93	3·31	3·18	2·14	2·90	3·88	3·87	3·48	3·22	37·21
Madras Presidency	1·9	1·6	1·6	1·6	1·6	1·7	1·8	1·9	1·0	1·9	1·8	2·1	21·4
Coorg . .	2·2	1·43	1·99	2·09	2·25	2·65	3·01	2·70	1·90	2·07	1·70	2·36	20·24
Bombay Presidency	3·70	3·15	3·18	3·03	2·58	2·14	2·13	2·43	2·48	2·39	2·44	2·59	31·84
Lower Burma	1·89	1·63	1·71	1·91	1·87	1·94	2·29	2·41	2·34	2·37	2·31	2·27	24·93
Upper Burma	1·74	1·38	1·52	1·58	1·59	1·58	1·90	2·03	2·04	2·07	2·50	2·54	22·46
Ajmer-Merwara . .	2·51	3·28	5·49	6·34	3·51	1·74	1·04	2·08	1·82	1·81	1·94	2·08	34·35

Year	Bengal*	Assam	United Provinces of Agra and Oudh	Punjab	(a) N.-W. Frontier Province	Central Provinces	Berar	Madras	Coorg	Bombay	Lower Burma	Upper Burma †	Ajmer-Merwara	Rajputana	Central India	Hyderabad (Cantonment stations)	Mysore
1877	155,305	11,157	31,170	29	...	3,413	842	357,430	†	57,228	7,176	...	11	60	926	7,414	2,902
1878	95,192	6,730	22,221	215	...	40,085	34,306	47,161	40	46,743	6,259	...	210	2,293	8,047	6,694	773
1879	130,363	17,415	36,892	26,115	...	27,575	221	13,396	...	6,037	1,855	...	120	918	2,734	6	11
1880	39,643	2,083	71,546	274	...	330	1	603	...	64	2,638	...	3	...	209	...	25
1881	79,180	5,010	25,365	5,207	...	9,140	3,404	9,446	3	16,694	5,239	...	16	197	581	1,721	25
1882	182,352	21,755	89,332	30	...	11,032	3,573	23,604	31	57,054	7,177	...	289	1,317	1,362	130	893
1883	99,439	14,098	18,60	150	...	16,135	21,897	36,384	...	57,054	2,185	...	87	797	1,740	4,017	174
1884	134,421	22,276	30,143	614	...	149	87	75,476	...	13,804	5,335	...	227	1,297	1,018	2,179	330
1885	173,707	7,753	63,457	1,936	...	21,898	5,133	58,109	...	31,087	7,585	...	100	1,615	4,721	1,387	2,677
1886	118,368	20,188	31,505	12	...	16,673	926	12,417	...	169	4,027	...	763	173	290	499	10
1887	112,578	7,941	200,028	8,804	...	13,576	14,396	28,159	3	25,711	4,049	...	381	2,612	8,863	2,831	532
1888	111,391	9,613	13,704	14,938	...	921	395	53,017	2	36,500	15,682	...	13	32	191	2,057	1,015
1889	171,103	15,383	48,474	2,845	...	52,588	10,925	76,020	5	32,431	3,210	...	55	6,903	3,344	1,128	1,599
1890	145,585	15,3961	9,295	3,401	...	4,287	847	35,288	7	3,259	1,976	...	408	2,246	3,132	3,462	1,326
1891	236,575	23,882	16,023	10,107	...	21,312	7,053	98,773	7	17,830	2,100	...	532	2,046	13,472	3,802	1,204
1892	259,368	21,552	194,886	75,089	...	39,077	2,030	79,933	53	47,900	6,209	...	2,352	26,750	8,584	53	5,497
1893	156,976	21,849	12,154	697	...	587	1,186	37,299	9	18,553	2,303	...	3	314	127	165	630
1894	296,150	13,497	178,079	113	...	7,043	3,452	47,289	8	33,658	7,428	...	...	2	5,710	1,862	318
1895	177,087	13,962	51,362	540	...	15,506	11,019	21,172	...	9,590	54,139	...	289	1,049	6,043	467	2,334
1896	226,824	17,642	69,147	5,146	...	52,085	12,564	81,847	49	35,404	2,059	...	12	3,797	15,796	535	2,100
1897	196,247	33,240	44,208	622	...	57,131	10,122	143,445	103	57,109	8,358	...	19	1,496	13,202	1,019	4,748
1898	65,030	11,149	2,508	335	...	7	...	65,444	8	4,368	2,972	...	...	6	2	6	1,103
1899	107,658	8,350	8,142	1,816	...	766	541	29,082	...	8,379	4,842	2,059	475	479	...	...	121
1900	345,898	23,761	84,960	28,260	...	63,114	18,375	60,662	58	163,889	3,449	41	4,842	28,712	20,450	3,813	779
1901	110,753	7,468	53,095	150	...	49	17	81,370	58	13,000	3,552	++	59	6	72	1	11,351
1902	159,971	12,658	25,160	331	117	28	16	29,269	...	3,230	1,841	57	32	1,319	12	...	218
1903	203,405	8,360	47,159	14,683	...	437	...	27,303	...	1,835	5,216	2,887	...	236	1,410	...	98
1904	137,701	8,588	6,617	716	1	2,661	...	23,405	...	13,156	7,477	908	...	1	150	...	471
1905	146,339	141,312**	121,790	2,197	300	1,217	...	16,888	...	5,396	3,311	1,836	...	3	27	64	626

* Excluding Calcutta from 1877 to 1892.
† Statistics from 1877 to 1898 not available.
‡ Statistics not available.
§ Including 30 deaths in cantonments.

‖ Excluding Zamindaris.
† Including Berar (from 1901).
** Eastern Bengal and Assam.
(a) 1877-1900 included in the Punjab.

STATEMENT II.—*Deaths from CHOLERA in British Provinces, by months, during the year 1905.*

PROVINCE.	January.	February.	March.	April.	May.	June.	July.	August.	September.	October.	November.	December.	TOTAL.	RATIO PER 1,000 OF POPULATION. 1905.	RATIO PER 1,000 OF POPULATION. 1904.
Bengal . .	12,040	5,403	4,895	5,674	3,618	2,929	5,159	9,911	16,798	26,274	30,977	22,671	146,339	2·93	1·63
E. B. and A. .	9,179	1,439	1,230	2,479	3,098	1,659	1,285	1,664	9,938	28,649	39,982	41,710	142,312	4·77	2·07
U. P. of Agra and Oudh.	56	28	64	472	1,893	2,315	2,718	3,495	12,215	28,418	46,923	23,593	121,790	2·55	·14
Punjab .	...	...	...	...	13	36	324	688	909	174	51	2	2,197	·11	·04
N.-W. F. Province.	...	...	...	...	...	...	...	...	169	126	5	...	300	·15	·001
C. P. and B. .	...	...	...	...	2	64	195	249	500	170	12	25	1,217	·10	·24
Madras Presidency.	26	16	20	5	42	449	1,268	4,191	3,717	3,680	1,499	7,975	16,888	·5	·6
Coorg . .	...	...	...	...	...	...	...	...	...	...	...	...	...	...	...
Bombay Presidency.	51	49	133	777	871	1,980	879	340	225	62	27	2	5,396	·29	·71
Burma { Lower	331	413	292	271	216	44	68	113	326	571	481	385	3,511	63	·45
Burma { Upper	...	1	10	70	31	59	125	194	134	211	580	421	1,896	·63	·17
Ajmer-Merwara.	...	...	...	...	...	...	...	...	...	...	...	...	...	...	...
TOTAL .	21,683	7,349	6,634	9,748	9,784	9,535	11,621	20,845	44,931	88,335	120,537	90,784	441,786	1·96	·85

STATEMENT III.—*Details of the distribution and occurrence of CHOLERA during the year 1905.*

PROVINCE.	Mortality in 1905.	Mean mortality of previous 5 years.	Urban mortality.	Rural mortality.	Percentage of villages attacked.	Maximum mortality in any one district ct excluding towns.	Maximum mortality in any one town.	Month of maximum mortality.
Bengal	2·93	2·77	3·85	2·86	13·90	6·41	12·94	November.
Eastern Bengal and Assam . .	4·77	2·10	4·52	4·77	20·81	10·60	17·66	December.
United Provinces of Agra and Oudh . .	2·55	·92	·64	2·70	7·84	25·30	13·48	November.
Punjab	·11	·41	·32	·09	·90	1·26	13·78	September.
N.-W. Frontier Province .	·15	·48	·38	·13	1·30	·44	7·40	September.
Central Provinces and Berar .	·10	Not available	·07	·11	·46	·53	4·57	September.
Madras Presidency .	·5	1·2	1·2	·4	4·24	3·2	10·1	August.
Coorg . . .	...	·06	...	...	...	...	...	...
Bombay Presidency .	·29	2·08	·20	·31	1·84	1·95	6·26	June.
Burma { Lower .	·63	·63	1·04	·57	4·16	4·59	18·98	October.
Burma { Upper .	·63	·25	1·47	·53	3·48	1·39	9·10	November.
Ajmer-Merwara . . .	...	2·07	...	...	...	...	...	...

TABLE I.—*Small-pox mortality.*

Provinces, Districts, Towns.	Bengal.	Eastern Bengal and Assam.	United Provinces.	Punjab.	North-West Frontier Province.	Central Provinces and Berar.	Lower Burma.	Upper Burma.	Madras Presidency.	Bombay Presidency.	Ajmer-Merwara.	Coorg.	Registration India.
I.—Mortality by Provinces :—													
A.—Deaths by months—													
January	619	304	107	306	81	379	162	19	1,333	3,354	65	3	5,695
February	544	252	149	330	66	625	151	9	1,313	3,838	78	3	7,333
March	748	393	195	219	46	786	527	33	1,627	3,883	60	15	8,540
April	833	395	316	278	45	1,314	731	84	1,453	3,033	42	12	8,457
May	1,190	690	472	427	43	1,315	858	137	1,459	1,532	25	11	8,227
June	756	751	475	515	35	1,107	717	78	1,336	643	4	18	6,453
July	533	620	337	438	63	811	641	46	1,274	309	...	25	5,099
August	313	313	251	331	14	552	442	31	1,562	204	...	13	4,703
September	325	325	155	219	15	389	291	29	1,417	171	1	6	3,723
October	262	311	86	163	26	195	258	46	1,716	237	...	10	3,264
November	303	191	107	486	28	283	310	30	1,730	331	...	8	3,357
December	791	232	462	1,115	204	650	405	8	2,306	470	2	10	6,565
Total	7,213	4,723	3,373	4,733	571	8,314	5,623	537	17,540	16,985	277	132	70,562
B.—Annual death ratios—													
Ratio per 1,000 of population, 1905	'14	'15	'07	0'23	0'39	0'70	1'01	'18	0'5	'63	'38	'73	'31
Ratio per 1,000 of population, 1904.	'22	'26	'15	'43	'78	'16	'24	'17	'3	'33	'37	'09	'24
Difference	—'08	—'11	—'08	—'25	—0'49	+0'54	+0'77	+'01	+'02	+'69	+'31	+'64	+'07
Mean ratio per 1000, during 1900—1904	'54	'37	'15	0'52	0'77	Not available.	'36	'41	0'6	0'36	1'35	'93	0'40
Difference	—'40	—'12	—'08	—'29	—'43	...	+'65	—'23	—'1	+'65	—'57	—'20	—'09
II.—District mortality excluding towns :—													
Number of districts affected	30	33	48	23	4	23	14	10	22	24	9	5	339
Highest district ratio	1'89	2'52	1'63	0'77	0'63	3'62	3'24	1'52	1'4	5'34	1'34	1'53	5'74
Name of that district	Puri.	Kamrup.	Jhansi.	Feroze-pore.	Hazara.	Saugor.	Hantha-waddy.	Mandalay.	Madura.	Khandesh.	Pokhar.	Yedenaik-tad.	Khandesh.
Lowest district ratio	'002	'004	'001	0'01	0'024	0'05	0'002	'01	0'02	0'02	'05	'18	'001
Name of that district	Ranchi.	Malda.	Saharan-pur.	Kangra.	Dera-Ismail Khan.	Chhind-wara.	Akyab.	Kyaukse.	Vizagapa-tam.	Belgaum.	Sawar.	Naujaraj-patan.	Saharan-pur.
Number of districts without mortality.	1	None.	None.	1	1	None.	4	1	None.	None.	8	None.	16
District death-rate per 1,000 of population.	'13	'15	'07	0'20	0'29	0'65	'82	'18	0'5	'72	'16	'73	0'27
III.—Town mortality :—													
Number of towns affected	59	17	33	76	6	70	35	3	105	43	2	3	443
Highest town ratio	3'93	6'34	'61	9'07	0'59	10'65	13'20	4'05	8	8 33	3'00	3'54	13'20
Name of that town	Kishan-ganj.	Tezpur.	Benares.	Ludhiana	Abbotta-bad.	Mohpa.	Danol yu.	Pyicmana	V. T. C. Dindigal.	Shikarpur	Ajmer Town.	Virajen-drapet.	Danubyu.
Lowest town ratio	'03	'04	'006	0'01	0'04	0'07	'c8	'01	0'01	'02	'23	'41	'005
Name of that town	South-suburban.	Narayan-ganj.	Allahabad	Delhi.	Dera Ismail Khan.	Jubbulpur	Akyab.	Mandalay	M. T. C. Calicut.	Broach.	Pisangan Town.	Mercara.	Allahabad.
Number of towns without mortality	68	37	71	69	3	32	14	9	127	20	3	2	456
Town death-rate per 1,000 of population.	'22	'37	'05	0'52	0'20	1'10	3'34	'31	0'5	3'18	1'71	'93	0'78
IV.—Infantile mortality :—													
Children under 1 year	1,250	802	1,336	1,195	118	2,923	402	31	7,177	6,413	68	...	21,553
Children 1-10 years	5,247	3,223	1,683	2'064	238	3,163	1,487	196	6,513	8,127	173	...	39,503
Percentage of children in tot l small-pox mortality.	63'73	64'05	92'42	81'71	82'36	73'93	33'59	43'75	73'84	84'35	92'66	...	73'58

TABLE II.—*Fever mortality.*

Provinces, Districts and Towns.	Bengal.	Eastern Bengal and Assam.	United Provinces.	Punjab.	North-West Frontier Province.	Central Provinces and Berar.	Lower Burma.	Upper Burma.	Madras Presidency.	Bombay Presidency.	Ajmer-Merwara.	Coorg.	Registration India.
I.—Mortality by Provinces :—													
A.—Deaths by months—													
January	102,012	31,708	105,821	33,210	3,570	13,057	4,615	1,985	23,332	24,759	871	318	395,875
February	70,120	32,328	96,657	30,474	3,721	14,037	3,545	1,345	19,510	23,381	1,003	230	325,905
March	57,131	45,070	101,595	33,505	3,335	13,009	2,514	1,046	19,052	23,730	1,302	295	522,867
April	88,811	49,777	101,931	30,002	3,166	16,130	3,815	1,633	20,067	22,558	1,175	316	329,836
May	107,131	66,637	107,076	31,255	3,256	19,652	3,512	1,832	20,016	19,391	1,007	384	352,118
June	91,631	54,875	106,619	31,142	3,643	19,482	3,774	1,407	22,274	15,910	839	410	351,315
July	75,875	50,945	94,037	28,471	3,987	13,502	4,545	1,541	22,309	15,589	615	410	312,683
August	91,075	46,160	85,289	28,902	2,976	15,097	4,729	8,431	22,416	17,113	811	382	317,376
September	100,586	45,176	100,305	25,549	3,019	20,094	4,404	1,462	21,741	17,117	650	980	344,481
October	135,087	65,080	131,363	28,355	3,815	22,440	4,143	1,616	22,374	18,324	680	256	432,411
November	156,636	63,121	126,570	23,731	3,308	21,305	4,445	1,928	22,010	22,526	705	254	441,000
December	128,302	75,782	124,863	26,382	4,504	18,503	4,901	2,194	26,063	24,578	748	344	458,792
Total	1,214,457	705,200	1,294,164	370,047	41,190	207,195	49,055	19,728	265,044	245,373	10,326	3,876	4,417,655
B.—Annual death ratios—													
Ratio per 1,000 of population, 1905.	24'34	23'68	26'92	18'40	20'69	17'43	8'97	6'75	7'2	13'38	21'67	21'46	19'57
Ratio per 1,000 of population, 1904.	20'49	23'91	23'92	18'83	22'30	13'54	9'91	7'05	8'0	13'60	21'83	22'45	18'09
Difference	+ 3'85	+ '08	+ 3'00	— '43	— 1'61	+ 3'89	— '94	— '29	— 0'8	— 0'32	— 0'16	— 0'99	+ 1'48
Mean ratio per 1,000 during 1900–1904.	21'14	23'23	24'63	25'05	19'01	Not available.	9'68	7'32	8'2	17'40	25'33	25'43	19'60
Difference	+ 3'20	+ 0'45	+ 2'33	— 6'65	+ 1'68	...	— 0'71	— 0'46	— 1'0	— 4'12	— 16'71	— 3'97	— 0'03
II.—District mortality excluding towns :—													
Number of districts affected	31	22	48	29	5	23	18	11	22	24	17	5	255
Highest district ratio	39'10	36'24	38'01	23'44	25'23	29'81	15'50	13'94	18'6	27'84	33'37	28'95	39'81
Name of that district	Shahabad.	Dinajpur.	Saharanpur.	Attock	Peshawar	Nimar.	Akyab	Mandalay.	Vizagapatam,	Ahmednoad.	Ghegal,	Nanjarajpat-a.	Nimar.
Lowest district ratio	7'56	5'23	18'32	4'62	17'32	7'65	4'90	3'92	1'7	5'54	14'62	16'87	1'7
Name of that district	Puri .	Sylhet	Almera	Simla	Hazara	Buldana.	Amherst.	Meiktila	Bellary	Belgaum	Dewair	Vedenaik-nad.	Bellary.
Number of districts without mortality	None	None	None	None	None	None	None	None	None	None	None	None	None
District death rate per 1,000 of population.	25'10	23'92	25'92	18'47	21'08	19'03	9'39	6'85	7'4	13'90	21'81	22'17	20'11
III.—Town mortality :—													
Number of towns affected	117	54	107	145	10	93	39	13	228	63	6	5	895
Highest town ratio	41'40	33'75	50'05	31'98	13'66	31'59	21'01	9'40	30'1	53'36	39'35	23'75	53'36
Name of that town	Murshidabad.	Mangaldai.	Brindaban	Shahiwal	Haripur	Wadi-gaon.	Myaung-mya.	Menywa	T. C. Sruega-vaarapu-keta. ,	Umarkot.	Beawar Town,	Fraserpet.	Umarkot.
Lowest town ratio	2'78	1'71	11'33	2'00	6'35	1'64	2'89	1'73	0'1	'35	12'98	'70	0'1
Name of that town	Ulubaria.	Jhalakhati	Konch,	Phildour.	Kulachi.	Denigaon Raja.	Moulmein	Ninbu.	T. C. Ellya-puram.	Deolali Cantonment.	Aumer Town,	Virajend-rapet.	Ellya-puram.
Number of towns without mortality	None	None	None	None	None	None	None	None	4	None	None	None	4
Town death rate per 1,000 of population.	13'15	11'82	26'58	17'70	16'57	13'46	6'80	6'17	5'2	9'13	21'31	13'77	13'51

TABLE III.—*Dysentery and Diarrhœa mortality.*

Provinces, Districts and Towns.	Bengal.	Eastern Bengal and Assam.	United Provinces.	Punjab.	North-West Frontier Province.	Central Provinces and Berar.	Lower Burma.	Upper Burma.	Madras Presidency.	Bombay Presidency.	Ajmer-Merwara.	Coorg.	Registration India.
I.—Mortality by Provinces.—													
A.—Deaths by months.—													
January	4,065	1,751	1,811	759	2	1,645	620	44	4,119	4,853	24	5	19,774
February	3,149	954	1,467	742	25	1,536	492	33	2,670	4,111	21	3	15,363
March	3,209	910	1,338	729	31	1,477	444	50	3,613	4,065	33	1	15,265
April	3,162	1,311	1,721	840	16	1,999	503	75	3,243	4,164	75	5	17,623
May	3,704	2,082	2,613	1,510	43	2,464	605	73	3,013	4,853	43	13	21,853
June	2,833	2,849	2,659	1,119	47	2,171	650	135	4,851	4,835	31	14	23,416
July	2,793	2,130	3,055	1,373	47	2,321	884	176	5,57-	5,76)	23	11	23,774
August	3,028	1,712	2,512	1,341	85	4,110	814	221	5,261	6,738	38	17	27,230
September	4,215	1,150	2,782	1,903	75	6,137	854	131	4,304	5,775	32	7	27,426
October	4,017	2,709	2,191	1,271	57	5,109	669	117	4,170	4,111	23	9	25,074
November	4,349	3,142	2,176	1,143	24	3,455	891	156	3,843	3,347	31	7	23,509
December	4,465	2,252	2,130	1,129	51	2,706	711	104	4,216	3,412	31	4	22,505
Total	47,270	25,582	26,838	13,712	523	35,579	7,946	1,295	51,299	55,488	424	96	2,64,134
B.—Annual death ratios—													
Ratio per 1,000 of population, 1905.	·90	·85	·16	0·65	0·13	3·02	1·43	·44	1·4	3·00	·80	·53	1·17
Ratio per 1,000 of population, 1904.	·83	·62	·50	·69	·31	2·14	1·27	·31	1·3	3·33	·78	·50	1·06
Difference	+ ·07	+ ·23	+ ·06	+ 0·c8	— 0·03	↑ 0·53	+ 0·16	+ ·13	+ 0·1	—0·33	+ ·12	+ ·03	+ ·11
Mean ratio per 1,000 during 1900-1,904.	·99	·65	·63	0·82	0·25	Not available.	1·48	·33	1·2	4·87	4·44	·84	1·37
Difference	— ·09	+ ·20	— ·07	— ·14	…	…	— ·05	+ ·11	+ ·3	— 1·87	—3·55	— ·31	— 0·33
II.—District mortality excluding towns.—													
Number of districts affected.	31	28	48	29	5	23	13	11	22	24	11	5	219
Highest district ratio	4·19	6·17	11·51	3·22	0·65	11·34	2·90	·65	4·3	9·73	1·77	·67	11·50
Name of that district	Patna.	Lakhimpur.	Garhwal.	Simla.	Dera Ismail Khan.	Akola.	Tavoy.	Myingyan.	The Nilgiris.	Khandesh.	Pokhar	Kiggatnad	Garhwal.
Lowest district ratio	·03	·03	·01	0·14	0·03	0·15	·57	·05	0·2	·03	·03	·07	·01
Name of that district	Purnea.	Dinajpur	Gonda.	Gujranwala.	Peshawar.	Bhandara.	Toungoo	Kyauksé.	Vizagapatam.	Larkhana	Masuda	Padinalknad Taluk.	Gonda.
Number of districts without mortality.	None	None	None	None	None	None	None	None	None	None	6	None	6
District death-rate per 1,000 of population.	·73	·83	·42	0·53	0·19	2·93	1·11	·43	1·1	2·91	·22	·31	1·00
III.—Town mortality—													
Number of towns affected.	123	54	95	111	10	97	39	12	207	60	4	2	844
Highest town ratio	8·50	16·02	10·93	10·82	2·85	2·15	7·55	5·04	11·2	10·87	4·52	7·00	20·26
Name of that town	Hooghly and Chinsurah	North-Lakhimpur.	Ballia.	Kalabagh.	Kulachi.	Jalgaon.	Thayetmyo.	Pyinmana.	M. T. C. Madras.	Poona.	Ajmer suburbs	Virajendtapet,	Jalgaon.
Lowest town ratio	·11	·07	·03	0·15	0·22	0·12	·16	·10	0·1	·19	·14	1·12	·01
Name of that town	Baraset	Pirapur	Meerut	Eminabad	Kohat	Morsi	Danubyu	Shwebo	T. C. Sivagiri.	Jacobabad.	Kekri Town.	Kadlipet.	Meerut.
Number of towns without mortality.	5	None	12	4	None	2	None	None	25	3	2	1	55
Town death-rate per 1,000 of population.	3·13	1·90	2·19	2·11	1·23	3·77	3·57	·60	4·0	3·61	2·63	2·93	3·06

Appendix B to Section VI.—Chief Diseases—continued.

TABLE IV.—Plague deaths.

Province or State.	January.	February.	March.	April.	May.	June.	July.	August.	September.	October.	November.	December.	Total 1905.	Total 1904.
British Provinces :—														
Bengal	30,523	23,515	40,002	25,495	8,801	1,512	142	245	403	814	1,255	7,728	135,084	77,413
Eastern Bengal and Assam	...	...	2	1	3	...	...	...	...	...	...	...	0	3
United Provinces of Agra and Oudh.	63,929	65,540	116,259	91,055	25,007	3,131	355	441	545	705	1,858	4,854	293,801	179,082
Punjab	23,613	27,253	61,650	112,374	90,683	9,030	548	193	129	235	1,107	1,370	374,807	306,357
North-West Frontier Province.	...	...	2	...	1	...	...	...	...	...	...	...	3	1
Central Provinces and Berar	2,040	2,000	7,073	1,044	212	14	1	137	1,192	2,108	1,016	868	12,706	41,866
Burma { Lower . . .	1	61	173	573	356	405	521	325	232	151	127	105	3,045	3
Burma { Upper . . .	...	...	...	1	...	...	92	76	41	40	157	234	647	...
Madras Presidency . .	1,700	1,016	458	239	52	31	94	436	859	405	359	222	5,988	30,125
Bombay	11,069	9,161	10,525	8,405	4,353	1,573	7,046	4,459	7,258	6,549	3,516	2,459	77,373	233,057
Ajmer-Merwara . . .	33	170	665	1,273	315	...	1	4	1	1	18	3	2,480	158
Coorg	...	...	...	...	...	...	...	...	...	...	...	...	...	25
Total { 1905	132,318	138,125	231,957	241,450	140,145	15,772	9,814	6,318	10,651	10,725	9,282	15,353	940,811	938,010
Total { 1904	75,925	100,488	170,841	304,952	121,992	17,526	8,185	22,152	35,977	10,030	50,445	79,053	...	...
Native States, etc. :—														
Punjab Native States .	5,021	6,684	13,049	18,739	11,226	171	2	...	...	...	104	272	54,568	38,319
Jammu and Kashmir State.	69	84	229	955	1,110	95	1	...	...	7	20	89	2,659	8,692
Native States in Central Provinces.	...	...	...	...	...	...	...	...	...	...	...	...	...	101
Madras Native States .	...	...	...	...	...	...	...	...	...	...	...	...	...	4
Bombay Presidency Native States.	3,091	3,500	3,560	3,477	1,071	731	307	1,315	3,571	3,575	1,853	1,253	23,040	62,347
Baluchistan	...	...	...	...	...	...	...	...	...	...	...	...	...	25
United Provinces (Tehri Garhwal.)	...	...	...	...	...	...	...	...	...	...	24	8	32	...
Rajputana	2,548	1,077	5,310	11,925	7,416	536	19	275	1	43	50	43	30,453	16,268
Central India . .	273	353	495	273	60	1	10	60	146	250	490	217	2,826	26,645
Mysore	953	911	423	174	47	45	125	95	196	177	177	228	3,459	31,622
Bangalore Civil and Military Station.	321	196	101	60	57	70	73	90	158	273	164	258	1,854	2,574
Hyderabad State .	3,658	1,725	1,776	794	292	...	27	201	273	437	358	275	9,008	25,596
Total { 1905	14,575	11,920	24,437	35,421	22,259	1,225	640	2,374	3,417	3,708	3,170	2,681	128,310	305,983
Total { 1904	17,313	12,891	25,074	40,300	10,573	3,573	4,417	12,171	18,005	20,013	15,427	12,703	...	...
Grand Total { 1905	144,371	142,715	252,274	257,574	151,633	16,997	4,355	5,760	11,715	14,155	12,752	15,034	1,059,140	...
Grand Total { 1904	54,212	120,073	191,275	345,171	132,175	20,465	12,511	3,175	54,550	7,073	65,852	91,735	1,143,023	...
Calcutta City .	321	443	2,451	3,078	1,066	159	52	47	63	79	65	81	7,171	1,689
Bombay City .	1,069	1,371	2,173	3,744	2,033	673	121	205	101	77	45	58	14,171	13,504
Madras City .	...	4	1	2	7	2	1	...	...	...	...	5	22	5

Appendix B. to Section VI—Chief diseases—concluded.
TABLE V.—Respiratory disease.

Provinces, Districts and Towns.	Bengal.	Eastern Bengal and Assam.	United Provinces.	Punjab.	North-West Frontier Province.	Central Provinces and Berar.	Lower Burma.	Upper Burma.	Madras Presidency.	Bombay Presidency.	Ajmer-Merwara.	Coorg.	Registration India.
I.—Mortality by Provinces :—													
A—Deaths by months.													
January	...	209	3,379	5,495	51	1,503	309	124	1,018	5,884	48	...	17,933
February	...	140	2,431	5,895	91	1,716	351	160	1,777	5,811	68	3	18,335
March	...	153	2,643	6,345	67	1,388	366	105	1,646	5,596	138	2	18,148
April	...	142	2,577	5,307	93	1,503	323	91	1,511	5,943	89	6	6,873
May	...	133	1,911	5,748	73	1,365	245	98	1,445	4,359	40	4	15,316
June	...	131	1,479	4,709	91	977	521	105	1,471	3,573	31	1	13,890
July	...	103	1,301	4,358	113	847	358	110	1,513	3,460	19	2	13,182
August	...	99	1,421	4,335	92	881	365	138	1,673	3,904	19	3	13,050
September	...	142	1,674	3,900	111	1,113	481	185	1,514	3,938	17	2	13,077
October	...	152	1,531	4,398	88	1,395	426	166	1,723	3,738	30	2	13,430
November	...	170	1,704	5,071	94	1,408	489	205	1,581	4,182	35	1	14,047
December	...	259	2,038	5,860	164	1,513	349	231	1,916	5,038	40	6	17,457
Total	11,189*	1,829	23,171	64,353	1,138	15,377	4,083	1,643	19,657	54,527	561	30	1,04,7148
B.—Annual death ratios :—													
Ratio per 1,000 of population, 1905.	·33	·06	·49	3·05	0·57	1·70	·73	·56	0·5	2·95	1·18	·17	0·86
Ratio per 1,000 of population, 1904.	·16	†	·41	2·66	·39	1·38	†	†	0·5	3·22	†	†	†
Difference	+ ·06	†	+ ·08	+ ·39	+ ·18	+ ·63	†	†	0·0	− ·27	†	†	†
Mean ratio per 1,000 during 1901-1904.	†	†	†	2·61	0·35	†	†	†	0·3	†	·42	·21	†
Difference	†	†	†	+ ·44	+·22	†	†	†	+0·2	†	+ ·76	+ 0·06	†
II.—District mortality excluding towns.													
Number of districts affected	30	33	48	27	5	23	17	11	21	24	12	1	243
Highest district ratio	1·32	·94	5·25	11·68	0·67	4·23	·93	·74	1·2	9·86	2·13	·17	11·68
Name of that district	Patna.	Lakhimpore.	Hamirpore.	Gurdaspore.	Dera-Ismail-Khan.	Buldana.	Heozada.	Lower Chindwin	Cuddapah	Katra	Pokhar.	Yedenalknad.	Gurdaspore.
Lowest district ratio	·001	·008	·01	0·13	0·09	0·08	·01	·03	0·002	·05	·04	—	0·001
Name of that district	Furnea.	Chittagong.	Basti.	Multan.	Kohat.	Narsingh-pore.	Amherst.	Kyaukse.	Nellore.	Upper Sind Frontier.	Beawar Rural District.	—	Furnea.
Number of districts without mortality.	1	None	None	None	None	None	1	None	1	None	5	4	13
District death-rate per 1,000 of population.	·12	·05	·36	2·63	0·37	1·16	·43	·33	0·4	2·19	·39	·03	0·64
III.—Town mortality :—													
Number of towns affected	63	23	93	144	10	76	23	5	162	59	4	3	670
Highest town ratio	4·63	1·66	13·03	16·38	4·56	8·30	10·30	15·65	C·6	16·20	5·79	4·07	19·62
Name of that town	Calcutta.	Maldo.	Benares.	Narowal.	Peshawar	Balapur.	Toungoo.	Monywa.	Vallam.	Jambusar.	Ajmer Suburb.	Virarae-rapet.	Benares.
Lowest town ratio	·03	·05	·03	0·12	0·19	0·08	·05	·53	0·1	·07	·13	·74	·03
Name of that town	Serampore.	Comilla.	Ghazipore	Pindighab	Lakki.	Hinganghat.	Akyab.	Sagaing.	Ambur.	Karad.	Pisangan Town.	Mercara.	Ghazipur.
Number of towns without mortality.	64	21	8	1	None	23	15	7	70	4	2	3	339
Town death-rate per 1,000 of population.	1·63	·22	3·44	6·30	5·65	2·53	2·91	5·63	1·3	8·03	5·35	1·64	3·35

* Distribution by months not available.
† Not available.
‡ Include 11,189 deaths in Bengal of which the monthly distribution is not given.

STATEMENT I.—*Total Primary and Re-vaccinations, successful cases among the children, cost of the Special Vaccination Department, etc., during the official year 1905-06.*

PROVINCE.	NUMBER OF PERSONS VACCINATED BY THE SPECIAL AND DISPENSARY STAFFS COMBINED.		PERCENTAGE OF SUCCESSFUL CASES* TO TOTAL OPERATIONS.		NUMBER OF CHILDREN SUCCESSFULLY VACCINATED BY THE SPECIAL AND DISPENSARY STAFFS COMBINED		Average number of operations performed in each vaccinator of the Special Staff.	Total cost of the Special Department.	Average cost of each successful case vaccinated by the Special Department.
	Primary.	Re-vacci-nations.	Primary.	Re-vacci-nations.	Under one year.	1 to 6 years.			
								Rs.	Rs. A. P.
Bengal	1,024,625	136,605	99·13	66·29	809,606	973,461	1,031	1,57,803	0 1 3
Eastern Bengal and Assam . .	1,385,538	43,483	98·28	71·71	414,101	804,591	1,110	80,105	0 0 11
United Provinces of Agra and Oudh .	1,591,905	93,511	97·03	81·78	930,885	512,398	1,850	1,53,241	0 1 6
Punjab	638,310	110,420	99·19	83 38	517,638	103,636	2,810	97,702	0 2 3
North-West Frontier Province . . .	87,771	13,053	98·55	84·45	49,320	23,224	3,162	11,544	0 2 0
Central Provinces and Berar .	551,140	74,998	98·59	72 30	418,160	103,918	2,089	66,358	0 1 11
Madras Presidency	1,357,745	107,245	92·67	76 02	462,317	611,549	†1,660	2,24,419	0 3 8
Coorg	9,306	1,985	93·21	76 26	844	5,150	1,208	2,740	0 4 6
Bombay Presidency .	610,585	45,133	97·36	76·13	448,634	99,951	1,522	3,01,305	0 5 3
Burma . .	426,735	61,962	90·80	53·55	89,131	208,026	1,956	1,14,464	0 4 9
Ajmer-Merwara .	14,310	425	95·50	81·65	11,203	2,726	956	2,849	0 3 1
TOTAL .	8,556,770	691,125	97·20	74·20	4,161,967	3,474,929	1,480	1,282,636	0 2 5

* Excluding those the results of which were not known.
† Excludes average of work done by each medical subordinate.

STATEMENT II.—*Vaccination operations performed by the Special and Dispensary Establishments separately, deaths from small-pox, etc., during the official year 1905-06.*

PROVINCE.	Population.	NUMBER OF PERSONS VACCINATED (PRIMARY AND RE-VACCINATIONS COMBINED.)			Ratio of successful vaccinations per 1,000 of population.	Percentage of annual estimated births at 40 per 1,000 of population successfully vaccinated.	DEATHS FROM SMALL-POX*.	
		By Special Department.	By Dispensary Staff.	Total.			Number.	Ratio per 1,000 of population
Bengal	51,656,383	1,919,125	122,105	2,041,250	37·94	39·18	7,213	·14
Eastern Bengal and Assam . .	30,788,131	1,401,291	27,735	1,429,026	44 86	33·62	4,723	·15
United Provinces of Agra and Oudh .	47,960,074	1,691,761	6355	1,695,416	33·79	49·57	3,272	·07
Punjab	20,293,834	747,237	1,402	748,739	31 31	62·54	4,723	·23
North-West Frontier Province .	2,092,713	58,021	2,8459	100,866	44 11	58 92	571	·29
Central Provinces and Berar . .	13,021,159	597,348	28,790	626,138	44 83 ‡	76 75	8,364	·70
Madras Presidency .	38,108,531	1,462,890	2,100	1,464,990	33·75	30·00	18,540	·5
Coorg . .	180,607	10,878	413	11,201	55·45	11·68	132	·73
Bombay Presidency .	21,438,760	652,469	3,869	656,338	27·35	52·32	16,985	·92
Burma	10,477,508	471,420	17,507	488,927	37·84	21·27	6,161	·73
Ajmer-Merwara .	4 6,912	14,935	...	14,935	30 63	53·25	277	·58
Total .	237,395,617	9,070,475	207,471	9,277,856	36·30	43·83	70,961	·31

* For the Calendar year.
† Includes 10,010 persons enumerated at the Ajodhya fair.
‡ No special vaccinators are attached to dispensaries. Operations performed by Medical Subordinates.
§ There were no special vaccinators attached to the dispensaries. Operations were performed by Civil Surgeons and Medical subordinates.

STATEMENT III.—*The number of persons primarily vaccinated and the number of those who were successfully vaccinated in His Majesty's European and Native Army in India, during 1905.*

COMMANDS AND DIVISIONS.	EUROPEAN ARMY.														NATIVE ARMY.													
	Officers.		Officers' wives.		Officers' children.		Warrant and Non-Commissioned Officers and men.		Women.		Children.		Total.		European Officers.		European Officers' Wives.		European Officers' Children		Native Commissioned, Non-Commissioned Officers and men.		Women.		Children		Total.	
	Primary.	Successful.	Primary.	Successful.	Primary.	Successful.	Primary.	Successful.	Primary.	Successful.	Primary.	Successful.	Primary.	Successful.	Primary.	Successful.	Primary.	Successful.	Primary.	Successful.	Primary.	Successful.	Primary.	Successful.	Primary.	Successful.	Primary.	Successful.
Northern Command	...	...	...	...	7	7	40	33	1	1	331	287	379	328	8	5	...	...	5	5	2,170	1,917	226	303	2,452	1,097	4,561	4,127
Western Command	...	...	...	...	4	2	...	...	...	...	301	203	395	205	...	...	...	...	19	15	2,186	1,485	142	89	2,389	1,039	4,736	3,528
Eastern Command	...	...	...	...	11	11	3	2	1	1	301	213	316	257	...	...	...	...	7	6	1,405	772	163	147	1,159	995	2,714	1,920
Secunderabad Division	...	...	...	...	1	1	26	21	4	4	146	115	177	141	...	...	...	...	10	9	359	226	26	23	932	873	1,327	1,075
Burma Division	...	...	...	...	2	1	...	...	...	...	39	39	41	33	...	...	...	...	...	...	66	40	3	3	138	117	207	160
INDIA	...	...	...	...	25	22	69	56	6	6	1,118	880	1,218	964	8	5	...	...	41	35	6,186	4,434	560	465	6,730	5,881	13,545	10,810

ANNUAL RETURNS

OF THE

EUROPEAN ARMY OF INDIA

OF THE

NATIVE ARMY AND OF THE JAIL POPULATION

FOR THE YEAR

1905.

———————

COMPILED AND SYSTEMATICALLY ARRANGED FROM THE ORIGINAL DOCUMENTS

BY

A. C. MacGILCHRIST., M.A., M.B., Captain, I.M.S.,

Offg. Statistical Officer to the Government of India in the Sanitary and Medical Departments

CONTENTS.

I.—EUROPEAN TROOPS 1905.

A.—MEN.

B.—WOMEN.

C.—CHILDREN.

II.—NATIVE TROOPS, 1905.

* Omitted for the present by order of Government.
† There being no cases of cholera the table is blank.

III.—PRISONERS 1905.

(European, Eurasian, native ; male, female; adult, juvenile.)

IV.—TROOPS AND PRISONERS, 1905.

NOTE—In the tables for European troops, Native troops, and for prisoners, the months mentioned are calendar months.

TABLE G.

Grouping of Diseases in the Main Tables for 1905.[*]

HEAD OF DISEASE.	Includes or includes also
CHOLERA	Choleraic diarrhœa.
HEAT-STROKE	Sunstroke and Heat-Apoplexy.
ALCOHOLISM	Delirium tremens. Alcoholic Poisoning.
TUBERCLE OF THE LUNGS	Tubercular Phthisis, and Hæmoptysis due to tubercle.
OTHER RESPIRATORY DISEASES.	Includes Hæmoptysis and Cirrhosis of the lung not due to tubercle, and excludes Pneumonia and Tubercular Phthisis.
ANÆMIA AND DEBILITY	Old age (Tables for men and women). Immaturity at birth (Tables for children).
DIARRHŒA	Epidemic Diarrhœa.
HEPATIC CONGESTION AND INFLAMMATION.	Congestion of liver, Hepatitis, Perihepatitis ; but excludes Cirrhosis of liver.
VENEREAL DISEASES	Syphilis, Gonorrhœa, and Soft Chancre, which include also their sequelæ.
GUINEA-WORM AND OTHER ENTOZOA .	The entozoa numbered from 1 to 56, 67 to 81 : also Nos. 105 and 106.
PHAGEDÆNA, SLOUGH, AND GANGRENE.	Nomenclature of 1896, Nos. 25 *a* and *b*, 8. o. and 847. } These two headings appear only in jail tables.
ABSCESS, ULCER, AND BOIL .	Nomenclature of 1896, Nos. 799, 843, and 845. }
ABORTION AND PUERPERAL AFFECTIONS.	Nomenclature of 1896, Nos. 700 and 706 to 718, and any other diseases stated by medical officers to have been puerperal.
OTHER DISEASES PECULIAR TO WOMEN.	Nomenclature of 1896, No. 426, Vomiting of Pregnancy, Nos. 632 to 699, 701 to 705, and 719 to 730.

[*] For details of individual diseases, see Table LIII.

.

I.—EUROPEAN TROOPS, 1905.
A.—MEN.

Stations.	Height above sea level in feet.	Authority for height.	Stations.	Height above sea level in feet.	Authority for height.	Stations.	Height above sea level in feet.	Authority for height.
NORTHERN COMMAND :—			**WESTERN COMMAND :—contd.**			**EASTERN COMMAND :—contd.**		
Peshawar	1,165	S. G.	Ahmedabad	170	S. G.	Barrackpore	24	S. G.
Nowshera	1,100	M. O.	Kamptee	941	,,	Dinapore	171	,,
Rawalpindi	1,707	S. G.	Sitabaldi	1,236	,,	Allahabad	298	,,
Campbellpur	1,200	M. O.	Nasirabad	1,461	,,	Fort Allahabad	298	,,
Attock	891	S. G.	Neemuch	1,613	,,	Benares	250	,,
Sialkot	829	,,	Deesa	468	,,	Cawnpore	417	,,
Mian Mir	706	,,	Jhansi	800	,,	Fyzabad	336	,,
Fort Lahore	706	,,	Nowgong	770	I. B.	‡ Lebong	6,000	I. B.
Mooltan	402	,,	Jubbulpore	1,306	S. G.	‡ Ranikhet	5,983	S. G.
Ferozepore	645	,,	Saugor	1,753	,,	‡ Chaubuttia	6,912	,,
Jullundur	900	,,	Poona	1,900	,,	‡ Chakrata	6,885	,,
Amritsar	756	,,	Kirkee	1,837	,,	‡ Landour Convalescent Depôt	7,362	,,
Ambala	903	,,	Colaba (Bombay)	20	,,	‡ Naini Tal ,, ,,	6,400	,,
‡ Cherat	4,520	,,	Deolali Depôt	1,829	,,	‡ Darjeeling ,, ,,	7,168	,,
‡ Khyragully	8,746	,,	‡ Mount Abu Sanitarium	3,060	,,			
‡ Baragully	7,800	M. O.	‡ Pachmarhi	3,481	,,	**SECUNDERABAD DIVISION :—**		
‡ Kuldunnah	7,040	S. G.	‡ Purandhur	4,561	,,	Secunderabad	1,732	S. G.
‡ Kalabagh	7,036	I. D.	‡ Khandalla	2,000	M. O.	Bellary	1,483	,,
‡ Gharial	5,112	S. G.	Ahmednagar	2,123	S. G.	Bangalore	3,021	,,
‡ Barian Camp	7,133	I. B.				Madras	13	,,
‡ Upper Topa	7,000	M. O.	Belgaum	2,473	,,	St. Thomas' Mount	250	,,
‡ Lower Topa	7,321	I. B.	Aden	26	,,			
‡ Khan Spur	7,500	M. O.				Cannanore	47	,,
‡ Dagshai	5,692	S. G.	**EASTERN COMMAND :—**			Calicut	27	M. D.
‡ Solon	5,406	,,	Meerut	731	S. G.	Mallapuram	500	M. O.
‡ Subathu	4,124	,,	Delhi	715	,,	‡ Ramandrug	3,150	S. G.
‡ Jutogh	6,771	,,	Muttra	576	,,	‡ Wellington	6,160	,,
‡ Murree Convalescent Depôt	7,098	,,	Agra	554	,,	Poonamallee Depôt	50	M. O.
‡ Dalhousie ,, ,,	6,732	,,	Bareilly	560	,,			
‡ Kasauli ,, ,,	5,971	,,	Shahjahanpur	507	,,	**BURMA DIVISION :—**		
			Roorkee	884	,,	Fort Dufferin (Mandalay)	249	S. G.
WESTERN COMMAND :—			Lucknow	400	,,	Shwebo	600	M. O.
‡ Quetta	5,511	S. G.	Sitapur	449	,,	Bhamo	351	S. G.
Kurrachee	28	,,	Fatehgarh	444	I. B.	Meiktila	298	,,
Hyderabad (Sind)	134	I. B.	Fort William (Calcutta)	17	S. G.	Thayetmyo	145	,,
Mhow	1,903	S. G.	Fort Fulta	18	,,	Rangoon	14	,,
Indore	1,806	,,	Fort Chingrikhal	...	...	Port Blair	85	,,
‡ Taragarh Sanitarium	2,855	,,	Dum-Dum	...	...	‡Maymyo	3,500	,,

* These heights are usually those of the survey-marks or of the mercury-surface in barometer-cisterns of meteorological observatories.
† S. G. = Surveyor-General of India ; I. B. = Intelligence Branch of the Quartermaster-General's Department ; M. D. = Meteorological Department ; M. O. = Medical Officers in charge of Station Hospitals in their Sanitary Reports.
‡ These are the official " Hill Stations."

TABLE I.

RATIOS OF COMMANDS.

The ratios of admissions and deaths to strength are taken from Table III. The actuals will be found in Table IV.

	Northern Command.	Western Command.	Eastern Command.	Secunderabad Division.	Burma Division.	India.[*]
	RATIOS PER 1,000 OF THE AVERAGE STRENGTH.					
I.—STRENGTH	18,724	19,537	18,983	8,389	3,812	71,343
II.—† CONSTANTLY-SICK-RATE OF EACH MONTH—						
January	64·9	57·9	62·5	82·8	36·6	59·4
February	48·1	54·4	57·7	59·2	58·0	52·3
March	39·7	57·2	57·2	59·1	52·6	47·0
April	37·9	45·8	47·5	50·8	49·1	44·8
May	48·7	44·5	45·6	60·8	51·6	48·7
June	51·3	50·5	50·4	51·0	50·5	51·1
July	54·8	52·1	53·4	55·2	52·3	54·1
August	52·3	61·0	52·6	62·2	58·8	57·3
September	46·0	66·7	54·8	66·0	54·7	55·3
October	43·2	59·9	54·5	65·1	48·3	52·4
November	52·4	50·2	61·8	58·9	59·1	54·7
December	58·3	56·7	52·1	57·3	59·1	55·4
OF THE YEAR	49·3	55·0	55·6	61·6	54·0	52·3
II.—ADMISSION-RATE OF THE YEAR—						
Influenza	33·4	7·1	11·2	·1	1·3	14·2
Cholera	...	·1	·1	·1	...	·1
Small-pox	·3	3·9	·1	1·2	·8	1·4
Enteric Fever	17·7	17·5	16·9	16·8	2·4	16·1
Intermittent Fever	69·6	162·8	113·2	77·7	114·4	111·4
Remittent Fever	·5	·8	3·6	·1	10·5	2·3
Simple Continued Fever	63·4	42·4	48·0	200·0	80·5	47·9
Tubercle of the lungs	1·7	3·0	1·8	1·1	3·9	2·1
Pneumonia	5·7	4·2	3·2	1·4	2·9	4·1
Other Respiratory Diseases	20·2	26·4	22·5	20·1	30·7	23·0
Dysentery	6·3	17·2	12·6	13·7	15·5	13·4
Diarrhœa	14·0	16·2	19·6	5·8	17·1	15·2
Hepatic Abscess	1·3	2·9	2·4	2·7	·5	2·1
„ Congestion and Inflammation	23·4	14·4	19·8	9·3	11·3	17·1
Venereal Diseases	116·1	155·4	165·8	194·1	207·0	153·7
ALL CAUSES	800·5	896·4	820·2	812·8	970·9	834·3
IV.—DEATH-RATE OF THE YEAR—						
Cholera	...	·10	·11	...	1·31	·13
Small-pox	...	·05	...	...	...	·01
Enteric Fever	3·12	3·89	3·11	2·62	·79	2·99
Intermittent Fever	·05	·20	·11	·12	...	·11
Remittent Fever	...	...	·05	...	...	·01
Simple Continued Fever	...	...	...	...	...	...
Heat-stroke	1·15	·51	1·00	·48	·26	·77
Circulatory Diseases	·49	·67	·63	·24	·52	·56
Tubercle of the lungs	·27	·41	·37	...	...	·28
Pneumonia	·82	·67	·47	·36	1·05	·63
Other Respiratory Diseases	...	...	·21	...	...	·06
Dysentery	·05	·72	·42	·60	1·31	·46
Diarrhœa	...	...	...	...	...	...
Hepatic Abscess	·44	1·54	1·42	1·79	·26	1·18
ALL CAUSES	9·08	11·62	10·90	8·46	9·71	10·05
V.—PERCENTAGE IN 100 ADMISSIONS—						
Influenza	4·42	·81	1·36	·01	·14	1·70
Cholera	...	·01	·01	·01	...	·02
Small-pox	·04	·44	·01	·15	·08	·17
Enteric Fever	2·21	1·95	2·06	2·07	·24	1·93
Intermittent Fever	8·69	18·16	13·80	9·56	11·78	13·35
Remittent Fever	·07	·31	·12	·01	1·08	·27
Simple Continued Fever	7·92	4·73	5·86	2·46	8·30	5·74
Tubercle of the lungs	·21	·34	·22	·13	·41	·25
Pneumonia	·71	·53	·39	·18	·30	·50
Other Respiratory Diseases	2·53	2·94	2·74	2·48	3·16	2·76
Dysentery	·81	1·92	1·54	2·30	1·59	1·61
Diarrhœa	1·75	1·80	2·40	·72	1·76	1·83
Hepatic Abscess	·16	·33	·29	·34	·05	·26
„ Congestion and Inflammation	2·03	1·60	2·42	1·14	1·16	2·05
Venereal Diseases	14·51	17·34	20·21	23·87	21·32	18·42
VI.—PERCENTAGE IN 100 DEATHS—						
Cholera	...	·9	1·0	...	13·5	1·3
Small-pox	...	·4	...	...	...	·1
Enteric Fever	34·3	31·7	28·5	31·0	8·1	29·7
Intermittent Fever	·6	1·8	1·0	1·4	...	1·1
Remittent Fever	...	...	·5	...	...	·1
Simple Continued Fever	...	...	...	...	...	...
Heat-stroke	13·7	4·4	9·2	5·6	2·7	7·7
Circulatory Diseases	5·4	5·7	6·3	2·8	5·4	5·6
Tubercle of the lungs	3·0	3·5	3·4	...	...	2·8
Pneumonia	9·0	5·7	4·3	4·3	10·8	6·3
Other Respiratory Diseases	...	...	1·9	...	...	·6
Dysentery	·6	6·2	3·9	7·0	13·5	4·6
Diarrhœa	...	...	...	...	...	...
Hepatic Abscess	4·8	14·1	13·0	21·1	2·7	11·7

[*] Including troops on the line of march, and the Field Force. For complete detail of diseases see Table LIII.
† Worked on the aggregates.

TABLE II.

RATIOS of GEOGRAPHICAL GROUPS.

The ratios of admissions and deaths to strength are taken from Table III.   **The actuals will be found in Table IV.**

RATIOS PER 1,000 OF THE AVERAGE STRENGTH.

	I Burma Coast and Bay Islands.	II Burma inland.	IV Bengal and Orissa.	V Gange-tic Plain and Chutia Nagpur.	VI Upper Sub-Hima-laya.	VII N.W., Frontier Indus Valley, and N.W. Rajpu-tana.	VIII S.E. Rajpu-tana, Central India, and Gujarat.	IX Deccan.	X Western Coast.	XI South-ern India.	XIIa Hill Stations.	XIIb Hill Conva-lescent Depôts, and Sanita-ria.	India.
I.—STRENGTH	1,259	2,063	1,690	7,315	13,724	5,068	6,562	10,434	1,591	3,738	10,508	3,512	71,343
II.—†CONSTANTLY-SICK-RATE OF EACH MONTH—													
January	59·4	52·9	86·6	50·6	69·6	54·6	61·8	54·5	70·2	108·8	55·5	66·2	56·4
February	57·4	55·3	81·9	50·6	50·8	48·9	58·0	52·1	63·0	56·1	46·1	68·4	52·3
March	55·0	50·0	75·6	42·2	40·4	43·4	51·1	50·3	75·7	47·6	36·4	63·6	47·0
April	49·8	43·4	70·7	46·8	44·2	39·2	45·0	44·9	45·3	47·7	36·6	52·0	44·8
May	40·4	51·9	59·5	45·6	51·1	63·7	48·3	44·8	41·8	51·1	48·7	59·6	48·7
June	38·9	47·0	60·7	55·9	52·2	62·0	47·4	45·5	48·8	53·7	45·6	61·9	51·7
July	46·9	49·3	71·1	56·4	58·7	74·8	43·9	46·9	59·4	55·1	47·3	58·3	54·1
August	64·5	51·0	103·3	51·5	59·6	80·7	47·0	57·8	65·4	50·7	48·3	59·8	57·3
September . . .	49·8	52·4	100·8	54·4	52·9	69·6	52·8	60·4	64·3	53·9	44·2	59·8	55·5
October . . .	34·3	54·7	78·8	51·7	49·1	58·2	43·4	58·9	73·9	51·8	45·4	59·1	57·4
November . . .	64·3	49·2	76·4	56·8	51·5	57·6	61·4	56·2	43·5	52·5	70·6	78·7	54·7
December . . .	63·0	53·8	63·7	55·8	57·5	57·3	60·1	52·4	55·1	50·8	50·4	63·2	53·4
OF THE YEAR .	51·7	50·0	77·3	51·4	52·5	57·2	51·7	52·1	58·3	54·7	45·4	60·4	52·3
III.—ADMISSION-RATE OF THE YEAR—													
Influenza	·8	1·9	60·4	1·9	15·5	93·7	5·0	6·1	7·5	...	5·5	3·1	14·2
Cholera	...	...	...	·3	...	...	...	·2	...	·3	...	...	·2
Small-pox . . .	2·4	...	...	·3	·3	1·0	3·8	3·6	5·7	2·4	·1	·3	1·4
Enteric Fever . .	·8	2·9	7·7	17·4	23·3	18·9	10·5	22·6	1·3	13·1	10·0	9·7	16·0
Intermittent Fever .	37·0	135·6	255·6	95·4	74·6	190·8	161·5	107·0	152·1	91·0	60·0	121·0	111·4
Remittent Fever .	26·2	3·4	4·7	7·5	2·2	·2	6·4	·5	3·8	...	·3	·9	2·3
Simple Continued Fever .	135·8	65·0	21·3	104·4	53·6	64·5	8·7	41·9	1·9	15·2	53·6	13·1	47·9
Rheumatic Fever .	·8	...	·6	·3	·3	·4	·5	·3	...	...	·7	·6	·4
Tubercle of the lungs	8·7	1·5	3·6	2·2	1·6	2·4	·3	2·6	1·3	·3	1·4	2·3	2·1
Pneumonia . . .	2·4	3·4	4·7	1·8	5·4	7·1	4·4	3·5	3·1	·5	5·8	1·1	4·1
Other Respiratory Diseases	31·8	27·6	42·0	19·5	19·1	35·6	18·1	19·5	19·5	14·7	30·3	35·9	23·0
Dysentery . . .	35·7	5·8	21·3	12·9	9·0	7·7	12·3	19·5	6·3	16·6	10·6	13·7	13·4
Diarrhœa . . .	23·8	14·5	23·7	11·5	13·2	10·3	14·9	12·7	17·6	6·7	26·7	18·8	15·2
Hepatic {Abscess .	...	...	3·6	3·1	1·3	1·0	2·6	3·5	3·8	·1	1·3	3·7	2·1
Congestion and Inflammation .	7·9	11·6	26·0	20·9	28·9	10·7	14·9	12·9	15·1	10·4	14·6	12·2	17·1
Venereal Diseases . .	230·7	157·5	244·4	152·3	139·0	130·2	173·7	165·3	243·2	200·5	122·5	140·1	157·7
ALL CAUSES	1,220·8	823·1	1,160·4	800·8	815·5	1,037·7	820·2	806·2	824·0	793·3	754·5	813·9	834·3
IV.—DEATH-RATE OF THE YEAR—													
Cholera	...	2·42	...	·27	...	...	...	·19	...	...	...	...	·13
Small-pox . . .	...	...	...	...	...	...	·20	...	...	...	...	...	·01
Enteric Fever . .	·79	·48	·59	2·87	4·23	3·75	4·88	3·55	·63	1·87	2·76	·85	2·99
Intermittent Fever .	...	...	·59	·14	...	·20	...	·10	...	·27	...	...	·11
Remittent Fever .	...	...	...	...	·07	...	...	...	...	...	...	...	·01
Simple Continued Fever .	...	...	...	...	...	...	...	...	...	...	...	...	...
Heat-stroke . .	...	·48	1·18	1·23	1·02	1·78	1·07	·29	1·89	·27	·10	...	·77
Circulatory Diseases	...	...	·96	·58	·59	·61	1·0	·63	·54	·67	·28	·55	·58
Tubercle of the lungs	...	...	·55	...	·20	·46	·79	...	...	·10	·23	...	·38
Pneumonia . .	·79	1·45	1·78	·14	·86	1·5	·30	·58	·63	·27	1·14	·37	·63
Other Respiratory Diseases	...	...	·59	...	·07	...	...	·06	...	...	·10	·23	·06
Dysentery . . .	3·18	...	1·78	·27	·15	...	·30	...	...	...	·67	·85	·46
Diarrhœa . . .	...	...	...	...	...	...	...	...	...	...	...	...	...
Hepatic Abscess . .	...	...	2·96	1·50	·44	·59	1·68	1·73	1·26	1·07	·86	2·28	1·18
ALL CAUS	7·15	8·73	17·75	10·25	10·27	10·06	11·89	10·16	9·43	6·42	8·95	9·11	10·05
V.—PERCENTAGE IN 100 ADMISSIONS—													
Influenza . . .	·07	·24	5·20	·24	1·90	8·53	·61	·76	·88	...	·73	·38	1·70
Cholera . . .	...	...	...	·03	...	...	...	·02	...	·03	...	...	·02
Small-pox . . .	·20	...	...	·03	·04	·09	·46	·45	·66	·30	·01	·01	·17
Enteric Fever . .	·07	·35	·66	2·17	2·86	1·73	2·04	3·05	·15	1·65	1·45	1·18	1·93
Intermittent Fever .	2·21	18·90	22·03	11·92	9·15	17·40	19·70	13·26	17·70	11·44	7·95	14·78	13·35
Remittent Fever .	2·15	·41	·41	·34	·27	·02	·73	·06	·44	...	·10	·10	·27
Simple Continued Fever .	11·13	7·89	1·84	13·04	6·58	5·90	1·06	5·19	·23	1·92	7·10	1·60	5·74
Rheumatic Fever .	·07	...	·05	·03	·04	·04	·06	·04	...	...	·09	·07	·05
Tubercle of the lungs	·72	·18	·31	·27	·20	·2	·22	·32	·15	·17	·19	·23	·25
Pneumonia . . .	·20	·41	·41	·12	·66	·65	·51	·44	·37	·07	·77	·14	·50
Other Respiratory Diseases	2·60	3·36	3·62	2·44	2·34	2·56	2·41	2·41	2·28	1·85	4·01	4·38	2·76
Dysentery . . .	2·03	·71	1·84	1·60	1·11	·70	1·50	2·41	·74	2·00	1·40	1·68	1·61
Diarrhœa . . .	1·95	1·77	2·04	1·43	1·62	·91	1·82	1·57	2·06	·84	3·54	2·29	1·83
Hepatic {Abscess Congestion and Inflammation .	·61	...	·31	·39	·16	·09	·3	·43	·44	·17	·18	·49	·26
Venereal Diseases . .	19·39	1·41 19·14	2·24 21·06	2·61 19·02	3·55 17·05	·08 11·92	1·82 21·18	1·60 20·49	1·76 28·46	1·31 26·34	1·93 16·23	1·50 17·11	2·05 18·42
VI.—PERCENTAGE IN 100 DEATHS—													
Cholera . . .	...	27·8	...	2·7	...	...	...	1·9	...	...	...	...	1·3
Small-pox . . .	...	...	...	...	...	...	1·9	...	...	...	...	...	·1
Enteric Fever .	11·1	5·6	3·3	28·0	41·1	35·2	41·0	34·9	6·7	29·2	30·9	9·4	29·7
Intermittent Fever .	...	...	3·3	1·3	...	·9	...	·9	...	4·2	...	...	1·1
Remittent Fever .	...	...	...	...	·7	...	...	...	...	...	...	...	·1
Simple Continued Fever .	...	...	...	...	...	...	...	...	...	...	...	...	...
Heat-stroke . .	...	5·6	6·7	12·0	9·9	16·7	10·3	2·8	20·0	4·2	1·1	...	7·7
Circulatory Diseases	...	...	5·4	·93	5·7	5·5	5·1	1·0	6·7	8·3	7·4	3·1	5·6
Tubercle of the lungs	...	...	6·7	5·3	1·4	1·9	5·8	2·8	...	...	1·1	3·1	3·8
Pneumonia . .	11·1	16·7	10·0	1·3	6·4	5·0	2·6	5·7	6·7	4·2	12·8	6·3	6·3
Other Respiratory Diseases	...	...	3·1	...	·7	...	...	·6	...	...	1·1	3·1	·6
Dysentery . . .	44·4	...	10·0	2·7	1·4	...	2·6	9·4	...	...	7·4	6·3	4·6
Diarrhœa . . .	...	...	...	...	...	...	...	...	...	...	...	...	...
Hepatic Abscess .	...	...	16·7	14·7	4·3	5·6	14·1	17·0	13·3	16·7	9·6	25·1	11·7

* For complete detail of diseases, see Table LIII.   † Worked on the aggregates.

TABLE III.

RATIOS of STATIONS, GROUPS, and COMMANDS.   For actuals see Table IV.

Stations and Groups.	Average annual strength.	Influenza.	Cholera.	Small-pox.	Enteric Fever.	Intermittent Fever.	Remittent Fever.	Simple Continued Fever.	Rheumatic Fever.	Heat-stroke.	Circulatory Diseases.	Tubercle of the lungs.	Pneumonia.	Other Respiratory Diseases.	Dysentery.	Diarrhœa.	Hepatic Abscess.	Hepatic Congestion and Inflammation.	Venereal Diseases.	All Causes.	Constantly Sick Rate.	Syphilis.	Soft Chancre.	Gonorrhœa.
												1. Admission-rate. 2. Death-rate.												
Port Blair	140	...	...	...	21·4	135·7	35·7	...	...	...	7·1	...	23·6	...	...	...	...	7·1	100·0	893·9	31·9	57·1	7·1	35·7
		...	...	...	...	...	...	...	...	...	...	...	...	...	...	...	...	...	...	...		...	...	...
Rangoon	1,119	·9	...	2·7	·9	27·7	...	148·3	·9	...	7·1	8·9	2·7	32·2	40·2	26·8	...	8·0	253·8	1,261·5	54·2	50·0	76·0	127·8
		...	...	...	·89	...	...	...	...	...	·89	...	3·57	...	...	...	...	·89	8·04			·80	...	...
Group I.—Burma Coast and Bay Islands.	1,25·	·8	...	...	·3	27·0	26·2	135·8	·8	...	6·4	8·7	2·4	31·8	35·7	23·8	...	7·9	236·7	1,220·8	† 51·7	50·8	68·3	117·6
		...	...	...	·79	...	...	...	...	...	...	...	·79	...	3·18	...	...	...	·79	7·15		·79	...	...
Thayetmyo	483	...	...	...	6·1	500·0	...	...	...	2·0	4·1	4·1	10·2	36·9	8·2	20·5	...	6·1	139·3	1,030·7	45·9	26·6	22·5	90·2
		...	10·25	...	2·05	...	...	...	...	2·05	...	6·15	...	...	...	...	...	...	...	20·49			...	...
Meiktila	370	...	...	...	13·2	5·3	23·2	...	...	...	10·6	...	2·6	31·7	7·9	21·1	...	5·3	153·0	807·7	55·9	29·0	66·0	58·0
		...	...	...	...	...	...	...	...	...	...	...	...	...	...	...	...	...	...	2·64			...	...
Fort Dufferin	379	...	...	...	29·0	...	97·6	...	...	...	5·3	2·6	...	34·3	5·3	21·1	...	2·6	160·9	926·1	65·7	29·0	63·3	68·6
		...	...	...	...	...	...	...	...	...	...	...	...	...	...	...	...	...	...	2·64			...	...
Shwebo	535	6·8	...	...	3·4	40·5	3·4	13·6	...	3·4	6·8	...	1·7	20·4	5·1	...	...	23·9	175·2	596·9	42·9	54·0	59·5	64·6
		...	...	...	...	...	...	...	...	...	...	...	...	...	...	...	...	...	...	6·80			...	...
Bhamo	229	...	...	...	4·4	161·6	13·1	4·4	...	...	8·7	...	...	8·7	...	17·5	...	4·4	152·8	821·0	46·2	39·3	34·9	78·6
		...	...	...	...	...	...	...	...	...	...	...	...	...	...	...	...	...	...	8·73			...	...
Group II.—Burma Inland.	2,063	1·9	...	...	2·0	155·6	3·4	65·0	...	1·5	6·8	1·5	3·4	27·6	5·8	14·5	...	11·6	157·5	823·1	† 50·6	35·9	49·9	71·7
		...	2·42	...	·48	...	...	...	...	·48	...	...	1·45	...	...	...	...	...	...	8·73			...	...
Fort William	1,044	92·9	...	...	3·8	292·1	3·9	15·3	1·0	...	23·0	3·8	2·9	36·4	16·3	21·1	1·0	11·5	283·3	1,297·8	92·7	61·3	90·0	132·2
		...	...	...	...	·96	...	...	1·92	...	·96	...	·96	...	...	...	3·83	...	...	15·33			...	...
,, Fulta	27	...	...	...	...	148·1	...	37·0	...	...	...	...	37·0	37·0	...	...	...	...	222·2	555·6	7·0	...	74·1	148·1
		...	...	...	...	...	...	...	...	...	...	...	...	...	...	...	...	...	...	...			...	...
,, Chiogrikhal.	51	...	...	...	...	176·5	...	...	...	...	19·6	...	...	...	...	19·6	...	...	333·3	882·4	8·6	98·0	137·3	98·0
		...	...	...	...	...	...	...	...	...	...	...	...	...	...	...	...	...	...	...			...	...
Dum-Dum	281	17·8	...	...	10·7	135·8	7·1	...	...	...	3·6	7·1	17·8	10·7	7·1	10·7	3·6	67·6	89·0	782·9	50·3	3·6	28·5	56·9
		...	...	...	...	...	...	...	...	...	...	7·12	7·12	...	3·56	...	...	...	...	28·47			...	...
Barrackpore	286	...	...	...	21·0	262·2	10·5	66·4	...	14·0	...	...	10·5	55·9	49·0	10·5	45·5	241·3	1,132·2	66·7	69·9	76·9	94·4
		...	...	...	3·50	...	...	...	...	...	...	...	3·50	3·50	...	3·50	...	...	...	20·98			...	...
Group IV.—Bengal and Orissa.	1,690	60·4	...	...	7·7	255·6	4·7	21·3	·6	...	17·8	3·6	4·7	42·0	21·3	23·7	3·6	26·0	244·4	1,160·4	† 77·3	53·3	78·7	112·4
		...	...	...	·59	·51	...	...	...	1·18	...	1·18	1·78	·59	1·78	...	2·90	...	...	17·75			...	...
B Dinapore	609	...	1·6	...	32·8	83·7	19·7	11·5	...	1·6	4·9	4·9	6·6	23·0	8·2	3·3	8·2	85·4	174·1	715·9	42·9	27·6	46·0	98·5
		...	1·64	...	...	...	...	...	...	...	...	...	...	...	...	...	6·57	...	...	11·49			...	...
Benares	191	...	...	...	5·2	67·0	...	168·8	...	15·5	15·5	...	...	25·8	...	10·3	...	10·3	97·9	701·0	34·8	10·3	25·8	61·9
		...	...	...	5·15	...	...	...	...	5·15	...	...	...	...	...	...	...	...	...	20·62			...	...
Allahabad	976	3·1	1·0	...	6·1	137·3	1·0	77·9	...	8·2	4·1	...	2·0	10·2	3·1	6·1	...	5·1	167·0	703·9	58·0	41·0	38·0	87·1
		...	1·02	...	1·02	1·02	...	...	...	1·02	3·07	...	1·02	...	...	1·02	...	...	...	11·27			...	...
Fort Allahabad.	216	...	...	...	...	407·4	...	101·9	...	13·9	...	4·6	...	23·1	...	...	...	4·6	157·4	1,027·8	53·8	18·5	55·6	83·3
		...	...	...	...	...	...	...	...	4·63	...	...	...	...	...	...	...	...	...	4·63			...	...
Fyzabad	987	...	...	...	9·1	90·2	...	97·3	1·0	7·1	11·1	...	1·0	24·3	10·1	17·2	2·0	6·1	164·1	827·7	52·3	33·4	49·6	81·1
		...	...	...	1·01	...	...	...	1·01	2·02	...	...	...	...	...	...	...	...	...	6·08			...	...
Sitapur	582	...	...	...	3·4	55·0	...	144·3	...	1·7	10·3	5·2	1·7	34·4	...	3·4	...	3·4	55·0	651·2	30·8	17·2	3·4	34·4
		...	...	...	1·72	...	...	...	...	...	...	...	...	...	...	...	...	...	...	3·44			...	...
Lucknow	2,556	1·6	...	...	27·8	80·2	1·2	167·1	·39	7·4	11·0	3·1	2·0	21·1	27·4	14·5	5·1	26·6	167·4	943·7	61·4	35·6	23·5	108·2
		...	...	...	4·69	...	...	...	...	1·17	·78	·7	...	...	...	...	1·96	...	...	10·95			...	...
Cawnpore	1,039	1·0	...	1·0	3·8	60·6	...	14·4	...	50·0	9·6	1·0	...	9·6	4·8	9·6	1·9	12·5	136·7	603·5	39·0	16·4	28·9	91·4
		...	...	...	·96	...	...	...	...	1·92	...	1·92	...	...	·96	...	·96	...	...	10·59			...	...
Fatehgarh	157	38·2	...	...	82·7	146·5	12·7	38·2	...	...	12·7	...	...	6·4	6·4	51·0	...	25·5	178·3	936·3	43·4	12·7	38·2	127·4
		...	...	...	25·48	...	...	...	...	...	...	...	...	...	6·37	...	...	...	...	31·85			...	...
Group V.—Gangetic Plain and Chutia Nagpur.	7,315	1·9	·3	·1	17·4	95·1	2·5	104·4	·3	12·0	9·2	2·2	1·8	19·5	12·9	11·5	3·1	20·9	152·3	800·8	† 51·4	29·7	31·4	91·2
		...	·27	...	2·87	·14	...	...	...	1·23	·96	·55	·14	...	·27	...	1·50	...	...	10·25			...	...

* Derived from the aggregates.   † Worked on the aggregates.

TABLE III—continued.

RATIOS of STATIONS, GROUPS, and COMMANDS.

For actuals see Table IV.

STATIONS AND GROUPS.	Average annual strength.	Influenza.	Cholera.	Small-pox.	Enteric Fever.	Intermittent Fever.	Remittent Fever.	Simple Continued Fever.	Rheumatic Fever.	Heat-stroke.	Circulatory Diseases.	Tubercle of the lungs.	Pneumonia.	Other Respiratory Diseases.	Dysentery.	Diarrhœa.	Hepatic Abscess.	Hepatic Congestion and Inflammation.	Venereal Diseases.	ALL CAUSES.	CONSTANTLY SICK RATE.	Syphilis.	Soft Chancre.	Gonorrhœa.
					1. ADMISSION-RATE.								2. DEATH-RATE.											
A Shahjahanpur.	489	... ...	... ...	... ...	8·2 ...	36·8 ...	... ...	10·2 ...	... ...	... ...	6·1 ...	2·0 ...	... ...	16·4 ...	6·1 ...	6·1 ...	... ...	2·0 ...	151·3 ...	627·8 2·04	43·8	24·5 ...	24·5 ...	102·2 ...
Barcilly	1,149	... ...	... ...	... ...	10·4 1·74	94·9 ...	... ...	·9 ...	... ...	9·6 ...	6·1 ·87	... ...	7·0 ...	18·3 ·67	6·1 ...	12·2 ...	... ...	49·6 ...	120·1 ...	768·c 4·35	47·2	26·1 ...	15·7 ...	78·3 ...
Roorkee	393	2·5 ...	... ...	... ...	12·7 5·09	81·4 ··	... ...	2·5 ...	... ...	7·6 ...	7·6 2·54	2·5 ...	... ...	12·7 ...	28·0 2·54	17·8 ...	... ...	12·7 ...	221·4 ...	786·3 10·18	48·8	33·1 ...	81·4 ...	106·9 ...
Meerut	1,884	9·0 ...	... ...	... ...	57·3 11·86	118·4 ...	·5 ·53	21·8 ...	1·1 ...	2·1 ·53	14·9 ·53	2·7 ...	4·8 ·53	21·8 ·53	19·6 ·53	18·6 ...	2·1 1·06	20·7 ...	171·4 ·53	833·0 16·99	63·2	43·0 ·53	62·1 ...	66·3 ...
Delhi	327	30·6 ...	... ...	... ...	12·2 6·12	529·1 ...	61·2 ...	12·2 ...	... ...	24·5 3·06	18·3 ...	... ...	... ...	15·3 ...	6·1 ...	6·1 ...	... ...	... ...	192·7 ...	1,250·5 15·29	80·0	58·1 ...	79·5 ...	55·0 ...
Ambala	2,314	1·7 ...	... ...	... ...	5·2 2·16	31·5 ...	... ...	27·2 ...	... ...	1·7 ·43	7·3 ·43	1·7 ...	11·7 2·16	17·7 ...	2·6 ...	11·7 ...	2·2 ·43	12·1 ·43	152·1 ...	720·4 9·08	45·1	33·7 ...	33·7 ...	84·7 ...
B Jullundur	675	8·9 ...	... ...	... ...	8·9 ...	47·4 ...	... ...	13·3 ...	... ...	7·4 1·48	10·4 ...	1·5 ...	1·5 1·4	13·3 ...	5·9 ...	8·9 ...	1·5 1·48	4·4 ...	103·7 ...	545·2 5·93	34·9	17·8 ...	13·3 ...	72·6 ...
Ferozepore	823	1·1 ...	... ...	4·5 1·12	20·2 ...	92·9 —	...	134·4 ...	... ...	11·2 1·12	5·6 ...	3·4 ...	3·4 ...	15·7 ...	11·2 ...	16·8 ...	1·1 ...	11·2 ...	135·5 ...	771·6 5·60	55·9	44·8 ...	15·7 ...	75·0 ...
Amritsar	204	... ...	... ...	24·5 4·90	27·4 ...	... ...	63·7 ...	... ...	39·2 4·90	19·6 ...	... ...	... ...	4·9 ...	9·8 ...	4·9 ...	... ...	4·9 ...	137·3 ...	750·0 14·71	45·2	9·8 ...	24·5 ...	102·9 ...
Mian Mir	773	137·8 ...	... ...	22·1 2·77	42·9 ...	2·6 ...	149·9 ...	... ...	30·4 6·22	6·9 2·77	1·4 1·38	5·3 ...	24·9 ...	27·7 ...	23·5 ...	1·4 ...	18·0 ...	168·7 ...	1,354·1 16·60	68·8	23·5 ...	41·5 ...	103·7 ...
Fort Lahore	121	16·5 ...	... ...	47·6 ...	90·9 ...	... ...	247·9 ...	... ...	74·4 ...	... ...	24·8 ...	16·5 ...	8·3 ...	41·3 ...	57·9 ...	... ...	41·3 ...	157·0 ...	1,363·6 ...	44·1	24·8 ...	33·1 ...	99·2 ...
Sialkot	1,208	47·2 ...	... ...	7·5 1·66	49·7 ...	... ...	99·3 ...	·8 ...	10·8 ...	13·2 ...	... ...	·8 ...	18·2 ...	1·7 ...	17·4 ...	2·5 ·83	13·2 ...	66·2 ...	814·6 4·14	40·3	20·7 ...	8·3 ...	37·5 ...
Rawalpindi	2,923	6·5 ...	... ...	23·9 3·42	37·6 ...	... ...	25·0 ...	·3 ·34	4·4 1·03	147 ·66	1·0 ...	5·8 ·34	24·3 ...	3·4 ...	6·2 ...	·7 ·34	67·1 ...	131·4 ...	783·1 8·69	55·0	42·4 ...	19·2 ...	69·8 ...
Campbellpur	226	... ...	... ...	190·3 35·40	39·8 ...	31·0 ...	8·8 ...	... ...	48·7 4·42	4·4 ...	—	4·4 ...	22·1 ...	22·1 ...	8·8 ...	... ...	61·9 ...	97·3 ...	854·0 57·52	56·3	17·7 ...	8·8 ...	70·8 ...
Attock	196	... ...	... ...	10·2 5·10	275·5 ...	... ...	5·1 ...	... ...	15·3 ...	... ...	5·1 5·10	... ...	... ...	30·6 ...	5·1 5·10	45·9 ...	127·6 ...	903·1 25·51	43·3	20·4 ...	30·6 ...	76·5 ...
GROUP VI.— UPPER SUB-HIMA-LAYA.	* 13,724	15·3 ...	... ...	·3 4·23	23·3 ...	74·6 ·0	2·7 ...	53·6 ...	·3 ·07	9·0 1·02	10·6 ·58	1·6 ·15	5·4 ·66	19·1 ·07	9·0 ·15	13·2 ...	1·3 ·44	28·9 ·07	130·0 ·07	815·5 10·27	† 52·5	33·8 ·07	30·5 ...	74·7 ...
A Nowshera	720	22·2 ...	... ...	1·4 1·39	54·2 ...	... ...	93·1 ...	1·4 ...	23·6 2·78	16·7 1·39	2·8 ...	17·5 1·39	23·6 ...	5·6 ...	8·3 ...	1·4 ...	1·4 ...	127·8 ...	812·5 8·33	49·1	38·9 ...	20·8 ...	68·1 ...
Peshawar	1,663	257·4 ·60	... ...	42·7 7·82	269·4 ·60	...	6·128·1 ...	·6 ...	47·5 3·61	4·2 ...	1·2 ·60	10·2 ·60	18·0 ...	4·8 ...	16·8 ...	·6 ...	16·8 ...	58·9 ...	1191·2 18·01	59·7	9·6 ...	15·0 ...	34·3 ...
Mooltan	953	... ...	... ...	1·0 2·10	4·2 ...	112·3 ...	... ...	38·8 ...	... ...	4·2 1·05	19·9 ...	3·1 ...	4·2 1·05	24·1 ...	19·9 ...	5·2 ...	... ...	4·2 ...	150·1 ...	884·6 5·25	43·4	30·4 ...	30·4 ...	89·2 ...
C Hyderabad	424	... ...	... ...	6·1 2·02	6·1 2·02	139·7 ...	... ...	14·2 ...	... ...	2·0 ...	14·2 ...	4·0 ...	8·1 ...	26·3 ...	12·1 ...	6·1 ...	2·0 2·02	14·2 ...	236·8 ...	950·3 8·10	57·2	32·4 ...	85·0 ...	119·4 ...
Kurrachee	1,238	25·0 ...	... ...	·8 1·62	13·7 ...	245·6 ...	... ...	2·4 ...	... ...	19·4 1·62	35·5 ...	2·4 ...	1·6 ...	47·7 ...	1·6 ...	8·1 ...	1·6 1·62	11·3 ...	169·6 ...	1,336·3 7·27	69·2	14·5 ...	70·3 ...	84·8 ...
GROUP VII.—N.-W. FRONTIER, INDUS VALLEY, AND N.-W. RAJ-PUTANA.	* 5,078	93·7 ·20	... ...	1·0 ·20	18·9 3·75	190·8 ·20	...	64·5 ...	·4 ...	24·7 1·78	17·6 ·51	2·4 ·20	7·1 ·59	25·0 ...	7·7 ...	10·3 ...	1·0 ·59	10·7 ...	130·2 ...	1,092·7 10·66	† 57·7	21·1 ...	39·1 ...	70·0 ...
A Deesa	217	... ...	... ...	... ...	92·2 ...	... ...	9·2 ...	... ...	... ...	... ...	... ...	13·8 ...	4·6 ...	... ...	... ...	17·3·7 ...	737·3 4·61	40·c	32·3 ...	55·3 ...	92·2 ...
Ahmedabad	241	... ...	... ...	8·3 4·15	809·1 ...	... ...	... ...	... ...	8·3 4·15	8·3 ...	... ...	12·4 ...	4·1 ...	12·4 ...	... ...	4·1 ...	232·4 ...	1,510·1 8·36	60·c	66·4 ...	87·1 ...	78·8 ...
B Neemuch	328	... ...	... ...	3·0 3·05	176·8 ...	... ...	... ...	... ...	... ...	3·0 ...	18·3 ...	6·1 ...	30·5 ...	3·0 3·05	... ...	82·3 ...	689·0 6·10	31·2	9·1 ...	33·5 ...	39·6 ...
Nasirabad	926	... ...	... ...	1·1 ...	4·3 ...	173·9 ...	... ...	2·2 ...	2·2 ...	... ...	5·4 ...	... ...	3·2 ...	8·0 ...	9·7 ...	16·2 ...	1·1 ...	6·5 ...	200·9 ...	867·3 2·16	49·3	38·9 ...	78·8 ...	83·2 ...
Muttra	508	5·9 ...	... ...	... ...	3·0 ...	82·7 ...	... ...	7·9 ...	... ...	7·0 1·97	15·7 ...	... ...	2·0 ...	9·8 ...	5·9 ...	9·8 ...	... ...	9·8 ...	173·2 ...	732·3 3·94	38·8	51·2 ...	27·6 ...	94·5 ...
Agra	1,146	12·2 ...	... ...	16·6 5·24	68·1 ...	... ...	1·7 ...	·9 ...	21·8 ·36	4·4 ·87	1·7 ·87	10·5 ·87	21·8 ...	·9 ...	·9 ...	1·7 1·75	32·3 ...	204·2 ·87	733·0 16·58	64·2	22·7 ·87	47·7 ...	131·8 ...

* Derived from the aggregates. † Worked on the aggregates.

Stations and Groups	Average annual strength	Influenza	Cholera	Small-pox	Enteric Fever	Intermittent Fever	Remittent Fever	Simple Continued Fever	Rheumatic Fever	Circulatory Diseases	Tubercle of the lungs	Pneumonia	Other Respiratory Diseases	Dysentery	Diarrhœa	Hepatic Abscess	Hepatic Congestion and Inflammation	Venereal Diseases	All Causes	Constantly Sick	Syphilis	Soft Chancre	Gonorrhœa
Jhansi	1,046	...	...	4·8	26·8 / 6·69	260·0	1·0	29·6	...	2·9 / 1·91	9·6 / ·90	1·9	2·0	28·7	22·0 / 1·91	14·3	4·8 / 3·82	16·3	328·5	984·7 / 21·03	67·5	66·0	61·2	101·3
Nowgong	322	...	...	...	28·0 / 9·32	130·4	6·2	34·2	...	6·2	6·2 / 3·11	3·11	...	34·2	24·8 / 18·6	6·2	34·2 / 3·11	81·0	819·9 / 15·53	50·4	18·6	9·3	59·0
Indore	136	7·4	...	...	44·1 / 14·71	44·1	...	22·1	...	...	7·4	...	...	14·7	22·1 / 7·4	7·4	22·1	135·0	588·2 / 29·41	36·1	22·1	66·2	36·8
Mhow	1,693	8·9	...	11·2	23·0 / 7·09	109·9	23·0	1·2	...	1·2 / 1·18	16·0	3·0	5·3 / ·59	15·4	17·7 / 24·8	3·0	10·6 / 1·77	133·6	734·6 / 11·23	48·4	25·4	51·2	55·0
Group VIII.—South-East Rajputana, Central India and Gujarat	6,362	5·0	...	3·8	16·8 / 4·68	161·5	6·4	8·7	·5	5·5 / 1·22	9·1 / ·61	1·8 / ·46	4·4 / ·30	18·1	12·3 / ·30	14·9	2·6 / 1·68	14·9	173·7 / ·15	820·2 / 11·89	52·7	35·8 / ·15	53·9	84·0
A Saugor	273	10·8	...	...	18·0 / 3·60	158·3	...	43·2	...	3·6	...	3·6	...	32·4 / 3·60	10·8	25·2	7·2	32·4	172·7 / 7·19	805·8 /	51·2	10·8	86·3	75·5
Jubbulpore	841	...	2·4 / 2·38	2·4	19·0 / 4·76	76·1	3·6	172·4	...	2·4	15·5	1·2	2·4 / 3·57	29·7	29·7 / 2·38	20·2	3·6	22·6	151·0 / 15·46	948·9 / 15·46	50·9	21·4	33·3	95·3
Kamptee	952	...	...	5·3	8·4 / 2·10	200·6	...	92·4	...	8·4 / 1·05	8·4 / 1·05	1·1	...	16·8 / 1·05	12·6	2·4	3·2 / 3·15	14·7	175·4 / 16·50	973·7 / 16·50	51·4	39·9	76·7	58·8
Sitabaldi	76	26·3	...	...	13·2	184·2	13·2	421·1	...	...	39·5	13·2	118·4	...	39·5	92·1 / 1,684·2	20·3	65·8	...	26·3
B Secunderabad	3,228	...	...	·3	27·3 / 4·63	40·3	...	32·8	·3	2·5 / ·63	6·5	·3	2·5 / ·31	16·4 / 1·24	22·9	...	2·2 / 1·55	7·7	185·3 / 10·23	755·0 / 10·23	37·9	32·8	81·8	70·6
Belgaum	1,110	1·8	...	·9	9·9 / ·90	88·3	...	5·4	...	1·8 / ·90	4·5 / ·90	2·7 / ·90	1·8 / ·90	12·6	19·8	7·2	3·6 / ·90	5·4	191·9 / 7·21	785·6 / 7·21	54·2	37·8 / ·90	92·8	61·3
Poona	2,132	17·4	...	12·2	43·2 / 4·09	178·7	...	·7	·3	·9 / ·47	14·5	6·6 / ·47	4·7	23·9	21·6 / ·94	27·7	5·2 / ·94	13·5	127·1 / 8·91	833·5 / 8·91	51·1	16·4 / ·47	50·2	60·5
Kirkee	751	26·6	...	4·0	24·0 / 1·33	43·9	1·3	21·3	...	2·7 / ·33	2·7	4·0	9·3 / 1·33	17·3	13·3 / 3·99	10·7	5·3 / 3·99	12·0	185·1 / 14·65	737·7 / 14·65	47·2	46·6	46·6	91·9
Ahmednagar	1,065	...	...	...	16·9 / 2·82	151·2 / ·94	...	6·6	·9	...	9·4 / ·94	2·8 / 2·82	7·5	17·8	0·4 / ·94	·9	1·9 / ·94	16·0	145·5 / 9·32	658·2 / 9·32	42·0	24·4 / ·94	37·6	83·6
Group IX.—Deccan	10,434	6·1	·2 / ·19	3·6	24·6 / 3·55	107·0 / ·10	·5	41·9	·3	2·4 / ·29	8·6 / ·10	2·6 / ·29	3·1 / ·38	19·5	19·5 / ·96	12·7	3·3 / 1·73	129	165·3 / 10·16	806·8 / 10·16	52·1	29·5 / ·29	64·6	71·2
Colaba	1,218	8·7	...	7·1	1·6 / ·79	183·8	4·7	...	...	1·6 / 2·37	10·5 / ·79	1·6 / ·79	3·9	14·2	5·5 / ·79	4·2	3·2 / ·79	18·0 / ·79	265·0 / 9·46	863·3 / 9·46	61·3	67·8	88·3	108·8
Cannanore	96	...	...	...	41·7	...	31·2	...	...	10·4	...	20·8 / 10·42	10·4	...	208·3 / 10·42	645·8 / 20·83	56·9	41·7 / 10·42	83·3	83·3
Calicut	92	...	...	...	10·9	...	...	...	21·7	...	43·5 / ·99	21·7	87·0	10·9	...	195·7	902·2	50·9	32·6	54·3	108·7
Mallapuram	136	7·4	...	...	29·4	...	...	7·4	22·1	...	51·5 / 7·35	...	14·7	7·4 / 7·35	95·6	838·2 / 7·35	36·2	14·7	36·8	44·1
Group X.—Western Coast	1,591	7·5	...	5·7	1·3 / ·63	152·1	3·8	1·9	...	1·9 / 1·83	11·9 / 1·26	1·3 / ·63	·3	19·5	6·3 / ·63	17·6	3·8 / ·63	15·1 / ·63	243·2 / 9·43	854·8 / 9·43	38·3	59·7 / ·63	81·7	101·8
A Bellary	512	...	...	3·9	17·6 / 3·91	189·5	...	...	...	2·0	19·5	3·9	...	11·7	...	2·0 / 1·95	29·3	263·7 / 7·51	943·4 / 7·51	94·6	62·5	121·1	80·1
Mangalore	2,201	...	...	1·3	17·3 / 2·27	94·0 / ·45	...	14·5	...	·5	7·3	1·4	·5	15·0	19·5 / ·91	0·5	·9 / ·91	3·2	189·9 / 5·45	685·6 / 5·45	52·5	76·3	41·8	71·5
B St. Thomas' Mount	354	...	...	2·8	16·9	70·6	...	11·3 / 2·82	2·8 / 2·82	...	8·5	28·2	8·5	2·8 / 1·49	14·1	144·1 / 8·47	827·7 / 8·47	44·4	19·8	37·5	84·7
Madras	671	...	1·5	3·0	3·0 / 1·49	44·7	...	1·5 / 1·49	26·8	...	1·5 / 1·49	13·4	1·5	1·5 / 1·49	17·9	266·8 / 7·45	1,025·3 / 7·45	36·8	64·1	67·1	133·6
Group XI.—Southern India	3,738	...	·3	2·4	13·1 / 1·87	91·0 / ·27	...	15·2	...	1·0 / ·27	12·0 / ·54	1·3	·5 / ·27	14·7	16·6	6·7 / 1·07	1·9	10·4	209·5 / 6·42	795·3 / 6·42	54·7	66·9	57·0	85·6

TABLE III—*continued.*

RATIOS of STATIONS, GROUPS, and COMMANDS.

For actuals see Table IV.

Stations and Groups.	Average annual strength.	1. ADMISSION-RATE.																2. DEATH-RATE.			Constantly Sick-rate.	Syphilis.	Soft Chancre.	Gonorrhœa.
		Influenza.	Cholera.	Small-pox.	Enteric Fever.	Intermittent Fever.	Remittent Fever.	Simple Continued Fever.	Rheumatic Fever.	Heat-stroke.	Circulatory Diseases.	Tubercle of the lungs.	Pneumonia.	Other Respiratory Diseases.	Dysentery.	Diarrhœa.	Hepatic Abscess.	Hepatic Congestion and Inflammation.	Venereal Diseases.	All Causes.				
Ranikhet	1,087	...	...	...	3'7 '92	62'6	...	12'9	...	'9	27'6	...	1'8	24'8	13'8	108'6	1'8 1'84	16'6	106'7	753'4 2'76	38'1	16'6	5'3	84'6
Chaubattia	414	...	...	...	2'4	33'8	...	7'2	2'4	...	2'4	...	2'4	36'2	9'7	33'8	2'4 2'42	2'4	118'4	514'5 2'42	25'3	12'1	14'5	91'8
Chakrata	1,036	1'9	...	...	8'7 '97	66'6	2'9	23'2	...	...	9'7	1'0	4'8 1'93	28'0 '97	16'4	8'7	1'0	6'8	166'0 '97	725'9 11'58	48'9	35'0 '97	58'9	54'1
Lebong	728	56'3	...	...	4'1	109'9	...	5'5	4'1	...	6'9	...	2'7 1'37	26'1 1'37	2'7	49'4	2'7 1'37	2'7 1'37	229'4 1'37	894'2 9'62	54'5	90'7 1'37	45'5	93'4
Solon	250	8'0	...	...	16'0 8'00	28'0	...	8'0	...	...	8'0	...	4'0	4'0	...	...	...	12'0	132'0	440'0 8'00	26'1	36'0	16'0	80'0
Dagshai	663	...	...	...	3'0	16'6	...	7'5	...	...	6'0 1'51	1'5 1'51	9'0 1'51	16'6	...	42'2	...	9'0	116'1	701'4 4'52	44'6	43'7	42'2	30'2
Subathu	420	9'5	...	2'4	26'2 9'52	42'9	...	54'8	4'8	...	2'4	...	4'8	28'6	2'4	14'3	2'4	7'1	121'4	754'8 11'90	48'1	35'7	9'5	76'2
Jutogh	221	...	...	...	...	9'0	...	...	...	...	9'0 4'52	...	9'0 4'52	9'0	...	4'5	...	40'7	149'3	565'0 9'05	33'8	...	122'2	27'1
Khyragully	68	...	...	...	14'7	14'7	...	...	...	...	14'7	...	...	...	...	...	...	29'4	88'2	647'1	20'6	44'1	...	44'1
Baragully	51	...	...	...	...	...	...	...	...	...	19'6	...	...	...	...	...	...	137'3	58'8	568'6	22'2	...	196	39'2
Kuldunnah	372	...	...	...	2'7	26'9	...	...	2'7	...	37'6 2'69	2'7	5'4	48'4	...	16'1	...	34'9	158'6	739'2 3'38	46'3	26'9	10'	121'0
Kalabagh	52	...	...	...	...	19'2	...	...	...	...	...	...	...	...	...	...	...	192'3	403'8		24'0	57'7	57'7	115'4
Camp Gharial.	528	1'9	...	...	1'9	41'7	...	24'6	...	...	13'3	...	...	9'5	...	...	...	15'2	77'6	462'1	26'8	43'6	11'4	32'7
Camp Barian.	465	15'1	...	...	6'5 4'30	83'9	...	6'5	...	4'3	19'4	4'3	2'2 2'15	15'1	2'2	21'5	...	25'8	68'8	700'7 6'45	38'8	12'9	15'1	40'9
Camp Upper Topa.	177	5'6	...	...	5'6	16'9	...	11'3	...	5'6	...	...	...	...	...	...	...	45'2	96'0	576'3	24'0	16'9	...	79'1
Camp Lower Topa.	45	...	...	...	22'2	66'7	...	...	...	...	44'4	...	...	44'4	...	...	...	...	377'8	1,088'9	55'0	22'2 177'8	177'8
Khan Spur	459	...	...	...	17'4 2'18	69'7	...	58'8	...	...	2'2	...	2'2	24'0	...	4'4	2'2 2'18	26'1	58'8	712'4 6'54	32'6	21'8	6'5	30'5
Cherat	384	...	...	...	26'0	18'2	...	161'5	...	...	7'8	...	2'6	2'6	2'6	5'2	5'2 2'60	5'2	39'1	744'8 2'60	36'5	15'6	26	20'8
Quetta	2,590	...	...	...	20'5 6'56	62'5	...	146'2	9'7 '39	51'7 '77	3'5	13'1 2'32	53'3	25'9 1'54	17'0	'8 '77	12'0 '77	74'9	886'5 15'06	55'4	13'5	16'6	44'3
Ramandrug	8	...	...	...	...	...	...	...	...	...	...	...	1 5'0 125'00	...	...	...	...	...	875'0 125'00	55'0			
Maymyo	491	...	...	...	4'1 2'04	165'0	...	4'1	...	...	12'2 4'07	2'0	2'0	40'7	4'1 2'04	10'2	4'1 2'04	18'3 4'07	335'1 2'04	949'1 20'37	74'5	99'8 183'3	110'0
Group XII a— Hill Stations.	10,508	5'5	...	'1	10'9 2'76	60'0	'3	53'6	7	2'8 '10	22'2 '67	1'4 '14	5'8 '10	30'3 '67	10'6	26'7	1'3 '86	14'6 '29	122'5 '38	754'5 8'95	45'4	33'1 '38	29'1	60'2
Darjeeling	472	8'5	...	...	2'1	137'7	...	10'6	...	...	4'2	4'2	...	10'6	10'6	6'4 2'12	6'4 2'12	6'4	165'3 8'47	611'6 8'47	38'3	65'7 2'12	40'3	59'3
Naini Tal	188	16'0	...	...	26'6 5'32	74'5	...	5'3	10'6	...	10'6 5'32	...	16'0	5'3	5'3	5'3 5'32	10'6	69'1	558'5 15'96	47'7	21'3	10'6	37'2
Landour	157	6'4	...	...	25'5 6'37	210'2	...	19'1	...	...	12'7	...	31'8 6'37	12'7	6'4	6'4 6'37	101'9	974'5 25'48	51'0	31'8	12'7	57'3
Kasauli	368	...	...	...	16'3	51'6	...	...	8'2	19'0	5'4	...	49'0	43'5	8'2 5'43	27'2	214'7	1,024'9 13'59	115'3	32'6 103'3	78'8
Dalhousie	822	2'4	...	...	12'2 1'21	36'5	...	19'5	...	1'2	15'8	2'4	1'2	30'4 1'22	10'5	36'0	...	6'1	75'4	491'5 2'43	38'2	12'2	12'2	51'1
Murree	111	9'0	...	...	18'0 9'01	27'0	...	36'0	...	...	...	9'0 9'01	9'01	54'1	...	...	...	36'0	437'4 27'03	164'3	18'	9'0	9'0
Taragarh	26	...	...	...	...	76'9	...	230'8	...	...	...	...	38'5 115'4	...	...	38'5	115'4 1,230'8	88'5	...	38'5	76'9
Mount Abu	105	...	...	...	...	209'5	...	...	...	...	19'0	...	10'0 9'52	10'0	...	28'6	104'8	761'9 19'05	41'2	57'	28'6	19'0
Pachmarhi	124	...	...	...	...	103'5	16'1	121'0	...	40'3	...	...	56'5	16'1	32'3	...	40'3	137'1	1,258'1 8'00	54'1	72'6	32'3	32'3

* Derived from the aggregates.   † Worked on the aggregates.

STATIONS AND COMMANDS	Average annual strength	1. ADMISSION-RATE.									2. DEATH-RATE.						3. CONSTANTLY SICK-RATE.							
		Influenza.	Cholera.	Small-pox.	Enteric Fever.	Intermittent Fever.	Remittent Fever.	Simple Continued Fever.	Rheumatic Fever.	Heat-stroke.	Circulatory Diseases.	Tubercle of the lungs.	Pneumonia.	Other Respiratory Diseases.	Dysentery.	Diarrhœa.	Hepatic Abscess.	Hepatic Congestion and Inflammation.	Venereal Diseases.	ALL CAUSES.	CONSTANTLY SICK.	Syphilis.	Soft Chancre.	Gonorrhœa.
Parædbur .	130	...	...	7·7	15·4	284·6	...	...	...	...	...	...	...	30·8	38·5	15·4	...	15·4	2,46·2	1,030·8	64·3	107·7	61·5	76·9
Khandalla .	47	...	...	...	...	361·7	...	...	...	...	...	...	21·3	85·1	...	...	...	63·8	85·1	1,340·4	76·4	42·6	21·3	21·3
Wellington	961	...	...	...	4·2	167·5	1·0	...	...	...	8·3	1·0	3·1 1·04	48·9	7·3	11·4	6· 4·16	8·3	180·0 1·04	1,045·8 8·32	62·2	72·8 1·04	48·9	58·3
GROUP XII 8.— Hill Convalescent Depôts and Sanitaria.	3,512	3·1	...	·3	9·7 ·85	121·0	·9	13·1	·6	1· ·25	12·8 ·28	2·3 ·28	1·1 ·57	35·9 ·28	13·4 ·37	18·8	4·0 2·36	12·2 ·57	140·1	818·9 9·11	60·4	47·0 ·57	38·7	54·4
Troops marching, India.	1,999	3·8	2·5	1·0	4·5	49·5	2·5	20·5	...	·3 1·00	5·5	...	7·0	19·0	13·0	12·8	·5	4·5	117·1	474·7 3·00	2·4	23·0	35·0	53·0
Aden Columbo Field Force, (Dthalia).	349	...	...	...	...	455·6	...	...	...	...	20·1	...	5·7 2·87	14·3	65·9	8·6	2·9 2·85	14·3	25·8	983·5 8·60	30·7	5·7	2·9	17·2
eolali Depôt	452	44·2 4·42	...	2·2	2·2	254·4	...	...	4·4	...	11·1 2·21	8·8 6·64	2·2	33·2	8·8	6·6	2·2 2·21	·33	265·5 15·49	1,044·2	92·8	70·8	49·7	146·0
Poonamallee Depôt.	131	...	...	...	...	91·6	...	15·3	...	15·3 7·63	61·6	15·3	...	7·6	76·3	22·9	22·9 7·63	45·8	175·6	1,053·4 13·27	383·4	99·2	...	76·3
EXTRA INDIA Aden .	947	...	...	...	6·3 3·17	286·2 3·17	...	43·3	...	16·9 3·17	10·6 2·11	3·2	15·8	33·8 1·06	6·3	3·2 3·17	20·1	50·7	948·3 24·29	66·4	15·8	16·9	18·0
India .	71,343	14·2 ·04 ·4	·1 ·13	1·4 ·01 ·1	16·1 2·99 3·1	111·4 ·11 4·3	2·3 ·01	47·9 ·1 1·7	·4 ·01 ·1	6·6 ·77 ·3	12·5 ·56 1·2	2·1 ·28 ·4	4·1 ·63 ·3	23·0 ·00 1·11	13·4 ·40 1·0	15·2 ·5	2·1 1·18 1·0	17·1 ·07 ·5	153·7 ·12 15·3	834·3 10·05 52·3	35·6 52·3 ·18 4·8	43·3 ·06 4·0	74·8 6·5
NORTHERN COMMAND	18,274	35·4 0·5	...	·3	17·7 3·12	69·6 ·05	·2	63·4	·05 1·15	11·1 ·42	13·1 ·27	1·7 ·82	5·7 ·05	20·2	6·5	14·0	1·3 ·44	23·4 ·05	116·1	800·5 9·08	49·3	28·7	23·3	64·1
WESTERN COMMAND	19,537	7·3 ·10	·1 ·10	3·9 ·05	17·5 3·6	162·8 ·20	2·8	42·4	·3	4·7 ·67	17·4 ·41	3·0	4·7 ·72	26·4	17·2	16·2 1·64	2·9 ·15	14·4 ·15	155·4 11·62	896·4	55·0	31·6 ·15	52·8	71·0
EASTERN COMMAND.	18,983	11·2	·1 ·11	·1	16·9 3·11	113·2 ·11	2·6 ·0.	48·0	·6	7·0 1·00	11·0 ·63	1·8 ·37	3·2 ·47	22·5 ·21	12·0 ·42	16·6	2·4 1·42	19·8 ·05	165·8 ·20	820·2 10·90	153·6	36·0 ·26	40·3	58·4
SECUNDERABAD DIVISION.	8,389	·1	·1	1·2	16·8 2·62	77·7 ·12	·1	20·0	·1	2·1 ·45	11·0 ·24	1·1	1·4 ·36	20·1	18·7 ·60	5·8	2·7 1·79	9·3	194·1 ·24	812·8 8·46	61·6	53·4 ·24	64·6	75·1
BURMA DIVISION	3,812	1·5	...	·8 1·31	2·4 ·79	114·4	10·5	80·5	·3	·2 ·26	7·3 ·52	3·9	2·9 1·05	30·7	15·5 1·31	17·1	·5 ·20	11·3	207·0 ·70	970·0 9·71	154·0	47·1 ·70	66·1	91·8
Lucknow	2,556	·1	...	·0	6·0	3·0	·0	5·3	·2	·4	1·5	·5	·2	1·0	2·7	·5	·9	1·4	18·4	61·4	61·4	5·1	2·4	10·7
Ambala	2,314	·1	...	·1	2·6	1·6	·1	·8	...	·0	·0	·1	·5	1·1	·0	·3	·2	·8	16·0	48·1	48·1	4·4	3·3	8·3
Rawalpindi	2,923	·1	...	·0	4·7	2·0	...	1·1	·1	1·7	·3	·8	1·0	·6	·2	·3	2·6	15·4	35·0	55·0	6·7	1·3	7·4
Secunderabad	3,228	...	...	·0	4·5	3·1	...	1·9	·1	·2	·5	·0	·2	·0	·1	·3	·6	15·3	57·9	57·9	5·6	4·0	7·8
Poona	2,132	·0	...	1·0	7·3	5·4	...	·3	·1	·0	·9	1·2	·3	1·1	1·8	·3	1·0	11·3	51·1	51·1	1·8	5·1	4·6
Bangalore	2,201	...	...	·1	3·4	4·0	...	1·2	...	·0	·1	·0	·0	1·0	1·7	·5	·3	·4	20·1	52·5	52·5	10·3	3·0	6·5
Quetta	2,390	...	...	...	3·7	3·5	...	4·6	...	1·0	4·4	·4	1·0	2·4	2·	·4	·1	7·9	55·4	55·4	2·5	1·4	4·0

* Derived from the aggregates.   † Worked on the aggregates.   ‡ Constantly sick-rate per 1,000 by diseases at the largest station.

TABLE IV.

ACTUALS of STATIONS, GROUPS, and COMMANDS on which the ratios in Tables I—III have been calculated.

Stations and Groups.	Average annual strength.	1. ADMISSIONS.														2. DEATHS.						3. CONSTANTLY SICK.							
		Influenza.	Cholera.	Small-pox.	Enteric Fever.	Intermittent Fever.	Remittent Fever.	Simple Continued Fever.	Rheumatic Fever.	Heat-stroke.	Circulatory Disease.	Tubercle of the lungs.	Pneumonia.	Other Respiratory Diseases.	Dysentery.	Diarrhœa.	Hepatic Abscess.	Congestion and Inflammation.	Venereal Diseases.	All Causes.	Syphilis.	Soft Chancre.	Gonorrhœa.	Tænia.	Other En 100.		
Port Blair	140					3	23	5						1		14	125	8		1	5						
Rangoon	1,119					1	166		10		3	36		28	1,419	26	85	143									
GROUP I.—BURMA COAST AND BAY ISLANDS.	1,239					24	33	171		8	11		40	30	398	1,537	61	86	148								
Thayetmyo	488				3	244			2	3	5	15		10	68	503	13	11	44								
Meiktila	379				3		83			3	13	59	305	11	23	28											
Fort Dufferin	379				11	37			3	12	8	6	351	11	24	25											
Shwebo	588	4			21	3	8		3	12	17	100	351	30	35	58											
Bhamo	329				37	3				8	35	188	9	7	18												
GROUP II.—BURMA INLAND.	2,063		5		6	231	7	134	3	14	3	37	13	30	24	335	1,696	74	103	148	10						
Fort William	1,044	97			305	3	16		2	35	17	22	12	396	1,357	64	94	138									
Fort Falta	27				4					6	15																
Fort Chingrikhali	58				6					1	17	45	5	7	5												
Dum-Dum	284	5			3	39	8			39	3	19	35	220	8	16											
Barrackpore	276				6	75	3	10		4	15	14	3	69	324	20	23	37									
GROUP IV.—BENGAL AND ORISSA.	1,690	102			13	422	8	36	2	30	6	78	31	40	413	1,861	90	133	190								
B Dinapore	602		1		20	51	11	7	1	3	4	14	5	53	106	430	18	25	60								
Benares	194				1	13		31	3	5	13	136	2	5	15												
Allahabad	975	3	1		6	134	1	78	8	4	2	3	6	5	163	657	40	31	85								
Fort Allahabad	216					86		22		3	1	34	232	4	11	18											
Fyzabad	587				9	89		95	7	11	4	10	17	10	51	373	33	49	60								
Sitapur	551				3	38		84	1	6	3	37	10	2	30												
Lucknow	2,556				71	305	3	423	19	28	5	70	37	13	84	413	3,412	91	60	277	6						
Cawnpore	1,032				6	15	53	10	5	10	22	13	147	637	17	30	95										
Fatehgarh	151	6			14	23		6		1	8	28	147	2	6	30											
GROUP V.—GANGETIC PLAIN AND CHOTTA NAGPUR.	7,313		3		117	761	19	784	2	94	84	23	153	3,214	5,845	317	130	667	14	6							

* Derived from the aggregates.

Stations and Groups.	Average annual strength.	Influenza.	Cholera.	Small-pox.	Enteric Fever.	Intermittent Fever.	Remittent Fever.	Simple Continued Fever.	Rheumatic Fever.	Heat-stroke.	Circulatory Diseases.	Tubercle of the lungs.	Pneumonia.	Other Respiratory Diseases.	Dysentery.	Diarrhœa.	Hepatic Abscess.	Congestion and Inflammation.	Venereal Diseases.	All Causes.	Syphilis.	Soft Chancre.	Gonorrhœa.	Tænia.	Other Entozoa.
A Shahjehanpur	489																								
Bareilly	1,149																								
Roorkee	393																								
Meerut	1,584																								
Delhi	317																								
Ambala	3,314																								
B Jullundur	675																								
Ferozepore	893																								
Amritsar	704																								
Allan Mir	733																								
Fort Lahore	122																								
Sialkot	1,708																								
Rawalpindi	2,973																								
Campbellpur	226																								
Attock	196																								
Group VI.—Upper Sub-Himalaya.	13,674																								
A Nowshera	720																								
Peshawar	1,663																								
Mooltan	953																								
C Hyderabad	474																								
Karrachee	1,236																								
Group VII.—N.W. Frontier, Indus Valley, and N.W. Rajpotana.	5,068																								

*Derived from the aggregates.

TABLE IV—*continued.*

ACTUALS of STATIONS, GROUPS, and COMMANDS on which the ratios in Tables I—III have been calculated.

STATIONS AND GROUPS.	Average annual strength.	Influenza.	Cholera.	Small-pox.	Enteric Fever.	Intermittent Fever.	Remittent Fever.	Simple Continued Fever.	Rheumatic Fever.	Heat-stroke.	Circulatory Diseases.	Tubercle of the lungs.	Pneumonia.	Other Respiratory Diseases.	Dysentery.	Diarrhœa.	Hepatic Abscess.	Hepatic Congestion and Inflammation.	Venereal Diseases.	All Causes.	Syphilis.	Soft Chancre.	Gonorrhœa.	Tænia.	Other Entozoa.
														1. ADMISSIONS.			2. DEATHS.			3. CONSTANTLY SICK.					
Deesa A	317	...	...	...	...	20	...	3	...	...	...	...	...	3	...	...	...	...	39	130	7	11	20	...	1
		...	...	...	...	·04	...	·06	...	...	...	...	...	·05	·04	...	...	...	3·77	8·33	·71	1·07	1·99	...	·21
Ahmedabad B	241	...	...	...	1	105	...	2	...	...	2	·01	...	3	1	...	...	1	50	264	10	11	19	1	...
		...	...	...	·23	6·20	...	...	...	...	·33		...	·20	·08	·06	...	·04	3·47	14·45	1·24	1·01	1·33	·05	...
Neemuch	328	...	...	...	1	58	...	...	...	...	...	...	...	6	2	10	1	...	37	228	3	11	13	...	...
		...	...	...	·04	1·74	...	...	...	...	...	·06	·24	·07	·20	·28	...	...	3·31	10·24	·23	1·21	·90	...	·03
Nasirabad	936	...	1	...	1	161	...	3	2	...	5	...	3	8	9	15	1	0	186	807	36	73	77	2	...
		...	...	·25	·84	4·05	...	·14	·29	...	·57	...	·16	·25	·07	·57	·24	·48	13·13	45·63	3·58	8·10	6·44	·04	...
Muttra	505	3	...	...	1	42	...	4	...	4	...	1	5	3	1	...	...	5	88	372	16	14	48	3	...
		·05	...	...	·72	1·04	...	·16	...	·13	·85	...	·06	·48	·19	·20	...	·25	7·10	19·69	2·11	2·06	3·91	·05	·03
Agra	1,146	14	...	...	6	72	...	3	...	25	5	2	13	25	...	1	2	27	234	840	25	57	151	7	...
		1·84	...	...	6·00	7·94	...	1·9	·09	·43	·51	·30	2·13	...	·06	·03	...	2·3·03	23·03	73·57	4·66	7·29	11·68	·33	...
Jhansi	1,010	...	...	5	28	271	2	3	1	...	3	10	·9	3	32	23	15	5	17	239	1,030	69	61	106	...	...
		...	·74	...	4·81	9·47	·03	1·37	...	·07	1·35	·06	·20	1·33	1·82	·53	·76	·79	21·74	70·63	8·99	5·84	7·91	...	...
Nowgong	323	...	...	...	9	42	2	11	...	...	1	1	8	1	3	6	1	11	25	284	6	3	19	1	...
		...	...	...	2·95	1·73	·23	·85	...	·13	·16	·03	...	·26	·30	·24	·18	·87	1·79	16·23	·73	·10	·93	·03	...
Indore	136	...	...	...	...	1	2	...	...	...	1	...	...	2	1	...	2	...	17	80	...	...	9	...	...
		·04	...	...	·49	...	·30	...	...	·05	...	...	·03	·00	·02	·21	·27	·97	4·91	...	·21	·54	·29	...	...
Mhow	1,693	15	...	19	35	185	39	1	...	9	27	5	9	26	30	43	5	18	236	1,343	43	90	93	1	...
		·59	...	1·39	7·36	8·05	3·43	·03	...	·03	2·95	·84	7·0	1·9	3·10	1·41	·47	1·33	21·32	81·90	5·73	7·31	8·24	·01	·16
Group VIII—S.-E. RAJPUTANA, CENTRAL INDIA, AND GUJARAT.	6,562*	33	...	25	110	1,010	43	37	3	36	64	13	57	119	83	96	27	99	1,140	5,382	233	354	551	16	...
		1·47	...	2·48	23·01	41·01	3·68	3·40	·28	·85	5·25	1·21	3·40	6·43	5·59	3·25	3·45	5·53	103·15	316·09	21·13	33·43	43·00	·30	·79
Saugor A	378	3	...	...	5	44	...	13	...	...	1	...	1	...	9	3	7	9	48	814	3	24	21	...	...
		·14	...	...	·92	1·44	...	·78	...	·04	·35	...	0	...	·23	·18	·37	3·94	14·91	·55	3·08	1·30	...	...
Jubbulpore	841	...	3	...	16	64	3	145	...	2	13	1	3	25	15	17	2	19	137	725	15	29	81	1	...
		·01	·01	·01	2·11	3·12	·02	5·37	...	·03	1·83	·01	·01	1·02	2·23	1·24	1·01	1·07	9·75	47·75	2·14	3·50	4·03	·02	·01
Kamptee	952	...	...	5	8	101	...	83	...	8	8	1	...	16	12	23	5	11	157	937	39	23	50	1	...
		...	...	·38	1·30	5·11	...	3·30	...	·25	·53	·01	...	·37	·93	73	...	·87	16·97	48·22	4·94	8·83	3·18	·01	...
Sitabaldi	76	...	...	...	1	11	...	33	...	1	...	...	...	...	...	1	...	2	7	128	5	...	2	...	...
		·01	...	...	·01	·16	·01	·31	...	...	...	...	·02	·08	...	·16	...	...	·05	1·54	·04	...	...	...	...
Secunderabad B	3,318	...	...	1	85	130	...	106	...	8	21	1	8	53	74	...	7	25	595	3,427	105	264	235	...	1
		...	...	·01	14·44	9·91	...	6·28	·43	·65	1·71	...	·65	1·93	3·48	...	1·06	1·53	53·11	137·00	15·18	15·03	3·06	...	·01
Belgaum	1,110	2	...	1	11	95	...	6	...	2	...	1	22	8	...	4	6	...	213	873	42	103	68	4	...
		·15	...	·11	·90	5·48	...	·20	...	·09	·57	·74	·27	·68	·62	·25	·97	...	22·55	62·13	6·76	11·23	5·3	·18	...
Poona	2,131	27	...	10	92	361	...	15	...	3	31	14	10	51	46	57	1	33	271	1,777	33	107	130	...	...
		·21	...	3·00	5·03	11·45	...	·72	·34	·03	3·63	·66	2·70	2·63	1·91	2·53	1·03	3·16	34·46	105·59	3·84	12·91	9·71	·31	...
Kirkee	751	10	...	3	18	33	1	16	...	2	1	1	7	12	10	8	2	9	131	551	33	21	60	...	...
		·32	...	·20	3·80	2·24	·02	·35	...	·01	·05	1·05	·25	·41	1·40	·07	·35	·35	15·07	35·4	5·25	3·43	6·20	·0	1
Ahmednagar	1,035	...	...	18	101	...	7	...	...	13	2	8	13	1	2	3	...	135	701	20	42	89	...	...
		...	...	2·37	5·43	...	·23	...	·11	...	1·31	·33	1·15	·76	·01	·25	3·16	...	13·43	44·09	2·56	2·77	8·16	...	...
Group IX.—DECCAN	10,424	64	...	28	317	1,116	3	432	...	23	90	9	32	203	103	132	35	125	1,725	8,415	301	674	743	16	2
		1·02	·01	3·60	41·35	41·4	·01	17·01	·58	1·40	7·5	5·00	3·07	8·47	14·55	4·31	5·73	9·30	166·15	54·73	44·31	52·73	63·08	·54	·03
Colaba	1,168	...	...	...	13	...	...	2	...	1	3	2	5	3	1	...	2	3	3,0	1,101	20	111	135	...	...
		·38	...	·0	·51	0·78	·51	...	...	·23	1·74	·33	·53	·58	·9	·47	·04	·05	1·75	34·43	77·71	11·07	9·05	13·13	...	...
Cannanore	96	...	...	...	...	13	...	...	...	...	1	...	...	...	2	...	...	3	9	61	...	4	8	...	...
		...	...	...	...	·91	...	...	...	...	·03	...	...	·13	·07	·64	·04	...	4·03	8·3	...	1·91	1·27	...	...
Calicut	63	...	...	...	...	11	...	...	...	...	1	...	...	1	...	...	...	...	13	51	3	11	10	...	...
		...	...	...	...	·91	...	...	...	...	·13	·7	·64	·04	...	...	...	...	2·15	8·59	·37	...	·59	...	...
Mallapuram	136	...	1	...	...	11	...	...	...	...	3	...	...	3	1	...	...	...	13	114	1	5	11	...	...
		·71	...	...	·48	·96	...	...	...	...	·31	...	...	·39	·01	·04	...	...	4·01	...	·14	·73	·45	...	...
Group X —WESTERN COAST.	1,501*	13	...	1	13	48	1	3	...	1	10	2	5	21	3	3	3	6	3·1	1,370	95	130	165	...	...
		·33	...	·7	73	0·91	·31	·01	...	·33	1·54	3	39	·24	1·33	·23	·23	...	20·61	91·77	11·01	12·05	15·47	...	...

*Derived from the aggregate.

STATIONS AND GROUPS.	Average annual strength.	Influenza.	Cholera.	Small-pox.	Enteric Fever.	Intermittent Fever.	Remittent Fever.	Simple Continued Fever.	Rheumatic Fever.	Heat-stroke.	Circulatory Diseases.	Tubercle of the lungs.	Pneumonia.	Other Respiratory Diseases.	Dysentery.	Diarrhœa.	Hepatic Abscess.	Hepatic Congestion and Inflammation.	Venereal Diseases.	All Causes.	Syphilis.	Soft Chancre.	Gonorrhœa.	Tænia.	Other Entozoa.

A

Bellary ... 517

Bangalore ... 2,301

B

St. Thomas' Mount ... 354

Madras ... 671

Group XI.—Southern India ... 2,738

Ranikhet ... 1,087

Chaubuttia ... 414

Chakrata ... 1,016

Lebong ... 723

Solon ... 250

Dagshai ... 663

Subathu ... 410

Jutogh ... 221

Khyragully ... 68

Baragully ... 51

Kuldunnah ... 370

Kalabagh ... 52

Camp Ghatial ... 523

  „ Hariso ... 465

  „ Upper Topa ... 177

  „ Lower Topa ... 45

Khan Spar ... 459

Chirat ... 384

* Derived from the aggregates.

19

C

TABLE IV—continued.

ACTUALS of STATIONS, GROUPS, and COMMANDS, on which the ratios in Tables I—III have been calculated.

Stations and Groups.	Average annual strength.	1. ADMISSIONS. Influenza	Cholera	Small-pox	Enteric Fever	Intermittent Fever	Remittent Fever	Simple Continued Fever	Rheumatic Fever	Heat-stroke	Circulatory Diseases	Tubercle of the lungs	Pneumonia	Other Respiratory Diseases	Dysentery	Diarrhœa	Hepatic Abscess	Hepatic Congestion and Inflammation	Venereal Diseases	2. DEATHS / 3. CONSTANTLY SICK All Causes	Syphilis	Soft Chancre	Gonorrhœa	Tænia	Other Entozoa
Quetta	2,550																								
Ramandrug	8																								
Maymyo	491																								
Group XIIa.—Hill Stations.	10,508																								
Darjeeling	473																								
Naini Tal	188																								
Landour	157																								
Kasauli	358																								
Dalhousie	832																								
Murree	111																								
Taragarh	26																								
Mount Abu	105																								
Pachmarhi	114																								
Purandhar	130																								
Khandala	47																								
Wellington	961																								
Group XIIb.—Hill Convalescent Depôts, and Sanitaria.	3,513																								
Troops marching, India.	1,992																								
Aden Column Field Force (Uthalla).	349																								
Deolali Depôt	457																								
Poonamallee Depôt	131																								

* Derived from the aggregates.   † Details by diseases are not available.

Stations and Commands.	Average annual strength.	Influenza.	Cholera.	Small-pox.	Enteric Fever.	Intermittent Fever.	Remittent Fever.	Simple Continued Fever.	Rheumatic Fever.	Heat-stroke.	Circulatory Diseases.	Tubercle of the lungs.	Pneumonia.	Other Respiratory Diseases.	Dysentery.	Diarrhœa.	Hepatic Abscess.	Hepatic Congestion and Inflammation.	Venereal Diseases.	All Causes.	Syphilis.	Soft Chancre.	Gonorrhœa.	Tænia.	Other Entozoa.
EXTRA INDIA. Aden	947	...	...	...	6	27	...	4	...	16	10	3	...	15	32	6	3	19	48	898	15	16	17	4	1
NORTHERN COMMAND	18,274	647	...	6	324	1,271	10	1,156	7	205	202	31	104	370	118	251	23	428	2,122	14,983	524	420	1,172	65	5
WESTERN COMMAND	12,53	143	2	77	342	2,162	55	899	6	52	340	89	92	515	336	316	57	281	2,037	17,512	618	1,032	1,317	44	4
EASTERN COMMAND	18,083	212	8	...	321	2,148	50	972	12	157	209	34	61	427	210	373	26	375	3,247	15,560	720	768	1,679	41	21
SECUNDERABAD DIVISION.	8,359	1	1	10	141	652	...	168	1	18	92	9	13	169	157	49	23	78	1,628	6,815	448	542	621	7	3
BURMA DIVISION	3,813	...	5	...	425	409	307	1	28	11	117	53	65	2	43	759	3,701	1,871	252	350	13	...

Groups and Commands.		1. Strength. 2. Constantly sick.											Total.
	Jan.	Feb.	March.	April.	May.	June.	July.	Aug.	Sept.	Oct.	Nov.	Dec.	
GROUP I.—BURMA COAST AND BAY ISLANDS.	1,023	1,111	1,236	1,280	1,285	1,268	1,256	1,234	1,230	1,410	1,360	1,384	15,107
	60·77	63·79	67·93	63·80	51·90	49·37	58·88	79·65	60·04	49·36	87·47	88·06	781·02
„ II.—BURMA INLAND	1,947	1,899	1,964	2,029	2,114	2,306	2,245	2,247	2,167	2,173	1,925	1,726	24,752
	103·03	105·07	98·18	88·39	109·63	108·45	103·91	114·65	113·60	118·82	94·76	92·89	1,251·44
„ IV.—BENGAL AND ORISSA.	1,821	1,950	1,660	1,543	1,614	1,618	1,565	1,580	1,575	1,585	1,774	1,593	20,278
	157·73	159·74	125·48	109·04	96·01	98·19	111·26	163·11	158·77	124·87	135·58	126·99	1,566·77
„ V.—GANGETIC PLAIN AND CHUTIA NAGPUR.	7,911	8,119	8,278	7,334	7,020	6,949	6,911	6,937	6,992	6,855	7,202	7,182	87,780
	400·16	410·56	349·49	343·26	319·97	388·69	390·01	359·23	330·16	354·18	414·10	400·71	4,510·52
„ VI.—UPPER SUB-HIMALAYA.	15,092	19,342	20,407	15,287	10,394	9,802	9,541	9,591	9,728	11,340	17,300	16,861	164,691
	1,049·70	982·54	823·80	674·99	531·03	511·82	560·32	572·10	514·27	557·39	590·11	970·28	8,638·35
„ VII.—N.-W. FRONTIER, INDUS VALLEY, AND WESTERN RAJPUTANA.	5,714	6,579	6,695	6,013	4,423	4,125	4,038	3,943	3,967	4,462	5,521	5,341	60,821
	312·16	321·93	290·50	236·02	281·78	255·77	302·22	318·32	276·03	259·71	318·20	306·19	3,478·83
„ VIII.—SOUTH-EASTERN RAJPUTANA, CENTRAL INDIA, AND GUJARAT.	6,167	6,713	7,142	6,827	6,558	6,411	6,323	6,413	6,505	6,813	6,700	6,177	78,749
	430·25	389·63	365·17	313·14	316·68	304·00	277·29	301·15	343·49	329·50	411·40	371·29	4,153·19
„ IX.—DECCAN	10,508	10,504	9,858	10,253	10,038	10,126	10,186	10,705	10,989	10,844	10,435	10,664	125,212
	572·74	546·75	425·52	451·16	449·31	460·29	478·02	618·52	658·06	638·93	586·02	569·15	6,524·47

Note.—Constantly sick × 365 = total annual loss of service.
* Derived from the aggregates.
† Remaining + admitted = total treated; remaining + admitted + died out of hospital = total cases.

TABLE IV—*concluded.*

GROUPS AND COMMANDS.		1. STRENGTH.											2. CONSTANTLY SICK.	TOTAL.
		Jan.	Feb.	March.	April.	May.	June.	July.	Aug.	Sept.	Oct.	Nov.	Dec.	
GROUP X.—WESTERN COAST	{	1,574	1,660	1,724	1,558	1,576	1,507	1,465	1,484	1,558	1,476	2,107	1,404	19,093
		110·48	104·56	130·51	70·36	65·83	69·07	87·01	97·01	100·21	109·06	91·70	77·42	1,113·2 5
,, XI.—SOUTHERN INDIA	{	1,960	3,320	3,50	4,021	3,992	4,132	4,167	4,173	4,044	4,074	3,849	3,575	44,851
		213·20	186·36	170·87	191·93	203·94	222·04	229·80	211·43	217·93	220·44	201·95	181·61	2,451·55
,, XIIa.—HILL STATIONS	{	4,205	4,237	4,626	10,108	16,024	16,988	17,071	17,145	16,662	11,321	4,179	3,534	126,100
		233·31	195·52	168·26	370·44	619·39	774·90	807·20	823·76	735·70	514·43	295·24	177·96	5,720·30
,, XIIb.—HILL CONVALESCENT DEPÔTS, AND SANITARIA.	{	1,338	1,352	1,817	3,811	5,545	5,872	5,581	5,229	4,979	3,722	1,500	1,392	42,144
		88·51	92·64	115·61	198·32	330·60	363·57	325·55	312·79	292·56	220·15	118·00	87·93	2,546·32
INDIA	{	69,150	71,084	72,095	73,228	72,712	72,889	72,161	72,477	72,574	70,390	69,574	67,779	856,116
		3,858·12	3,718·09	3,355·31	3,777·01	3,542·88	3,766·75	3,904·51	4,151·23	4,028·93	3,689·65	3,803·07	3,616·75	44,781·80
NORTHERN COMMAND	{	14,328	17,872	19,293	19,297	19,444	20,028	19,761	19,680	19,492	17,903	17,181	15,010	219,294
		930·68	859·94	765·59	732·22	945·35	1,026·54	1,082·86	1,029·20	896·38	774·31	900·29	875·79	10,820·08
WESTERN COMMAND	{	19,178	19,657	20,118	19,760	19,531	19,479	19,359	19,406	19,698	20,052	19,426	18,776	234,440
		1,110·23	1,069·42	1,050·46	905·48	928·10	991·31	1,007·64	1,182·84	1,235·73	1,201·97	1,149·08	1,064·80	12,897·12
EASTERN COMMAND	{	17,772	19,653	19,817	19,786	20,292	20,121	19,858	20,069	19,875	17,152	16,258	17,190	227,793
		1,108·29	1,114·27	914·72	939·87	924·78	1,014·73	1,060·61	1,136·08	1,088·60	934·84	1,005·01	964·67	12,206·47
SECUNDERABAD DIVISION	{	6,265	7,60	7,010	8,652	8,631	8,810	8,761	8,886	8,745	8,628	8,845	8,934	100,668
		515·94	449·91	443·57	491·62	524·86	519·55	527·32	552·33	577·00	562·68	521·37	512·31	6,201·31
BURMA DIVISION	{	3,447	3,502	3,821	3,979	4,016	4,045	4,016	4,035	3,947	4,161	3,607	3,170	45,748
		195·06	206·25	201·17	195·26	207·18	204·42	209·98	237·39	216·04	201·83	209·70	187·30	2,471·58

TABLE V.

ABSTRACT of the CANTONMENT SANITARY REPORTS of the most UNHEALTHY STATIONS, SANITARY DEFECTS, IMPROVEMENTS, SUGGESTIONS, etc.

(The ratios of sickness and mortality will be found in Table III.)

NORTHERN COMMAND.

CAMPBELLPORE.—No defects reported.

ATTOCK.—No defects reported.

MURREE.—No defects reported.

KASAULI.—The following defects have been brought to notice :—(1) The incinerators have worked badly. (2) Minor defects exist in the bazaar.

The General Officer Commanding the Brigade states in regard to the above defects:—(1) A system of trenching is being tried experimentally. (2) Are being remedied locally.

WESTERN COMMAND.

INDORE.—No defects reported.

ADEN :—No defects reported.

DEOLALI.—The following defects have been brought to notice :—(1) The Station Hospital is too small for requirements during the trooping season. (2) The enlargement of the Cantonment boundaries on sanitary grounds is essential.

The General Officer Commanding the Brigade says as regards the defects numbered above :—(1) An approximate estimate for enlargement has been submitted. The needs of the Depôt are now considerably diminished. (2) The Cantonment Committee has reported on this question. The matter is now in the hands of the Civil authorities (Collector of Nasik).

The Lieutenant-General Commanding the Forces remarks on the above defects :—(1) No action is necessary in view of the proposals for the reduction of the Depôt into a Rest Camp. (2) Is under consideration.

EASTERN COMMAND.

DUM-DUM.—The following defects have been brought to notice :—(1) Water should be laid on to the Guard Room Lavatory and the Canteen. (2) The bottom of the main drain behind the Mall Road should be properly levelled and made pucca. This is one of the principal outlets to the Salt Lakes of the Cantonment drainage system. (3) Gauge protection to the latrine seats in the Barracks is needed. (4) The want of a disinfecting room in the Station Hospital.

The General Officer Commanding the Brigade says as regards the defects numbered above :—(1) A requisition for this has been called for. (2) The requirement has already been discussed and proposals will be put forward shortly. (3) Hardly necessary. (4) Not necessary.

The Major General Commanding the Forces has no remarks to offer.

FATEHGARH.—No defects reported.

LANDOUR.—No defects reported.

SECUNDERABAD DIVISION.

CANNANORE.—No defects reported.

RAMANDROOG.—No defects reported.

POONAMALLEE.—The following defect has been brought to notice :—The need which exists for supplying improved means for draining the moat surrounding the hospital, as the water collects and stagnates during the rains, affording breeding places for mosquitoes.

The Colonel Commanding the Madras Brigade states that the matter has been brought to the notice of the Officer Commanding Royal Engineer for action.

The Lieutenant-General Commanding the Division remarks that the matter is receiving attention.

TABLE V—*concluded.*

ABSTRACT of the CANTONMENT SANITARY REPORTS of the most UNHEALTHY STATIONS, SANITARY DEFECTS, IMPROVEMENTS, SUGGESTIONS, etc.

BURMA DIVISION.

THAYETMYO.—No defects reported.

MAYMYO.—The following defects have been brought to notice :—(1) Imperfect drainage of the swamps in and around Cantonments. (2) The construction of impermeable floors for latrines and urinals in the Infantry Barracks and Hospital is essential. (3) The need of wire-netting in latrines in the Infantry Barracks and Hospital to render them fly proof.

The Lieutenant-General Commanding the Forces remarks on the above defects :—(1) R7,000 has been sanctioned and the acquisition of the land is in hand. An estimate amounting to R8,342 for the drainage of the swamps has also been sanctioned. (2) and (3) Estimates have been prepared for the construction of impermeable floors and the provision of wire-netting.

TABLE VI.

INFLUENZA by months, stations, groups, and commands.

TABLE VII.

CHOLERA by months, stations, groups, and commands.

Stations and Groups.	Jan.	Feb.	Mar.	Apr.	May	June	July	Aug.	Sept.	Oct.	Nov.	Dec.	TOTAL	Jan.	Feb.	Mar.	Apr.	May	June	July	Aug.	Sept.	Oct.	Nov.	Dec.	TOTAL
	Admissions from Influenza in each month.													Admissions from Cholera in each month.												
Rangoon	...	...	...	...	...	1	...	...	...	...	...	...	1	...	...	...	...	...	...	...	...	...	...	...	...	...
Group I.—Burma Coast and Bay Islands	...	...	...	...	...	1	...	...	...	...	...	...	1	...	...	...	...	...	...	...	...	...	...	...	...	...
Shwebo	...	...	...	...	...	...	...	3	...	1	...	...	4	...	...	...	...	...	...	...	...	...	...	...	...	...
Group II.—Burma Inland	...	...	...	...	...	...	...	3	...	1	...	...	4	...	...	...	...	...	...	...	...	...	...	...	...	...
Fort William	...	...	...	...	...	...	3	42	46	6	...	...	97	...	...	...	...	...	...	...	...	...	...	...	...	...
Dum Dum	...	...	...	...	...	...	2	3	...	...	...	...	5	...	...	...	...	...	...	...	...	...	...	...	...	...
Group IV.—Bengal and Orissa	...	...	...	...	...	...	5	45	46	6	...	...	102	...	...	...	...	...	...	...	...	...	...	...	...	...
B																										
Dinapore	...	...	...	...	...	...	...	...	...	...	...	...	...	...	...	...	...	...	...	...	...	1	...	1	...	1
Allahabad	...	...	...	...	...	...	...	...	2	1	...	...	3	...	...	...	...	...	...	...	1	...	...	...	...	1
Lucknow	...	...	...	...	2	...	...	1	...	...	...	1	4	...	...	...	...	...	...	...	...	...	...	...	...	...
Cawnpore	...	...	...	1	...	...	...	...	...	...	...	...	1	...	...	...	...	...	...	...	...	...	...	...	...	...
Fatehgarh	...	2	3	1	...	...	...	...	...	...	...	...	6	...	...	...	...	...	...	...	...	...	...	...	...	...
Group V.—Gangetic Plain and Chutia Nagpur	...	2	4	1	2	...	...	1	2	1	1	...	14	...	...	...	...	...	...	...	1	...	1	...	...	2
A																										
Roorkee	...	...	1	...	...	...	...	...	...	...	...	...	1	...	...	...	...	...	...	...	...	...	...	...	...	...
Meerut	...	...	3	2	...	...	2	...	3	1	6	...	17	...	...	...	...	...	...	...	...	...	...	...	...	...
Delhi	...	...	...	...	...	...	...	...	...	3	7	...	10	...	...	...	...	...	...	...	...	...	...	...	...	...
Ambala	...	3	1	...	...	...	...	...	...	...	...	...	4	...	...	...	...	...	...	...	...	...	...	...	...	...
B																										
Jullundur	...	...	...	...	1	4	1	...	...	...	...	...	6	...	...	...	...	...	...	...	...	...	...	...	...	...
Ferozepore	...	...	...	...	...	...	...	1	...	...	...	...	1	...	...	...	...	...	...	...	...	...	...	...	...	...
Mian Mir	9	2	1	26	51	7	...	...	...	...	...	...	96	...	...	...	...	...	...	...	...	...	...	...	...	...
Fort Lahore	1	...	...	...	...	1	...	...	...	...	...	...	2	...	...	...	...	...	...	...	...	...	...	...	...	...
Sialkot	9	4	4	...	39	1	...	...	...	...	...	...	57	...	...	...	...	...	...	...	...	...	...	...	...	...
Rawalpindi	14	4	...	1	...	...	...	...	...	...	...	...	19	...	...	...	...	...	...	...	...	...	...	...	...	...
Group VI.—Upper Sub-Himalaya	33	13	10	28	92	12	4	...	3	1	10	7	213	...	...	...	...	...	...	...	...	...	...	...	...	...
A																										
Nowshera	...	...	...	...	12	3	1	...	...	...	...	...	16	...	...	...	...	...	...	...	...	...	...	...	...	...
Peshawar	...	...	...	...	317	108	3	...	...	...	...	...	428	...	...	...	...	...	...	...	...	...	...	...	...	...
C																										
Kurrachee	...	...	...	1	15	1	...	3	2	5	4	...	31	...	...	...	...	...	...	...	...	...	...	...	...	...
Group VII.—North-West Frontier, Indus Valley, and North-Western Rajputana	...	...	...	...	330	126	5	...	3	2	5	4	475	...	...	...	...	...	...	...	...	...	...	...	...	...
B																										
Mutjra	1	...	2	...	...	...	...	...	...	...	...	...	3	...	...	...	...	...	...	...	...	...	...	...	...	...
Agra	...	...	5	4	...	2	...	...	3	...	...	...	14	...	...	...	...	...	...	...	...	...	...	...	...	...
Indore	...	...	1	...	...	...	...	...	...	...	...	...	1	...	...	...	...	...	...	...	...	...	...	...	...	...
Mhow	2	7	1	1	1	...	...	1	...	1	...	...	15	...	...	...	...	...	...	...	...	...	...	...	...	...
Group VIII.—South-Eastern Rajputana, Central India, and Gujarat	3	7	8	6	1	2	...	1	3	1	1	...	35	...	...	...	...	...	...	...	...	...	...	...	...	...
A																										
Saugor	...	...	1	...	...	2	...	...	...	...	...	...	3	...	...	...	...	...	...	...	...	...	...	...	...	...
Jubbulpore	...	...	...	...	...	...	...	...	2	...	...	...	2	...	...	...	...	...	...	...	...	...	2	...	...	2
Sitabaldi	...	...	...	...	...	...	...	...	2	...	...	...	2	...	...	...	...	...	...	...	...	...	...	...	...	...
B																										
Belgaum	...	...	...	...	...	...	...	2	...	...	...	...	2	...	...	...	...	...	...	...	...	...	...	...	...	...
Poona	...	9	13	2	1	3	1	...	...	6	2	...	37	...	...	...	...	...	...	...	...	...	...	...	...	...
Kirkee	...	6	12	1	...	1	...	...	...	...	...	...	20	...	...	...	...	...	...	...	...	...	...	...	...	...
Group IX.—Deccan	...	15	26	3	1	6	1	...	...	2	8	2	64	...	...	...	...	...	...	...	...	...	2	...	...	2

* Stations where neither Influenza nor Cholera occurred are not shown in these tables. For the annual ratios, see Table III.

TABLE VI—concluded.

INFLUENZA by months, stations, groups, and commands.

TABLE VII—concluded.

CHOLERA by months, stations, groups, and commands.

Stations and Groups	ADMISSIONS FROM INFLUENZA IN EACH MONTH													ADMISSIONS FROM CHOLERA IN EACH MONTH												
	Jan.	Feb.	Mar.	Apr.	May	June	July	Aug.	Sep.	Oct.	Nov.	Dec.	Total	Jan.	Feb.	Mar.	Apr.	May	June	July	Aug.	Sep.	Oct.	Nov.	Dec.	Total
Colaba	...	...	...	...	...	...	...	...	1	6	4	...	11	...	...	...	...	...	...	...	...	...	...	...	...	...
Mallapuram	...	1	...	...	...	...	...	...	...	...	...	...	1	...	...	...	...	...	...	...	...	...	...	...	...	...
Group X.—Western Coast	...	1	...	...	...	...	...	...	1	6	4		12	...	...	...	...	...	...	...	...	...	...	...	...	...
D																										
Madras	...	...	...	...	...	...	...	...	...	...	...	...	...	...	...	...	...	...	...	...	1	...	...	...	...	1
Group XI.—Southern India	...	...	...	...	...	...	...	...	...	...	...	...	...	...	...	...	...	...	...	...	1	...	...	...	...	1
Chakrata	2	1	5	1	...	...	1	4	...	...	23	4	41	...	...	...	...	...	...	...	...	...	...	...	...	...
Lebong	...	...	...	...	...	...	...	...	...	...	...	...		...	...	...	...	...	...	...	...	...	...	...	...	...
Solon	...	...	...	2	...	...	...	...	...	...	...	...	2	...	...	...	...	...	...	...	...	...	...	...	...	...
Subathu	...	...	...	...	1	3	...	...	...	...	...	...	4	...	...	...	...	...	...	...	...	...	...	...	...	...
Camp Gharial	...	...	...	1	...	...	...	...	...	...	...	...	1	...	...	...	...	...	...	...	...	...	...	...	...	...
„ Thobba	...	...	...	...	...	...	...	1	5	1	...	...	7	...	...	...	...	...	...	...	...	...	...	...	...	...
„ Upper Topa	...	...	...	...	...	1	...	...	...	...	...	...	1	...	...	...	...	...	...	...	...	...	...	...	...	...
Group XIIa.—Hill Stations	2	1	5	4	1	5	2	5	5	1	23	4	58	...	...	...	...	...	...	...	...	...	...	...	...	...
Darjeeling	...	...	4	...	...	...	...	...	...	...	...	...	4	...	...	...	...	...	...	...	...	...	...	...	...	...
Naini Tal	...	...	...	...	...	2	1	...	...	...	...	...	3	...	...	...	...	...	...	...	...	...	...	...	...	...
Landour	...	...	...	...	...	...	...	1	...	...	...	...	1	...	...	...	...	...	...	...	...	...	...	...	...	...
Dalhousie	...	...	...	1	1	...	...	...	...	...	...	...	2	...	...	...	...	...	...	...	...	...	...	...	...	...
Murree	...	...	...	...	...	...	...	...	1	...	...	...	1	...	...	...	...	...	...	...	...	...	...	...	...	...
Group XIIb.—Hill Convalescent Depôts, and Sanitaria	...	...	4	1	2	1	...	1	1	...	...		11	...	...	...	...	...	...	...	...	...	...	...	...	...
Troops, Marching, India	5	2	...	...	...	...	...	...	...	...	...	...	7	...	...	...	...	...	...	...	...	...	...	5	...	5
Deolali Depôt	...	14	4	...	...	...	...	...	...	2	...	...	20	...	...	...	...	...	...	...	...	...	...	...	...	...
INDIA	43	41	71	47	428	154	18	53	1	27	52	19	1,014	...	...	...	...	...	...	...	1	1	2	6	...	10
Northern Command	33	13	6	30	423	127	6	1	5	2	1	...	647	...	...	...	...	...	...	...	...	...	...	...	...	...
Western Command	2	22	41	9	3	21	2	...	5	12	18	7	142	...	...	...	...	...	...	...	...	...	2	...	...	2
Eastern Command	3	3	24	8	2	5	10	49	51	12	33	12	212	...	...	...	...	...	...	...	1	...	1	...	...	2
Secunderabad Division	...	1	...	...	...	...	...	...	...	...	...	...	1	...	...	...	...	...	...	...	1	...	...	...	...	1
Burma Division	...	...		...	...	1	...	3	...	1	...	...	5	...	...	...	...	...	...	...	...	...	...	...	...	...

TABLE VIII.
ENTERIC FEVER by months, stations, groups, and commands.

TABLE IX.
SIMPLE CONTINUED FEVER by months, stations, groups, and commands.

ADMISSIONS FROM ENTERIC FEVER IN EACH MONTH.

STATIONS and GROUPS.	January	February	March	April	May	June	July	August	September	October	November	December	TOTAL
Port Blair	...	...	...		...	...	...	...	...	...	...	...	...
Raugoon	...	...	...	1	...	...	...	...	...	...	...	...	1
GROUP I.—BURMA COAST AND BAY ISLANDS	...	...	...	1	...	...	...	...	...	...	...	...	1
Thayetmyo	...	...	...	...	...	...	1	...	...	...	2	...	3
Meiktila	...	...	...	...	...	...	...	...	...	...	...	...	...
Fort Dufferin	...	...	...	...	...	...	...	...	...	...	...	...	...
Shwebo	...	...	...	1	1	...	...	...	1	...	...	...	2
Bhamo	...	...	...	...	...	...	...	1	...	...	...	...	1
GROUP II.—BURMA INLAND	...	...	...	1	1	...	1	...	1	...	2	...	6
Fort William	...	1	...	1	...	...	...	...	1	...	1	...	4
„ Felta	...	...	...	...	...	...	...	...	...	...	...	...	...
Dum-Dum	...	...	...	...	...	...	...	...	2	1	...	...	3
Barrackpore	...	2	...	1	1	...	1	...	...	...	...	...	6
GROUP IV.—BENGAL AND ORISSA	...	3	...	1	...	1	1	...	2	2	2	1	13
B													
Dinapore	1	...	...	...	...	...	...	1	...	18	...	...	20
Benares	...	...	...	...	...	...	1	...	...	...	...	...	1
Allahabad	1	...	...	...	...	3	...	2	...	...	...	...	6
Fort Allahabad	...	...	...	...	...	...	...	...	...	...	...	...	...
Fyzabad	...	4	1	...	2	...	...	1	...	1	...	...	9
Sitapur	2	...	...	...	...	...	...	...	...	...	...	...	2
Lucknow	1	1	...	3	9	9	8	8	11	3	7	6	71
Cawnpore	...	...	...	...	3	...	...	2	1	...	1	...	4
Fatehgarh	...	...	...	...	...	...	...	...	1	10	...	...	14
GROUP V.—GANGETIC PLAIN AND CHUTIA NAGPUR	5	5	1	8	14	9	8	13	14	8	35	7	127
A													
Shahjahanpur	1	...	...	...	1	...	...	1	...	...	...	...	4
Bareilly	...	...	1	...	6	...	3	2	...	...	...	...	12
Roorkee	...	...	...	...	...	...	...	...	1	1	...	...	3
Meerut	2	3	...	15	15	3	18	1	3	34	14	...	103
Delhi	1	...	...	...	...	...	...	2	1	...	...	...	4
Ambala	...	...	...	1	1	3	3	...	1	2	...	...	12
B													
Jullundur	...	...	...	...	1	...	2	2	...	1	...	...	6
Ferozepore	2	3	1	...	5	3	1	2	...	...	...	1	18
Amritsar	...	...	...	...	1	...	...	...	...	1	3	...	5
Mian Mir	...	...	...	...	1	4	3	...	...	3	4	1	16
Fort Lahore	1	...	...	...	...	...	...	1	1	2	...	...	6
Sialkot	...	...	...	...	3	...	1	2	...	1	2	...	9
Rawalpindi	1	...	...	1	3	13	9	7	4	7	10	10	70
Campbellpur	...	...	...	...	21	10	...	...	12	...	...	...	43
Attock	...	...	...	1	...	...	...	1	...	...	...	...	2
GROUP VI.—UPPER SUB-HIMALAYA	8	6	2	3	59	48	21	41	13	26	53	40	320
A													
Nowshera	...	...	...	...	1	...	...	...	...	...	...	...	1
Peshawar	2	...	1	1	8	42	7	3	1	5	...	...	71
Mooltan	...	...	...	1	1	...	...	1	...	...	1	...	4
C													
Hyderabad	1	...	1	...	1	...	...	...	...	...	...	...	3
Kurrachee	2	3	1	...	1	1	...	...	1	...	2	6	17
GROUP VII.—NORTH-WEST FRONTIER, INDUS VALLEY, AND NORTH-WESTERN RAJPUTANA	5	3	3	2	3	11	43	7	4	2	7	6	96
A													
Deesa	...	...	...	...	...	...	...	...	...	...	...	...	...
Ahmedabad	...	...	...	...	...	...	...	...	...	1	...	...	...

ADMISSIONS FROM SIMPLE CONTINUED FEVER IN EACH MONTH.

STATIONS and GROUPS.	January	February	March	April	May	June	July	August	September	October	November	December	TOTAL
Port Blair	1	1	1	1	...	...	...	...	1	...	...	...	5
Raugoon	3	5	1	3	2	7	19	21	10	23	1	53	166
GROUP I.—BURMA COAST AND BAY ISLANDS	4	6	2	4	2	7	19	21	11	23	19	53	171
Thayetmyo	...	...	...	...	...	...	...	...	...	...	...	...	...
Meiktila	1	2	6	5	12	4	5	2	11	35	6	1	88
Fort Dufferin	...	4	...	2	...	2	...	5	...	3	8	6	5	37
Shwebo	...	1	...	1	2	...	...	...	...	...	3	1	8
Bhamo	...	...	...	...	...	...	...	...	...	...	1	...	1
GROUP II.—BURMA INLAND	1	7	6	6	16	6	7	7	14	43	14	7	134
Fort William	1	1	...	...	...	...	5	9	...	...	...	...	16
„ Felta	...	...	...	...	...	...	...	1	...	...	...	...	1
Dum-Dum	...	...	...	...	...	...	...	...	...	...	...	...	...
Barrackpore	...	...	2	...	1	...	...	...	6	4	3	3	19
GROUP IV.—BENGAL AND ORISSA	1	1	2	...	1	5	9	1	6	4	3	3	36
B													
Dinapore	...	...	...	...	...	1	...	...	6	...	...	...	7
Benares	...	...	2	2	8	5	8	3	...	1	...	...	37
Allahabad	...	...	4	3	14	6	10	4	8	8	...	...	76
Fort Allahabad	...	...	1	1	4	8	2	4	1	1	...	...	22
Fyzabad	...	6	12	24	17	7	12	11	6	...	...	...	96
Sitapur	...	5	25	6	11	12	16	6	2	1	...	...	84
Lucknow	12	7	13	19	19	87	103	71	63	17	8	6	427
Cawnpore	...	...	...	...	6	3	2	3	...	1	...	...	15
Fatehgarh	...	...	...	1	3	1	...	...	...	...	...	1	6
GROUP V.—GANGETIC PLAIN AND CHUTIA NAGPUR	13	7	23	57	46	155	154	116	130	39	26	7	764
A													
Shahjahanpur	1	1	...	2	...	...	...	...	...	...	...	...	5
Bareilly	...	...	1	...	...	...	...	...	...	...	...	...	1
Roorkee	...	...	...	3	21	...	6	3	...	...	...	6	41
Meerut	...	...	...	2	1	...	...	...	...	...	...	...	4
Delhi	3	1	2	...	4	...	13	16	2	11	7	4	63
Ambala	...	...	...	...	...	...	...	...	...	...	...	...	...
B													
Jullundur	...	...	1	2	5	...	1	...	...	...	...	...	9
Ferozepore	...	3	17	30	30	12	19	2	2	1	...	...	120
Amritsar	...	1	5	4	2	1	2	...	...	...	...	...	13
Mian Mir	...	4	8	52	95	47	23	5	15	4	...	...	253
Fort Lahore	...	...	3	6	8	9	...	...	3	...	...	...	30
Sialkot	8	11	7	16	57	13	5	3	...	...	...	...	120
Rawalpindi	1	1	...	...	18	11	23	17	1	...	...	...	73
Campbellpur	1	...	1	...	...	...	...	...	...	...	...	...	2
Attock	...	...	...	...	...	...	...	...	...	...	...	...	1
GROUP VI.—UPPER SUB-HIMALAYA	5	2	6	17	48	122	176	164	79	30	51	16	736
A													
Nowshera	...	...	...	2	3	9	6	12	20	4	4	3	67
Peshawar	2	1	3	4	11	6	44	20	22	34	41	19	213
Mooltan	...	...	...	...	4	8	8	3	3	3	5	6	37
C													
Hyderabad	...	...	...	...	...	...	6	1	...	...	...	...	7
Kurrachee	...	...	...	1	...	...	...	...	...	...	...	2	3
GROUP VII.—NORTH-WEST FRONTIER, INDUS VALLEY, AND NORTH-WESTERN RAJPUTANA	3	1	5	7	21	16	64	60	30	41	49	30	327
A													
Deesa	...	2	...	...	...	...	...	...	...	...	...	...	2
Ahmedabad	...	...	...	...	...	...	...	...	...	...	...	...	...

* Stations where neither Enteric Fever nor Simple Continued Fever occurred are not shown in these tables. For the annual ratios, see Table III.

TABLE VIII—*concluded.*

ENTERIC FEVER by *months, stations, groups, and commands.*

TABLE IX—*concluded.*

SIMPLE CONTINUED FEVER by *months, stations, groups, and commands.*

STATIONS AND GROUPS.	January.	February.	March.	April.	May.	June.	July.	August.	September.	October.	November.	December.	TOTAL.	January.	February.	March.	April.	May.	June.	July.	August.	September.	October.	November.	December.	TOTAL.
B																										
Neemuch	1	...	...	...	...	...	...	...	...	...	...	...	1	...	...	...	...	...	...	...	...	...	...	...	...	...
Nasirabad	...	1	1	...	...	...	2	...	...	1	...	...	4	1	...	1	...	...	...	...	...	...	...	...	...	2
Muttra	...	...	...	...	2	...	...	...	...	...	...	...	2	...	...	...	...	1	3	...	...	...	...	...	...	4
Agra	3	...	1	2	1	4	3	1	...	1	2	1	19	...	...	...	...	...	1	...	1	...	...	...	...	3
Jhansi	3	5	3	...	1	...	...	...	5	6	1	4	28	...	...	3	3	...	3	3	...	4	...	8	7	31
Nowgong	2	...	...	1	...	...	...	...	5	...	...	...	9	...	...	...	...	1	1	3	1	...	5	...	...	11
Indore	...	...	...	...	2	...	1	1	...	...	2	...	6	...	...	...	1	...	...	...	2	...	...	...	...	3
Mhow	6	5	4	7	3	2	1	2	3	2	...	4	39	...	...	...	...	1	...	1	...	...	...	...	...	3
GROUP VIII.—SOUTH-EAST-ERN RAJPUTANA, CENTRAL INDIA, AND GUJARAT	13	11	9	10	7	8	5	4	15	10	6	10	110	3	...	4	3	1	7	7	4	8	...	13	7	57
A																										
Saugor	...	...	...	2	1	...	...	1	1	...	...	...	5	...	...	...	6	6	...	...	...	...	...	...	...	12
Jubbulpore	...	...	2	2	...	...	1	5	3	1	1	1	16	...	...	...	5	8	8	10	19	35	32	14	4	145
Kamptee	...	...	...	2	...	1	...	1	1	2	...	...	5	1	...	10	10	5	13	8	12	13	5	7	4	88
Sitabaldi	...	...	...	...	1	...	...	...	...	...	...	...	1	...	...	...	1	1	2	6	2	9	7	3	1	32
B																										
Secunderabad	1	1	2	4	1	4	2	33	21	16	2	1	88	...	3	4	11	12	9	4	15	14	13	15	6	106
Belgaum	...	1	1	...	4	...	3	...	2	...	...	...	11	4	1	...	...	...	...	...	...	...	1	...	...	6
Poona	2	2	...	1	1	1	3	55	22	15	...	...	92	...	2	...	1	...	2	...	...	2	2	9	6	25
Kirkee	...	...	...	1	1	...	...	7	4	2	1	1	18	...	...	...	...	...	3	3	7	2	1	...	...	16
Ahmednagar	...	...	5	...	...	2	...	4	6	...	...	...	18	...	...	...	...	...	1	...	1	3	...	3	4	7
GROUP IX.—DECCAN	3	4	10	12	9	9	9	95	70	37	5	3	257	5	6	14	34	35	34	31	51	68	81	53	25	437
Colaba	1	...	...	...	...	...	...	...	1	...	...	...	2	...	...	...	...	...	...	...	...	...	...	...	...	...
Cannanore	...	...	...	...	...	...	...	...	...	...	...	...	...	1	...	...	...	...	...	...	...	...	...	...	2	3
GROUP X.—WESTERN COAST	1	...	...	...	...	...	...	...	1	...	...	...	2	1	...	...	...	...	...	...	...	...	...	...	2	3
A																										
Bellary	1	...	...	...	...	...	...	...	...	1	...	...	9	...	...	...	...	...	...	...	...	...	...	...	...	...
Bangalore	...	...	3	4	2	1	6	4	2	9	4	3	38	...	...	1	4	3	7	1	3	3	...	2	8	32
B																										
St. Thomas' Mount	...	...	...	...	...	...	...	...	...	...	...	...	...	...	3	...	...	1	...	2	5	4	5	2	3	25
Madras	...	1	...	...	1	...	...	...	...	...	...	...	2	...	...	...	...	...	...	...	...	...	...	...	...	...
GROUP XI.—SOUTHERN INDIA	1	1	3	6	4	3	6	4	2	11	5	3	49	...	3	1	4	7	3	8	7	5	4	11	57
Ranikhet	...	...	...	1	...	1	1	1	...	...	...	...	4	...	...	2	2	4	4	1	1	...	...	...	...	14
Chaubuttia	...	...	...	...	...	...	...	...	3	...	...	...	3	...	...	1	...	1	1	...	...	...	...	...	...	3
Chakrata	...	...	1	...	2	6	...	...	...	...	...	...	9	...	...	...	7	6	4	4	2	1	...	...	...	24
Lebong	...	...	...	...	...	3	...	...	...	...	...	...	3	...	...	...	1	2	...	...	...	...	...	...	...	4
Solon	...	...	...	...	3	1	...	...	...	...	...	...	4	...	...	...	3	...	...	...	...	...	...	...	...	4
Dagshai	...	...	...	1	1	...	...	...	...	...	...	...	2	...	...	...	3	1	...	1	...	...	...	...	...	5
Subathu	...	...	...	...	...	5	6	...	...	...	...	...	11	...	...	1	2	4	6	4	1	1	...	...	...	23
Khyragully	...	...	...	...	...	...	...	...	...	...	...	...	1	...	...	...	...	...	...	...	...	...	...	...	...	...
Kuldunnah	...	...	...	...	...	...	...	...	...	...	...	...	1	...	...	...	...	...	...	...	...	...	...	...	...	...
Camp Gharial	...	...	...	1	...	...	...	...	...	...	...	...	1	...	...	...	...	5	4	2	2	...	...	...	...	13
,, Thobba	...	...	...	3	...	...	...	...	...	...	...	...	3	...	...	3	...	...	...	...	...	...	...	...	...	3
,, Upper Topa	...	...	...	...	...	1	...	...	...	...	...	...	1	...	...	2	...	...	...	...	...	...	...	...	...	2
,, Lower Topa	...	...	...	...	1	...	...	...	...	...	...	...	1	...	...	...	...	...	...	...	...	...	...	...	...	...
Khan Spur	...	...	...	2	6	...	...	...	...	...	...	...	8	...	...	10	7	3	3	3	...	...	...	...	...	77
Cherat	...	...	...	1	1	...	2	1	5	...	...	...	10	...	...	10	21	11	9	8	2	...	...	...	...	62
Quetta	2	...	1	6	1	...	13	13	5	5	5	2	53	1	...	2	7	12	20	18	18	156	75	18	...	379
Maymyo	...	...	...	...	...	...	1	...	...	...	...	...	1	...	...	...	...	1	...	...	...	...	...	1	...	2
GROUP XIIa.—HILL STATIONS	2	...	1	9	9	13	27	22	16	9	5	2	115	1	...	2	13	47	67	54	93	179	83	21	2	563
Darjeeling	...	...	...	...	...	...	...	...	...	...	...	...	...	...	...	...	...	...	...	...	...	...	...	...	...	5
Naini Tal	...	...	...	...	1	...	2	1	...	1	...	...	5	...	...	...	...	...	...	...	...	...	...	...	...	1
Landour	...	...	...	...	...	1	1	1	1	...	...	...	5	...	...	...	...	1	...	...	...	...	...	...	...	3
Kasauli	...	...	...	...	2	2	2	...	...	...	...	...	6	...	...	...	...	1	...	...	...	...	...	...	...	...
Dalhousie	...	...	...	...	4	2	...	1	1	...	...	...	10	...	...	...	4	7	4	1	...	...	...	...	...	16
Murree	...	...	...	...	1	...	...	...	...	...	...	...	2	...	...	...	3	...	...	...	...	...	...	...	...	4
Tara-arh	...	...	...	...	...	...	...	...	...	...	...	...	...	...	...	...	2	...	...	...	...	...	...	...	...	2
Pachmarhi	...	...	...	...	...	...	1	...	...	...	...	...	1	...	...	1	...	...	...	1	2	4	3	1	...	15
Puranhor	...	...	...	...	...	...	...	...	...	...	...	...	...	...	...	...	...	...	...	...	...	...	...	...	...	...
Wellington	...	...	...	...	1	...	2	...	1	...	...	...	4	...	...	...	...	...	...	...	...	...	...	...	...	...
GROUP XIIb.—HILL CON-VALESCENT DEPÔTS AND SANITARIA	...	...	1	2	5	8	4	4	1	1	...	...	34	...	...	3	4	13	6	2	4	2	...	...	...	46

STATIONS AND COMMANDS.	January.	February.	March.	April.	May.	June.	July.	August.	September.	October.	November.	December.	TOTAL.	January.	February.	March.	April.	May.	June.	July.	August.	September.	October.	November.	December.	TOTAL.
Troops, marching, India	2	...	...	...	...	...	...	...	...	1	1	5	9	13	...	...	1	14	...	...	...	...	5	7	1	41
Deolali Depôt	...	...	...	...	...	...	...	...	1	...	...	...	1	...	...	...	...	...	...	...	...	...	...	...	...	...
Poonamallee ,,	...	...	...	...	...	...	...	...	...	...	...	...	...	1	...	...	...	...	1	...	...	...	...	...	...	2
EXTRA INDIA.																										
Adeu	...	...	1	...	...	1	...	...	3	1	...	...	6	...	...	...	...	...	19	13	6	2	1	...	...	41
INDIA	42	33	30	53	116	110	129	180	145	109	122	77	1,146	52	33	68	150	248	456	542	515	528	378	261	164	3,415
Northern Command	6	3	2	6	52	56	70	35	21	27	22	24	324	7	2	10	30	103	154	253	234	126	97	45	41	1,158
Western Command	20	17	20	22	16	9	23	70	75	37	14	19	342	9	3	17	33	40	70	64	124	223	146	70	30	829
Eastern Command	12	11	3	14	39	37	24	38	25	17	76	25	321	16	9	28	60	57	202	186	126	133	45	36	13	912
Secunderabad Division	2	2	5	10	6	8	10	37	23	27	7	4	141	2	6	5	15	16	17	7	23	21	18	19	19	168
Burma Division	...	...	...	1	3	...	2	...	1	...	2	...	9	5	13	8	11	13	13	26	23	25	66	34	60	307

TABLE X.

INTERMITTENT FEVER by months, stations, groups, and commands.

ADMISSIONS FROM INTERMITTENT FEVER IN EACH MONTH.

Stations and Groups.	January	February	March	April	May	June	July	August	September	October	November	December	TOTAL
Port Blair	...	...	...	...	...	...	...	...	2	...	...	...	3
Rangoon	5	5	3	3	2	4	1	2	4	1	...	...	31
Group I.—Burma Coast and Bay Islands	5	5	3	3	2	4	1	2	4	3	...	2	34
Thayetmyo	22	12	13	55	66	11	15	14	7	18	10	2	244
Meiktila	...	...	2	...	...	...	...	1	1	...	1	...	5
Fort Dufferin	2	1	...	1	1	...	1	1	1	1	...	2	13
Shwebo	...	...	1	...	4	6	3	1	...	1	3	5	24
Bhamo	5	4	...	...	3	6	3	7	3	6	2	...	37
Group II.—Burma Inland	29	17	15	56	74	23	20	24	12	26	16	9	321
Fort William	21	16	12	8	12	10	24	32	39	40	56	35	305
„ Fulta	...	...	...	...	...	...	...	...	...	4	...	...	4
„ Chingrikhal	1	...	...	...	...	...	1	...	...	4	3	...	9
Dum-Dum	5	1	...	...	2	...	...	1	6	7	11	...	39
Barrackpore	11	4	2	4	4	10	7	7	4	2	15	5	75
Group IV.—Bengal and Orissa	38	21	14	12	18	20	31	41	47	49	90	49	432
B													
Dinapore	13	9	15	2	3	1	...	1	6	...	...	1	51
Benares	...	...	...	1	2	...	2	...	4	3	...	1	13
Allahabad	5	...	2	3	1	5	6	5	22	32	25	7	134
Fort Allahabad	2	1	...	...	3	8	2	4	15	28	24	7	88
Fyzabad	3	2	...	6	8	9	5	4	15	13	17	7	80
Sitapur	3	...	...	1	...	1	3	2	1	3	11	7	32
Lucknow	8	8	1	16	22	44	23	36	11	5	21	10	205
Cawnpore	1	1	...	1	5	6	7	8	8	15	7	4	63
Fatehgarh	3	2	...	2	1	7	2	...	...	2	1	3	23
Group V.—Gangetic Plain and Chutia Nagpur	38	24	18	32	45	73	48	62	78	122	109	47	698
A													
Shahjahanpur	2	...	...	2	4	3	4	2	...	...	...	...	18
Bareilly	6	4	4	5	9	14	13	31	10	5	4	...	109
Roorkee	...	1	...	...	1	9	12	...	2	3	1	...	52
Meerut	14	7	16	8	12	16	20	17	21	7	65	20	223
Delhi	10	4	2	7	30	30	29	4	19	7	1	...	173
Ambala	9	3	1	1	1	8	7	14	15	10	3	...	73
B													
Jullundur	2	4	1	2	1	...	2	4	7	4	1	...	32
Ferozepore	4	4	8	11	6	2	6	6	8	12	4	3	83
Amritsar	...	1	...	...	...	...	1	...	3	...	1	...	6
Mian Mir	6	1	3	3	...	1	2	3	3	5	4	...	31
Fort Lahore	1	...	...	3	...	2	1	3	...	...	2	...	11
Sialkot	7	9	4	7	3	5	4	9	2	9	6	...	60
Rawalpindi	10	5	5	12	3	2	9	11	21	10	13	...	110
Campbellpur	...	...	...	4	2	...	...	2	...	...	...	...	9
Attock	7	9	4	7	6	2	...	...	3	8	8	...	54
Group VI.—Upper Sub-Himalaya	73	52	49	60	82	83	109	131	79	109	136	61	1,024
A													
Nowshera	2	3	1	1	1	6	6	1	3	6	8	...	39
Peshawar	10	11	6	15	15	13	20	83	42	57	118	46	448
Moultan	6	2	9	6	8	4	25	5	7	12	14	3	107
C													
Hyderabad	8	2	5	3	3	3	3	...	...	1	15	26	69
Kurrachee	15	20	13	11	31	31	36	22	34	10	16	43	304
Group VII.—N.W. Frontier, Indus Valley, and N.W. Rajputana	41	47	36	41	59	57	102	111	83	95	171	127	967

TABLE XI.

REMITTENT FEVER by months, stations, groups, and commands.

ADMISSIONS FROM REMITTENT FEVER IN EACH MONTH.

Stations and Groups.	January	February	March	April	May	June	July	August	September	October	November	December	TOTAL
Port Blair	...	...	...	2	2	8	8	8	...	...	3	1	33
Rangoon	...	...	...	...	...	...	...	...	...	...	...	...	...
Group I.—Burma Coast and Bay Islands	...	...	...	2	2	8	8	8	...	3	1	1	33
Thayetmyo	...	...	...	...	...	...	...	...	...	...	...	...	...
Meiktila	...	...	...	...	...	...	...	...	1	1	...	...	2
Fort Dufferin	...	...	...	...	...	...	...	...	...	...	...	...	1
Shwebo	...	...	...	...	1	...	...	...	...	...	...	1	2
Bhamo	...	...	...	...	1	...	1	...	...	...	1	...	3
Group II.—Burma Inland	...	...	...	1	1	1	...	...	1	2	...	1	7
Fort William	...	...	...	...	...	3	...	...	...	...	...	...	3
„ Fulta	...	...	...	...	...	...	...	...	...	...	...	...	...
„ Chingrikhal	...	...	...	...	...	...	...	...	...	...	2	...	2
Dum-Dum	...	...	...	...	...	...	...	...	...	2	...	...	2
Barrackpore	...	...	...	...	...	...	...	...	2	1	...	...	3
Group IV.—Bengal and Orissa	...	...	...	...	...	...	...	...	3	2	3	...	8
B													
Dinapore	8	...	1	...	...	...	...	...	3	...	...	...	12
Benares	...	...	...	...	...	...	...	...	...	...	...	...	...
Allahabad	1	...	...	...	...	...	...	...	...	...	...	...	1
Fort Allahabad	...	...	...	...	...	...	...	...	...	...	...	...	...
Fyzabad	...	...	...	...	...	...	...	...	...	...	...	...	...
Sitapur	...	...	...	...	...	...	...	...	...	...	...	...	...
Lucknow	...	...	...	...	...	...	...	...	3	...	...	...	3
Cawnpore	...	...	...	...	...	...	...	...	...	...	...	...	...
Fatehgarh	...	...	...	...	...	1	1	...	...	...	...	...	2
Group V.—Gangetic Plain and Chutia Nagpur	9	...	1	...	1	1	...	3	3	...	...	...	18
A													
Shahjahanpur	...	...	...	...	...	...	...	...	...	...	...	...	...
Bareilly	...	...	...	...	...	...	...	...	...	...	...	...	...
Roorkee	...	...	...	...	...	...	...	...	...	...	...	...	...
Meerut	...	...	...	...	1	...	...	...	...	...	...	...	1
Delhi	...	...	...	1	5	10	1	3	...	...	...	...	20
Ambala	...	...	...	...	...	...	...	...	...	...	...	...	...
B													
Jullundur	...	...	...	...	...	...	...	...	...	...	...	...	...
Ferozepore	...	...	...	...	...	...	...	...	...	...	...	...	...
Amritsar	...	...	...	...	...	...	...	...	...	...	...	...	...
Mian Mir	...	...	...	...	...	...	...	...	...	2	...	...	2
Fort Lahore	...	...	...	...	...	...	...	...	...	...	...	...	...
Sialkot	...	...	...	...	...	...	...	...	...	...	...	...	...
Rawalpindi	...	...	...	...	...	...	...	...	...	...	...	...	...
Campbellpur	...	...	...	...	...	...	...	1	...	...	2	3	7
Attock	...	...	...	...	...	...	...	...	...	...	...	...	...
Group VI.—Upper Sub-Himalaya	...	...	...	1	6	11	1	3	...	4	3	1	30
A													
Nowshera	...	...	...	...	...	...	...	...	...	...	...	...	...
Peshawar	...	...	...	...	...	...	1	...	...	...	...	...	1
Moultan	...	...	...	...	...	...	...	...	...	...	...	...	...
C													
Hyderabad	...	...	...	...	...	...	...	...	...	...	...	...	...
Kurrachee	...	...	...	...	...	...	...	...	...	...	...	...	...
Group VII.—N.W. Frontier, Indus Valley, and N.W. Rajputana	...	...	...	...	...	...	1	...	...	...	...	...	1

* Stations where neither Intermittent Fever nor Remittent Fever occurred are not shown in these tables. For the annual ratios, see Table IIL.

| Stations and Groups | \multicolumn{13}{c}{Admissions from Intermittent Fever in each month.} | | | | | | | | | | | | | \multicolumn{13}{c}{Admissions from Remittent Fever in each month.} | | | | | | | | | | | | |

Stations and Groups	Jan.	Feb.	Mar.	Apr.	May.	June.	July.	Aug.	Sept.	Oct.	Nov.	Dec.	Total.	Jan.	Feb.	Mar.	Apr.	May.	June.	July.	Aug.	Sept.	Oct.	Nov.	Dec.	Total.
A																										
Deess	...	1	1	1	1	1	3	1	4	1	3	3	20	...	...	...	...	...	...	...	...	...	...	...	...	...
Ahmedabad	9	3	5	13	15	9	6	13	25	23	38	36	195	...	...	...	...	...	...	...	...	...	...	...	...	...
B																										
Neemuch	5	...	3	5	9	2	13	5	4	5	5	2	58	...	...	...	...	...	...	...	...	...	...	...	...	...
Nasirabad	15	14	11	12	9	12	9	13	9	19	31	7	161	...	...	...	...	...	...	...	...	...	...	...	...	...
Muttra	...	1	1	3	9	4	3	9	2	7	3	..	42	...	...	...	...	...	...	...	...	...	...	...	...	...
Agra	6	5	6	4	5	8	11	7	15	15	5	1	78	...	...	...	...	...	...	...	...	...	...	...	...	...
Jhansi	23	10	5	14	15	24	30	10	51	54	18	9	272	...	...	...	...	...	...	...	...	...	...	...	...	1
Nowgong	1	1	...	2	11	7	1	2	11	6	...	...	42	...	...	...	...	...	...	...	2	...	...	...	...	2
Indore	...	...	...	1	...	...	...	3	1	...	1	...	6	...	...	...	...	...	...	...	...	...	...	...	...	...
Mhow	13	6	9	21	13	5	5	12	36	21	15	30	186	2	...	5	1	...	...	5	8	4	9	2	3	39
Group VIII.—S.-E. Rajputana, Central India, and Gujarat	72	41	41	76	67	72	81	84	148	151	119	88	1,060	2	1	5	1	...	...	5	8	6	9	2	3	42
A																										
Saugor	...	1	...	...	3	1	3	4	16	11	5	...	44	...	...	...	...	...	...	...	...	...	...	...	...	...
Jubbulpore	4	3	2	...	2	...	2	6	15	12	12	6	64	...	1	...	...	...	1	...	1	...	...	...	...	3
Kamptee	10	14	10	10	5	3	6	10	47	54	16	6	191	...	...	...	...	...	...	...	...	...	...	...	...	...
Sitabaldi	1	1	2	1	2	3	2	...	2	...	1	...	14	...	...	...	...	...	...	...	...	...	1	...	...	1
B																										
Secunderabad	21	10	22	7	17	9	7	7	6	8	11	2	130	...	...	...	...	...	...	...	...	...	...	...	...	...
Belgaum	8	5	3	4	8	6	12	11	8	10	16	7	93	...	...	...	...	...	...	...	...	...	...	...	...	...
Poona	30	22	17	11	13	41	83	43	37	24	38	22	381	...	...	...	...	...	...	...	...	...	...	...	...	...
Kirkee	1	...	...	1	3	...	8	6	1	2	5	6	33	...	...	...	...	...	...	1	...	...	...	...	...	1
Ahmednagar	11	19	15	17	26	8	12	17	11	7	6	12	161	...	...	...	...	...	...	...	...	...	...	...	...	...
Group IX.—Deccan	89	75	71	51	79	70	135	104	143	128	110	61	1,116	...	1	...	...	...	2	...	1	...	...	...	...	5
Colaba	21	15	2	12	13	20	14	42	37	23	27		233	...	...	...	...	2	1	1	2	...	...	...	...	6
Cannanore	...	1	...	1	...	1	...	...	...	...	...	...	4	...	...	...	...	...	...	...	...	...	...	...	...	...
Callicut	...	...	...	...	...	...	1	...	...	...	...	...	1	...	...	...	...	...	...	...	...	...	...	...	...	...
Mallapuram	...	...	...	...	4	...	...	...	...	--	...		4	...	...	...	...	...	...	...	...	...	...	...	...	...
Group X.—Western Coast	21	16	2	6	12	18	21	15	42	37	25	27	242	...	...	...	...	2	1	1	2	...	...	...	...	6
A																										
Bellary	11	4	2	3	9	5	9	10	10	8	11	13	97	...	...	...	...	...	...	...	...	...	...	...	...	...
Bangalore	4	12	17	22	34	35	20	15	21	14	11	7	207	...	...	...	...	...	...	...	...	...	...	...	...	...
B																										
St. Thomas' Mount	...	...	2	...	...	1	...	1	...	2	...	...	6	...	...	...	...	...	...	...	...	...	...	...	...	...
Madras	...	...	2	2	7	3	3	8	1	3	...	1	30	...	...	...	...	...	...	...	...	...	...	...	...	...
Group XI.—Southern India	15	16	18	29	50	44	37	34	32	27	22	21	340	...	...	...	...	...	...	...	...	...	...	...	...	...
Ranikhet	...	...	...	9	5	10	13	10	7	8	3	...	68	...	...	...	...	...	...	...	...	...	...	...	...	...
Chaubuttia	...	...	...	...	1	4	5	3	1	...	...	...	14	...	...	...	...	...	...	...	...	...	...	...	...	...
Chakrata	...	...	4	6	2	15	13	15	9	2	...	4	69	...	...	...	...	1	...	...	...	...	2	...	...	3
Lebong	3	4	5	4	...	1	3	2	...	...	37	15	80	...	...	...	...	...	...	...	...	...	...	...	...	...
Solon	...	...	...	...	...	1	3	2	1	...	...	...	7	...	...	...	...	...	...	...	...	...	...	...	...	...
Dagshai	...	...	...	3	1	2	2	2	1	...	...	...	11	...	...	...	...	...	...	...	...	...	...	...	...	...
Subathu	...	...	...	...	4	3	5	1	3	2	...	...	18	...	...	...	...	...	...	...	...	...	...	...	...	...
Jutogh	...	...	...	1	...	...	...	...	...	...	...	...	2	...	...	...	...	...	...	...	...	...	...	...	...	...
Khyragully	...	...	...	1	...	...	...	1	...	...	...	...	1	...	...	...	...	...	...	...	...	...	...	...	...	...
Kuldunnah	...	...	...	2	2	2	1	3	...	...	...	...	10	...	...	...	...	...	...	...	...	...	...	...	...	...
Kalabagh	...	...	...	...	...	...	...	1	...	...	...	...	1	...	...	...	...	...	...	...	...	...	...	...	...	...
Camp Gharial' Thobba	...	...	...	9	2	4	5	1	1	...	...	...	22	...	...	...	...	...	...	...	...	...	...	...	...	...
,, Upper Topa	...	...	...	13	6	8	5	6	1	...	...	...	39	...	...	...	...	...	...	...	...	...	...	...	...	...
,, Lower Topa	...	...	...	...	...	...	3	...	...	...	...	...	3	...	...	...	...	...	...	...	...	...	...	...	...	...
Khan Spur	...	...	...	c	3	...	1	...	...	...	...	...	3	...	...	...	...	...	...	...	...	...	...	...	...	...
Cherat	...	...	2	...	3	...	4	11	1	...	...	...	32	...	...	...	...	...	...	...	...	...	...	...	...	...
Quetta	...	5	6	19	15	14	17	15	10	23	15	18	162	...	...	...	...	...	...	...	...	...	...	...	...	...
Maymyo	8	2	7	4	1	6	22	12	4	8	6	1	81	...	...	...	...	...	...	...	...	...	...	...	...	...
Group XIIa.—Hill Stations	16	13	18	44	65	78	104	86	58	51	64	35	630	...	...	...	...	1	...	...	...	...	2	...	...	3

TABLE X—concluded.

INTERMITTENT FEVER by months, stations, groups, and commands.

TABLE XI—concluded.

REMITTENT FEVER by months, stations, groups, and commands.

STATIONS, GROUPS, AND COMMANDS.	ADMISSIONS FROM INTERMITTENT FEVER IN EACH MONTH.													ADMISSIONS FROM REMITTENT FEVER IN EACH MONTH.											
	January.	February.	March.	April.	May.	June.	July.	August.	September.	October.	November.	December.	TOTAL.	January.	February.	March.	April.	May.	June.	July.	August.	September.	October.	November.	December.	TOTAL.
Darjeeling	...	...	10	7	14	6	3	3	3	3	15	1	65	...	...	...	...	...	...	...	...	...	...	...	...	...
Naini Tal	...	...	...	...	1	2	4	3	2	2	...	...	14	...	...	...	...	...	...	...	...	...	...	...	...	...
Landour	...	...	...	6	3	2	8	4	9	1	...	...	33	...	...	...	...	...	...	...	...	...	...	...	...	...
Kasauli	...	...	...	1	11	3	2	2	...	...	...	...	19	...	...	...	...	...	...	...	...	...	...	...	...	...
Dalhousie	...	...	...	...	1	8	...	9	...	5	3	...	30	...	...	...	...	...	...	...	...	...	...	...	...	...
Murree	...	...	...	...	4	...	...	...	...	2	1	...	3	...	...	...	...	...	...	...	...	...	...	...	...	...
Mount Abu	...	...	...	1	1	...	8	3	...	1	1	5	22	...	...	...	...	...	...	...	...	...	...	...	...	...
Pachmarhi	...	...	...	1	3	6	3	5	6	...	...	...	24	...	...	...	...	1	...	...	1	...	...	...	...	2
Purandhur	11	2	4	3	1	3	3	3	2	...	2	3	37	...	...	...	...	...	...	...	...	...	...	...	...	...
Khandalla	1	2	1	1	...	...	...	...	...	1	6	5	17	...	...	...	1	...	...	...	...	...	...	...	...	...
Wellington	8	11	8	17	21	42	29	13	3	2	3	4	161	...	...	...	1	...	...	...	...	...	...	...	...	1
GROUP XIIb.— HILL CONVALESCENT DEPÔTS, AND SANITARIA	20	15	24	36	59	73	63	42	33	14	31	15	425	...	...	...	1	1	...	...	1	...	...	...	...	3
Troops marching, India	39	10	5	...	2	...	...	...	...	10	13	20	99	4	...	1	...	...	...	...	...	...	...	...	...	5
Aden Column Field Force (Dthalla)	12	9	5	32	17	21	17	8	23	11	10	4	159	...	...	...	...	...	...	...	...	...	...	...	...	...
Deolali Depôt	5	8	5	9	14	11	12	7	7	14	11	12	115	...	...	...	...	...	...	...	...	...	...	...	...	...
Poonamallee Depôt	...	...	...	...	...	...	2	1	3	1	2	3	12	...	...	...	...	...	...	...	...	...	...	...	...	...
EXTRA INDIA. Aden	18	16	28	66	25	39	33	13	45	11	3	4	271	...	...	...	...	...	...	...	...	...	...	...	...	...
INDIA	531	383	352	536	684	688	800	765	809	859	932	580	7,945	15	2	7	6	14	24	19	29	13	21	6	6	161
Northern Command	59	55	43	64	103	70	130	165	126	163	199	88	1,271	...	...	...	...	...	1	1	...	...	4	3	1	10
Western „	215	176	150	239	240	249	325	244	392	356	308	286	3,180	2	2	5	1	3	3	6	12	6	10	2	3	55
Eastern „	117	71	76	104	162	215	319	254	204	244	342	140	2,148	9	...	1	1	8	11	4	8	6	2	...	...	50
Secunderabad Division.	47	36	48	54	88	100	71	57	44	38	38	30	652	...	...	...	1	...	...	...	...	...	...	...	...	1
Burma Division	42	24	25	63	77	33	43	35	20	37	22	12	436	...	...	...	3	3	9	8	1	5	1	2		40

TABLE XII.

PNEUMONIA by months, stations, groups, and commands.

TABLE XIII.

DYSENTERY by months, stations, groups, and commands.

Stations and Groups.	ADMISSIONS FROM PNEUMONIA IN EACH MONTH.													ADMISSIONS FROM DYSENTERY IN EACH MONTH.												
	Jan.	Feb.	Mar.	Apr.	May	Jun.	Jul.	Aug.	Sep.	Oct.	Nov.	Dec.	Total	Jan.	Feb.	Mar.	Apr.	May	Jun.	Jul.	Aug.	Sep.	Oct.	Nov.	Dec.	Total
Rangoon	...	...	...	...	...	...	...	1	2	...	...	...	3	4	...	3	3	9	3	12	4	3	1	...	5	45
Group I.—Burma Coast and Bay Islands	...	...	...	...	...	...	...	1	2	...	...	...	3	4	...	3	3	9	3	12	4	3	1	...	3	45
Thayetmyo	...	...	1	2	...	...	...	...	2	...	...	5		...	...	...	1	2	...	...	...	1	...	...	4	
Meiktila	...	...	...	...	...	...	...	...	...	...	1		...	2	...	...	...	...	...	1	...	...	3		
Fort Dufferin	...	...	...	1	...	...	...	...	...	...	2		1	...	...	...	...	...	1	1	...	...	2		
Shwebo	...	...	...	...	...	...	...	...	...	...	3		...	...	...	...	...	...	...	...	1	...	3		
Group II.—Burma Inland	...	...	2	3	...	...	...	...	2	...	...	7	1	2	...	1	2	...	...	2	2	1	...	1	12	
Fort William	...	...	...	...	...	...	...	...	...	1	1	3	5	3	3	1	...	1	...	2	...	1	...	1	17	
„ Fulta	...	...	...	...	...	...	...	...	...	...	...		...	...	...	...	...	...	1	...	...	...	1		
Dum-Dum	...	...	...	...	1	2	2	...	...	...	5		...	...	...	...	...	1	...	1	...	1	...	2	
Barrackpore	...	...	...	...	...	...	...	...	...	...	...		...	1	1	...	...	1	4	5	...	1	1	2	16
Group IV.—Bengal and Orissa	1	...	...	...	1	2	2	...	...	1	1	8	5	4	4	1	...	2	4	9	...	3	1	3	36	
B																										
Dinapore	2	...	...	...	...	...	1	...	...	1	...	4	1	...	...	...	...	...	...	4	...	...	...	5	
Allahabad	...	1	...	...	...	...	1	...	...	1	...	3	2	1	...	...	...	2	...	...	...	...	3		
Fyzabad	1	...	...	...	...	...	...	...	...	1		1	1	...	...	3	...	...	...	1	2	1	10		
Sitapur	...	...	...	...	...	...	...	...	1		...	...	...	...	...	...	...	...	...	...	1			
Lucknow	1	...	...	...	2	...	...	...	2	5	4	1	3	5	5	6	5	9	10	7	13	2	70		
Cawnpore	...	...	...	...	...	...	...	...	...		...	...	...	1	...	2	...	...	1	1	...	5			
Fatehgarh	...	...	...	...	...	...	...	...	...		...	...	...	...	...	...	...	...	...	1		1			
Group V.—Gangetic Plain and Chutia Nagpur	4	1	...	...	...	3	1	...	...	2	2	13	7	2	5	5	9	8	11	14	8	15	5	94		
A																										
Shahjahanpur	...	...	...	...	...	...	...	...	...	...		...	...	1	...	...	...	...	...	...	1	3			
Bareilly	...	...	...	...	...	...	1	1	...	6	8		...	...	2	1	...	...	...	1	2	1	2	7		
Roorkee	...	...	...	...	...	...	...	...	...		...	1	...	1	...	2	...	...	...	2	3	2	11		
Meerut	...	2	1	...	...	2	...	...	...	4	9	2	...	1	1	11	2	2	3	3	2	6	4	37		
Delhi	...	...	...	...	...	...	...	...	...	...		...	...	...	...	1	...	...	1	...	...	2			
Ambala	2	4	...	1	...	...	...	...	...	1	19	27	2	1	...	...	...	...	1	...	...	1	...	6		
B																										
Jullundur	...	1	...	...	...	...	...	...	...	1	1	...	1	...	1	...	...	...	...	...	4				
Ferozepore	...	...	...	...	...	...	...	...	...	3	...	...	...	2	...	2	...	...	1	4	1	10			
Amritsar	...	...	...	...	...	...	...	...	...		1	...	...	...	...	...	...	1	...	...	2			
Mian Mir	2	...	...	1	...	...	...	1	...	4	...	3	...	2	1	...	...	2	1	3	2	6	20			
Fort Lahore	...	...	...	...	...	...	...	1	1	2	...	...	1	...	...	...	3	...	...	...	...	5			
Sialkot	...	...	...	...	...	...	...	...	...	1	1	1	...	...	...	...	...	...	...	1	...	3			
Rawalpindi	2	2	...	...	1	1	1	...	...	1	9	17	...	1	...	...	...	...	3	2	...	2	2	10		
Campbellpur	...	...	...	1	...	...	...	...	...	1		...	...	...	...	1	3	...	...	...	...	5			
Attock	...	...	...	...	...	...	...	1	...	1		...	...	...	...	...	...	...	...	...	...	...			
Group VI.—Upper Sub-Himalaya	6	9	2	3	1	3	1	1	1	3	3	41	74	6	7	5	10	17	7	9	9	6	13	18	17	124
A																										
Nowshera	2	1	1	1	...	...	...	...	1	2	9	2	...	...	...	...	...	...	...	...	...	2	4			
Peshawar	1	1	6	3	2	1	...	1	...	...	2	17	...	...	1	...	3	...	...	...	2	1	1	8		
Mooltan	...	2	1	...	...	...	...	...	...	1	4	3	...	1	4	1	2	4	...	2	1	1	19			
C																										
Hyderabad	1	...	...	1	...	...	...	1	...	4	...	...	...	...	...	1	3	1	...	...	6				
Kurrachee	1	...	...	...	...	...	...	...	...	2	...	...	2	...	...	...	...	...	...	...	2				
Group VII.—N.-W. Frontier, Indus Valley, and N.-W. Rajputana	5	4	8	5	2	1	...	1	1	1	7	36	2	3	3	2	7	1	2	6	3	3	3	4	39
A																										
Deesa	...	...	...	...	...	...	...	...	...	...		...	...	...	...	...	...	1	...	...	...	1			
Ahmedabad	...	...	...	...	...	...	...	...	...		...	...	...	...	...	...	...	1	...	...	1				
B																										
Neemuch	...	...	1	...	...	...	...	...	...	1	...	...	...	...	...	...	...	...	1	...	1	2			
Nasirabad	...	...	...	...	...	...	...	...	3	3	1	...	...	2	...	1	1	...	3	...	9				
Muttra	...	...	...	...	...	...	...	1	1	1	1	...	...	...	...	...	1	...	...	3				
Agra	...	1	...	1	2	2	1	...	1	2	12	...	1	...	...	...	...	...	...	...	...	1			
Jhansi	...	...	...	...	...	1	...	...	1	3	5	...	1	4	1	...	...	1	4	3	1	3	23			
Nowgong	...	...	...	...	...	...	...	...	...	2	2	3	...	...	...	...	...	...	...	...	8			
Indore	...	...	...	...	1	...	...	...	...		...	1	...	...	...	...	...	...	...	...	3			
Mhow	2	1	2	...	...	1	...	...	2	...	9	1	1	...	3	1	...	2	2	7	7	3	...	4	30	
Group VIII.—S.-E. Rajputana, Central India, and Gujarat	2	2	3	1	2	3	1	2	1	2	3	7	29	9	4	5	9	5	2	4	11	13	9	2	8	81

* Stations where neither Pneumonia nor Dysentery occurred are not shown in these tables. For the annual ratios, see Table III.

E

TABLE XII—*concluded.*

PNEUMONIA by months, stations, groups, and commands.

TABLE XIII—*concluded.*

DYSENTERY by months, stations, groups, and commands.

ADMISSIONS FROM PNEUMONIA IN EACH MONTH.

Stations, and Groups.	Jan.	Feb.	Mar.	Apr.	May.	June.	July.	Aug.	Sept.	Oct.	Nov.	Dec.	Total
A													
Saugor	...	...	...	...	...	...	...	...	...	...	...	...	...
Jubbulpore	1	...	...	...	1	...	...	...	...	...	...	...	2
Kamptee	...	...	...	...	...	...	...	...	...	...	...	...	...
Sitabaldi	...	...	...	...	...	...	...	...	...	...	...	...	...
B													
Secunderabad	...	...	2	...	...	3	...	...	2	1	...	...	8
Belgaum	...	1	...	...	...	...	...	...	...	1	...	...	2
Poona	1	2	...	2	1	...	...	1	...	...	...	...	3
Kirkee	...	1	1	...	...	...	1	...	...	...	...	4	7
Ahmednagar	...	...	...	4	...	...	...	...	...	...	4	8	
Group IX.—Deccan	2	4	3	6	2	3	2	...	3	1	...	11	37
Colaba	...	...	...	...	1	...	...	1	...	1	2	...	5
Cannanore	...	...	...	...	...	...	...	...	...	...	...	...	...
Calicut	...	...	...	...	...	...	...	...	...	...	...	...	...
Group X.—Western Coast	...	...	...	1	...	...	1	...	1	2	...		5
A													
Bangalore	...	...	...	1	...	...	...	...	...	...	...		1
B													
St. Thomas' Mount	...	...	...	...	...	...	...	...	...	...	...	...	...
Madras	...	...	...	...	...	...	...	1	...	...	...		
Group XI.—Southern India	...	...	...	1	...	...	...	1	...	...			2
Ranikhet	...	...	...	...	1	...	...	1	...	...	...	...	2
Chaubuttia	...	...	...	1	...	...	...	...	...	...	...	...	1
Chakrata	...	...	...	2	1	...	2	...	...	...	...	...	5
Lebong	...	...	...	...	...	...	2	...	...	...	...	...	2
Solon	...	...	1	1	...	1	...	1	...	...	...	...	6
Darghai	...	...	1	...	1	...	1	...	...	...	...	...	2
Subathu	...	...	...	1	...	1	...	...	...	...	...	...	2
Jutogh	...	...	...	1	...	...	1	...	...	...	...	...	2
Kuldunnah	...	...	...	2	...	...	...	...	...	...	...	...	2
Camp Thobba	...	...	...	1	...	...	...	...	...	...	...	...	1
Khan Spur	...	...	...	...	...	...	...	...	...	...	...	...	1
Cherat	...	...	1	...	...	...	...	...	...	...	...	...	1
Quetta	3	3	6	4	2	...	1	...	5	7	1	2	34
Ramandrug	...	...	...	1	...	...	...	...	...	...	...	...	1
Maymyo	...	...	...	1	...	...	...	...	...	...	...	...	1
Group XII a.—Hill Stations	3	3	6	9	9	5	4	7	9	1	2		61
Darjeeling	...	...	...	...	...	...	...	...	...	...	...	...	...
Naini Tal	...	...	...	...	...	...	...	...	...	...	...	...	...
Landour	...	...	...	...	...	...	...	...	...	...	...	...	...
Kasauli	...	...	...	...	...	...	...	...	...	...	...	...	...
Dalhousie	...	...	...	...	...	...	1	...	...	...	...	...	1
Taragarh	...	...	...	...	...	...	...	...	...	...	...	...	...
Mount Abu	...	...	...	...	...	...	...	...	...	...	...	...	...
Pachmarhi	...	...	...	...	...	...	...	...	...	...	...	...	...
Purandhur	...	...	...	...	...	...	...	...	...	...	...	...	...
Khandalla	...	...	...	...	...	...	...	...	...	...	1	...	1
Wellington	1	...	...	1	...	...	...	...	...	...	...	...	2
Group XII b.—Hill Convalescent Depôts, and Sanitaria	1	...	...	...	...	...	1	1	...	...	1		4
Troops, Marching, India; Aden Column Field Force	5	1	...	1	...	...	...	...	1	1	5		14
(Dchalia)	...	1	...	...	...	...	1	...	1	...	1		
Daulali Depôt	...	...	...	...	...	...	...	...	...	...	...	...	...
Poonamallee Depôt	...	...	...	...	...	...	...	...	...	...	...	...	...
Extra India.													
Aden	...	...	...	...	...	...	...	...	...	1	...	...	...
INDIA	21	25	21	21	18	13	10	12	16	24	13	17	206
Northern Command	9	11	9	11	7	5	3	3	...	5	4	30	104
Western Command	9	8	10	11	5	1	3	3	7	8	7	20	92
Eastern Command	5	4	1	1	5	10	4	5	4	5	1	16	61
Secunderabad Division	1	...	2	1	1	3	...	...	2	2	...		12
Burma Division	...	...	2	4	...	1	...	...	1	...	...		11

ADMISSIONS FROM DYSENTERY IN EACH MONTH.

Stations, and Groups.	Jan.	Feb.	Mar.	Apr.	May.	June.	July.	Aug.	Sept.	Oct.	Nov.	Dec.	Total
A													
Saugor	...	...	...	1	...	...	...	1	...	...	...	...	3
Jubbulpore	...	...	...	...	...	...	2	2	6	6	6	3	25
Kamptee	2	...	...	1	...	...	...	2	3	4	...	...	12
Sitabaldi	...	...	1	...	...	...	...	...	...	...	...	...	1
B													
Secunderabad	6	6	1	5	5	4	6	12	13	6	5	5	74
Belgaum	...	...	1	4	1	1	4	3	1	3	3		22
Poona	1	1	...	...	1	3	7	14	10	4	3	2	46
Kirkee	...	...	...	1	...	...	4	3	...	1	1		10
Ahmednagar	...	1	...	2	...	...	...	...	1	5	1		10
Group IX.—Deccan	9	7	5	7	13	8	16	39	39	26	20	14	203
Colaba	...	...	...	...	...	1	1	...	...	1	3	1	7
Cannanore	...	...	...	...	...	...	...	...	...	1	...	...	1
Calicut	...	...	1	...	...	...	...	...	...	...	...	...	2
Group X.—Western Coast	...	...	1	1	...	...	3	...	3	1			10
A													
Bangalore	2	5	4	6	4	2	3	5	6	1	2	3	43
B													
St. Thomas' Mount	...	...	...	1	1	1	2	...	2	2	1	...	10
Madras	...	1	...	...	1	...	1	3	...	2	1		9
Group XI.—Southern India	2	6	4	7	5	4	5	6	11	3	5	4	62
Ranikhet	1	...	...	1	2	6	3	...	1	...	...	...	15
Chaubuttia	...	...	...	...	2	1	1	...	...	...	...	...	4
Chakrata	...	...	4	...	1	11	...	1	...	...	...	...	17
Lebong	...	...	1	...	...	...	...	1	...	...	...	...	2
Solon	...	...	...	...	...	...	...	...	...	...	...	...	...
Darghai	...	...	...	...	...	1	...	...	...	...	...	...	1
Subathu	...	...	...	...	...	...	...	...	...	...	...	...	...
Jutogh	...	...	...	...	...	...	...	...	...	...	...	...	...
Kuldunnah	...	1	...	...	...	...	...	...	...	...	...	...	...
Camp Thobba	...	...	...	1	...	...	...	...	...	...	...	...	...
Khan Spur	...	...	...	1	...	...	...	...	...	...	...	...	1
Cherat	...	...	...	...	...	...	...	...	...	...	...	...	...
Quetta	...	...	2	4	14	11	12	11	10	2	1		67
Ramandrug	...	...	...	...	...	...	...	...	...	...	...	...	1
Maymyo	...	1	...	...	...	...	...	...	...	...	...	...	...
Group XII a.—Hill Stations	1	...	9	9	23	26	13	14	10	3	2		111
Darjeeling	1	...	...	1	...	...	...	1	1	1	...		5
Naini Tal	...	...	...	...	...	1	...	...	...	...	...		1
Landour	...	...	...	...	...	1	1	...	...	...	...		2
Kasauli	...	...	2	1	1	...	...	...	...	...	...		4
Dalhousie	...	1	2	5	7	...	1	...	...	...	...		16
Taragarh	...	...	1	...	1	...	1	...	...	...	...		3
Mount Abu	...	1	...	...	1	...	...	...	...	...	...		2
Pachmarhi	...	...	...	1	...	2	...	...	...	...	...		3
Purandhur	1	...	1	...	...	1	...	...	...	...	...		5
Khandalla	...	...	...	...	...	...	...	...	...	1	1		...
Wellington	...	1	2	...	1	...	...	1	1	...	...		7
Group XII b.—Hill Convalescent Depôts, and Sanitaria	2	1	3	9	6	9	8	4	1	3	1		47
Troops, Marching, India; Aden Column Field Force	5	2	...	...	...	...	...	...	1	8	10		26
(Dchalia)	1	2	...	...	...	2	6	8	2	...	1		23
Daulali Depôt	...	...	...	...	...	1	2	...	1	...	...		4
Poonamallee Depôt	...	...	...	...	2	1	1	2	3	...	...		10
Extra India.													
Aden	3	1	...	4	1	5	7	2	1	5	1		32
INDIA	57	41	40	68	79	72	102	124	119	94	89	74	959
Northern Command	0	9	2	10	14	12	10	12	...	11	11	12	118
Western Command	15	6	13	19	20	23	31	61	57	43	27	21	336
Eastern Command	17	7	14	20	23	30	24	37	17	27	15		240
Secunderabad Division	8	13	7	14	11	10	13	19	27	12	14	9	137
Burma Division	5	2	4	4	12	3	13	6	5	7	...	4	59

TABLE XV.

A.—STRENGTH, ADMISSIONS from ALL CAUSES, ADMISSIONS from ENTERIC FEVER, of the Army of India 1905, in relation to AGE and LENGTH of RESIDENCE in INDIA.

	BY AGE.						BY LENGTH OF RESIDENCE.						
	Under 20.	20 and less than 25.	25 and less than 30.	30 and less than 35.	35 and less than 40.	40 and upwards.	Under 1 year.	1 and less than 2.	2 and less than 3.	3 and less than 4.	4 and less than 5.	5 and less than 10.	10 years and upwards.
Strength	2,446	34,243	23,079	7,876	1,967	397	17,287	12,887	15,452	7,614	5,570	8,284	2,914
Per cent. of total . . .	3	49	33	11	3	1	25	18	22	11	8	12	4
1900-1904	3	44	32	11	3	1	15	15	15	14	13	23	5
Admissions from all causes .	1,383	33,449	18,322	4,829	959	210	12,207	10,545	12,225	8,482	4,353	8,779	2,561
Admissions from Enteric Fever .	33	800	255	48	7	3	470	267	205	101	26	65	12
All causes per 1,000 . . .	565·4	976·8	793·9	613·1	487·5	529·0	706·1	818·3	791·2	1,114·0	781·5	1,059·8	878·9
Enteric Fever per 1,000 . .	13·5	23·4	11·0	6·1	3·6	7·6	27·2	20·7	13·3	13·3	4·7	7·8	4·1
Liability to Enteric Fever (The above column expressed in percentages) .	20·71	35·89	16·87	9·36	5·52	11·66	29·86	22·72	14·60	14·60	5·16	8·56	4·50
Enteric Fever per cent. of all causes .	2·39	2·39	1·39	·99	·73	1·43	3·85	2·53	1·68	1·19	·60	·74	·47

B.—CHANGE of PERSONNEL, YOUTHFULNESS, RECENT ARRIVAL, and MARRIAGE, in relation to VENEREAL DISEASE and ENTERIC FEVER.

YEAR.	ARRIVED IN INDIA.*		YEAR.	MEN.									
				PER CENT. OF STRENGTH.					RATIO PER 1,000.			RATIO PER CENT. OF TOTAL ADMISSION.	
	Men.	Women.		Age. Under 25 years.	Length of residence. Under 5 years.	Married. ‡	Strength.	Admissions.				Venereal Diseases.	Enteric Fever.
								All causes.	Venereal Diseases.	Enteric Fever.		
1875-76 . . .	7,368	752	1875	36	...	10·80	58,400	1,337·8	205·1	2·8	15·33	·21
1876-77 . . .	8,170	591	1876	33	...	10·37	57,858	1,361·5	189·9	4·6	13·95	·34
1877-78 . . .	9,113	482	1877	33	56	9·70	57,260	1,257·3	208·5	4·1	16·59	·32
1878-79 . . .	13,113	575	1878	35	60	7·39	56,475	1,651·3	271·3	8·5	16·43	·51
1879-80 . . .	13,342	612	1879	39	61	6·63	59,082	1,871·2	234·8	8·0	12·55	·43
1880-81 . . .	13,165	664	1880	41	65	6·36	59,717	1,754·2	240·7	7·9	14·23	·45
1881-82 . . .	9,895	349	1881	43	70	5·94	58,728	1,604·6	260·5	5·6	16·23	·35
1882-83 . . .	9,748	325	1882	41	72	5·43	57,269	1,444·9	265·2	6·2	18·35	·43
1883-84 . . .	12,525	433	1883	41	75	5·20	55,525	1,335·7	270·3	7·7	20·23	·58
1884-85 . . .	11,822	393	1884	45	75	5·05	54,996	1,513·4	293·9	11·7	19·42	·77
1885-86 . . .	17,706	508	1885	48	73	4·23	56,967	1,532·7	347·7	11·2	22·36	·73
1886-87 . . .	11,645	372	1886	52	75	3·90	61,015	1,513·9	389·5	18·1	25·73	1·20
1887-88 . . .	11,729	459	1887	52	73	3·84	63,515	1,369·7	361·2	12·7	26·37	·93
1888-89 . . .	12,407	506	1888	50	76	3·65	68,887	1,381·7	370·6	13·6	26·82	·99
1889-90 . . .	12,270	532	1889	49	78	3·60	69,266	1,498·0	481·5	22·9	32·14	1·53
1890-91 . . .	14,046	542	1890	50	80	3·70	67,823	1,520·2	503·5	18·5	33·12	1·22
1891-92 . . .	15,456	529	1891	51	79	3·36	67,030	1,379·1	400·7	20·4	29·06	1·48
1892-93 . . .	15,894	540	1892	51	80	3·29	68,137	1,517·3	409·9	22·1	27·01	1·46
1893-94 . . .	15,090	482	1893	53	79	3·20	70,091	1,414·9	466·0	20·0	32·94	1·41
1894-95 . . .	15,937	517	1894	54	81	...†	71,082	1,508·0	511·4	20·9	33·91	1·38
1895-96 . . .	14,346	654	1895	55	83	...	71,031	1,461·8	522·3	26·3	35·73	1·80
1896-97 . . .	14,805	545	1896	56	82	...	70,484	1,386·7	511·6	25·5	36·89	1·84
1897-98 . . .	16,227	543	1897	55	84	...	68,395	1,555·9	485·7	32·4	31·20	2·08
1898-99 . . .	16,911	648	1898	54	81	...	67,741	1,435·9	362·9	36·9	25·26	2·57
1899-1900 . . .	3,369	168	1899	53	78	...	67,697	1,148·7	313·4	20·6	27·28	1·79
1900-01 . . .	5,058	185	1900	45	69	...	60,553	1,143·2	298·1	16·0	26·07	1·40
1901-02 . . .	18,594	438	1901	42	63	...	60,838	1,104·3	276·0	12·8	24·99	1·16
1902-03 . . .	24,840	961	1902	43	68	...	60,540	1,078·4	281·4	16·7	26·09	1·55
1903-04 . . .	15,126	758	1903	51	76	...	70,445	1,023·4	247·0	19·6	23·90	1·90
1904-05 . . .	16,366	820	1904	52	80	...	71,083	900·4	198·5	19·6	22·05	2·18
1905-06 . . .	15,178	804	1905	52	84	...	71,343	834·3	153·7	16·1	18·42	1·93

*In ordinary years the departures plus the deaths nearly balance the arrivals.   † Return abolished.   ‡ On the 1st May of each year.

E

TABLE XVI.

RELATION of MORTALITY to AGE and LENGTH of RESIDENCE in

	A.—AGE.										B.—LENGTH OF R
	(a) DIED PER 1,000.					(b) LIABILITY IN PERCENTAGES.					(c) DIED PER 1,000.
CAUSES OF DEATH.	Under 20.	20 and less than 25.	25 and less than 30.	30 and less than 35.	35 and less than 40.	40 and upwards.	Under 20.	20 and less than 25.	25 and less than 30.	30 and less than 35.	35 and less than 40.	40 and upwards.	Under 1 year.	1 and less than 2.	2 and less than 3.	3 and less than 4.	4 and less than 5.	5 and less than 10.
---	---	---	---	---	---	---	---	---	---	---	---	---	---	---	---	---	---	---
Enteric Fever	·82	4·32	2·34	1·14	...	...	10	50	27	13	...	...	4·34	3·34	3·24	3·02	1·44	1·57
Cholera	...	·15	·04	·13	·51	...	...	17	10	15	58	...	·12	·10	...	·26	...	·36
Dysentery	...	·50	·33	·71	1·02	...	...	19	13	29	39	...	·58	·47	·20	·39	·18	·72
Intermittent and Remittent Fevers	...	·16	·13	...	...	...	...	58	42	...	...	...	...	·23	·06	·34	...	·24
Alcoholism	...	...	·13	·51	1·02	5·04	...	...	2	8	15	75	...	·08	·26	·26	...	·24
Tubercle of the lungs	...	·20	·22	·76	1·02	...	...	9	10	35	46	...	·17	·31	·06	·26	...	·85
Nervous Diseases	·41	·32	·30	·38	·51	...	21	17	16	20	27	...	·23	·23	·39	·66	·18	·24
Circulatory Diseases	...	·15	·65	1·78	2·03	5·04	...	2	7	18	21	52	·12	·31	·25	·53	·36	1·81
Pneumonia	·41	·47	·61	1·14	2·03	...	9	10	13	24	44	...	·35	·47	·45	·66	·54	1·45
Other Respiratory Diseases	...	·03	...	·25	·51	...	...	4	...	32	65	...	·12	...	...	...	...	·24
Abscess of the liver	...	1·05	1·30	1·52	2·03	2·52	...	12	15	20	24	30	·50	·70	1·29	1·97	1·08	1·81
Urinary Diseases	...	·06	·09	·25	...	2·52	...	2	3	9	...	56	...	...	·13	·30	·18	·12
TOTAL	1·64	7·42	6·20	8·51	10·65	15·11	3	15	13	17	22	30	6·56	6·29	6·34	8·67	3·95	9·66
Heat-stroke	·41	·50	·69	1·65	2·54	7·56	3	4	5	12	19	57	·87	·78	·58	·39	·36	1·57
Suicide	...	·15	·52	·51	1·02	2·52	...	3	11	11	22	53	·12	·16	·13	·74	·18	·85
Other injuries	...	·33	·82	·89	·51	2·52	...	7	16	17	10	49	·35	·62	·45	·39	...	1·33
All Causes	3·68	9·72	9·45	13·59	16·75	15·20	4	11	11	15	19	40	8·56	9·08	8·41	11·95	6·64	16·18

	(c) NUMBER OF DEATHS.						(d) COMPOSITION OF 100 DEATHS AT EACH AGE.						(i) NUMBER OF DEATHS.
Enteric Fever	2	148	54	9	...	...	22	44	25	8	...	...	75	43	50	23	8	13
Cholera	...	5	2	1	1	...	...	2	1	1	3	...	2	2	...	2	...	3
Dysentery	...	17	8	6	2	...	...	5	4	6	6	...	10	6	4	3	1	6
Intermittent and Remittent Fevers	...	6	3	...	...	...	...	2	1	...	...	...	...	3	1	3	...	2
Alcoholism	...	...	3	4	2	2	...	...	1	4	6	14	...	1	4	2	...	2
Tubercle of the lungs	...	7	5	6	2	...	...	2	2	6	6	—	3	4	1	2	...	7
Nervous Diseases	1	11	7	3	1	...	11	3	3	3	3	1	4	3	6	5	1	2
Circulatory Diseases	...	5	15	14	4	2	...	2	7	13	12	14	2	4	4	4	2	15
Pneumonia	...	16	14	9	4	...	11	5	6	8	12	...	6	6	7	5	3	12
Other Respiratory Diseases	...	1	...	2	1	...	...	0	...	2	3	...	2	...	...	...	...	2
Abscess of the liver	...	36	30	12	4	1	...	11	14	11	12	7	10	9	20	15	6	15
Urinary Diseases	...	2	2	2	...	1	...	1	1	2	...	7	...	...	2	2	1	1
TOTAL	4	254	143	68	21	6	44	76	66	64	63	43	114	81	99	66	22	80
Heat-stroke	1	17	16	13	5	3	11	5	7	12	15	21	15	10	9	3	2	13
Suicide	...	5	12	4	2	1	...	2	6	4	6	7	2	2	2	6	1	7
Other injuries	...	13	19	7	1	1	...	4	9	7	3	7	6	8	7	3	...	11
All Causes	9	333	218	107	33	14	100	100	100	100	100	100	148	117	130	91	37	134

	(e) NUMBER OF DEATHS.						(f) PERCENTAGE AT EACH AGE TO TOTAL NUMBER.						(k) NUMBER OF DEATHS.
Enteric Fever	2	148	54	9	...	...	1	69	25	4	...	...	75	43	50	23	8	13
Cholera	...	5	2	1	1	...	...	55	22	11	11	...	2	2	...	2	...	3
Abscess of the liver	...	36	30	12	4	1	...	43	36	14	5	1	10	9	20	15	6	15
Suicide	...	5	12	4	2	1	...	21	50	17	8	4	2	2	2	6	1	7
All Causes	9	333	218	107	33	14	1	47	31	15	5	2	148	117	130	91	37	134

TABLE XVII.

RELATION of INVALIDING to AGE and LENGTH of RESIDENCE in INDIA.

	A.—AGE.												B.—LENGTH OF RESIDENCE IN INDIA.											

(a) INVALIDED PER 1,000. / (b) LIABILITY IN PERCENTAGES. / (g) INVALIDED PER 1,000. / (h) LIABILITY IN PERCENTAGES.

CAUSES OF INVALIDING.	Under 20.	20 and less than 25.	25 and less than 30.	30 and less than 35.	35 and less than 40.	40 and upwards.	Under 20.	20 and less than 25.	25 and less than 30.	30 and less than 35.	35 and less than 40.	40 and upwards.	Under 1.	1 and less than 2.	2 and less than 3.	3 and less than 4.	4 and less than 5.	5 and less than 10.	10 and upwards.	Under 1.	1 and less than 2.	2 and less than 3.	3 and less than 4.	4 and less than 5.	5 and less than 10.	10 and upwards.
Dysentery	·41	·70	·52	...	...	...	25	43	32	...	...	...	·06	·62	·65	·79	·90	·83	...	2	16	17	20	23	22	...
Intermittent and Remittent Fevers	1·23	1·49	2·25	1·90	2·54	...	13	16	24	20	27	...	·52	·93	·97	1·97	2·69	6·04	3·43	3	6	6	12	16	36	21
Venereal Diseases	1·31	1·60	·89	·51	...	...	27	39	22	12	...	...	·29	·70	1·04	1·31	1·62	3·84	·69	3	7	11	14	17	41	7
Debility	1·64	1·69	1·91	3·53	5·08	20·15	5	5	6	10	15	59	·69	1·40	1·62	3·28	1·26	3·86	11·32	3	6	7	14	5	16	48
Rheumatism	·47	·78	1·27	·51	...	...	15	26	42	17	...	...	·33	·16	·52	·39	·36	2·05	2·40	6	3	8	6	6	33	30
Tubercle of the lungs	1·23	1·72	1·56	1·65	2·54	...	14	20	18	19	28	...	·93	1·63	1·75	1·84	1·44	2·41	3·43	7	12	13	14	11	18	26
Mental Diseases	·91	1·13	·76	1·02	...	...	24	30	20	27	...	...	·56	·93	·45	1·18	1·62	1·51	1·03	8	12	6	15	21	24	14
Epilepsy	·41	·50	·43	·76	1·53	2·52	7	8	7	12	25	41	·23	·31	·53	·39	·36	·72	3·43	4	5	10	6	6	12	57
Other Nervous Diseases	...	·35	·56	1·02	4·58	2·52	...	4	6	11	51	28	·35	·39	·52	·13	·36	1·09	4·12	5	6	7	2	5	10	50
Eye, ear, and nose Diseases	2·04	2·02	1·39	1·02	...	...	31	31	21	16	...	...	1·16	1·86	1·36	2·36	1·26	2·66	·69	10	16	12	21	11	23	6
Palpitation	·83	1·28	1·08	·25	1·02	2·52	12	18	16	4	15	36	·40	1·40	·97	1·53	1·44	1·45	1·37	5	16	11	18	17	17	16
Valvular disease of the heart	...	1·17	1·34	2·29	3·05	...	...	15	17	29	39	...	·65	1·32	1·10	·92	1·08	2·78	4·46	6	11	9	7	9	23	36
Other Circulatory Diseases	...	·47	·30	·51	·51	...	...	26	17	23	28	...	·06	·47	·39	·53	·36	·72	1·03	2	13	11	15	10	20	29
Respiratory Diseases	...	·23	·52	·51	...	...	...	18	41	40	...	...	·17	·31	·19	·39	·36	·97	·34	6	11	7	14	13	36	12
Congestion, inflammation and Abscess of the liver	...	·81	1·21	1·52	1·53	2·52	...	11	16	20	20	33	·12	·47	1·29	1·18	1·44	2·17	3·09	1	5	13	12	12	22	32
Locomotive Diseases	...	·85	·91	·51	...	2·52	...	18	19	11	...	53	·12	·85	·58	1·13	1·26	1·81	·69	2	13	9	18	19	23	11
Injuries	·41	·99	·74	1·65	2·03	...	7	17	13	28	35	...	·17	·70	·78	1·31	1·26	2·17	3·43	2	9	8	13	13	22	35
All Causes	8·59	20·44	22·05	24·89	33·05	42·82	5	13	15	16	22	28	3·21	18·31	18·06	24·03	24·60	43·94	57·31	4	9	9	12	13	23	29

(c) NUMBER INVALIDED. / (d) COMPOSITION OF 100 INVALIDINGS AT EACH AGE. / (i) NUMBER INVALIDED. / (j) COMPOSITION OF 100 INVALIDINGS IN EACH PERIOD OF RESIDENCE.

Dysentery	1	24	12	—	...	...	5	3	2	...	...	...	1	8	10	6	5	7	...	1	3	4	3	4	2	...
Intermittent and Remittent Fevers	3	51	52	15	5	...	14	7	10	8	8	...	9	12	15	15	15	50	10	6	5	5	8	11	14	6
Venereal Diseases	...	38	37	7	1	...	...	5	7	4	2	...	5	9	11	10	9	32	2	3	4	5	5	7	9	1
Debility	4	58	44	28	10	8	19	8	9	14	15	47	12	18	25	25	7	32	33	8	8	9	14	5	9	20
Rheumatism	...	16	18	10	1	...	...	2	4	5	2	...	6	2	8	3	2	17	7	4	1	3	2	1	5	4
Tubercle of the lungs	3	59	36	13	5	...	14	8	7	7	8	...	16	21	27	14	8	20	10	11	9	10	8	6	5	6
Mental Diseases	...	31	26	6	2	...	...	4	5	3	3	...	10	12	7	9	9	15	3	7	5	3	5	7	4	2
Epilepsy	1	17	10	6	3	1	5	2	2	3	5	6	4	4	9	3	2	6	9	4	2	3	2	1	2	6
Other Nervous Diseases	...	12	13	8	9	1	...	2	3	4	14	6	6	5	8	1	2	9	12	4	2	3	1	1	2	7
Eye, ear, and nose Diseases	5	69	32	8	...	...	24	10	6	4	...	...	20	34	21	18	7	22	2	14	10	8	10	5	6	1
Palpitation	2	44	25	2	1	1	10	6	5	1	3	6	7	18	15	12	8	12	4	5	8	5	7	6	3	2
Valvular disease of the heart	...	40	31	18	6	...	...	6	6	9	9	...	12	17	17	7	6	23	13	8	7	6	4	4	6	8
Other Circulatory Diseases	...	16	7	4	1	...	...	2	1	2	3	...	1	6	6	4	2	6	3	1	3	2	2	1	2	2
Respiratory Diseases	...	8	12	4	...	...	...	1	2	2	...	...	3	4	3	3	2	8	1	2	2	1	2	1	2	1
Congestion, inflammation and Abscess of the liver	...	28	28	12	3	1	...	4	6	6	5	6	2	6	20	9	8	18	9	1	3	7	5	6	5	5
Locomotive Diseases	...	29	21	4	...	1	...	4	4	2	...	6	2	11	9	7	7	15	2	1	5	3	5	5	4	1
Injuries	1	34	17	13	4	...	5	5	3	7	6	...	3	9	12	10	7	18	10	2	4	5	5	5	5	6
All Causes	21	700	509	196	65	17	100	100	100	100	100	100	142	236	279	183	137	364	167	100	100	100	100	100	100	100

(e) NUMBER INVALIDED. / (f) PERCENTAGE AT EACH AGE TO TOTAL NUMBER. / (k) NUMBER INVALIDED. / (l) PERCENTAGE IN EACH PERIOD OF RESIDENCE TO TOTAL NUMBER.

Intermittent and Remittent Fevers	3	51	52	15	5	...	2	40	41	12	4	...	9	12	15	15	15	50	10	7	10	12	12	12	40	8
Venereal Diseases	...	38	37	7	1	...	...	46	45	8	1	...	5	9	16	10	9	32	2	6	11	19	12	11	39	2
Debility	4	58	44	28	10	8	3	33	29	18	7	5	12	18	25	25	7	32	33	8	12	16	16	5	21	22
All Causes	21	700	509	196	65	17	1	46	34	13	4	1	142	236	279	183	137	364	167	9	16	19	12	9	24	11

TABLE XVIII.

STATISTICS OF OFFICERS.

A.—SICKNESS and MORTALITY among OFFICERS of the BRITISH ARMY in 1905. (From the Medical Returns of the Army.)

	Northern Command.		Western Command.		Eastern Command.		Secunderabad Division.		Burma Division.		India.*	
STRENGTH	574		687		581		301		101		2,323	
CASES REMAINING FROM 1904	24		16		10		10		1		61	
	Ratios.	Actuals.	Ratios.	Actuals.	Ratios.	Actuals.	Ratios.	Actuals.	Ratios.	Actuals.	Ratios.	Actuals.
CONSTANTLY SICK	29·1	16·72	26·9	18·49	35·1	20·38	25·8	7·77	28·4	2·83	28·8	66·81
INVALIDS	40·07	23	58·22	40	67·13	39	49·83	15	59·41	6	52·95	123
ADMISSIONS.												
Influenza	52·3	30	14·6	10	46·5	27	10·0	3	...	...	30·1	70
Cholera	...	...	...	...	...	...	...	...	...	...	...	...
Small-pox	1·7	1	1·5	1	...	...	...	...	...	...	·9	2
Enteric Fever	22·6	13	24·7	17	17·2	10	23·3	7	9·9	1	21·5	50
Intermittent Fever	83·6	48	90·2	62	82·6	48	56·5	17	138·6	14	87·0	202
Remittent Fever	3·5	2	5·8	4	...	...	3·3	1	...	...	3·0	7
Simple Continued Fever	61·0	35	40·8	28	51·6	30	29·9	9	99·0	10	48·2	112
Tubercle of the lungs	1·7	1	...	...	3·4	2	...	...	...	...	1·3	3
Pneumonia	...	...	7·3	5	3·4	2	...	...	—	...	3·0	7
Other Respiratory Diseases	13·9	8	18·9	13	22·4	13	16·6	5	49·5	5	19·4	45
Dysentery	1·7	1	45·6	32	20·7	12	33·2	10	...	...	23·7	55
Diarrhœa	25·1	15	21·8	15	60·2	35	13·3	4	49·5	5	31·9	74
Hepatic Abscess	...	...	1·5	1	...	...	...	...	...	...	·4	1
„ Congestion and Inflammation	22·6	13	14·6	10	43·0	25	6·6	2	9·9	1	22·0	51
Venereal Diseases	3·5	2	1·5	1	8·6	5	3·3	1	...	...	3·9	9
ALL CAUSES	695·1	399	691·4	475	764·2	444	568·1	171	851·5	86	692·2	1608
DEATHS												
Cholera	...	...	—	...	—	...	...	...	...	...	—	...
Small-pox	...	...	...	...	—	...	...	...	...	...	—	...
Enteric Fever	3·48	2	1·46	1	...	...	6·64	2	9·90	1	2·58	6
Intermittent Fever	...	...	...	...	...	...	...	...	...	...	...	...
Remittent Fever	...	...	...	...	...	...	...	...	...	...	...	...
Simple Continued Fever	...	...	...	...	...	...	...	...	...	...	...	...
eat-stroke	...	...	...	...	...	...	...	...	...	...	...	...
Circulatory Diseases	...	...	...	...	...	...	...	...	...	...	...	...
Tubercle of the lungs	1·74	1	...	—	...	...	...	...	...	...	·43	1
Pneumonia	...	...	...	...	...	...	...	...	...	...	...	...
Other Respiratory Diseases	...	...	...	...	...	...	...	...	...	...	...	...
Dysentery	...	...	...	...	...	...	...	...	...	...	...	...
Diarrhœa	...	...	...	—	...	...	...	...	...	...	...	...
Hepatic Abscess	...	...	...	...	...	...	...	...	...	...	...	...
All causes	10·45	6	7·28	5	8·61	5	9·97	3	9·90	1	9·04	21
DIED OUT OF HOSPITAL	...	...	4·37	3	5·16	3	...	...	9·90	1	4·4	11

* Including officers on the line of march and with the Field Force. (*Vide* Table E.—Detail of Diseases.)

B.—CAUSES of DEATH among OFFICERS of the BRITISH and INDIAN ARMIES in 1905. (From non-medical sources.)

ARMIES.	Strength in India, whether on leave or not, on the 1st of July 1905	Strength in Europe or beyond sea on 1st July 1905, whether on furlough or sick leave.	IN INDIA.																
			Cholera.	Small-pox.	Enteric Fever.	Intermittent Fever.	Remittent Fever.	Simple Continued Fever.	Heat-stroke.	Circulatory Diseases.	Tubercle of the lungs.	Pneumonia.	Other Respiratory Diseases.	Dysentery.	Diarrhœa.	Hepatic Abscess.	TOTAL.	Deaths in England and other countries.	Deaths at sea.	GRAND TOTAL.	Ratio per 1,000.
BRITISH	3,043	570	...	...	6	...	...	...	...	—	1	...	...	...	...	...	21	...	1	22	6·07
INDIAN	3,259	595	...	...	9	...	...	1	...	1	...	...	1	1	1	...	28	10	...	38	9·86

STATIONS* AND GROUPS.	Average annual strength.	January.	February.	March.	April.	May.	June.	July.	August.	September.	October.	November.
		NUMBER OF ADMISSIONS FROM ENTERIC FEVER IN EACH MONTH.										
Meiktila	8	...	...	...	...	...	...	...	...	1	...	...
GROUP II.—BURMA INLAND	49	...	...	...	...	...	...	...	...	1	...	...
Barrackpore	9	...	...	...	...	...	1	...	...	...	...	...
GROUP IV.—BENGAL AND ORISSA	61	...	...	...	...	...	1	...	...	...	...	...
B												
Lucknow	73	1	...	...	...	...	...	...	...	...	1	...
Cawnpore	25	...	...	1	...	...	...	...	...	...	...	...
Fatehgarh	10	...	...	...	...	...	...	...	...	...	...	1
GROUP V.—GANGETIC PLAIN AND CHUTIA NAGPUR	214	1	...	1	...	...	...	...	...	...	1	1
A												
Delhi	8	...	...	...	...	...	...	1	...	...	...	...
B												
Ferozepore	28	...	1	...	...	1	...	...	...	...	1	...
Mian Mir	30	1	...	...	...	...	...	...	...	...	...	...
Rawalpindi	95	...	...	...	...	...	...	...	...	...	...	1
GROUP VI.—UPPER SUB-HIMALAYA	441	1	1	...	...	1	...	1	...	...	1	1
A												
Nowshera	21	1	...	...	...	...	...	...	...	...	...	...
Peshawar	43	...	...	...	...	...	...	...	...	...	2	...
C												
Kurrachee	43	...	...	...	...	...	...	...	...	...	...	...
GROUP VII.—NORTH-WEST FRONTIER, INDUS VALLEY, AND NORTH-WESTERN RAJPUTANA	147	1	...	...	...	...	...	...	...	...	2	...
B												
Mhow	57	...	...	2	...	...	...	...	...	...	...	...
GROUP VIII.—SOUTH-EASTERN RAJPUTANA, CENTRAL INDIA, AND GUJARAT	196	...	...	2	...	...	...	...	...	...	...	...
A												
Kamptee	30	...	...	...	...	...	...	...	...	...	1	...
B												
Secunderabad	104	1	...	...	...	1	...	...	...	2	...	...
Belgaum	31	...	...	...	...	...	...	...	...	...	...	1
Poona	98	...	1	2	...	...	...	...	...	2	1	...
Kirkee	39	...	...	1	...	...	...	...	...	...	1	...
Ahmednagar	29	...	...	...	...	...	...	...	...	1	...	...
GROUP IX.—DECCAN	370	1	1	3	...	1	...	...	...	5	4	1
A												
Bellary	20	...	...	...	...	...	...	...	...	...	...	...
GROUP XI.—SOUTHERN INDIA	144	...	...	...	...	...	...	...	...	...	...	...
Ranikhet	30	...	...	...	...	...	1	...	...	...	...	...
Dagshai	19	...	...	...	...	...	...	...	...	...	1	...
Khan Spur	8	...	...	...	...	...	1	...	...	...	...	...
Quetta	92	1	...	...	...	...	...	1	...	...	...	...
GROUP XIIa.—HILL STATIONS	303	1	...	...	...	1	1	1	...	...	1	...
Landour	5	...	...	...	...	...	...	1	...	...	...	...
Kasauli	15	...	...	...	...	...	...	1	...	...	...	...
Murree	20	...	...	...	...	1	...	...	...	...	...	...
GROUP XIIb.—HILL CONVALESCENT DEPÔTS AND SANITARIA	168	...	...	...	...	1	...	2	...	...	1	...
Troops, marching, India	70	1	...	...	...	1	...	...	...	...	...	...
INDIA	2,314	6	2	6	...	5	2	4	...		10	3
NORTHERN COMMAND	574	2	1	...	...	3	...	1	...	...	4	1
WESTERN „	687	1	1	5	...	...	...	1	...	3	2	1
EASTERN „	551	...	...	1	...	...	2	2	...	...	2	...
SECUNDERABAD DIVISION	301	1	...	...	...	1	...	...	...	2	2	...
BURMA DIVISION	101	...	...	...	...	...	...	...	...	1	...	...

* Stations where Enteric Fever did not occur are not shown in this table.

39

TABLE XVIII—continued.

STATISTICS OF OFFICERS.
E.—DETAIL of DISEASES.

DISEASES.	British Officers attached to European Troops — India.* Admissions	Deaths	Invalids	Field Service Admissions	Deaths	British Officers attached to Native Troops — India. Admissions	Deaths	Invalids.†
Small-pox	2	...	...	...	...	...	...	...
Cow-pox	1	...	...	...	...	...	...	...
Chicken-pox	3	...	...	...	...	1	...	...
Measles	3	...	...	...	...	7	...	...
Rubella	4	...	...	...	...	2	...	...
Scarlet fever	2	...	...	...	...	...	...	...
Plague	1	...	...	...	...	...	...	...
Dengue	15	...	...	...	...	...	...	...
Influenza	70	...	1	...	...	20	...	...
Mumps	...	...	...	...	...	3	...	...
Diphtheria	1	1	...	...	...	...	...	...
Simple continued fever	112	...	1	...	...	34	...	...
Enteric fever	50	6	29	...	...	33	8	...
Dysentery	55	...	8	...	...	55	1	...
Ague	191	...	10	...	...	157	...	...
Remittent fever	7	...	1	...	...	10	...	...
Erysipelas	...	...	...	...	...	1	...	...
Septicæmia	1	1	...	...	...	...	...	...
Tubercle of lung	3	1	1	...	...	3	...	...
„ of testicles	...	...	...	...	...	1	...	...
Syphilis	...	...	...	...	...	2	...	...
Gonorrhœa	9	...	1	...	...	2	...	...
Hydrophobia	...	1	...	...	...	...	...	...
Tænia solium	7	...	...	...	...	...	...	...
Tænia mediocanellata	1	...	...	...	...	...	...	...
Thread-worm	1	...	...	...	...	...	...	...
Scabies	2	...	...	...	...	...	...	...
Ringworm	2	...	...	...	...	...	...	...
Rheumatic fever	1	...	...	...	...	...	...	...
Rheumatism	19	...	...	...	...	10	...	...
Gout	2	...	...	...	...	7	...	...
Chondroma	1	...	...	...	...	...	...	...
Diabetes mellitus	...	...	...	...	...	1	1	...
Congenital phimosis	1	...	...	...	...	...	...	...
Debility	39	...	13	...	...	14	...	...
Neuritis	3	...	...	...	...	4	...	...
Spinal meningitis	...	...	...	...	...	1	...	...
Softening of the brain	...	...	...	...	...	1	...	...
Sanguineous apoplexy	...	...	...	...	...	1	1	...
Apoplexy	...	...	...	...	...	1	...	...
Hemiplegia	1	...	...	...	...	...	...	...
Epilepsy	1	...	1	...	...	1	...	...
Vertigo	1	...	...	...	...	...	...	...
Megrim	...	...	...	...	...	1	...	...
Headache	...	...	...	...	...	1	...	...
Neuralgia	10	...	...	...	...	2	...	...
Nervous weakness	4	...	3	...	...	6	...	...
Mania	1	1	...	...	...	...	...	...
Melancholia	1	...	1	...	...	...	...	...
Conjunctivitis	8	...	...	...	...	7	...	...
Œdema of the Conjunctiva	1	...	...	...	...	...	...	...
Keratitis	...	...	...	...	...	1	...	...
Ulcerative keratitis	...	...	...	...	...	1	...	...
Iritis	1	...	...	...	...	1	...	...
Optic neuritis	1	...	...	...	...	...	...	...
Stye	2	...	...	...	...	1	...	...
Inflammation of the external ear	4	...	...	...	...	1	...	...
Inflammation of middle ear	2	...	...	...	...	...	...	...
Inflammation of middle ear suppurative	1	...	...	...	...	...	...	...
Perforation of membrana tympani	1	...	1	...	...	...	...	...
Rhinitis	1	...	...	...	...	...	...	...
Coryza	4	...	...	...	...	2	...	...
Valvular disease of the heart	4	...	...	...	...	1	...	...
Degeneration of the mus. sub. of the heart	...	...	...	...	...	...	...	...
Dilatation of the heart	1	...	1	...	...	1	...	...
Disordered action of the heart	2	...	...	...	...	3	...	...
Phlebitis	4	...	...	...	...	...	...	...
Varix	2	...	...	...	...	...	...	...
Laryngitis	3	...	...	...	...	4	...	...
Bronchitis	34	...	1	...	...	34	...	...
Spasmodic asthma	1	...	1	...	...	1	...	...
Congestion of the lung	1	...	...	...	...	...	...	...
Pneumonia	7	...	1	...	...	1	...	...
Pleurisy	3	...	1	1	...	1	...	...
Stomatitis	1	...	...	...	...	...	...	...
Inflammation of the dental periosteum	1	...	...	...	...	2	...	...

DISEASES.	British Officers attached to European Troops — India.* Admissions	Deaths	Invalids	Field Service Admissions	Deaths	British Officers attached to Native Troops — India. Admissions	Deaths	Invalids.†
Gum-boil	6	...	...	...	...	1	...	...
Sore throat	11	...	...	...	...	4	...	...
Ulceration of the tonsils	2	...	...	...	...	...	...	...
Tonsillitis	10	...	...	...	...	...	...	...
„ Follicular	36	...	...	...	...	28	...	...
Quinsy	4	...	...	...	...	...	...	...
Inflammation of the pharynx	...	...	...	...	...	5	...	...
Gastritis	16	...	...	...	...	15	...	...
Indigestion	12	...	2	11	...	4	...	...
Enteritis	14	...	2	...	...	6	...	...
Typhlitis	7	...	3	...	...	4	...	...
Colitis	15	...	2	...	...	3	...	...
Catarrhal inflammation of the intestines	8	...	...	...	...	...	...	...
Fæcal accumulation in the intestines	1	...	...	...	...	...	...	...
Sprue	2	...	1	...	...	1	...	...
Hernia	2	...	...	...	...	3	...	...
Valvulus	1	1	...	...	...	...	...	...
Obstruction of intestines	1	...	...	...	...	...	...	...
Colic	12	...	...	...	...	1	...	...
Diarrhœa	74	...	1	...	...	33	...	...
Neuralgia	...	...	...	...	...	1	...	...
Abscess of the rectum and anus	...	...	...	...	...	1	...	...
Fissure of the anus	1	...	...	...	...	4	...	...
Fistula in ano	1	...	...	...	...	8	...	...
Piles	1	...	...	...	...	1	...	...
Hepatitis	17	...	3	...	...	13	...	...
Abscess of the liver	1	...	1	...	...	3	1	...
Cirrhosis of the liver	...	...	...	...	...	2	...	...
Perihepatitis	1	...	...	...	...	...	...	...
Congestion of liver	33	...	2	...	...	4	...	...
Jaundice	20	...	1	...	...	8	...	...
Cholecystitis	2	...	...	...	...	1	...	...
Peritonitis	5	...	3	...	...	1	...	...
Splenitis	1	...	...	...	...	...	...	...
Inflammation of lymph glands	18	...	1	...	...	10	...	...
Acute nephritis	4	...	1	...	...	1	...	...
Pyelitis	1	...	...	...	...	...	...	...
Calculus in kidney	...	...	...	...	...	2	...	...
Nephralgia	1	...	...	...	...	1	...	...
Hæmaturia	...	...	...	...	...	1	...	...
Albuminuria	2	...	...	...	...	...	...	...
Lithuria	...	...	...	...	...	1	...	...
Inflammation of the bladder	3	...	...	...	...	6	...	...
Calculus in the bladder	1	...	...	...	...	1	...	...
Urethritis	...	...	...	...	...	1	...	...
Inflammation of the prostate	...	...	...	...	...	1	...	...
Calculus in prostate	1	...	...	...	...	...	...	...
Phimosis	1	...	...	...	...	1	...	...
Balanitis	1	...	...	...	...	1	...	...
Hydrocele of the spermatic cord	1	...	1	...	...	1	...	...
Hæmatocele	...	...	...	...	...	1	...	...
Varicocele	2	...	...	...	...	1	...	...
Orchitis	12	...	...	...	...	4	...	...
Epididymitis	4	...	...	...	...	1	...	...
Periostitis	2	...	1	...	...	4	...	...
Synovitis	42	...	3	...	...	24	...	...
Dislocation of articular cartilage	6	...	1	...	...	2	...	...
Loose body	...	...	...	...	...	1	...	...
Inflammation of the spine	1	...	1	...	...	...	...	...
Myalgia	7	...	...	...	...	6	...	...
Leuco-synovitis	1	...	...	...	...	...	...	...
Inflammation of bursæ	4	...	...	...	...	1	...	...
Abscess of bursæ	1	...	...	...	...	1	...	...
Hammer Toe	3	...	...	...	...	...	...	...
Inflammation of connective tissue	36	...	...	1	...	8	...	...
Abscess of connective tissue	33	...	...	...	...	10	...	...
Roseola	1	...	...	...	...	...	...	...
Urticaria	1	...	1	...	...	...	...	...
Eczema	5	...	...	...	...	3	...	...
Impetigo	1	...	...	...	...	1	...	...
Psoriasis	1	...	...	...	...	...	...	...
Pemphigus	1	...	...	...	...	...	...	...
Sycosis	1	...	...	...	...	...	...	...
Ulcer	12	...	...	...	...	15	...	...

* Excluding Field Service. † Information not available.

DISEASES.	BRITISH OFFICERS ATTACHED TO EUROPEAN TROOPS.					BRITISH OFFICERS ATTACHED TO NATIVE TROOPS.			DISEASES.	BRITISH OFFICERS ATTACHED TO EUROPEAN TROOPS.					BRITISH OFFICERS ATTACHED TO NATIVE TROOPS.		
	INDIA.*			FIELD SERVICE.		INDIA.				INDIA.*			FIELD SERVICE.		INDIA.		
	Admissions.	Deaths.	Invalids.	Admissions.	Deaths.	Admissions.	Deaths.	Invalids. †		Admissions.	Deaths.	Invalids.	Admissions.	Deaths.	Admissions.	Deaths.	Invalids. †
Boil	28	—	...	...	...	18	...	...	Rupture of muscles	2	...	1	...	...	...	...	...
Carbuncle	5	...	...	...	...	...	...	...	Concussion of the brain	28	—	6	...	...	6	1	...
Whitlow	2	...	...	...	...	...	...	...	Laceration „ „	1	...	...	...	...	...	...	...
Onychia	8	...	...	...	...	1	...	...	Compression of the brain	1	1	...	...	...	...	.	...
Corn	1	...	...	...	...	...	...	...	Rupture of viscera	—	...	...	...	...	1	...	...
Delhi boil	...	...	...	...	...	2	...	...	Murdered	...	2	...	...	...	...	...	...
Effects of heat	...	...	...	..	...	3	...	...	Poison, cyanide of potassium	...	1	...	...	...	...	...	...
Heat-stroke	13	...	2	...	...	2	...	...	Poison, sulphonal	1	...	...	...	...	...	...	...
Sun-stroke	7	...	2	...	...	3	...	...	„ ptomaines	2	...	...	...	...	...	...	...
Suffocation from compression	...	...	...	—	...	...	1	...	„ putrid exhalation	1	...	...	...	...	...	...	...
Suffocation from submersion	...	4	...	...	...	...	...	...	Poisoned wound, not defined	1	...	...	...	...	1	...	...
Contusions	67	...	...	...	...	41	...	...	Poisoned wound by dog	1	...	...	...	...	...	...	...
Strains and sprains	49	...	...	—	...	38	...	...	„ stinging insect	7	...	...	...	...	...	...	...
Dislocation of other bones	15	...	...	—	...	8	...	...	„ by dyed breeches	1	—	...	...	...	...	...	...
Gunshot wounds	2	...	1	...	...	3	...	...	No appreciable disease	1	—	...	...	...	...	...	...
Wounds	37	...	...	—	...	19	—	...	Not yet diagnosed	—	...	...	...	...	1	...	...
Burns and scalds	1	...	...	...	...	2	...	...									
Abrasion	12	—	...	...	—	11	...	—									
Fracture of vault of skull	1	...	1	...	...	...	...	...									
Fracture of base of skull	2	1	...	...	...	2	1	...									
Fracture of other bones	43	...	4	—	...	13	—	...	TOTAL	1,595	21	123	13	...	907	17	...

* Excluding Field Service.

† Information not available.

B.—WOMEN.

TABLE XIX.

RATIOS AND ACTUALS OF COMMANDS.

	Northern Command.		Western Command.		Eastern Command.		Secunderabad Division.		Burma Division.		India.*		
	Ratios.	Actuals.	Ratios.	Actuals.	Ratios.	Actuals.	Ratios.	Actuals.	Ratios.	Actuals.	Ratios.	Actuals.	Remaining from 1904.
Strength	838		974		873		518		172		3,375		
Constantly sick	29·7	24·85	26·9	26·16	24·7	21·59	33·2	17·20	25·1	4·32	27·9	94·12	
ADMISSIONS—													
Influenza	15·5	...	·1	3	8·0	7	...	...	...	...	6·8	23	...
Cholera	...	...	...	...	...	...	...	...	...	...	...	...	...
Small-pox	2·4	2	8·2	8	2·3	2	5·8	3	5·8	1	4·7	16	...
Enteric Fever	9·5	8	9·2	9	6·9	6	11·6	6	...	...	8·6	29	4
Intermittent Fever	22·7	19	51·3	50	61·9	54	57·9	30	58·1	10	48·3	163	4
Remittent Fever	3·6	3	...	...	6·9	6	...	...	...	...	2·7	9	1
Simple Continued Fever	16·7	14	21·6	21	20·6	18	9·7	5	29·1	5	18·7	63	1
Tubercle of the lungs	6·0	5	7·2	7	4·6	4	3·9	2	...	...	5·3	18	...
Pneumonia	2·4	...	...	...	...	...	1·9	1	...	...	9	3	...
Other Respiratory Diseases	8·4	7	17·5	17	13·7	12	23·2	12	5·8	1	14·5	49	2
Dysentery	4·8	4	26·7	26	4·6	4	13·5	7	5·8	1	12·4	42	1
Diarrhœa	9·5	8	14·4	14	28·6	25	23·2	12	5·8	1	17·8	60	1
Anæmia and Debility	353·2	296	278·2	271	262·3	229	175·7	91	180·2	31	272·0	918	30
Abortion and Puerperal Affections	53·7	45	26·7	26	47·0	41	42·5	22	52·3	9	42·4	143	9
Other Diseases peculiar to women	46·5	39	37·0	36	30·9	27	36·7	19	40·7	7	37·9	128	3
ALL CAUSES	668·3	560	625·3	609	752·6	657	530·9	275	476·7	82	646·8	2,183	72
DEATHS—													**Deaths out of hospital.**
Cholera	...	...	...	...	...	...	...	...	...	...	...	...	...
Small-pox	...	...	2·05	2	...	...	...	...	5·81	1	·89	3	...
Enteric Fever	...	...	2·05	2	3·44	3	...	...	...	...	1·48	5	...
Intermittent Fever	1·19	1	...	...	...	...	...	...	...	...	·30	1	...
Remittent Fever	...	...	...	...	...	...	...	...	...	...	...	...	...
Simple Continued Fever	...	...	...	...	...	...	...	...	...	...	...	...	...
Tubercle of the lungs	...	...	2·05	2	1·15	1	...	...	...	...	·89	3	...
Pneumonia	...	...	...	...	...	...	...	...	...	...	...	...	...
Other Respiratory Diseases	...	...	...	...	...	...	...	...	...	...	...	...	...
Dysentery	1·19	1	1·03	1	...	...	1·93	1	...	...	·8	3	...
Diarrhœa	...	...	...	...	...	...	...	...	...	...	...	...	...
Hepatic Abscess	...	...	...	...	...	...	...	...	...	...	...	...	...
Childbirth and Abortion	1·19	1	1·03	1	2·29	2	1·93	1	...	...	1·48	5	1
ALL CAUSES	9·55	8	14·37	14	11·45	10	8·79	3	11·63	2	10·96	37	3
PERCENTAGE IN 100 ADMISSIONS—													
Influenza	2·32		·40		1·07		...		...		1·05		
Cholera	...		...		...		...		...		...		
Small-pox	·36		1·31		·30		1·09		1·22		·73		
Enteric Fever	1·43		1·48		·91		2·18		...		1·33		
Intermittent Fever	3·39		8·21		8·22		10·91		12·20		7·47		
Remittent Fever	·54		...		·91		...		...		·41		
Simple Continued Fever	2·50		3·45		2·74		1·82		6·10		2·89		
Tubercle of the lungs	·89		1·15		·61		·73		...		·82		
Pneumonia	·36		...		...		·36		...		·14		
Other Respiratory Diseases	1·25		2·79		1·83		4·16		1·22		2·24		
Dysentery	·71		4·27		·61		2·55		1·22		1·93		
Diarrhœa	1·43		2·30		3·81		4·16		1·22		2·75		
Anæmia and Debility	52·86		41·30		34·86		33·02		37·80		42·05		
Abortion and Puerperal Affections	8·04		4·27		6·24		8·00		10·98		6·55		
Other Diseases peculiar to women	6·96		5·91		4·11		6·91		8·54		5·86		
PERCENTAGE IN 100 DEATHS—													
Cholera	...		...		...		...		...		...		
Small-pox	...		14·3		...		...		50·0		8·1		
Enteric Fever	...		14·3		30·0		...		...		13·5		
Intermittent Fever	12·5		...		...		...		...		2·7		
Remittent Fever	...		...		...		...		...		...		
Simple Continued Fever	...		...		...		...		...		...		
Tubercle of the lungs	...		14·3		10·0		...		...		8·1		
Pneumonia	...		...		...		...		...		...		
Other Respiratory Diseases	...		...		...		...		...		...		
Dysentery	12·5		7·1		...		33·3		...		8·1		
Diarrhœa	...		...		...		...		...		...		
Hepatic Abscess	...		...		...		...		...		...		
Childbirth and Abortion	12·5		7·1		20·0		33·3		...		13·5		

* For complete detail of diseases, see Table LIII.

TABLE XXI.

ENTERIC FEVER by months, stations, groups, and commands.

Stations* and Groups.	Average annual strength.	January.	February.	March.	April.	May.	June.	July.	August.	September.	October.	November.	December.	Total Admissions.	Admission-rate per 1,000 of strength.	Total deaths.	Death-rate per 1,000 of strength.
B																	
Dinapore	24	...	...	...	...	...	...	...	...	1	...	...	...	1	41·7	...	...
Sitapur	22	...	...	...	...	1	...	...	...	...	...	...	...	1	45·5	1	45·45
Group V.—Gangetic Plain and Chutia Nagpur	296	...	...	...	...	1	...	...	...	1	...	...	...	2	6·9	1	3·38
A																	
Bareilly	44	...	...	...	1	...	...	...	...	...	...	...	...	1	22·7	1	22·73
erut	92	...	...	...	...	...	1	...	...	...	...	1	...	2	21·7	1	10·87
B																	
Rawalpindi	130	...	...	...	...	...	...	...	...	...	...	...	1	1	7·7	...	...
Group VI.—Upper Sub-Himalaya	619	...	...	...	1	...	1	...	...	...	...	1	1	4	6·5	2	3·23
A																	
Ahmedabad	14	...	...	...	...	...	...	...	...	1	...	...	...	1	71·4	1	71·43
B																	
Muttra	17	...	...	...	...	...	...	1	...	...	...	...	...	1	58·8	...	...
Jhansi	46	...	...	...	...	...	...	...	...	...	...	...	1	1	21·7	...	...
Mhow	89	...	...	...	...	...	...	...	...	...	...	1	...	1	11·2	...	...
Group VIII.—South-Eastern Rajputana, Central India, and Gujarat	263	...	...	...	...	...	...	1	...	1	...	1	1	4	15·2	1	3·80
B																	
Secunderabad	163	...	1	...	...	...	...	1	1	1	...	...	...	4	24·2	...	...
Belgaum	37	...	...	...	...	...	...	1	...	...	...	...	...	1	27·0	...	...
Poona	90	...	...	...	...	...	...	...	...	1	1	...	1	3	33·3	...	...
Ahmednagar	44	...	...	...	...	...	...	...	1	...	...	...	...	1	22·7	...	...
Group IX.—Deccan	514	...	1	...	...	...	...	2	1	3	1	...	1	9	17·5	...	...
A																	
Bangalore	135	...	...	...	...	...	1	1	...	...	...	...	...	2	14·8	...	...
Group XI.—Southern India	251	...	...	...	...	...	1	1	...	...	...	...	...	2	7·3	...	...
Dagshai	40	...	...	...	...	...	1	...	...	...	...	...	...	1	25·0	...	...
Khyragully	6	...	...	...	...	2	...	...	...	...	...	...	...	2	333·3	...	...
Group XIIa.—Hill Stations	438	...	...	...	...	2	1	...	...	...	...	...	...	3	6·8	...	...
Kasauli	37	...	...	...	...	1	2	1	...	...	...	...	...	4	108·1	...	...
Group XIIb.—Hill Convalescent Depôts, and Sanitaria	293	...	...	...	...	1	2	1	...	...	...	...	...	4	13·4	...	...
Extra India.																	
Aden	33	...	1	...	...	...	...	...	...	...	...	...	...	1	30·3	1	30·30
INDIA	3,375	...	2	...	1	4	5	5	1	5	1	2	3	29	8·6	5	1·48
Northern Command	838	...	...	...	...	3	3	1	...	...	...	...	1	8	9·5	...	...
Western „	974	...	1	...	...	...	...	1	...	3	1	1	2	9	9·2	2	2·05
Eastern „	873	...	...	...	1	1	1	1	...	1	...	1	...	6	6·9	3	3·44
Secunderabad Division	518	...	1	...	...	...	1	2	1	1	...	...	...	6	11·6	...	...
Burma Division	172	...	...	...	...	...	...	...	1	...	...	...	...	...	...	...	...

* Stations where Enteric Fever did not occur are not shown in this table.

C.–CHILDREN.

TABLE XXII.

RATIOS AND ACTUALS OF COMMANDS.

	Northern Command.		Western Command.		*Eastern Command.		Secunderabad Division.		Burma Division.		India.*		
	Ratios.	Actuals.	Ratios.	Actuals.	Ratios.	Actuals.	Ratios.	Actuals.	Ratios.	Actuals.	Ratios.	Actuals.	Remaining from 1904.
Strength	1,301		1,484		1,292		829		248		5,154		
Constantly sick	14'5	18'86	58'2	86'43	14'0	18'08	30'5	25'30	8'7	2'15	29'3	150'82	
ADMISSIONS—													
Influenza	2'3	3	1'3	2	2'3	3	...	...	...	...	1'6	8	1
Cholera	...	...	...	...	...	...	...	...	...	...	...	...	...
Small-pox	1'5	2	3'4	5	1'5	2	...	...	...	...	1'7	9	...
Measles	3'8	5	1'3	2	1'5	2	3'6	3	...	...	2'3	12	1
Whooping Cough . . .	5'4	7	5'4	8	4'6	6	2'4	2	...	...	4'5	23	...
Enteric Fever	5'4	7	10'1	15	2'3	3	1'2	1	...	...	5'0	26	2
Intermittent Fever . . .	23'1	30	53'2	79	54'2	70	83'2	69	36'3	9	49'9	257	6
Remittent Fever . . .	...	...	4'7	7	8	1	...	...	4'c	1	1'7	9	...
Simple Continued Fever . .	13'8	18	14'2	21	18'6	24	32'6	27	20'2	5	18'4	95	...
Tubercular Diseases . . .	2'3	3	6'1	9	...	...	1'2	1	...	...	2'5	13	1
Respiratory Diseases . .	36'1	47	49'2	73	70'4	91	86'9	72	16'1	4	55'7	287	9
Dysentery	4'6	6	2'0	3	7'7	10	13'3	11	8'1	2	6'2	32	2
Diarrhœa	43'8	57	51'2	76	38'7	50	21'7	18	24'2	6	40'2	207	4
Eye Diseases	3'8	5	10'1	15	4'6	6	82'0	68	4'0	1	18'4	95	4
ALL CAUSES .	319'0	415	319'7	519	370'3	490	737'0	611	229'8	57	495'9	2,092	56
DEATHS—													Deaths out of hospital.
Cholera	...	...	...	...	...	...	...	...	...	...	'78	4	...
Small-pox	...	...	2'02	3	'77	1	...	...	...	...	'78	4	...
Diphtheria and Croup . .	'77	1	...	...	2'32	3	...	...	...	...	'39	2	...
Enteric Fever	...	...	'67	1	'77	1	...	...	...	...	'39	2	...
Intermittent Fever . . .	'77	1	...	...	1'55	2	3'41	2	...	...	'97	5	...
Remittent Fever . . .	...	...	...	...	...	...	...	...	4'03	1	'19	1	...
Simple Continued Fever . .	...	...	...	...	...	...	...	...	...	...	...	...	...
Tubercular Diseases . . .	'77	1	3'37	5	...	...	1'21	1	...	...	1'36	7	...
Convulsions	3'07	4	4'72	7	1'55	2	2'41	2	...	...	2'91	15	1
Respiratory Diseases . .	7'69	10	4'72	7	8'51	11	3'62	3	4'03	1	6'21	32	1
Teething	'77	1	2'02	3	3'87	5	7'24	6	...	...	2'91	15	...
Dysentery	'77	1	...	...	'77	1	...	...	...	...	'39	2	...
Diarrhœa	1'54	2	9'43	14	6'07	9	...	...	...	...	4'85	25	1
Anæmia, Debility, and Immaturity .	7'69	10	4'72	7	10'81	14	7'24	6	4'03	1	7'37	38	2
ALL CAUSES .	32'28	42	39'76	59	50'31	65	37'39	31	12'10	3	38'80	200	6
PERCENTAGE IN 100 ADMISSIONS—													
Influenza	'72		'39		'61		...		...		'38		
Cholera	...		...		...		...		...		...		
Small-pox	'48		'96		'44		...		...		'43		
Measles	1'20		'39		'44		'49		...		'57		
Whooping Cough . . .	1'69		1'54		1'22		'33		...		1'10		
Enteric Fever	1'69		2'89		'61		'16		...		1'24		
Intermittent Fever . . .	7'73		15'22		14'29		11'29		15'79		12'28		
Remittent Fever . . .	...		1'35		'20		...		1'75		'43		
Simple Continued Fever . .	4'34		4'05		4'90		4'42		8'77		4'54		
Tubercular Diseases . .	'72		1'73		...		'16		...		'62		
Respiratory Diseases . .	11'32		14'07		18'57		11'78		7'02		13'72		
Dysentery	1'45		'58		2'04		1'80		3'51		1'53		
Diarrhœa	13'73		14'64		10'21		2'95		10'53		9'89		
Eye Diseases	1'20		2'89		1'22		11'13		1'75		4'54		
PERCENTAGE IN 100 DEATHS—													
Cholera	...		...		...		...		...		...		
Small-pox	...		5'1		1'5		...		...		2'0		
Diphtheria and Croup . .	2'4		...		4'6		...		...		2'0		
Enteric Fever	...		1'7		1'5		...		...		1'0		
Intermittent Fever . . .	2'4		...		3'1		6'5		...		2'5		
Remittent Fever . . .	...		...		...		...		33'3		'5		
Simple Continued Fever . .	...		...		...		...		...		...		
Tubercular Diseases . .	2'4		8'5		...		3'2		...		3'5		
Convulsions	9'5		11'9		3'1		6'5		...		7'5		
Respiratory Diseases . .	23'8		11'9		16'9		9'7		33'3		16'0		
Teething	2'4		5'1		7'7		19'4		...		7'5		
Dysentery	2'4		...		'5		...		...		1'0		
Diarrhœa	4'8		23'7		13'8		...		...		12'5		
Anæmia, Debility, and Immaturity .	23'8		11'7		21'5		19'4		33'3		19'0		

* For complete detail of diseases, see Table LIII.

TABLE XXIV.

ENTERIC FEVER by months, stations, groups, and commands.

STATIONS,* AND GROUPS.	Average annual strength.	Jan.	Feb.	Mar.	Apl.	May.	June.	July.	Aug.	Sept.	Oct.	Nov.	Dec.	Total admissions.	Admission-rate per 1,000 of strength.	Total deaths.	Death-rate per 1,000 of strength.
B																	
Fyzabad	53	...	...	...	...	...	...	...	...	...	1	...		1	18·9	1	18·87
Cawnpore	34	...	...	...	...	...	...	...	...	1	1	...		2	58·8	...	...
GROUP V.—GANGETIC PLAIN AND CHUTIA NAGPUR.	428	...	...	...	...	...	...	...	...	1	2	...		3	7·0	1	2·34
B																	
Ferozepore	93	...	...	1	...	...	..	...	...	...	...	...		1	10·8	...	
Campbellpur	36	...	...	...	...	2	...	...	...	...	...	...		2	55·6	...	...
GROUP VI.—UPPER SUB-HIMALAVA.	935	...	...	1	...	2	...	...	...	...	...	...		3	3·2	...	
A																	
Peshawar	75	...	...	...	...	...	...	...	...	...	...	1		1	13·3	...	...
GROUP VII.—N.-W. FRONTIER INDUS VALLEY, AND NORTH-WESTERN RAJPUTANA.	302	...	...	...	...	...	...	...	...	...	...	1		1	3·3	..	
B																	
Nasirabad	31	...	...	...	...	...	...	...	...	...	1	...		1	32·3	...	...
Mhow	125	1	1	...	...	...	...	...	1	2	...	1		6	48·0	1	8·00
GROUP VIII.—SOUTH-EASTERN RAJPUTANA, CENTRAL INDIA, AND GUJARAT.	386	1	1	...	...	...	..	...	...	1	2	1	1	7	18·1	1	2·59
B																	
Secunderabad	246	1	...	...	...	...	...	...	...	...	...	...		1	4·1	...	
Belgaum	45	...	...	...	..	...	1	...	1	...	...	...		2	44·4	...	
Poona	108	...	...	...	...	...	...	3	...	...	...	...		3	27·8	...	...
Kirkee	126	...	...	...	...	...	...	1	...	...	...	...		3	23·8	...	...
GROUP IX.—DECCAN	737	1	...	...	...	...	1	2	2	3	...	...	...	9	12·2	...	
Salon	6	...	...	...	...	...	1	...	...	...	...	...		1	166·7	...	
Jutogh	17	...	...	...	...	1	...	...	...	...	...	...		1	58·3	...	
Camp Bariac	37	...	...	...	...	...	...	...	...	...	...	...		1	27·0	...	...
GROUP XIIa.—HILL STATIONS	638	...	...	...	1	1	1	...	...	...	...	...		3	4·7	...	
INDIA	5,154	2	1	1	...	3	2	3	2	4	3	3	2	26	5·0	2	·39
NORTHERN COMMAND	1,301	...	...	1	...	3	1	1	...	...	...	...	1	7	5·4	...	...
WESTERN "	1,484	1	1	...	...	...	1	2	2	4	2	1	1	15	10·1	1	·67
EASTERN "	1,292	...	...	...	...	...	...	...	...	1	2	...		3	2·3	1	·77
SECUNDERABAD DIVISION	829	1	...	...	...	...	...	...	...	...	...	...		1	1·2	...	...
BURMA DIVISION	248	...	...	...	...	...	...	...	...	...	...	...		...	...	...	

* Stations where Enteric Fever did not occur are not shown in this table.

TABLE XXV.

DEATHS OF CHILDREN BY AGES AND CAUSES.

Age at Death.	Cholera.	Small-pox.	Diphtheria and Croup.	Enteric Fever.	Intermittent Fever.	Remittent Fever.	Simple Continued Fever.	Tubercular Diseases.	Convulsions.	Respiratory Diseases.	Teething.	Dysentery.	Diarrhœa.	Anæmia, Debility, and Immaturity at birth.	All Causes.	Average annual strength.	Death-rate per 1,000 of strength.	Liability. (The previous column expressed in percentages.)
Under 6 months	...	...	...	...	...	...	...	2	7	14	2	...	9	32*	85	512	162·84	45·77
Between 6 and 12 months‡	...	2	...	...	1	...	...	1	4	7	5	1	13	3	54	520	103·85	29·19
,, 12 and 18 ,,	...	2	1	...	...	...	...	2	...	3	6	1	3	...	24	524	45·80	12·87
,, 18 and 24 ,,	...	...	...	...	...	...	...	...	2	1	2	...	...	1	9	500	18·00	5·06
,, 2 years and 5 years	...	...	2	...	2	...	...	2	2	5	...	...	...	1	16	1,120	14·17	3·99
,, 5 ,, and 10 ,,	...	...	1	1	1	1	...	...	...	...	...	...	...	...	5	1,224	4·08	1·15
,, 10 ,, and 15 ,,	...	...	...	1	1	...	...	...	...	...	...	...	...	...	4	569	7·03	1·93
,, 15 ,, and upwards	...	...	...	...	...	...	...	...	...	...	...	...	...	...	...	123	...	...
Total	...	4	4†	2	5	1	...	7	15	30	15	2	25	37	197‡	5,111	38·54	100·00

* 24 immaturity.
† Diphtheria.
‡ Excluding 3 deaths and 43 average strength of children at Deolali for which details by ages are not available.

II.—NATIVE TROOPS, 1905.

51

TABLE H.
STATIONS by COMMANDS.

Stations	Height above the sea-level in feet.[*]	Authority for height.[†]
NORTHERN COMMAND :—		
Ambala	902	S. G.
Jullundur	900	,,
Ferozepore	645	,,
Mian Mir	706	,,
Amritsar	756	,,
Sialkot	829	,,
Jhelum	827	,,
Rawalpindi	1,707	,,
Attock	891	,,
Mardan	...	...
Nowshera	1,100	M. O.
Peshawar	1,165	S. G.
Fort Jamrud	1,610	,,
Kohat	1,765	,,
Thal	2,820	I. B.
Edwardesabad	1,279	,,
Dera Ismail Khan	571	S. G.
Jatta	1,000	I. B.
Draïand	1,600	,,
Fort Zam	1,350	,,
Dera Ghazi Khan	395	S. G.
Multan	402	,,
Jandola	2,400	I. B.
Simla	7,230	S. G.
Jutogh	6,371	,,
Dharmsala	6,111	,,
Bakloh	4,585	,,
Murree	7,098	,,
Khyragully	8,746	,,
Baragully	7,800	M.O.
Kalabagh	7,036	I. B.
Chitral	4,980	S.G.
Kila Drosh	4,250	I. B.
Malakand	3,589	S. G.
Dargai	...	...
Chakdara	3,903	S. G.
Abbottabad	4,152	,,
Cherat	4,520	,,
Fort Lockhart	6,567	I. B.
Hangu	3,650	,,
WESTERN COMMAND :—		
Bikaner	825	S. G.
Sibi	495	,,
Jacobabad	181	,,
Hyderabad	134	I. H.
Kurrachee	28	S. G.
Bhuj	...	...
Rajkot	417	S. G.
Deesa	468	,,
Ahmedabad	170	,,
Baroda	...	...
Alirajpore	977	S. G.
Sirdarpore	1,659	,,
Jhabwa	1,171	,,
Kherwara	1,050	,,
Kotra	1,013	,,
Udaipur	1,950	,,
Todgarh	2,300	,,
Erinpura	869	,,
Neemuch	1,613	,,
Deoli	1,122	,,

Stations	Height above the sea-level in feet.[*]	Authority for height.[†]
WESTERN COMMAND—contd.		
Tonk	...	...
Beawar	1,465	S. G.
Nasirabad	1,301	,,
Ajmer	1,627	,,
Sambhar	1,254	M. D.
Jaipur	1,582	S. G.
Gwalior	...	...
Jhansi	860	S. G.
Nowgong	770	I. B.
Goona	1,617	S. G.
Agar	1,671	,,
Schore	1,617	,,
Indore	1,805	,,
Mhow	1,901	I. B.
Asirgarh	2,283	S. G.
Saugor	1,753	,,
Sutna	1,040	M. D.
Jubbulpore	1,306	S. G.
Kamptee	941	,,
Sitabaldi	1,236	,,
Aurangabad	1,885	M. D.
Ahmednagar	2,125	S, G.
Belgam	2,473	,,
Satara	2,183	,,
Poona	1,903	,,
Kirkee	1,837	,,
Sirur	...	...
Bombay	20	S. G.
Santa Cruz	...	...
Mir Ali Khel	3,650	I. B.
Fort Sandeman	4,700	,,
Hindubagh	5,675	S, G.
Musa Khel	4,450	I. B.
Khan Mohamed Kot	3,431	S, G.
Kilia Saifulla	...	...
Murgha	5,100	I. B.
Loralai	4,450	S. G.
Gumbaz	3,000	I. B.
Quetta	5,511	S. G.
Peshin	5,157	,,
Shelabagh	7,700	I. H.
Spinwana	7,300	,,
Chaman	5,488	S. G
Mount Abu	3,960	,,
Chahbar	...	...
Jask	...	...
Muscat	...	...
Bushire	40	I. B.
Aden	26	S. G.
Khormaksar	50	I. B.
Sheikh Othman	...	...
Perim	249	I. B.
EASTERN COMMAND :—		
Manipur	2,619	S. G.
Sadiya	440	M. H. I.
Dibrugarh	342	S. G,
Fort William	17	S. G.
Alipore	21	I. B.
Barrackpore	24	S. G.
Buxa	2,457	,,
Cuttack	74	,,
Doranda (Ranchi)	2,166	,,
Dinapore	...	...
Benares	236	S. G.
Allahabad	298	,,
Fyzabad	336	,,

Stations	Height above the sea-level in feet.[*]	Authority for height.[†]
EASTERN COMMAND—contd.		
Lucknow	400	S. G.
Cawnpore	417	I. B.
Fatehgarh	444	,,
Bareilly	560	S. G.
Roorkee	884	,,
Dehra Dun	2,249	,,
Meerut	739	,,
Delhi	715	,,
Agra	554	,,
Kohima	4,500	I. B.
Shillong	4,937	S. G.
Gantak	5,000	I. B.
Almora	5,494	S. G.
Naini Tal	6,400	,,
Lansdowne	6,200	,,
SECUNDERABAD DIVISION :—		
Ellichpur	1,218	S. G.
Bulanum	...	S. G.
Secunderabad	1,732	,,
Cannanore	47	M. D.
Trivandrum	198	,,
Bellary	1,483	S. G.
Bangalore	3,021	,,
Trichinopoly	274	,,
Pallavaram	74	,,
St. Thomas' Mount	250	,,
Madras	15	,,
Vizianagram	191	,,
Ootacamund	7,216	,,
BURMA DIVISION :—		
Port Blair	85	S. G.
Rangoon	14	,,
Thayetmyo	145	,,
Keng Tung	2,773	,,
Fort Stedman	2,900	,,
Meiktila	298	,,
Fort Dufferin	249	,,
Bhamo	351	,,
Maymyo	3,600	,,

* These are usually the heights above sea-level of the survey-marks or of the mercury-surface in barometer-cisterns in the stations.

† S. G. = Surveyor-General of India: M, H. I. = Dr. Macnamara's "Himalayan India"; M. D. = Meteorological Department ; I. B. = Intelligence Branch of the Quarter-Master-General's Department ; M. O. = Medical Officers in charge of Station Hospitals in their Sanitary Reports.

TABLE XXVI.

RATIOS of COMMANDS.

The ratios of admissions and deaths to strength are taken from Table XXVIII. The actuals will be found in Table XXIX.

	Northern Command.	Western Command.	Eastern Command.	Secundera- bad Division.	Burma Division.	Army of India.* †
			RATIO PER 1,000 OF THE AVERAGE STRENGTH.			
I.—AVERAGE ANNUAL STRENGTH	36,696	33,932	21,823	11,794	4,793	123,434
II.—CONSTANTLY-SICK-RATE OF EACH MONTH—						
January	30·2	27·1	22·8	21·8	29·0	23·9
February	30·6	24·9	21·6	20·2	26·7	24·3
March	21·8	21·6	18·8	18·3	20·9	21·5
April	26·3	23·7	2·2	18·6	15·2	23·7
May	25·5	21·5	22·1	19·1	18·4	24·0
June	24·9	21·2	24·2	20·7	27·9	23·6
July	23·8	20·1	24·1	21·7	35·1	23·5
August	25·3	22·4	24·0	19·4	39·4	23·5
September	31·2	25·0	24·0	17·3	24·2	22·3
October	33·8	22·3	22·4	15·0	23·9	21·3
November	31·1	24·9	23·1	17·0	25·1	23·4
December	29·8	23·9	24·6	18·5	25·8	23·7
OF THE YEAR	26·5	23·1	23·7	19·1	25·6	23·2
III.—ADMISSION-RATE OF THE YEAR—						
Influenza	2·3	1·3	2·2	·3	1·0	1·5
Cholera	·0	·1	·2	·3	...	·1
Small-pox	·5	1·5	...	·3	...	·6
Enteric Fever	1·4	·5	2·0	1·1	·7	1·1
Intermittent Fever	142·6	185·0	159·4	80·4	167·1	17·12
Remittent Fever	11·1	3·7	10·7	1·8	7·3	8·4
Simple Continued Fever	10·7	8·4	10·2	42·9	15·9	14·3
Tubercle of the lungs	5·4	2·4	2·8	1·2	3·1	3·1
Pneumonia	20·4	11·4	9·0	5·5	3·1	12·5
Other Respiratory Diseases	35·4	29·6	23·1	15·4	22·7	28·5
Dysentery	34·6	24·9	30·8	25·0	37·0	32·5
Diarrhœa	6·7	7·9	10·7	2·2	5·2	7·5
Hepatic { Abscess	·1	·1	...	...	...	·2
{ Congestion and Inflammation	·5	·8	·7	·4	·8	·6
Scurvy	1·0	4·0	·6	·8	·4	1·7
Venereal Diseases	18·8	20·5	19·5	27·0	20·9	19·6
ALL CAUSES	651·8	602·4	571·5	466·1	614·2	617·1
IV.—DEATH-RATE OF THE YEAR—						
Cholera	·03	...	·18	·17	...	·06
Small-pox	...	...	...	·04	...	·01
Enteric Fever	·27	·21	·60	·34	...	·28
Intermittent Fever	·19	·41	·37	·34	·63	·37
Remittent Fever	·76	·32	·64	...	·63	·59
Simple Continued Fever	...	...	...	...	...	...
Circulatory Diseases	·16	·32	·18	·59	...	·26
Tubercle of the lungs	·76	·47	·55	·08	·83	·50
Pneumonia	3·32	1·74	1·33	1·27	·63	1·90
Other Respiratory Diseases	·32	·21	·37	·08	·21	·23
Dysentery	·10	·18	·18	·17	...	·18
Diarrhœa	·05	·09	·05	...	...	·05
Hepatic Abscess	...	·03	...	...	...	·01
Anæmia and Debility	...	·15	·05	...	·42	·09
ALL CAUSES	12·92	6·60	6·42	5·77	6·47	8·09
V.—PERCENTAGE IN 100 ADMISSIONS—						
Influenza	·36	·22	·38	·05	·17	·25
Cholera	·00	·01	·03	·03	...	·01
Small-pox	·08	·24	...	·05	...	·10
Enteric Fever	·22	·09	·35	·24	·03	·17
Intermittent Fever	21·88	30·70	27·89	17·25	27·21	28·11
Remittent Fever	1·70	·61	1·88	·38	1·19	1·38
Simple Continued Fever	2·56	1·39	1·78	9·21	2·58	2·36
Tubercle of the lungs	·82	·40	·50	·25	·51	·51
Pneumonia	3·13	1·89	1·58	1·18	·51	2·07
Other Respiratory Diseases	5·43	4·92	4·05	3·31	3·70	4·70
Dysentery	5·31	4·14	6·43	5·49	6·11	5·36
Diarrhœa	1·03	1·31	1·87	·47	·85	1·24
Hepatic { Abscess	·02	·02	·02	...	...	·03
{ Congestion and Inflammation	·07	·13	·13	·03	·14	·10
Scurvy	·16	·67	·10	·16	·07	·28
Venereal Diseases	2·88	3·40	3·41	5·80	3·40	3·23
VI.—PERCENTAGE IN 100 DEATHS—						
Cholera	·2	...	2·9	2·9	...	·7
Small-pox	...	...	...	1·5	...	·1
Enteric Fever	2·1	3·1	9·3	5·9	...	3·5
Intermittent Fever	1·5	6·3	5·7	5·9	9·7	4·6
Remittent Fever	5·9	4·9	10·0	...	9·7	6·9
Simple Continued Fever	...	...	...	...	...	...
Circulatory Diseases	1·3	4·9	2·9	10·3	...	3·2
Tubercle of the lungs	5·9	7·1	8·6	1·5	13·0	6·2
Pneumonia	25·7	26·3	20·7	22·1	9·7	23·5
Other Respiratory Diseases	2·1	3·1	5·0	1·5	3·2	2·8
Dysentery	1·5	2·7	2·9	2·9	...	2·2
Diarrhœa	·4	1·3	·7	...	...	·7
Hepatic Abscess	...	·4	...	...	...	·1
Anæmia and Debility	...	2·7	·7	...	6·5	1·1

* For complete detail of diseases see Table LIII.
† Excluding troops in China and Somaliland only.

H 2

TABLE XXVII.

RATIOS of GEOGRAPHICAL GROUPS.

The ratios of admissions and deaths to strength are taken from Table XXVIII.　　The actuals will be found in Table XXIX.

	I Burma Coast and Bay Islands.	II Burma Inland.	III Assam.	IV Bengal and Orissa.	V Gange-tic Plain and Chutia Nagpur.	VI Upper Sub-Hima-laya.	VII N.-W. Frontier, Indus Valley, and N.-W. Raj-putana.	VIII S.-E. Rajpu-tana, Central India, and Gujarat.	IX Dec-can.	X West-ern Coast.	XI South-ern India.	XII Hill Stations.	Army of India.†
I.—AVERAGE ANNUAL STRENGTH	1,195	2,916	949	2,558	7,046	16,649	17,946	13,651	15,977	1,978	5,967	20,224	123,434
II.—CONSTANTLY-SICK-RATE OF EACH MONTH—													
January	23·7	31·0	29·6	31·1	15·2	20·7	34·7	24·8	19·1	39·6	23·3	31·4	23·9
February	17·5	28·4	30·2	27·5	14·5	21·3	32·9	21·7	19·2	40·2	22·2	32·4	24·3
March	17·6	18·5	22·3	21·4	13·4	19·3	24·7	16·7	16·6	31·3	21·0	27·7	21·5
April	9·7	12·0	10·9	20·9	15·4	23·4	15·1	20·1	17·7	33·5	20·7	34·1	23·7
May	17·8	12·6	25·5	22·1	17·6	25·3	18·3	17·8	17·0	34·7	20·7	31·8	24·0
June	41·8	20·3	30·8	23·3	17·6	22·5	19·9	16·3	17·0	24·6	24·6	32·8	23·8
July	50·3	26·3	34·0	28·0	18·7	22·2	20·4	16·4	18·3	22·8	24·2	29·1	23·8
August	35·2	23·1	25·3	33·1	19·1	23·4	20·6	18·9	18·8	27·8	20·1	31·6	23·3
September	27·1	19·0	19·6	31·0	22·5	20·5	19·8	22·8	17·2	23·9	18·7	30·6	22·3
October	20·7	20·3	18·7	27·5	21·1	21·5	23·8	23·9	16·8	25·4	16·4	26·0	21·1
November	16·8	26·5	28·8	30·5	19·8	27·3	32·3	27·6	19·9	23·2	19·4	29·4	23·4
December	20·3	22·7	32·3	22·8	20·9	27·2	37·0	26·4	18·6	22·1	20·5	32·1	23·7
Of the Year	24·1	22·1	25·7	26·7	17·8	22·8	25·3	21·4	18·1	29·1	21·2	30·8	23·2
III.—ADMISSION-RATE OF THE YEAR—													
Influenza	4·2	...	...	11·3	1·1	2·0	4·0	1·0	·2	...	·5	·3	1·5
Cholera	...	...	...	...	·6	...	...	·2	...	·5	·0	·1
Small-pox	...	...	...	...	...	·4	·7	1·4	·3	·0	·2	·0	·6
Enteric Fever	...	...	...	...	·3	2·9	·7	·7	1·1	·5	·2	1·6	1·1
Intermittent Fever	168·2	159·5	213·9	353·8	119·8	114·4	170·6	211·5	73·2	192·6	122·2	170·3	171·2
Remittent Fever	5·0	8·6	2·1	24·6	9·2	8·8	9·9	1·4	3·4	·3	·8	11·5	8·4
Simple Continued Fever	52·7	4·5	...	21·1	13·1	5·3	7·0	6·8	20·7	10·6	57·3	23·4	14·3
Tubercle of the lungs	2·5	2·7	1·1	2·7	3·4	4·7	4·3	3·2	·6	3·5	1·7	4·6	3·1
Pneumonia	2·5	1·7	5·3	10·9	5·3	15·4	20·5	14·7	6·6	13·7	5·0	13·7	12·5
Other Respiratory Diseases	21·8	23·3	37·9	28·1	21·7	27·4	29·0	24·2	19·9	19·3	15·8	45·5	28·5
Dysentery	51·7	23·7	40·0	95·4	35·1	24·3	39·5	18·1	25·0	20·7	19·0	32·3	37·5
Diarrhœa	4·2	6·2	23·2	43·0	3·0	5·0	7·0	4·2	5·1	13·1	8·8	9·7	7·5
Hepatic { Abscess	...	...	...	...	·1	...	·1	...	·3	...	...	...	·2
{ Congestion and Inflammation	·8	·3	1·1	·4	·9	·7	·2	·9	·9	...	·5	·7	·6
Scurvy	...	·7	1·1	1·2	·6	·3	1·1	1·5	1·1	7·1	·3	4·2	1·7
Venereal Diseases	10·9	12·3	26·3	19·5	13·9	19·6	14·7	18·0	24·3	38·4	30·5	24·3	19·6
All Causes	625·1	539·4	751·3	877·2	505·5	512·2	664·0	588·2	444·6	606·2	551·7	729·3	607·1
IV.—DEATH-RATE OF THE YEAR—													
Cholera	...	...	...	...	·57	...	...	...	...	...	·34	·05	·06
Small-pox	...	...	...	...	...	...	·06	...	...	·17	...	·01
Enteric Fever	...	...	...	...	·14	...	·06	·29	·31	·51	...	·45	·28
Intermittent Fever	...	...	...	·78	·28	·78	...	·28	·51	·51	·50	·45	·27
Remittent Fever	·84	·69	1·05	·78	·14	·78	·50	·44	·13	·51	...	·84	·56
Simple Continued Fever	...	...	...	...	...	·06	...	...	...	...	...	...	...
Circulatory Diseases	...	...	...	...	·28	·06	·17	·22	·25	2·53	·50	·30	·26
Tubercle of the lungs	...	·34	...	...	·57	·36	·67	·51	·13	1·01	...	1·29	·50
Pneumonia	·84	·69	1·05	1·17	·99	2·76	3·23	1·98	·94	3·54	1·01	2·37	1·90
Other Respiratory Diseases	...	·34	...	...	·43	·18	·28	·37	·06	·51	...	·35	·23
Dysentery	...	·69	1·05	·39	·14	1·2	·06	·22	...	...	·34	·40	·18
Diarrhœa	...	...	·05	...	·06	...	·07	...	1·01	...	·06	·05	·06
Hepatic Abscess	...	...	...	...	...	...	...	...	·06	...	...	...	·01
Anæmia and Debility	·84	·34	...	·39	...	...	·15	...	·13	1·01	...	...	·09
All Causes	4·18	5·14	4·21	5·08	5·96	6·97	7·02	6·89	4·19	17·69	5·87	18·05	8·09
V.—PERCENTAGE IN 100 ADMISSIONS—													
Influenza	·67	...	...	1·29	·22	·52	·60	·17	·04	...	·09	·05	·25
Cholera	...	...	...	...	·11	...	...	·04	...	·09	·01	·02
Small-pox	...	...	...	...	...	·08	·11	·24	·30	·67	·03	·01	·10
Enteric Fever	...	...	...	...	·06	·57	·11	·12	·25	·08	·03	·22	·17
Intermittent Fever	26·91	29·56	28·47	40·33	23·63	21·75	25·72	36·00	16·47	31·78	22·14	23·35	28·11
Remittent Fever	·80	1·59	·28	2·81	1·82	1·72	1·49	·57	·53	·25	15	1·58	1·38
Simple Continued Fever	8·43	·83	...	2·41	2·58	1·03	1·06	1·10	4·65	1·75	10·39	3·21	2·36
Tubercle of the lungs	·40	·51	·14	·31	·67	·93	·68	·55	·13	·58	·30	·63	·51
Pneumonia	·40	·32	·70	1·25	1·04	3·00	3·09	2·49	1·48	2·25	·91	1·88	2·07
Other Respiratory Diseases	3·48	4·32	5·05	3·21	4·30	5·36	4·36	4·12	4·45	3·17	2·86	6·24	4·70
Dysentery	8·43	4·39	5·33	10·87	6·13	4·74	5·91	3·08	5·63	3·41	3·61	4·43	5·30
Diarrhœa	·67	1·14	3·09	4·90	·59	1·10	1·05	·72	1·14	2·17	·52	1·33	1·24
Hepatic { Abscess	...	...	...	...	·03	...	·01	·01	...	...	...	·03	·03
{ Congestion and Inflammation	·13	·06	·14	·04	·17	·13	·03	·15	·20	...	·09	·10	·10
Scurvy	...	·13	·14	·13	·11	·16	·17	·26	·24	1·17	·03	·57	·28
Venereal Diseases	1·74	2·29	3·51	2·23	2·75	3·83	2·22	3·06	5·48	6·34	5·53	3·33	3·23
VI.—PERCENTAGE IN 100 DEATHS—													
Cholera	...	...	...	...	9·5	...	...	...	...	...	5·7	·3	·7
Small-pox	...	...	...	...	...	...	·8	...	...	2·9	...	·1
Enteric Fever	...	...	...	...	2·4	11·2	4·3	7·5	2·9	...	2·5	3·5
Intermittent Fever	...	6·7	...	15·4	4·8	...	4·0	7·4	4·5	2·9	8·6	3·5	4·6
Remittent Fever	20·0	13·3	25·0	15·4	2·4	11·2	7·1	6·4	3·0	2·9	...	4·5	6·9
Simple Continued Fever	...	...	...	...	...	...	...	...	...	...	...	...	...
Circulatory Diseases	...	...	...	...	4·8	·9	2·4	3·2	6·0	14·3	8·6	1·6	3·2
Tubercle of the lungs	...	6·7	...	...	9·5	5·2	9·5	7·4	3·0	5·7	...	7·1	6·1
Pneumonia	20·0	13·3	25·0	23·1	16·7	39·7	46·8	28·7	22·4	20·0	17·1	13·2	23·5
Other Respiratory Diseases	...	6·7	...	...	7·1	2·6	4·0	5·3	1·5	2·9	...	1·9	2·8
Dysentery	...	...	...	7·7	2·4	1·7	·8	3·2	...	...	5·7	2·2	2·2
Diarrhœa	...	...	25·0	...	·9	...	1·1	...	...	...	...	·3	·7
Hepatic Abscess	...	...	...	...	...	...	...	...	1·5	...	...	...	·1
Anæmia and Debility	20·0	6·7	...	7·7	...	...	2·1	...	3·0	5·7	...	...	1·1

* See foot-note to Table XXVI.
†　　"　　"　　"　　"

54

TABLE XXVIII.

RATIOS of STATIONS, GROUPS, and COMMANDS.   For actuals see Table XXIX.

STATIONS AND GROUPS.	Average annual strength.	Influenza.	Cholera.	Small-pox.	Enteric Fever.	Intermittent Fever.	Remittent Fever.	Simple Continued Fever.	Circulatory Diseases.	Tubercle of the lungs.	Pneumonia.	Other Respiratory Diseases.	Dysentery.	Diarrhœa.	Hepatic Abscess.	Hepatic Congestion and Inflammation.	Scurvy.	Anæmia and Debility.	Venereal Diseases.	ALL CAUSES.	CONSTANTLY SICK.	Syphilis.	Soft Chancre.	Gonorrhœa.
		1. ADMISSION-RATE.																			2. DEATH-RATE.			
Port Blair	279	16·7 ...	... ...	... ...	... ...	605·4 ...	3·3 ...	13·4 ...	...	3·3 ...	3·3 3·34	26·8 ...	61·2 ...	... ...	... ...	3·3 ...	... ...	10·0 ...	6·7 ...	1006·7 3·34	33·4	6·7 ...	... ...	... ...
Rangoon	896	... ...	... ...	... ...	... ...	22·3 1·12	5·6 ...	65·8 ...	1·1	2·2 ...	2·2 ...	20·1 ...	30·2 ...	5·6 ...	... ...	... ...	... ...	15·6 1·12	12·3 ...	497·8 4·46	21·2	...	8·9	3·3
GROUP I.—BURMA COAST AND BAY ISLANDS.	1,195	4·2 ...	... ...	... ...	... ...	165·2 ·84	5·0 ·84	52·7 ...	·8	2·5 ...	2·5 ·84	21·8 ...	52·7 ...	4·2 ...	...	·8 ...	...	14·2 ·84	10·9 ...	625·1 4·18	24·1	1·7	6·7	2·5
Thayetmyo	19	... ...	... ...	... ...	... ...	52·6 ...	... ...	...	...	...	...	52·6 ...	52·6 ...	... ...	... ...	... ...	...	... ...	...	368·4	11·6	...	...	...
Keng Tung	55	... ...	... ...	... ...	... ...	72·7 ...	... ...	...	54·5	18·2 ...	...	... ...	... ...	... ...	... ...	... ...	...	36·4 ...	254·5	18·2	18·2	...	18·2
Fort Stedman	90	... ...	... ...	... ...	... ...	144·4 ...	11·1 ...	...	...	...	...	... ...	... ...	... ...	... ...	... ...	...	22·2 ...	66·7 ...	1,011·1	77·8	22·2	11·1	33·3
Meiktila	560	... ...	... ...	... ...	... ...	96·5 ...	19·6 ...	1·8	...	1·8 ...	...	25·0 ...	10·7 ...	17·9 ...	... ...	... ...	...	3·6 ...	18·0 ...	407·2 1·79	17·9	3·6	9·0	5·4
Fort Dufferin	1,347	... ...	... ...	... ...	... ...	164·8 ·74	8·2 ·74	3·2	2·2 ·74	2·2 1·48	31·9 ·74	34·1 ...	3·0 ...	... ...	·7 ...	... ...	...	26·7 ...	10·4 ...	597·6 6·68	22·3	3·0	2·2	5·2
Bhamo	845	... ...	... ...	... ...	... ...	202·4 1·18	2·4 1·18	10·7	1·2	1·2 ...	11·8 ...	18·9 ...	4·7 ...	... ...	2·4 1·18	... ...	...	27·2 ...	4·7 ...	506·5 5·92	18·9	1·2	1·2	2·4
GROUP II.—BURMA INLAND.	2,916	... ...	... ...	... ...	... ...	159·5 ·34	8·6 ·69	4·5	...	2·7 ·34	1·7 ·69	23·3 ·34	23·7 ...	6·2 ...	·3 ...	·7 ...	...	21·6 ·34	12·3 ...	539·4 5·14	22·1	3·4	3·4	5·5
Manipur	584	... ...	... ...	... ...	... ...	287·7 ...	1·7 ...	...	...	1·7 ...	53·1 ...	18·8 ...	18·9 ...	1·7 ...	1·7 ...	10·3 ...	39·4	98·1 1·71	...	981·2 1·71	32·5	6·3	15·4	17·1
Sadiya	69	... ...	... ...	... ...	... ...	130·4 ...	... ...	...	...	14·5 14·49	14·5 ...	72·5 ...	... ...	... ...	... ...	43·5 ...	...	420·3 14·49	...	420·3 14·49	14·5	...	...	...
Dibrugarh	296	... ...	... ...	... ...	... ...	87·8 3·38	3·4 ...	...	3·4	10·1 ...	13·5 ...	74·3 ...	37·2 3·38	... ...	... ...	3·4 ...	6·8	375·0 6·76	...	375·0 6·76	13·5	6·8	...	...
GROUP III.—ASSAM.	949	... ...	... ...	... ...	... ...	213·9 1·05	2·1 1·05	...	1·1	5·3 1·05	37·9 ...	40·0 ...	23·2 1·05	... ...	1·1 ...	1·1 ...	10·5	26·3 4·21	...	751·3 4·21	25·7	6·3	9·5	10·5
Fort William	618	40·3 ...	... ...	... ...	... ...	661·8 3·24	87·4 ...	19·4	...	9·7 1·62	45·3 ...	131·1 ...	93·9 ...	... ...	1·6 ...	12·9 1·62	17·8	1,401·3 11·33	...	1,401·3 11·33	43·7	11·3	4·9	1·6
Alipore	772	5·2 ...	... ...	... ...	... ...	202·7 1·30	1·3 1·30	51·8	2·6	9·1 1·30	16·8 ...	58·3 ...	33·4 ...	... ...	2·6 ...	6·5 ...	31·1	790·2 3·89	...	790·2 3·89	27·2	18·1	5·2	7·8
Barrackpore	586	... ...	... ...	... ...	... ...	307·2 1·71	10·2 ...	1·7	1·7	18·8 ...	22·2 ...	177·5 ...	30·7 ...	1·7 ...	... ...	41·0 ...	8·5	8259·5 1·71	...	8259·5 1·71	17·1	...	1·7	6·8
Buxa	264	... ...	... ...	... ...	... ...	219·7 ...	7·6 ...	...	15·2	15·2 ...	30·3 ...	26·5 ...	26·5 ...	... ...	... ...	41·7 ...	3·8	643·9 3·79	...	643·9 3·79	26·5	...	3·8	...
Cuttack	318	... ...	... ...	... ...	... ...	100·0 1·14	... ...	3·1	...	...	31·4 ...	22·0 ...	6·3 ...	... ...	... ...	28·3 ...	...	355·5 3·14	...	355·5 3·14	12·6	9·4	9·4	9·4
GROUP IV.—BENGAL AND ORISSA.	2,558	11·3 ...	... ...	... ...	... ...	353·8 ·76	24·6 ·78	21·1	...	2·7 1·17	10·9 ...	28·1 ·39	95·4 ...	43·0 ...	·4 ...	1·2 ·39	18·8	19·5 ...	877·2 5·08	26·7	9·4	4·7	5·5

* Worked on the aggregates.

55

TABLE XXVIII—continued.

RATIOS of STATIONS, GROUPS, and COMMANDS.   For actuals see Table XXIX

STATIONS AND GROUPS.	Average annual strength.	1. ADMISSION-RATE.														2. DEATH-RATE.									
		Influenza.	Cholera.	Small-pox.	Enteric Fever.	Intermittent Fever.	Remittent Fever.	Simple Continued Fever.	Circulatory Diseases.	Tubercle of the Lungs.	Pneumonia.	Other Respiratory Diseases.	Dysentery.	Diarrhœa.	Hepatic Abscess.	Hepatic Congestion and Inflammation.	Scurvy.	Anæmia and Debility.	Venereal Diseases.	ALL CAUSES.	CONSTANTLY SICK.	Syphilis.	Soft Chancre.	Gonorrhœa.
A. Deranda	403	2·5	...	...	...	235·7	...	...	7·4 2·48	5·0	...	39·7	9·9	2·5	...	2·5	2·5	9·9	533·5 9·93	27·3	7·4	2·5	...
B. Dinapore	750	...	2·7 2·67	...	7·7 1·33	69·3	9·5 1·33	37·3	4·0 1·33	9·3	12·0	52·0	...	...	...	1·3	5·3	13·7	410·7 8·00	17·3	8·0	4·0	6·7
Benares	714	...	...	...	...	68·6	5·6	25·2	1·4	1·4 1·40	23·8 1·40	12·6	...	...	1·4	...	2·8	4·2	395·0 7·00	12·6	2·8	1·4	...
Allahabad	1,199	·8	1·7 1·67	...	...	159·3	2·5	...	1·7	4·2 3·34	8·3 ·83	27·5 ·83	67·6 ·83	4·2	...	2·5	...	21·7	13·3	648·9 9·17	20·0	3·8	4·2	3·3
Fysabad	1,136	...	...	...	...	154·0	37·9	39·6 ·88	·9	3·5 ·82	·0	44·0	21·1	...	...	·9 ·8	7·0	20·2	743·8 3·52	22·9	7·9	10·6	1·8
Lucknow	1,632	...	...	...	...	87·0 ·61	3·7	...	·6 1·23	3·1	4·9	14·1 ·61	20·3	4·9	...	·6	7·4	11·0	316·2 3·68	12·3	4·3	2·5	4·3
Cawnpore	1,140	5·3	...	...	...	103·5 ·8	1·8	·9	·9 ·88	2·6	5·3 ·88	14·9	35·8	3·5	...	·9	6·1	16·7	479·3 4·39	19·3	8·8	1·8	6·1
Fatehgarh	72	...	...	...	...	305·6	...	...	...	13·9	41·7	41·7	...	...	...	...	13·0	152·8 13·8	27·8	...	...	13·9
GROUP V.—GANGETIC PLAIN AND CHUTIA NAGPUR.	7,046	1·1	·6 ·57	...	·3 ·14	119·8 ·28	9·2 ·14	15·1	·0 2·5	3·4 ·57	5·3 ·09	21·7 ·43	35·1 ·14	3·0	·1	·9	·6 ·14	8·5	13·9	505·5 5·96	17·8	6·2	4·0	3·7
A. Bareilly	1,074	·9	...	...	...	50·3	·6	...	1·9	9·4 2·79	20·5	21·4	3·7	...	·6	...	5·6	·7	211·4 4·66	9·3	·9	2·8	·9
Roorkee	776	...	...	...	...	23·2	3·9	...	...	2·6 2·55	16·5	45·1	3·9	...	...	...	5·2	12·9	234·5 3·58	10·3	2·6	6·4	3·9
Dehra Dun	2,848	3·5	...	...	...	14·0 3·86	243·7	16·2 ·70	4·9	4·6 ·70	9·1 1·05	21·0 ·70	23·2	7·4	...	·7	·4 ·35	36·2	4·0	733·3 11·59	37·2	9·5 ·35	16·9	9·8
Meerut	839	...	...	...	1·1 1·19	47·7	13·1 2·38	...	...	1·2	10·7	13·1	17·9	4·8	...	1·2	...	13·1 1·19	363·5 4·77	15·5	6·0 1·19	2·4	4·8
Delhi	791	...	...	...	...	183·3	11·4	...	...	72·8 5·06	40·3 1·26	22·8	2·5	...	1·3	5·1	11·4	574·0 6·32	22·8	6·3	...	5·1
Ambala	1,032	3·9	1·9	1·0	...	52·3	1·0	...	4·8	5·5 ·97	20·1	20·2	4·8	...	1·0	...	3·6	26·2	357·6 2·91	16·5	15·5	9·7	1·0
B. Jullundur	1,319	...	...	...	...	54·6	5·3	·8	·8	3·8 1·52	13·6	24·3	25·0	4·5	...	1·5	15·9	9·1	413·2 1·52	18·2	2·3	...	6·8
Ferozepore	1,316	15·2	...	·8	1·5	83·1	18·2 3·04	...	·8	4·6 4·56	14·4	21·3	27·4	·8	...	·8	7·6	15·2	513·7 9·88	25·1	4·6	3·8	6·8
Mian Mir	1,555	...	...	1·9	1·3 ·64	83·0	7·1	7·1	·6 ·64	6·4 3·86	19·3	36·0	32·2	5·1	·6	...	·6	10·3	10·3	463·7 7·07	19·3	5·1	3·9	1·3
Amritsar	134	...	...	...	...	52·2	7·5	...	...	52·2 7·46	14·9	14·9	7·5	...	...	...	29·9	...	388·1 14·93	29·9	...	...	...
Sialkot	1,491	...	...	...	·7	161·6	4·7 ·67	2·0	...	4·7 2·68	8·0	26·2	26·2	8·0	...	1·3	1·3	18·1	11·4	614·4 3·35	22·1	7·4	...	4·0
Jhelum	1,305	6·1	...	...	...	115·0	4·6	26·1	...	14·6 1·53	26·8 2·30	33·0	29·9	11·5	...	·8	16·9	26·1	625·0 6·00	22·2	10·7	4·6	10·7
Rawalpindi	2,684	·5	...	·5	1·0	56·1	9·6 1·92	12·0	1·0	5·3 ·96	25·0 5·28	28·3	25·4	6·7 ·48	...	1·4	2·9	10·6	30·2	544·1 10·56	25·4	9·6	7·7	13·1
Attock	85	...	...	...	...	164·7	...	...	...	...	11·8	...	11·8	...	...	...	...	...	458·8 ...	11·6	...	...	...
GROUP VI.— UPPER SUB-HIMALAYA.	16,649	2·6	...	·4	2·9 ·75	111·4	8·8 ·78	5·3	·5 ·0	4·7 ·36	15·4 2·76	27·4 ·18	24·3 ·17	5·0 ·0	·1	·7	·8	9·2	19·6 ·12	512·2 6·97	22·8	7·1 ·17	6·1	6·5
A Mardan	923	41·2	...	...	...	68·8 2·17	1·1	...	2·2	5·4 1·08	18·4 2·17	14·1 1·08	22·8	1·1	...	...	15·2	16·3	455·0 6·50	19·5	4·3	5·4	6·5
Nowshera	2,836	...	...	·4	...	113·7 ·35	14·5 ·35	...	1·1	2·5 7·1	19·3 2·12	42·7 ·71	35·3	4·2	...	...	19·7	4·6	556·4 5·98	22·9	2·3	1·4	·7
Peshawar	3,593	1·0	...	...	3·1 ·39	103·0	2·7 ·39	16·6	...	3·1 1·54	19·7	19·7	39·7 ·39	5·4	·4	·8	14·3	19·3	480·9 3·03	17·7	6·2	7·3	5·8

Stations and Groups.	Average annual strength.	Influenza.	Cholera.	Smallpox.	Enteric Fever.	Intermittent Fever.	Remittent Fever.	Simple Continued Fever.	Circulatory Diseases.	Tubercle of the lungs.	Pneumonia.	Other Respiratory Diseases.	Dysentery.	Diarrhœa.	Hepatic Abscess.	Hepatic Congestion and Inflammation.	Scurvy.	Anæmia and Debility.	Venereal Diseases.	All Causes.	Constantly Sick.	Syphilis.	Soft Chancre.	Gonorrhœa.
		1. Admission-rate.													**2. Death-rate.**									
Fort Jamrud	108 {	...	...	...	...	263·5	18·5	...	...	...	9·3	117·'	222·2	...	...	...	...	...	...	1,492·6 }	18·5	...	...	...
		...	...	...	...	...	...	...	...	...	7·7	2·68	·38	...	...	...	...	...	...					
Kohat	2,608 {	·4	...	1·2	·8	210·1	16·1	2·3	3·8	6·9	33·7	31·1	31·4	4·2	...	·8	3·8	27·6	12·7	768·0 }	33·4	4·2	2·7	5·8
		...	...	...	...	...	·38	...	...	7·7	2·68	·38	...	...	...	...	...	...	...	5·37 }				
Thal	207 {	...	...	...	...	478·3	...	...	...	...	9·7	96·6	62·8	4·8	...	...	4·8	24·2	14·5	1,009·7 }	24·2	...	...	14·5
		...	...	...	...	...	...	...	...	...	4·83	...	...	...	...	...	...	...	...	4·83 }				
Edwardesabad	1,961 {	·5	...	...	1·0	155·0	13·3	2·0	...	3·6	22·4	30·1	41·8	11·7	...	·5	17·8	18·9		798·1 }	29·6	3·1	9·2	6·6
		...	...	...	...	...	...	...	...	·51	3·57	·51	...	...	...	...	...	...	...	6·63 }				
Dera Ismail Khan	2,300 {	...	...	2·6	...	232·6	12·6	7·8	1·7	3·2	30·9	23·5	61·3	15·7	...	1·3	19·6	11·3		967·0 }	34·8	7·4	·9	3·0
		...	...	...	...	·43	2·17	...	·87	·43	9·57	...	...	...	...	...	...	...	...	6 52 }				
Jatta	61 {	...	...	...	...	213·1	...	...	16·4	...	31·8	32·5	32·8	...	...	...	...	32·8		672 1 }	16·4	16·4	...	16·4
		...	...	...	...	...	...	...	...	...	...	...	...	...	...	...	...	...	...	16·39 }				
Drazand	57 {	...	...	...	...	1,228·1	52·6	...	...	35·1	17·5	210·5	...	...	...	...	35·1			2,122·8 }	70·2	35·1	...	...
		...	...	...	...	...	...	...	...	...	...	...	...	...	...	...	...	...	...	17 54 }				
Fort Zam	51 {	...	...	...	...	588·2	78·4	...	...	...	117·6	39·2	...	...	...	...	...	...		1,196·1 }	19·6	...	...	...
		...	...	...	...	...	...	...	...	...	...	...	...	...	...	...	...	...	...					
Dera Ghazi Khan	630 {	...	...	...	...	106·3	4·8	...	...	4·8	11·1	23·8	41·3	3·2	...	1·6	4·8	7·9		3·6 81 }	15·9	4·8	...	3·2
		...	...	...	...	...	...	...	1·59	1·59	...	...	...	...	...	1·59	...	...	...	9·52 }				
Multan	1,406 {	...	...	·7	...	114·5	7·1	...	·7	12·1	13·5	22·0	14·9	2·1	...	...	12·8	16·4		466·6 }	19·2	3·6	8·5	4·3
		...	...	·1	...	·1	·71	...	·71	1·42	2·84	...	...	...	...	...	...	...	...	9·25 }				
Bikaner	30 {	...	...	...	...	...	...	...	...	...	...	...	...	...	...	...	...	...		}	...	...	...	...
		...	...	...	...	...	...	...	...	...	...	...	...	...	...	...	...	...	...					
B. Jandola	191 {	...	...	...	...	62·8	25·2	...	...	...	15·7	47·1	15·7	...	...	...	5·2	20·9		506·4 }	15·7	20·9	...	...
		...	...	...	...	...	...	...	...	...	...	...	...	...	...	...	...	...	...					
Sibi	277 {	...	...	...	...	711·2	...	...	...	38·9	3·6	83·0	10·8	...	...	3·6	32·5	21·7		1,314·1 }	28·9	...	...	21·7
		...	...	...	...	...	...	...	...	7·22	...	...	...	...	...	...	...	...	...	7·22 }				
C. Jacobabad	418 {	...	...	2·4	...	177·0	2·4	...	...	7·2	16·7	38·3	45·1	12·0	...	...	26·3	23·9		624·4 }	33·5	2·4	4·4	19·1
		...	...	...	...	...	...	...	...	4·78	...	...	...	...	...	...	...	...	...	7·18 }				
Hyderabad	703 {	...	...	1·4	...	83·9	...	...	...	2·8	7·1	13·5	15·6	1·4	...	...	...	32·7		361·3 }	15·6	2·8	15·6	14·2
		...	...	...	...	...	...	...	...	1·42	...	...	...	...	...	...	...	...	...	1·42 }				
Kurrachee	586 {	44·4	...	1·7	...	139·9	—	...	...	10·2	15·4	46·1	23·9	10·2	...	3·4	13·7	20·5		737·2 }	23·9	8·5	1·7	10·2
		...	...	...	...	...	...	...	3·41	...	...	...	...	...	...	...	...	...	...	3·41 }				
Group VII.—N.-W. Frontier, Indus Valley, and North-Western Rajputana.	17,946 {	4·0	...	·7	·7	170·8	9·9	7·0	1·2	4·5	20·5	29·0	39·5	7·0	·1	·3	1·1	17·5	14·7	664·0 }	23·3	4·7	4·5	5·6
		...	...	...	·06	·28	·50	...	·17	·67	3·29	·28	·96	...	...	...	·06	...	...	7·02 }				

TABLE XXVIII—*continued.*

RATIOS of STATIONS, GROUPS, and COMMANDS.   For actuals see Table XXIX.

Stations and Groups	Average annual strength	Influenza	Cholera	Small-pox	Enteric Fever	Intermittent Fever	Remittent Fever	Simple Continued Fever	Circulatory Diseases	Tubercle of the lungs	Pneumonia	Other Respiratory Diseases	Dysentery	Diarrhœa	Hepatic Abscess	Hepatic Congestion and Inflammation	Scurvy	Anæmia and Debility	Venereal Diseases	All Causes	Constantly Sick	Syphilis	Soft Chancre	Gonorrhœa
																1. ADMISSION-RATE.					2. DEATH-RATE.			
A.																								
Bhuj	163	…	…	…	…	116·6	…	…	…	6·1	…	67·5 / 6·13	6·1	…	…	…	…	30·7	12·3	67·5	5·67 / 0·13	24·5	30·7	42·9	…
Rajkot	746	…	…	…	…	140·8 / 1·34	12·1	…	…	4·0 / 2·68	22·8	18·8	5·4	…	…	1·3	5·4	20·1	30·8	528·2 / 12·08		22·8	2·7 / 1·34	12·1	16·1
Dessa	1,076	…	…	·9	…	245·4 / 2·79	·9	·9	…	1·9 / 3·72	17·7	15·8	13·0	…	…	·9	2·8 / ·93	13·0	13·9	652·4 / 10·22		22·3	7·4	3·7	2·8
Ahmedabad	677	…	…	…	5·9 / 1·48	1·5	636·7	…	…	4·4 / 1·48	14·8 / 2·05	65·0	47·3	3·0	…	…	1·5	7·4	19·2	1,214·2 / 5·91	38·2	11·8	4·4	3·0
Baroda	708	…	…	…	…	1,319·2 / 1·41	15·5 / 1·41	…	…	2·8 / 1·41	9·9	26·8	31·1	5·6	…	…	4·2	1·4	39·5	1,665·3 / 7·00	41·0	25·4	4·2	9·9
B.																								
Alirajpore	36	…	…	…	…	…	…	…	…	…	…	…	…	…	…	…	…	27·8	83·3 / 27·78	…	2·1	27·8	…	…
Sirdarpore	386	…	…	10·4	…	80·3 / 2·59	…	…	2·59	5·2 / 5·18	33·7	20·7 / 2·59	2·6	…	…	…	…	20·7	489·6 / 12·95		23·3	13·0	…	7·8
Jhabwa	37	…	…	27·0	…	135·1	…	…	…	…	…	…	…	…	…	…	…	…	216·2		6·6	…	…	…
Kherwara	340	…	…	2·9	…	83·4	…	…	…	8·8 / 2·94	51·2 / 11·76	82·4	17·0 / 2·94	2·9	…	…	…	14·7	20·6	911·8 / 26·47	50·0	11·8	2·9	5·9
Kotra	150	…	…	…	…	133·3	…	…	…	46·7 / 6·67	20·0	6·7 / 6·67	…	…	…	…	…	…	500·7 / 20·00		26·7	…	…	…
Udaipur	43	…	…	…	…	46·5	…	…	…	…	…	…	…	…	…	…	…	…	209·3		7·8	…	…	…
Todgarh	32	…	…	…	…	62·5	…	31·2 / 31·25	62·5 / 31·25	…	…	31·2	62·5	…	…	…	31·2 / 31·25	…	343·8 / 125·00		62·5	…	…	…
Erinpura	569	…	…	…	…	107·2 / 1·76	8·8	…	…	1·8 / 1·76	21·1	8·8	19·3	10·5	…	1·8	…	1·8	19·3	630·9 / 5·27	24·6	8·8	5·3	5·3
Neemuch	364	…	…	…	5·5 / 5·49	27·5	…	…	…	8·2 / 2·75	11·0	…	…	…	…	…	13·7	8·2	236·3 / 8·24		11·0	…	…	8·2
Deoli	578	…	…	…	…	51·9	12·1	3·5	…	1·7 / 1·73	8·7	22·5	12·1	…	…	1·7	…	1·7	13·8	430·8 / 1·73	19·0	5·2	5·2	3·5
Tonk	28	…	…	…	…	…	…	…	…	…	…	…	…	…	…	…	…	…	…	…	…	…	…
Beawar	46	…	…	…	…	…	…	…	…	65·2	…	21·7 / 21·74	…	…	…	…	…	21·7	130·4 / 21·74	21·7	…	21·7	…
Nasirabad	688	…	…	…	…	66·9 / 1·45	2·9 / 1·45	5·8	1·3 / 1·45	2·9 / 2·91	16·0	24·7	2·9	2·9	…	1·5	…	8·7	14·5	348·8 / 13·08	14·5	2·9	7·3	4·4
Ajmer	446	9·0	…	…	…	49·3	2·2	…	…	4·5	26·9	9·0	2·2	2·3	…	…	4·5	…	17·9	365·5		20·2	4·5	2·2	11·2
Sambhar	21	…	…	…	…	…	…	…	…	…	…	…	…	…	…	…	…	…	…	…	…	…	…
Jaipur	40	…	…	…	…	75·0	…	…	…	100·0	25·0	…	…	…	…	…	…	…	575·0		2·1	…	…	…
Agra	694	…	…	…	…	72·0 / 1·44	1·4	…	…	7·2 / 1·44	18·7	31·7	5·8	…	…	2·9	…	…	25·9	311·3 / 1·44	13·0	15·9	8·6	1·4

STATIONS AND GROUPS.	Average annual strength.	Influenza.	Cholera.	Small-pox.	Enteric Fever.	Intermittent Fever.	Remittent Fever.	Simple Continued Fever.	Circulatory Diseases.	Tubercle of the lungs.	Pneumonia.	Other Respiratory Diseases.	Dysentery.	Diarrhœa.	Hepatic Abscess.	Hepatic Congestion and Inflammation.	Scurvy.	Anæmia and Debility.	Venereal Diseases.	ALL CAUSES.	CONSTANTLY SICK.	Syphilis.	Soft Chancre.	Gonorrhœa.
Gwalior	30	...	...	...	...	...	33'3	...	...	...	...	...	...	...	...	...	...	...	...	133'3	11'6	...	...	...
Jhansi	1,926	1'0	1'6	...	1'6 '53	140'7	2'6	22'3	...	6'2 '52	5'7 '52	11'4 '52	30'1	2'1	...	...	...	6'7 ...	24'4	536'9 2'08	19'7	10'9	4'2	9'3
Nowgong	1,159	6'9	...	...	...	101'5 '86	2'6	12'9	...	1'7 '86	7'8 '86	7'8 ...	7'8	...	...	'9	...	3'5 ...	8'6	350'3 3'45	13'8	4'3	3'5	'9
Goona	415	...	...	...	...	26'5 ...	...	...	...	9'6 2'41	2'4 ...	...	...	...	...	...	...	4'8	113'3 2'41	4'8	4'8	...	...
Agar	314	...	...	...	...	35'0 ...	15'9 3'18	...	...	12'7	3'2	25'5	...	...	...	...	3'2	3'2	289'8 3'18	12'7	3'2	...	...
Sehore	711	...	...	2'8	...	225'0 ...	4'2 1'41	...	...	2'8	5'6	63'3	21'1	2'8	...	...	19'7 1'41	7'0	611'8 5'63	18'3	4'2	...	2'8
Indore	162	...	...	...	...	104'9	...	12'3	...	18'5 6'17	18'5	18'5	6'2	...	...	...	...	12'3	413'6 6'17	12'3	...	...	12'3
Mhow	1,066	...	...	5'6	3'8	226'1 '94	'9 '94	15'9	...	1'9 '94	11'3 2'81	34'7 '94	34'7 '94	26'3	...	4'7	1'9	6'6	13'1	264'5 8'44	23'5	3'8	3'8	5'6
GROUP VIII.— SOUTH-EASTERN RAJPUTANA, CENTRAL INDIA, AND GUJARAT	13,651	1'0 ...	'2 ...	1'4	'7 '29	211'8 '51	3'4 '44	6'8	'1 '22	3'2 '51	14'7 1'98	24'2 '37	15'1 '24	4'2 '07	...	'9 '07	1'5 '15	7'0 '07	18'0 '689	581'2	21'4	8'3 '07	4'0	5'8

A.

Asirgarh	10	...	...	...	...	...	...	...	...	...	...	...	...	...	...	...	...	...	...	...	...	...	...	...
Saugor	1,013	2'0 ...	...	—	1'0 ...	111'6	12'8 '09	38'5	1'0	1'0 1'97	8'9	21'7	7'9	4'9	...	2'0	9'9	17'8	533'1 5'93	20'7	8'9	5'9	3'0
Sutna	33	...	...	...	...	181'8	30'3 ...	...	...	60'6 30'30	30'3	30'3	...	...	...	...	60'6	697'0 30'30	30'3	60'6	...	...
Jubbulpore	930	...	...	1'1	2'2	95'7 1'08	2'2 ...	...	2'2	6'5	23'7	43'0	1'1	...	2'2	5'4 1'08	22'6	516'1 5'38	24'7	97	1'1	11'8
Kamptee	603	...	...	3'3	...	126'0	1'7	...	...	3'3	26'5	21'6	5'0	...	...	3'3	11'6	462'3 1'66	21'6	6'6	...	5'0
Sitabaldi	82	...	...	36'6	...	353'7	...	...	...	...	73'2	61'0	12'2	...	...	...	...	951'2	12'2	...	...	...

B.

Ellichpur	23	...	...	...	...	173'9	...	...	...	43'5	...	...	...	...	...	...	43'5	521'7	18'4	...	...	...
Aurangabad	1,738	...	...	2'3	'6	157'7	1'2 '58	1'2	2'3	...	7'8 '58	11'5	16'7	1'7	...	...	4'0	15'0	463'8 2'30	19'6	4'6	2'3	8'1
Ahmednagar	575	...	...	...	...	43'5	...	1'7 1'74	1'7 1'74	1'7	5'2	7'0	7'0	...	...	...	5'2	5'2	207'0 5'22	10'4	3'5	...	1'7
Bolarum	1,712	...	...	1'2	4'7 '58	13'4	27'1 1'17	1'2 '58	1'2 '58	3'5	21'6	38'0	'6	...	1'2	'6	18'1	29'8 '5	447'4 5'84	21'0	19'3 '58	1'2	9'3
Secunderabad	3,270	...	...	1'2 '92	37'0	4'6	9'8	1'5 '31	'3	7'6 1'53	8'6 '3	35'5	2'1	...	2'1	9'3	19'6 '31	322'6 4'59	13'1	8'0 '31	3'7	8'0
Belgaum	1,176	...	...	2'6 '85	53'6	...	17'0	...	...	11'1 1'70	14'5	17'0	5'1	...	1'7	6'0	34'9	506'8 3'40	16'2	9'4	6'8	18'7
Satara	615	...	...	1'6	73'2	1'6	...	1'6	3'3	1'6	8'1	29'3	1'6	...	1'6	1'6	45'5	590'2	27'6	16'3	19'5	9'8

TABLE XXVIII—*continued.*

RATIOS of STATIONS, GROUPS, and COMMANDS.   For actuals see Table XXIX.

Stations and Groups.	Average annual strength.	1. Admission-Rate.																		2. Death-Rate.				
		Influenza.	Cholera.	Small-pox.	Enteric Fever.	Intermittent Fever.	Remittent Fever.	Simple Continued Fever.	Circulatory Diseases.	Tubercle of the lungs.	Pneumonia.	Other Respiratory Diseases.	Dysentery.	Diarrhœa.	Hepatic Abscess.	Hepatic Congestion and Inflammation.	Scurvy.	Anæmia and Debility.	Venereal Diseases.	All Causes.	Constantly Sick.	Syphilis.	Soft Chancre.	Gonorrhœa.
Poona	2,057	·3	...	1·0	...	53·3 ·49	1·5	18·0	·5	...	5·3 ·49	15·1	26·3	9·7	·5 ·49	·5	1·0	5·8 ·49	34·0	401·5 7·29	19·0	19·4	7·8	6·8
Kirkee	1,600	...	...	3·8	...	10·16 ·62	...	41·9	2·5	...	7·5 1·2	68·1	13·1	16·9	·6	1·2	·6	11·9	31·2	614·4 1·88	16·5	11·9	11·2	8·1
Sirur	540	...	...	1·9	...	61·1	...	...	...	...	1·0	11·1	9·3	3·7	1·9	9·3	1·9	13·0	14·8	320·4	16·7	...	1·9	13·0
Group IX.—Deccan.	15,977	·2	...	1·3	1·1 ·31	73·3 ·19	2·4 ·13	20·7	1·2 ·25	·6 ·13	6·6 ·94	19·9 ·06	25·0	5·1	·3 ·06	·9	1·1	8·5 ·13	24·5 ·13	446·6 4·19	18·1	10·8 ·13	5·0	8·5
Bombay	720	...	...	9·7 1·90	1·4	376·4	2·7	27·ʼ	2·7 2·7	8·3 2·78	25·ʼ 5·36	34·7 1·30	36·ʼ	35·0 2·78	...	...	15·3	17·5 2·7	50·0	002·2 25·00	40·3	20·4	9·3	15·3
Santa Cruz	647	...	...	1·5	...	98·0	...	1·5	3·0	...	13·9 3·09	9·3	21·6	10·8	...	...	4·6	6·2 1·55	29·4	370·2 17·00	18·5 1·55	12·4	15·5	1·5
Cannanore	540	...	...	...	...	67·7 1·85	...	...	11·0 1·55	1·9	1·9 1·85	13·0	1·9	1·9	...	...	...	7·4	38·9	364·8 9·26	29·6	14·8	5·6	18·5
Trivandrum	71	...	...	...	...	140·8	14·1	...	...	...	...	...	...	...	...	...	...	...	...	549·31 14·08	14·1	...	...	...
Group X.—Western Coast.	1,978	...	...	4·0	·5 ·51	192·6 ·51	1·5 ·51	10·6	5·1 2·53	3·1 1·01	13·7 3·54	19·2 ·51	20·7	13·1 1·01	...	...	7·1	8·0 1·0	38·4 ·51	606·2 17·69	29·1	17·6 ·51	9·6	11·1
A.																								
Bellary	934	...	...	...	1·1	234·5 1·07	5·4	30·ʼ	6·4 1·07	...	6·4 1·07	16·1	15·0	...	...	...	...	10·7	61·0	704·5 4·28	27·8	31·0	11·8	19·2
Bangalore	2,762	1·1	...	...	1·3	131·8	...	50·5 ·72	2·2	1·4 1·09	7·6	18·8	21·0	2·5 ·35	...	·7	...	8·0	23·5	613·3 4·71	21·0	5·8	2·2	15·6
B.																								
Trichinopoly	636	...	...	1·6 1·57	...	14·7	...	23·6	1·6	...	4·7	4·7	1·6	...	...	...	...	...	23·6	267·3 6·29	12·6	17·3	...	6·3
Pallavaram	8	...	...	...	...	...	...	...	...	...	...	...	...	...	...	...	...	...	...	...	...	...	...	...
St. Thomas' Mount.	501	...	...	...	...	176·6 1·98	...	13·9	...	...	2·0	25·ʼ	45·6	7·9	...	...	2·0	29·8	19·8	732·1 7·94	27·8	6·0	2·0	11·9
Madras	767	...	3·ʼ 2·61	...	...	62·6 1·30	...	41·7	3·9	2·6	2·6 2·01	5·2	16·9 1·30	5·2	...	...	3·9	3·9	24·8	353·3 13·04	16·9	15·6	5·ʼ	3·9
C.																								
Vizianagram	356	...	...	...	...	28·1	5·6	2·8	...	19·7	22·5	2·8	...	2·8	...	...	14·0	44·9	355·ʼ	16·9	16·9	5·6	22·5
Group XI.—Southern India.	5,967	·5	·5 ·34	·2 ·17	·3	122·2 ·50	·8	57·3	2·8 ·50	1·7	5·0 1·0	15·8	19·9 ·34	2·8	·5	2	9·2	30·5	551·7 5·87	21·2	12·9	4·0	13·6
Maymyo	682	...	...	...	1·5	198·0 2·93	5·9	...	...	5·0 4·40	10·3	22·0	70·4	2·9	...	2·0	33·7	74·8	015·1 16·13	44·0	23·5	14·7	36·7
Kohima	153	...	...	...	...	379·1	...	...	...	26·1 6·54	111·1	19·6	13·1	...	...	6·5	52·6	45·8	1,183·0 6·54	39·2	32·7	...	13·1
Shillong	708	...	...	...	...	265·5 1·41	1·4	...	1·4 1·41	1·4	22·0 1·41	7·1	16·7	14·1	...	...	9·9	50·8	766·9 4·24	31·1	21·2	5·6	24·0
Gantok	65	...	...	15·4	...	92·3	30·8	15·4	...	...	15·4	76·9	46·2	...	...	...	15·4	...	...	553·8	0·8	...	...	...
Almora	732	...	...	...	...	39·6 1·37	6·9	71·0	...	2·ʼ	9·6	8·2	15·0	15·0	1·4	1·4	...	9·6	41·0	545·1 2·73	21·9	16·4	13·7	10·9

Stations and Groups	Average annual strength	1. Admission-rate																		2. Death-rate				
		Influenza	Cholera	Small-pox	Enteric Fever	Intermittent Fever	Remittent Fever	Simple Continued Fever	Circulatory Diseases	Tubercle of the lungs	Pneumonia	Other Respiratory Diseases	Dysentery	Diarrhœa	Hepatic Abscess	Hepatic Congestion and Inflammation	Scurvy	Anæmia and Debility	Venereal Diseases	All Causes	Constantly Sick	Syphilis	Soft Chancre	Gonorrhœa
Naini Tal	132	...	...	...	...	197·0	...	...	...	15·2 7·38	7·6 7·58	37·9	15·2	...	7·6	...	...	...	60·0	560·6 22·73	22·7	22·7	15·2	22·7
Lansdowne	2,458	...	...	...	...	89·1 ·51	9·0 2·44	3·7 ·41	1·6 2·03	4·3 ·81	8·5 ·41	30·5 8:	9·4	·4	·8	13·4	6·1	505·3 8·95	25·6	3·7	1·6	·8
Simla	148	...	...	...	...	20·3	...	13·5	...	6·8	37·0	...	13·5	...	...	...	6·8	33·8	207·5	13·5	13·5	20·3	...
Jutogh	160	...	...	...	...	...	...	...	...	...	6·2	25·0	...	18·8 0·25	...	...	...	6·2	306·2 6·25	19·8	...	...	...
Dharmsala	917	...	...	...	...	74·2	...	...	...	24·0 5·45	14·2 1·09	12·0 1·0)	24·0	3·3	...	...	9·8	39·5	832·8 215·92	60·0	20·7	6·5	12·0
Bakloh	1,170	4·3	...	...	...	170·1	38·5 3·42	...	1·7 ·35	7·7	18·8 1·71	21·4	21·9	5·1	...	2·6	4·3 ·85	10·3	47·0 10·26	739·3	41·0	16·2	10·3 ·85	20·
Murree	25	...	...	...	...	120·0	...	...	...	...	...	...	...	...	...	...	40·0	80·0	280·0	6·3	...	...	80·0
Khyragully	93	...	...	...	...	32·3	21·5	...	10·8	10·8	64·5	10·8	10·8	...	...	...	21·5	387·11 10·75	1·0	...	...	...
Baragully	68	...	...	...	...	44·1	...	...	...	...	...	...	14·7	...	...	...	411·8	14·7	...	...	...
Kalabagh	45	...	...	...	...	22·2	...	...	...	22·2 23·22	...	...	...	...	22·2	155·6 22·22	4·0	...	...	...
Chitral	820	...	...	...	...	177·3	4·5	140·9	...	4·5	4·5	13·6	36·	4·5	...	...	4·5	4·5	718·2	18·2	4·5	...	...
Kila Drosh	636	...	...	...	1·6 1·57	143·1 4·72	23·6	261·0	4·7	6·3 1·57	3·1	31·4	70·8	15·7	...	...	11·0	929·2 11·01	35·2	7·9	1·6	1·6
Malakand	779	...	...	...	...	56·5 1·2b	10·3	2·6	1·3 16·60	48·8 6·42	51·3 1·28	16·7	...	...	...	9·0	5·1	364·6 30·81	12·8	1·3	2·6	1·3
Dargai	356	...	...	...	...	82·0	2·7	2·7	...	2·7	16·4 2·73	5·5	5·5	2·7	...	...	2·7	8·2	314·2 2·73	8·2	2·7	2	2·7
Chakdara	556	...	...	...	...	143·1	9·0	...	3·6	1·3	7·2	10·9 1·80	23·4	...	1·8	...	1·8	10·8	9·0	426·3 3·60	9·0	1·8	3·6	3·6
Abbottabad	3,451	...	·3 ·29	...	8·1 2·03	231·8 ·23	19·4 ·58	59·8	...	5·5 2·03	18·0 2·32	106·1 ·20	43·5	12·2	·3	·6	6·3	9·3	40·0 9·27	1,103·7	45·8	11·0	9·0	20·9
Cherat	48	...	...	20·8	...	41·7	...	...	...	20·8	83·3	62·5	20·8	...	...	20·8	...	458·3	20·8	...	...	...
Fort Lockhart	568	...	...	...	...	216·5 1·76	1·8	1·8	...	7·0 1·76	19·4 7·04	28·2	22·9	...	...	1·8	8·8	12·3	498·2 10·56	22·9	5·3	1·8	5·3
Haogu	59	...	...	...	...	576·3	...	...	...	33·9	33·9	...	...	...	...	16·9	966·11	16·9	...	...	...
Mir Ali Khel	115	...	...	...	...	521·7	8·7	17·4 8·70	...	17·4	147·8	60·0	8·7	...	...	34·8	95·7	1,191·3 8·70	34·8	...	...	...
Fort Sandeman	604	1·7	...	...	...	278·1	33·1	...	3·3 1·66	9·9 3·31	13·2 1·65	51·3	26·5	14·9	...	...	8·3	29·8	16·6	307·2 6·62	29·8	6·6	9·9
Hindubagh	30	...	...	...	...	66·7	...	33·2	...	...	66·7	66·7	33·3	...	...	33·3	...	666·7	33·3	...	...	...
Musa Khel	28	...	...	...	...	642·9	35·7	...	...	...	...	214·3	...	...	...	...	1,178·6	35·7	...	...	...
Khan Mohamed Kot	33	...	...	...	...	272·7	30·3	...	...	...	121·2	121·2	151·5	...	...	30·3	121·2	30·3	...	...	30·3
Kilia Saifulla	31	...	...	...	...	351·8	...	...	...	64·5	64·5	129·0	32·3	...	...	32·3	903·2	32·3	...	...	...
Murgha	54	...	...	...	...	55·6	18·5	...	...	37·0	18·5	74·1	...	...	...	37·0	...	370·4	18·5	...	...	...

TABLE XXVIII—continued

RATIOS of STATIONS, GROUPS, and COMMANDS.

For actuals see Table XXIX.

STATIONS AND GROUPS.	Average annual strength.	Influenza.	Cholera.	Small-pox.	Enteric Fever.	Intermittent Fever.	Remittent Fever.	Simple Continued Fever.	Circulatory Diseases.	Tubercle of the lungs.	Pneumonia.	Other Respiratory Diseases.	Dysentery.	Diarrhœa.	Hepatic Abscess.	Hepatic Congestion and Inflammation.	Scurvy.	Anæmia and Debility.	Venereal Diseases.	All Causes.	Constantly Sick.	Syphilis.	Soft Chancre.	Gonorrhœa.
Loralai	856	...	...	...	...	195·3 / 1·17	1·2	...	...	1·2	7·0	32·7	18·7	21·0	...	1·2	18·7	28·0	3·5	789·7 / 7·01	30·4	...	1·2	2·3
Gumbaz	67	...	...	...	...	104·5	29·9	...	...	...	14·9	...	59·7	...	...	...	...	...	...	447·8 / 14·93	14·9	...	...	...
Quetta	3,079	·3	...	...	·3 / ·32	229·0	1·6 / ·32	·6	...	3·6 / ·32	7·8 / 1·30	66·6 / ·32	26·0 / ·13	13·0	...	1·0	13·6 / ·32	11·4	16·6	731·3 / 5·20	26·3	2·9	2·6	11·6
Peshin	30	...	...	...	...	33·3	33·3	...	...	...	...	33·3	...	...	...	...	...	...	...	133·3	...	...	...	...
Shelabagh	35	...	...	...	...	285·7	...	...	...	...	28·6 / 28·57	85·7	...	28·6	...	...	...	28·6	...	714·3 / 28·57	28·6	...	...	...
Spinwana	35	...	...	...	...	142·9	...	...	...	...	85·7	28·6	...	...	...	...	...	...	28·6	342·9	...	...	...	28·6
Chaman	698	...	...	...	...	88·8	25·8	...	1·4	...	20·1 / 5·73	43·0	21·5 / 1·43	...	...	4·3	4·3	12·9	409·7 / 10·03	13·6	2·9	1·4	8·6
Mount Abu	79	...	...	...	...	88·6	...	...	...	...	...	12·7	...	...	...	...	...	25·3	189·9	12·7	...	25·3	...
Ootacamund	211	...	...	...	...	118·5	...	...	...	9·5 / 9·4	75·8	4·7	...	...	...	...	...	4·7	644·5 / 9·48	19·0	4·7	...	...
GROUP XII.— HILL STATIONS.	20,224	·3	·0 / ·05	·0	1·6 / ·45	170·3 / ·45	11·5 / ·84	23·4	·6 / ·30	4·0 / 1·29	13·7 / 2·37	45·5 / ·35	33·3 / ·40	9·7 / ·05	2·1	·7	4·2 / ·05	12·5	24·3 / ·05	729·3 / 18·05	30·8	8·2	5·0 / ·05	11·1
Rawalpindi Manœuvres.	1,134	2·6	...	·9	·0	134·9	7·1	5·3	·9	1·8	56·4 / 3·53	31·7	44·1	3·5	...	...	...	5·3	18·5	601·9 / 3·53	12·3	6·2	3·5	8·8
Marching India	11,165	·3	...	·4	·2 / ·09	104·7 / ·09	7·3 / ·18	3·0	2·0 / ·18	1·0	11·8 / ·54	31·2 / ·18	31·7 / ·09	6·0 / ·09	·4	·2	·8	6·6	12·4	423·6 / 2·69	9·2	4·7	2·5	5·1
EXTRA INDIA. (a) In the Indian Command :—																								
Sikkim Tibet Mission Force.	358	...	...	...	...	217·9 / 19·55	108·9	22·3	2·8 / 2·79	5·6	2·8	41·9	92·2 / 2·79	50·3	2·8	...	...	2·8 / 2·79	41·9	818·4 / 33·52	33·5	5·6	11·2	25·1
Chabbar	52	...	...	...	...	1115·5	...	...	...	...	19·2	57·7	38·5	76·9	...	...	...	...	1634·6	38·5	...	...	...
Jask	52	...	...	...	...	38·5	19·2	...	...	...	38·5	57·7	...	...	96·2	...	...	...	519·2	38·5	...	...	...
Muscat	21	...	...	...	...	285·7	...	...	...	...	47·6	142·9	...	...	47·6	...	...	...	523·8	...	...	...	...
Bushire	65	...	...	...	...	76·9	...	...	...	15·4 / 15·33	15·4	...	15·4	...	30·8	...	...	...	200·0 / 15·38	...	...	...	...
Aden	571	...	...	1·8	...	569·2 / 5·23	1·8 / 1·75	...	3·5 / 1·75	1·8 / 1·75	5·3 / 1·75	33·3	157·6	28·0	...	...	17·5	56·0	15·8	1220·7 / 14·01	82·5	10·5	...	3·5

STATIONS AND COMMANDS	Average annual strength	1. ADMISSION-RATE																		2. DEATH-RATE			
		Influenza	Cholera	Small-pox	Enteric Fever	Intermittent Fever	Remittent Fever	Simple Continued Fever	Circulatory Diseases	Tubercle of the lungs	Pneumonia	Other Respiratory Diseases	Dysentery	Diarrhœa	Hepatic Abscess, Hepatic Congestion and Inflammation	Scurvy	Anæmia and Debility	Venereal Diseases	ALL CAUSES	CONSTANTLY SICK	Syphilis	Soft Chancre	Gonorrhœa
Khormaksar	62	...	...	...	...	435'5	...	...	...	...	...	80'6	96'5	...	...	...	...	...	983'9 / 49'39	48'4	...	...	...
Sheikh Othman	2	...	...	...	...	2500'0	...	...	...	...	...	...	500'0	500'0	...	...	...	...	4,500'0	...	...	...	...
Perim	23	...	...	...	...	87'0	...	...	...	...	43'5	...	52'17 / 43'45	...	...	...	...	...	1,130'4 / 173'91	43'5	...	...	...
Aden Column Field Force	695	...	...	...	...	687'8 / 1'44	2'9	...	5'8	...	2'9 / 1'44	25'9	107'9	48'9	...	1'4	2'9	8'6	18'7 / 1,306'5 / 3'98	36'0	11'5	...	7'2
(b) Not in the Indian Command:—																							
Mauritius	1,477	...	...	...	...	1715'0 / 5'42	61'6 / 2'71	...	6'8 / '68	...	'7	22'3	38'6	2'0	2'1	...	2'0	28'4	14'0 / 2163'2 / 8'80	66'4	10'2	'7	4'1
Singapore	701	...	...	...	...	205'4	...	35'7	1'4	4'3 / 1'43	1'4	12'8	138'4 / 1'43	11'4	...	...	108'4 / 1'43	5'7	777'5 / 571	37'1	2'0	1'4	1'4
ARMY OF INDIA	123,434	1'5	'1 / '05	'6 / '01	1'1 / '28	171'2 / '37	8'4 / '56	14'3	1'1 / '26	3'1 / '50	12'5 / 1'90	28'5 / '23	32'5 / '18	7'5 / '05	'2 / '01	'6	1'7 / '03	11'8 / '09	19'6 / '06 / 617'1 / 8'09	23'2	7'7 / '03	4'6 / '01	7'4
INDIA	120,203	1'6	'1 / '06	'6 / '01	1'1 / '29	149'0 / '31	7'5 / '4	14'4	1'0 / '25	3'2 / '31	12'9 / 1'93	28'7 / '23	31'2 / '17	7'2 / '06	'1 / '01	'6	1'7 / '03	11'1 / '07	19'7 / '06 / 582'3 / 8'04	22'5	7'6 / '05	4'7 / '01	7'4
NORTHERN COMD.	36,696	2'3	'0 / '03	'5	1'4 / '27	142'6 / '19	11'1 / '76	16'7	1'0 / '16	5'4 / '76	20'4 / 3'32	33'4 / '27	34'6 / '19	6'7 / '05	'1	'5	1'0 / '03	13'6 / '01	18'8 / '01 / 651'8 / 12'93	26'5	6'8 / '03	4'7	7'2
WESTERN	33,932	1'3	'1	1'5	'5 / '21	185'0 / '41	3'7 / '32	8'4	'6 / '33	2'4 / '47	11'4 / 1'74	29'6 / '21	24'0 / '18	7'9 / '09	'1 / '03	'8	4'0 / '06	9'8 / '18	20'5 / '06 / 602'4 / 6'6	23'1	8'0 / '06	4'6	7'8
EASTERN	21,823	2'2	'2 / '18	2'0 / '60	...	150'4 / '37	10'7 / '64	10'2	'5 / '18	2'8 / '55	9'0 / 1'33	23'1 / '32	36'8 / '18	10'7 / '09	'1 / '05	'7	'6 / '05	9'2 / '05	19'5 / '09 / 571'5 / 6'42	22'7	7'7 / '09	6'1	5'6
SECUNDERABAD DIVISION	11,794	'3	'3 / '17	'3 / '08	1'1 / '34	80'4 / '34	...	1'8	4'2 / '59	2'6 / '08	1'2 / 1'27	5'5 / '03	15'4 / '17	25'6	2'2	'4	'8	10'3 / '17	27'0 / 466'1 / 5'77	19'1	12'3 / '17	3'5	11'3
BURMA	4,793	1'0	...	...	'2	167'1 / '63	7'3 / '63	15'9	'2	3'1 / '83	3'1 / '63	22'7 / '21	37'6	5'2	'8	...	'4	21'5 / '42	20'9 / 614'2 / 6'47	25'6	5'8	5'8	9'2
China Garrison	1,753	43'9	...	...	'4	50'2	...	1'7	...	9'7 / 2'85	11'4	44'5	21'7	3'4	...	...	3'4	20'0	338'8 / 4'56	16'5	3'4	4'0	12'5
Somaliland Field Force	525	...	...	5'7	...	699'0 / 1'90	...	...	...	...	5'7	13'3	68'6	22'9	...	...	1'9	1'9 / '42	1,135'2 / 3'81	22'9	...	...	1'0

* See foot-note at the end of Table XXIX.
† „ „ „ „

TABLE XXIX.

ACTUALS of STATIONS, GROUPS, and COMMANDS, on which the ratios in Tables XXVI—XXVIII have been calculated.

Stations and Groups	Average annual strength	Influenza	Cholera	Small-pox	Enteric Fever	Intermittent Fever	Remittent Fever	Simple Continued Fever	Circulatory Diseases	Tubercle of the lungs	Pneumonia	Other Respiratory Diseases	Dysentery	Diarrhœa	Hepatic Abscess	Hepatic Congestion and Inflammation	Scurvy	Anæmia and Debility	Venereal Diseases	All Causes	Constantly Sick	Syphilis	Soft Chancre	Gonorrhœa	Dracunculus Madinensis	Other Entozoa
						1. ADMISSIONS.														**2. DEATHS.**						
Port Blair	299	5	...	...	...	131	1	4	...	1	1	8	18	...	...	1	...	3	2	301 / 1 } 10		2	...	...	...	...
Rangoon	895	...	...	...	...	20	5 / 1	59	1	2	2	18	45	5	...	...	...	14 / 1	11	446 / 4 } 19		...	8	3	...	1
GROUP I.—BURMA COAST AND BAY ISLANDS.	1,195	5	...	...	...	201	6	63	1	3	3	26	63	5	1	...	...	17 / 1	13	747 / 5 } 29		2	8	3	...	1
Thayetmyo	19	...	...	...	...	1	...	...	...	...	...	1	1	...	...	...	...	...	...	7 }		...	...	...	...	...
Keng Tung	55	...	...	...	...	4	...	...	3	1	...	...	...	...	...	...	...	2	...	14 } 1		1	...	1	...	...
Fort Stedman	90	...	...	...	...	13	1	...	...	1	...	...	...	...	...	...	...	2	6	91 / 1 } 7		2	1	3	...	...
Meiktila	560	...	...	...	...	54	11	3	1	...	...	14	6	10	...	...	...	2	10	228 / 1 } 10		2	5	3	...	...
Fort Dufferin	1,347	...	...	...	...	721	11	2	...	3 / 1	3 / 2	43	46	4	...	1	...	36 / 1	14	805 / 9 } 30		4	3	7	3	1
Bhamo	845	...	...	...	...	171	2	9	...	1 / 1	1	10	16	4	...	...	2	23 / 1	4	428 / 5 } 16		1	1	2	...	...
GROUP II.—BURMA INLAND	2,916	...	...	...	...	465 / 1	25 / 2	13	8	5 / 2	68	69	18	...	1	2	63 / 1	36	1,573 / 15 } 64		10	10	16	3	1
Manipur	384	...	...	...	...	168	1	...	...	1	31	11	11	...	1	1	6	23	573 / 1 } 19		4	9	10	3	...
Sadiya	69	...	...	...	...	9	...	...	...	1	1	5	...	...	...	...	3	...	29 / 1 } 1		1	...	...	...	...
Dibrugarh	296	...	...	...	...	26	1	...	1	3	4	22	11	...	...	...	1	2	111 / 2 } 4		2	...	...	...	...
GROUP III.—ASSAM	949	...	...	...	...	203	2	...	1	5 / 1	36	38	22	...	1	1	10	25	713 / 4 } 24		6	9	10	3	...
Fort William	615	25	...	...	...	400	54	12	...	6 / 2	28	81	38	...	...	1	8	11	866 / 7 } 27		7	3	1	1	...
Alipore	772	4	...	...	...	226 / 1	1	40	...	7	13	45	25	...	2	...	5	24	610 / 3 } 21		14	4	6	...	...
Barrackpore	586	...	...	...	...	180 / 1	6	1	...	11	13	104	18	1	...	...	24	5	484 / 1 } 10		...	1	4	...	...
Buxa	264	...	...	...	...	56 / 1	2	...	4	4	8	7	7	...	...	...	11	1	170 / 1 } 7		...	1	...	5	...
Cuttack	313	...	...	...	...	32	...	1	...	1	10	7	2	...	...	...	...	9	114 / 1 } 4		3	3	3	...	...
GROUP IV.—BENGAL AND ORISSA.	2,553	29	...	...	...	905 / 2	63 / 2	54	7	2 / 3	72	241	110	1	3	1	48 / 1	50	2,244 / 13 } 69		24	12	14	6	...

STATIONS AND GROUPS.	Average annual strength.	Influenza.	Cholera.	Small-pox.	Enteric Fever.	Intermittent Fever.	Remittent Fever.	Simple Continued Fever.	Circulatory Diseases.	Tubercle of the lungs.	Pneumonia.	Other Respiratory Diseases.	Dysentery.	Diarrhœa.	Hepatic Abscess.	Hepatic Congestion and Inflammation.	Scurvy.	Anæmia and Debility.	Venereal Diseases.	ALL CAUSES.	CONSTANTLY SICK.	Syphilis.	Soft Chancre.	Gonorrhœa.	Dracunculus Medinensis.	Other Entozoa.
A Doranda	403	1	...	...	...	95	...	...	...	3	2	...	10	4	...	1	1	...	4	215 } 4	11	3	1	...	8	...
B Dinapore	750	...	2 ... 2	2 ... 1	52	7	28	...	3	7	9	39	...	...	...	1	...	4	14	30 } 6	13	6	3	5	8	...
Benares	714	...	...	...	49	4	18	1	1	3	17	9	...	...	1	...	2	3	28 } 5	9	2	1	...	6	...
Allahabad	1,199	1	2	...	191	3	...	2	5	10	33	81	5	...	3	...	26	16	778 } 11	24	7	5	4	5	1
Fyzabad	1,136	...	...	...	175	43	45	1	4	1	51	24	...	...	1	8	23	845 } 4	26	9	12	2	8	7
Lucknow	1,632	...	...	...	143	6	...	1	5	8	23	33	...	...	12	18	516 } 0	20	7	4	7	2	...
Cawnpore	1,140	6	...	...	118	2	1	1	3	5	17	42	4	1	1	7	19	533 } 5	22	10	2	7	8	2
Fatehgarh	72	...	...	...	22	...	...	...	1	1	3	3	...	...	1	83 } 1	2	...	1	...	...
GROUP V.—GANGETIC PLAIN AND CHUTIA NAGPUR.	7,046	8	4 ... 2	844	65	92	6	24	37	153	247	21	1	6	4	60	98	3,562 } 42	127	44	28	26	45	10
A Bareilly	1,074	1	...	...	54	1	...	2	...	9	22	23	4	...	1	...	5	217 } 5	10	1	3	1	10	...
Roorkee	776	...	...	...	18	3	...	2	13	33	3	...	...	4	10	182 } 2	8	2	5	3	...
Dehra Dun	2,848	10	40	11	194	14	1	13	26	60	66	21	2	1	13	103	2,034 } 33	162	27	48	28	...	2
Meerut	839	...	1	40	11	...	1	9	11	15	4	1	...	11	305 } 4	13	5	2	4	13	2
Delhi	791	...	...	145	9	...	18	39	18	2	...	1	4	9	434 } 5	13	5	...	4	...
Ambala	1,032	4	2	1	54	1	...	5	6	30	27	3	...	4	27	369 } 3	17	16	10	1	...
B Jullundur	1,319	...	...	72	7	1	1	5	18	32	33	6	2	21	12	545 } 2	24	3	...	9	...
Ferozepore	1,316	20	1	2	116	24	...	6	19	28	36	1	1	10	20	696 } 13	33	6	5	9	5	2
Mian Mir	1,555	3	2	129	11	11	1	10	30	56	50	8	1	1	16	16	721 } 11	30	8	6	2	1
Amritsar	134	...	...	7	1	...	7	2	2	1	...	4	57 } 2	4	...	...	...
Sialkot	1,401	1	241	7	3	7	12	30	39	12	2	2	27	17	916 } 5	33	11	6	13	...
Jhelum	1,305	8	...	154	6	34	19	13	43	30	15	1	22	34	813 } 9	29	14	6	14	1	1
Rawalpindi	2,084	1	1	2	117	20	25	3	11	54	53	53	14	3	6	22	63	1,134 } 22	53	20	16	27	3	...
Attock	85	...	...	14	...	...	...	1	...	1	...	39 } 1	...	...	...	...
GROUP VI.—UPPER SUB-HIMALAYA.	16,649	44	7	49	1,855	147	38	8	70	256	457	404	94	11	14	153	327	8,527 } 116	379	118	101	108	46	7
A Mardan	923	38	420	82	1	...	2	5	17	13	21	...	1	14	15	420 } 6	19	4	5	6	...	...
Nowshera	2,836	...	...	321	41	1	3	7	37	121	100	12	...	56	13	1,578 } 17	65	7	4	2	3	1

65

TABLE XXIX—continued.

ACTUALS of STATIONS, GROUPS, and COMMANDS, on which the ratios in Tables XXVI—XXVIII have been calculated.

STATIONS AND GROUPS.	Average annual strength.	Influenza.	Cholera.	Small-pox.	Enteric Fever.	Intermittent Fever.	Remittent Fever.	Simple Continued Fever.	Circulatory Diseases.	Tubercle of the lungs.	Pneumonia.	Other Respiratory Diseases.	Dysentery.	Diarrhœa.	Hepatic Abscess.	Hepatic Congestion and Inflammation.	Scurvy.	Anæmia and Debility.	Venereal Diseases.	ALL CAUSES.	CONSTANTLY SICK.	Syphilis.	Soft Chancre.	Gonorrhœa.	Dracunculus Medinensis.	Other Entozoa.
									1. ADMISSIONS.										2. DEATHS.							
Peshawar	2,593	5	...	...	8	267	7	43	...	8	31	51	103	14	2	1	...	37	50	1,247 / 8	46	16	19	15	12	1
Fort Jamrud	108	...	...	...	...	29	2	...	...	...	1	12	24	...	...	...	...	...	...	118	2	...	...	...	...	1
Kohat	2,608	1	...	3	2	548	42	6	10	18	88	81	82	11	...	2	10	72	33	2,003 / 14	87	11	7	15	8	7
Thal	207	...	...	...	...	99	...	...	...	...	2	20	13	1	...	...	1	5	3	209 / 1	5	...	...	3	...	...
Edwardesabad	1,961	1	...	...	2	304	30	58	...	7	44	59	82	23	...	...	1	35	37	1,565 / 13	58	6	18	13	6	...
Dera Ismail Khan	2,300	...	...	6	...	650	29	18	4	5	71	54	141	36	...	...	3	45	26	2,224 / 38	80	17	2	7	3	...
Jatta	61	...	...	...	...	13	...	...	1	...	2	2	2	...	...	...	...	...	2	41 / 1	1	1	...	1	...	...
Drazand	57	...	...	...	...	70	3	...	...	...	2	1	17	...	...	...	...	...	2	121 / 1	4	2	...	...	...	...
Fort Zam	51	...	...	...	...	30	4	...	...	...	...	...	6	2	...	...	...	...	...	61	1	...	...	...	...	...
Dera Ghazi Khan	630	...	...	...	...	67	3	...	...	3	7	15	26	2	...	1	1	3	5	250 / 6	10	3	...	2	4	...
Multan	1,406	...	...	1	...	161	10	...	1	17	19	31	21	3	...	...	1	18	23	656 / 13	27	5	12	6	4	...
Bikaner	30	...	...	...	...	...	...	...	...	...	...	...	...	...	...	...	...	...	...	...	...	...	...	...	...	...
B. Jandola	191	...	...	...	...	12	5	...	...	...	...	3	9	3	...	...	...	1	4	112	3	4	...	...	...	...

STATIONS AND GROUPS.	Average annual strength.	Influenza.	Cholera.	Small-pox.	Enteric Fever.	Intermittent Fever.	Remittent Fever.	Simple Continued Fever.	Circulatory Diseases.	Tubercle of the lungs.	Pneumonia.	Other Respiratory Diseases.	Dysentery.	Diarrhœa.	Hepatic Abscess.	Hepatic Congestion and Inflammation.	Scurvy.	Anæmia and Debility.	Venereal Diseases.	ALL CAUSES.	CONSTANTLY SICK.	Syphilis.	Soft Chancre.	Gonorrhœa.	Dracunculus Medinensis.	Other Entozoa.
Sibi	277 {	...	...	...	...	197	...	...	...	...	8	1	23	3	...	...	1	9	6	364 2 } 8		...	...	6	...	...
		...	...	...	...	...	...	...	...	...	2	...	...	...	...	...	...	...	...			...	...	...	...	...
C. Jacobabad	418 {	...	...	1	...	74	1	...	...	3	7	16	18	5	...	...	...	11	10	261 3 } 14		1	1	8	4	1
		...	...	...	...	...	...	...	...	...	2	...	...	...	...	...	...	...	...			...	...	...	...	...
Hyderabad	703 {	...	...	1	...	59	...	...	...	2	5	13	11	1	...	...	...	...	23	254 1 } 11		2	11	10	...	...
		...	...	...	...	...	...	...	...	...	1	...	...	...	...	...	...	...	...			...	...	...	...	...
Kurrachee	536 {	26	...	1	...	82	...	...	...	6	9	27	14	6	...	...	2	8	12	432 2 } 14		5	1	6	9	43
		...	...	...	...	...	...	...	...	2	...	...	...	...	...	...	...	...	...			...	...	...	...	...
GROUP VII.—N.-W. FRONTIER, INDUS VALLEY, AND NORTH-WESTERN RAJPUTANA.	17,946 {	71	...	13	13	3,065	178	126	21	81	368	520	708	125	1	4	20	314	264	11,916 } 454		84	80	100	53	54
		...	...	...	1	5	9	...	3	12	59	5	1	...	...	...	1	...	...	126		...	...	...	...	...
A. Bhuj	163 {	...	...	...	...	19	...	...	...	1	...	11	...	1	...	...	5	2	11	94 1 } 5		7	...	4	2	5
		...	...	...	...	...	...	...	...	...	...	1	...	...	...	...	...	...	...			...	...	...	...	...
Rajkot	746 {	...	...	...	...	105	...	9	...	3	17	14	4	...	1	4	15	23	394 9 } 17		2	9	13	7	4
		...	...	...	...	1	...	...	...	2	...	...	...	...	...	...	...	...	...			...	...	...	...	...
Deesa	1,076 {	...	...	1	...	264	1	...	...	2	10	17	14	...	...	1	3	14	15	702 11 } 24		8	4	3	8	...
		...	...	...	...	3	...	...	...	...	4	...	...	...	...	...	...	...	...			...	...	...	...	...
Ahmedabad	677 {	...	...	4	1	437	...	...	...	3	10	44	32	2	...	...	1	5	13	822 4 } 26		8	3	2	7	3
		...	...	...	1	...	...	...	...	1	2	...	...	...	...	...	...	...	...			...	...	...	...	...
Baroda	708 {	...	...	...	...	934	11	...	...	2	7	19	22	4	...	...	3	1	28	1,179 5 } 29		18	3	7	11	...
		...	...	...	...	1	1	...	...	1	...	...	...	...	...	...	...	...	...			...	...	...	...	...
B. Alirajpore	36 {	...	...	...	...	...	...	...	...	...	...	...	...	...	...	...	...	...	1	3 1 } ...		1	...	...	...	...
		...	...	...	...	...	...	...	...	...	...	...	...	...	...	...	...	...	...			...	...	...	...	...
Sirdarpore	386 {	...	...	4	...	31	...	...	...	2	13	8	1	...	...	...	...	8	...	189 5 } 9		5	...	3	11	...
		...	...	...	...	1	...	...	...	...	2	1	...	...	...	...	...	...	...			...	...	...	...	...
Jhabwa	37 {	...	...	1	...	5	...	...	...	...	...	...	...	...	...	...	...	...	...	8 } ...		...	...	...	...	...
		...	...	...	...	...	...	...	...	...	...	...	...	...	...	...	...	...	...			...	...	...	...	...
Kherwara	340 {	...	...	1	...	26	...	...	...	3	31	25	6	1	...	...	5	7	310 9 } 17		4	1	2	93	...
		...	...	...	...	...	...	...	...	...	4	...	...	1	...	...	...	...	...			...	...	...	...	...
Kotra	150 {	...	...	...	...	23	...	...	...	7	3	1	...	...	...	...	...	...	76 3 } 4		...	...	9	...	...
		...	...	...	...	...	...	...	...	1	...	1	...	...	...	...	...	...	...			...	...	...	...	...
Udaipur	43 {	...	...	...	...	2	...	...	...	...	...	...	...	...	...	...	...	...	...	9 } ...		...	...	...	...	...
		...	...	...	...	...	...	...	...	...	...	...	...	...	...	...	...	...	...			...	...	...	...	...
Todgarh	32 {	...	...	...	...	2	...	...	1	1	...	...	...	2	...	...	...	1	...	11 4 } 2		...	...	1	...	...
		...	...	...	...	...	...	...	1	1	...	...	...	...	...	...	...	1	...			...	...	...	...	...
Erinpura	569 {	...	...	...	...	61	5	...	...	1	12	5	10	6	...	1	1	11	359 3 } 14		5	3	3	6	...
		...	...	...	...	...	...	...	...	1	1	...	...	...	...	...	...	...	...			...	...	...	...	...
Neemuch	364 {	...	...	...	2	10	...	...	...	...	3	4	...	...	...	...	5	3	86 3 } 4		...	...	3	1	3
		...	...	...	2	...	...	...	...	...	1	...	...	...	...	...	...	...	...			...	...	...	...	...
Deoli	573 {	...	...	...	...	30	7	2	...	1	5	13	7	...	...	1	1	8	249 4 } 11		3	3	2	11	...
		...	...	...	...	...	...	...	...	1	...	...	...	...	...	...	...	...	...			...	...	...	...	...
Tonk	28 {	...	...	...	...	...	...	...	...	...	...	...	...	...	...	...	...	...	...	} ...		...	...	...	...	...
		...	...	...	...	...	...	...	...	...	...	...	...	...	...	...	...	...	...			...	...	...	...	...
Beawar	46 {	...	...	...	...	...	...	...	...	3	...	...	1	...	...	...	...	1	6 1 } ...		1	...	...	...	...
		...	...	...	...	...	...	...	...	...	...	1	...	...	...	...	...	...	...			...	...	...	...	...
Nasirabad	688 {	...	...	...	...	46	2	4	1	2	11	17	2	2	...	1	6	10	240 9 } 10		2	5	3	13	1
		...	...	...	...	1	...	1	...	1	2	...	...	...	...	...	...	...	...			...	...	...	...	...
Ajmer	446 {	4	...	...	...	22	1	...	...	2	12	4	1	1	...	2	...	8	163 } 9		2	1	5	6	...
		...	...	...	...	...	...	...	...	...	1	...	...	...	...	...	...	...	...			...	...	...	...	...
Sambhar	21 {	...	...	...	...	...	...	...	...	1	...	...	...	...	...	...	...	...	...	} ...		...	...	...	...	...
		...	...	...	...	...	...	...	...	...	...	...	...	...	...	...	...	...	...			...	...	...	...	...

TABLE XXIX—continued.

ACTUALS of STATIONS, GROUPS, and COMMANDS, on which the ratios in Tables XXVI—XXVIII have been calculated.

STATIONS AND GROUPS.	Average annual strength.	Influenza.	Cholera.	Small-pox.	Enteric Fever.	Intermittent Fever.	Remittent Fever.	Simple Continued Fever.	Circulatory Diseases.	Tubercle of the lungs.	Pneumonia.	Other Respiratory Diseases.	Dysentery.	Diarrhœa.	Hepatic Abscess.	Hepatic Congestion and Inflammation.	Scurvy.	Anæmia and Debility.	Venereal Diseases.	ALL CAUSES.	CONSTANTLY SICK.	Syphilis.	Soft Chancre.	Gonorrhœa.	Dracunculus Medinensis.	Other Entozoa.
								1. ADMISSIONS.											2. DEATHS.							
Jaipur	40 {	...	...	...	...	3	...	...	...	...	4	...	1	...	...	...	...	...	...	...	15 } ...	...	...	...	...	...
		...	...	...	...		...	...	...	...		...	...	...	...	...	...	...	...			...	...	...	...	...
Agra	694 {	...	...	...	...	50	1	...	...	5	13	22	4	...	...	2	...	...	18	216 } 9		11	6	1	1	...
		...	...	...	...			...	...	...	1	...	...	...	...	...	...	...	...			...	...	...	...	...
Gwalior .	30 {	...	...	...	...	...	1	...	...	...	...	...	...	...	...	...	...	...	...	4 } ...		...	...	...	...	...
		...	...	...	...	...		...	...	...	...	...	...	...	...	...	...	...	...			...	...	...	...	...
Jhansi	1,926 {	2	3	...	3	271	5	43	...	12	11	22	59	4	...	...	...	13	47	1,034 } 38		21	8	18	2	5
		...	...	...	...		...	...	...	1	1	1			...	...	...		...			...	...	...	...	...
Nowgong . .	1,159 {	8	...	...	...	118	3	15	...	2	9	9	9	...	...	...	1	4	109	406 } 16		5	4	1	0	...
		...	...	...	...	1				1			...	...	...	...		...		4		...	...	...	...	...
Goona .	415 {	...	...	...	...	11	...	...	...	...	4	1	...	...	...	...	...	...	2	47 } 2		2	...	...	1	...
		...	...	...	...		...	...	...	...	1		...	...	...	...	...	...		1		...	...	...	...	...
Agar . .	314 {	...	...	...	...	11	5	...	...	...	4	1	8	...	...	...	1	1	91 } 4		1	...	...	3	1
		...	...	...	...			...	...	...		1	...	...	...	...			1			...	...	...	...	...
Sehore	711 {	...	...	2	...	160	3	...	...	2	4	45	15	2	...	...	14	5	435 } 13		3	...	2	...	...
		...	...	...	...		1	...	...	...		...	...	...	...	...	1		4		...	...	...	...	...
Indore . .	162 {	...	...	...	...	17	...	2	...	...	3	3	3	...	...	...	...	2	67 } 2		...	...	2	1	...
		...	...	...	...		...		...	...			...	...	...	...	...		1		...	...	...	...	...
Mhow .	1,066 {	...	6	4	...	241	...	17	...	2	12	37	37	28	...	5	2	7	14	815 } 25		4	4	6	18	...
		...	...	...	...		1	...	...	1	3	1			...					9		...	...	...	...	...
GROUP VIII.—SOUTH-EASTERN RAJPUT-ANA, CENTRAL INDIA, AND GUJA-RAT.	13,651 {	14	3	19 10	2,891	46	93	2	44	200	331	247	55	...	12	21	95	246	8,030 } 291		113	54	79	221	22
		4			7	6	...	3	7	27	5	3	1	...	...	1	2	1	94		...	...	...	...	...
A.																										
Asirgarh . .	10 {	...	...	...	...	...	...	...	...	...	...	...	...	...	...	...	...	...	... } ...		...	...	...	...	...
		...	...	...	...	...	...	...	...	...	...	...	...	...	...	...	...	...			...	...	...	...	...
Saugor .	1,013 {	2	...	1	113	13	39	1	1	9	22	8	5	2	2	10	18	540 } 21		9	6	3	8	...
		...	...	...		1				2	...	...	...	...	...			6		...	...	...	...	...
Satna .	33 {	...	...	...	6	1	...	...	2	1	1	...	...	...	...	2	23 } 1		2	...	...	...	...
		...	...	...			...	...		1	...	...	...	...	...		1		...	...	...	...	...
Jubbulpore .	930 {	...	...	1	89	2	...	2	6	22	40	1	...	2	5	21	480 } 23		9	1	11	...	22
		...	...	...	1		...	...		...		...	...	...	1	...	5		...	...	...	...	...
Kamptee . .	603 {	...	...	2	76	1	...	2	16	13	3	...	...	2	7	279 } 13		4	...	3	...	...
		...	...	...		...	...				...	...	...	...		1		...	...	...	...	...
Sitabaldi .	82 {	...	3	...	2	...	...	...	6	5	1	...	...	...	78 } ...		...	...	...	...	...
		...	...	...		...	...				...	...	...	...				...	...	...	...	...
B.																										
Ellichpur . .	23 {	...	...	...	4	...	...	...	...	...	...	...	...	12 } ...		...	...	...	...	...
		...	...	...		...	...			...	...	...	...			...	...	...	...	...
Aurangabad .	1,735 {	...	4	1	274	2	2	4	13	20	29	3	...	7	26	866 } 34		8	4	14	10	3
		...	...	...		1		...				...	...	1		4		...	...	...	...	...
Ahmednagar .	575 {	...	...	...	25	...	1	1	3	4	4	...	...	3	3	110 } 6		2	...	1	12	...
		...	...	...			1	1		1		...	...			3		...	...	...	...	...
Bolarum .	1,712 {	...	2	8	23	122	2	2	6	37	65	1	2	1	31	51	766 } 36		33	2	19	1 10	...
		...	...	...			...	2	1			...	...			1	10		1	...	...	...	...
Secundarabad .	3,270 {	...	...	4	121	15	35	5	25	28	116	7	...	7	31	64	1,055 } 43		20	12	26	5	2
		...	...	3				1	1			...	...			1	15		1	...	...	...	...

STATIONS AND GROUPS.	Average annual strength.	Influenza.	Cholera.	Small-pox. Enteric Fever.	Intermittent Fever.	Remittent Fever.	Simple Continued Fever.	Circulatory Diseases.	Tubercle of the lungs.	Pneumonia.	Other Respiratory Diseases.	Dysentery.	Diarrhœa.	Hepatic Abscess.	Hepatic Congestion and Inflammation.	Scurvy.	Anæmia and Debility.	Venereal Diseases.	ALL CAUSES.	CONSTANTLY SICK.	Syphilis.	Soft Chancre.	Gonorrhœa.	Dracunculus Medinensis.	Other Entozoa.
								1. ADMISSIONS.									2. DEATHS.								
Belgam	1,176	...	...	... 3 / 1	63	...	20	...	...	13/2	17	26	6 ... 2 ..		7	41	596/4 } 19			11	8	22	14	1
Satara	615	...	... 1	45	1	...	1	2	...	5	18	1 ...	1	1	28	363/3 } 17		10	12	6	15	1
Poona	2,057	1 ... 2 ..	108	3	37	1	...	11	31	54	20 1 1 2	12	70	830/15 } 39		40	16	14	32	22
Kirkee	1,600	... ... 6 ...	161	...	67	4	...	12 / 2	109	21	27 1 2 1	19	50	983/3 } 27		19	18	13	3	17
Sirur	540	... ... 1 ...	33	...	...	...	1	6	5	2 1 5 1	7	8	173/1 } 9		...	1	7	7	5
GROUP IX.—DECCAN	15,977	3 ... 21 18 / 5	1,170	38	330	19	6	105/15	318	400	81 4 14 17	136	389/2	7,103/67 } 289		173	80	136	107	83
Bombay	720	... ... 7 1	271	2	20	2 / 3	6 / 2	17/4	25	26	18 11	9	36	723/18 } 29		19	6	11	7	12
Santa Cruz	647	... ... 1 ...	64	...	1	... / 2	9/2	6	14	7 3	4	19/1	240/11 } 12		8	10	1	7	5
Cannanore	540	... ...	32	...	7	1	1	7	1	1	4	21	197/5 } 16		6	3	10	1	...
Trivandrum	71	... ...	10	...	...	1	...	...	...	... ...	...	...	3/1 }		...	...	...	1	...
GROUP X.—WESTERN COAST.	1,978	... ... 8 1	381	3	21	10	7	27	38	41	26 14	17	76	1,19.,/35 } 58		35	19	21	16	17
A. Bellary	934	 1	210	5	28	6	2	6	15	14	... ...	10	57	658/4 } 26		29	11	17	6	...
Bangalore	2,762	3	364	...	25	6 / 2	4	21/3	52	58	7 ... 2	22	65	1,694/13 } 55		16	6	43	5	1
B. Trichinopoly	636	... ... 1 ...	9	...	15	1	...	3	3	1	15	170/4 } 8		11	...	4	...	...
Pallavaram	8	... ...	...	...	...	...	...	...	...	... ...	...	... }		...	...	...	...	...
St. Thomas' Mount	1504	... ...	89	...	7	...	1	13	23	4 1	15	10	369/4 } 14		3	1	6	...	...
Madras	767	3	48	1	3	3	2	2	4	13	4 3	19	271/10 } 13		12	4	3	2	12
C. Vizianagram	356	... ...	...	...	10	2	1	7	8	1 ... 1	5	16	130/1 } 6		6	2	8	...	...
GROUP XI.—SOUTHERN INDIA.	5,967	3 3 1 1 / 2 11	729	5	342	17	10	30	94	115	17 ... 3 1	55	182/35	3,292/35 } 125		77	24	81	13	13
Maymyo	682	... ... 1 / 2	135	4	...	4 / 3	7	15	48	2 ... 2	23	51	624/11 } 30		16	10	25	...	...
Kohima	153	... ...	58	...	...	4	1	17	3	2 1	8	7	151/1 } 6		5	...	2	...	2
Shillong	708	... ...	180	1	1	1	16	5	26	10	7	36	543/3 } 22		15	4	17	...	...
Gantak	65	 1	6	2	1	...	...	1	5	3	1	...	36/2 } 2		...	...	...	3	...
Almora	73	... ...	29	8	52	2 / 1	7	6	11	11 1 1 ...	7	30	399/2 } 16		12	10	8	...	...

TABLE XXIX—continued.

ACTUALS of STATIONS, GROUPS, and COMMANDS, on which the ratios in Tables XXVI—XXVIII have been calculated.

STATIONS AND GROUPS.	Average annual strength.	Influenza.	Cholera.	Small-pox.	Enteric Fever.	Intermittent Fever.	Remittent Fever.	Simple Continued Fever.	Circulatory Diseases.	Tubercle of the lungs.	Pneumonia.	Other Respiratory Diseases.	Dysentery.	Diarrhœa.	Hepatic Abscess, Hepatic Congestion and Inflammation.	Scurvy.	Anæmia and Debility.	Venereal Diseases.	All Causes.	Constantly Sick.	Syphilis.	Soft Chancre.	Gonorrhœa.	Dracunculus Medinensis.	Other Entozoa.
Naini Tal	132	..	..	...	26	..	...	...	2	1	5	2	...	1	...	...	...	8	74 3	3	3	2	3	...	1
Lansdowne	2,458	...	...	...	219 2	2? 6	9	... 1	4 5	11 2	21 1	97 2	23	1	2	33	15	1,242 22	63	9	4	2	...	...
Simla	148	...	...	...	2	...	2	...	...	1	4	...	2	...	...	1	5	31	2	2	3	...	...	...
Jutogh	160	...	...	...	...	...	...	...	...	1	4	...	3 1	...	...	1	...	49 1	3	...	...	...	...	...
Dharmsala	917	...	...	...	68	...	...	...	22 5	13 1	11	22 1	3	...	...	9	36	78 198	55	19	6	11	...	...
Bakloh	1,170	5	...	...	199 4	45	...	2 1	9	22 2	25	28	6	3	5	12 1	53	865 12	48	19	12	24	...	...
Murree	25	...	...	...	3	...	...	...	...	...	...	...	...	...	...	1	2	7	...	...	...	2	...	...
Khyragully	93	...	...	...	3	2	...	...	1	...	6	1	...	...	...	2	...	36	1	...	...	...	...	...
Baragully	68	...	...	...	3	...	...	...	...	...	1	...	...	...	...	1	...	78	3	...	...	...	1	...
Kalahagh	45	...	...	...	1	...	...	...	...	1	...	...	...	...	...	1	...	7 1	...	...	...	...	...	...
Chitral	220	...	...	...	39	...	31	... 1	1	1	3	8 1	...	...	1	1	158	4	1	...	...	...	...
Kila Drosh	636	...	...	1	91	18 3	166	3 4	2 1	2	20	45	10	...	...	7	591 7	23	5	1	1	...	...
Malakand	779	...	...	...	44	8 1	2	1	...	38 13	40 5	13 1	...	...	7	4	284 24	10	1	2	...	7	1
Dargai	365	...	...	...	30	1	... 1	1	6	2	2	1	...	...	3	115	3	1	1	...	...
Chakdara	556	...	...	1	79	5	...	2 1	4	6	13	1	...	...	6	5	237 2	5	1	2	2	...	...
Abbottabad	3,451	1 7	19 1	800	67 2	203	... 1	19 7	62 8	366	150	42 1	2	2	32	141	3,809 32	158	38	31	72	6	...
Cherat	48	...	...	...	2	...	...	...	4	...	3	1	...	...	1	...	22	1	...	...	...	...	...
Fort Lockhart	568	...	...	...	123 1	1	1	4 1	11 4	16	13	...	...	5	7	283 6	13	3	1	3	...	...
Hangu	59	...	...	...	34	...	...	...	2	2	...	...	...	...	57	1	...	...	...	...	...

Stations and Groups.	Average annual Strength.	Influenza.	Cholera.	Small-pox.	Enteric Fever.	Intermittent Fever.	Remittent Fever.	Simple Continued Fever.	Circulatory Diseases.	Tubercle of the lungs.	Pneumonia.	Other Respiratory Diseases.	Dysentery.	Diarrhœa.	Hepatic Abscess.	Hepatic Congestion and Inflammation.	Scurvy.	Anæmia and Debility.	Venereal Diseases.	All Causes.	Constantly Sick.	Syphilis.	Soft Chancre.	Gonorrhœa.	Uracneulous Medinensis.	Other Entozoa.
					1. Admissions.															**2. Deaths.**						
Mir Ali Khel	115	...	...	...	60	1	2	...	...	2	17	8	1	...	...	4	11	...	1·0	4	...	...	...	2	...
Fort Sandeman	604	1	...	...	168	20	...	2	6	8	31	16	0	...	...	5	18	10	586	18	4	...	...	6	11	...
Hindubagh	30	...	...	...	2	...	1	...	...	2	2	1	...	...	1	...	...	...	20	1	...	...	...	...	...
Musa Khel	28	...	...	...	18	1	...	...	...	...	6	...	...	...	...	...	...	...	33	1	...	...	...	1	...
Khan Mohamed Kot	33	...	...	...	9	1	...	...	...	4	4	5	...	...	...	...	1	...	37	1	...	...	1	...	...
Killa Saifulla	31	...	...	...	11	...	...	...	...	2	2	4	1	...	...	1	...	...	28	1	...	...	...	...	...
Murgha	54	...	...	...	3	1	...	...	...	2	1	4	...	...	2	...	...	20	1	...	...	...	...	...
Loralai	856	...	...	...	168	1	...	1	6	28	16	18	...	1	16	24	3	676	6	26	...	1	2	17	...
Gumbaz	67	...	...	...	7	2	...	...	1	...	4	...	...	...	...	...	...	30	1	...	...	...	...	1
Quetta	3,079	1	...	1	705	5	2	11	24	205	82	40	...	3	42	35	51	7,221	81	9	8	34	19	13
						1		1	4	...	1	...	...	...	1	...	...	16						
Peshin	30	...	...	...	1	1	...	...	...	1	...	...	...	...	...	...	4	...	...	...	...	...	...
Shelabagh	35	...	...	...	10	...	...	1	3	...	1	...	...	1	...	...	25	1	...	...	...	...	...
Spinwana	35	...	...	...	5	...	...	3	1	...	...	...	...	...	...	1	12	...	...	1	...	...
Chaman	698	...	...	...	62	18	...	1	14	30	15	...	3	3	9	286	13	2	1	6	2	...
									4	...	1						7						
Mount Abu	79	...	...	...	7	...	...	...	...	1	...	...	...	...	2	15	1	...	2	...	...	...
Ootacamund	211	...	...	...	25	...	...	2	16	1	...	...	...	1	135	4	1	...	...	...	1
									2	...	...	...					2						
Group XII.—Hill Stations.	20,224	7	1	1	33	3,444	233	473	12	93	277	930	654	166	4	15	84	253	491	11,749	624	166	101	224	69	18
		1	1	...	9	9	17	...	6	26	48	7	8	1	...	1	...	1	365	...	1	...	...	
Rawalpindi Man-œuvres.	1,134	3	...	1	1	153	8	6	1	2	64	36	50	4	...	...	6	21	606	14	7	4	10	...	1
			...	...	...			...	4	...	...	...	...	...				4					
Marching, India	11,165	3	...	5	2	1,169	81	34	22	11	132	348	354	67	5	7	9	74	138	4,729	103	53	28	67	14	3
			...	1		1	2		2	...	6	2		1				...	30					
Extra India. *(a)* In the Indian Command :— Sikkim-Tibet Mission Force.	358	...	...	...	...	78	39	8	1	2	1	15	33	18	1	...	1	15	295	12	2	4	9	...	1
			...	...			7		1	...		...	1					1	...	12					
Chabbar	52	...	...	...	58	...	...	...	1	3	2	4	...	...	...	...	85	2	...	...	1	...
			...	...	...			...	...	...		...											
Jask	52	...	...	...	2	1	...	...	...	2	3	...	...	5	...	...	27	2	...	...	...	...

TABLE XXIX—continued.

ACTUALS of STATIONS. GROUPS. and COMMANDS. on which the ratios in Tables XXVI—XXVIII have been calculated.

STATIONS AND COMMANDS.	Average annual strength.	Influenza.	Cholera.	Small-pox.	Enteric Fever.	Intermittent Fever.	Remittent Fever.	Simple Continued Fever.	Circulatory Diseases.	Tubercle of the lungs.	Pneumonia.	Other Respiratory Diseases.	Dysentery.	Diarrhœa.	Hepatic Abscess.	Hepatic Congestion and Inflammation.	Scurvy.	Anæmia and Debility.	Venereal Diseases.	ALL CAUSES.	CONSTANTLY SICK.	Syphilis.	Soft Chancre.	Gonorrhœa.	Dracunculus Medi-mensis.	Other Entozoa.
Muscat	21	...	...	...	6	...	...	...	...	...	1	3	...	...	...	1	...	...	11	...	...	...	...	...	...
Bashire	65	...	...	...	5	...	...	...	1	1	...	1	...	1	...	2	...	...	13	...	...	...	...	...	...
Aden	571	...	...	1	325	1	...	2	1	3	19	90	...	16	...	10	32	9	697 / 8	47	6	1	2	9	...
Khormaksar	62	...	...	...	27	...	...	1	...	5	6	...	...	...	...	...	...	61 / 3	3	...	...	...	1	...
Sheikh Othman	2	...	...	...	5	...	...	...	...	1	1	...	...	...	...	...	9	...	...	...	...	...	...
Perim	23	...	...	...	2	...	...	1	...	12	...	...	...	...	...	26 / 4	1	...	...	...	...	...
Aden Column Field Force.	695	...	...	1	478	2	...	4	...	2	18	75	34	1	2	6	13	908 / 1	25	8	...	5	...	—
(b) Not in the Indian Command :— Mauritius	1,477	...	...	...	2,533 / 8	91 / 4	...	10	...	1	33	57	3	3	42	22	3,195 / 13	98	15	1	6	...	5
Singapore	701	...	...	...	144	...	23	1	3	1	9	97	8	...	76	4	545 / 4	25	2	1	1	2	9
ARMY OF INDIA	*Remaining from 1904. Admitted. Died. Died out of hospital.	123,434	20 ... 190 11	4 8 77 139 7 1 35	464 21,138 46 1	33 1,034 69 ...	19 1,768	10 137	35 385 32 11	112 1,548 62 ...	139 3,523 235 2	91 4,017 28 ...	15 929 22 ...	2 23 7 ...	5 72 1 ...	12 213 4 ...	77 1,458 11 ...	245 2,419 7 ...	2,795 74,940 998 231	2,867	113 945 6 ...	56 565 1 ...	76 909	17 609	31 245 1 ...
INDIA	*Remaining from 1904. Admitted. Died. Died out of hospital.	120,203	20 ... 190 11	4 8 77 130 7 1 35	457 17,095 37 1	33 902 58 ...	19 1,735	10 121	35 380 30 11	112 1,543 61 ...	137 3,448 234 2	90 3,755 28 ...	15 866 20 ...	2 16 7 ...	5 71 1 ...	12 205 4 ...	74 1,333 9 ...	238 2,365 7 ...	2,733 69,099 967 228	2,706	111 918 6 ...	55 559 1 ...	72 888	15 607	31 230 1 ...
NORTHERN COMMAND	30,690	86 1 19 52 5,232 407 612 35 197 74 1,298 1,271 247 4 17 38 499 689 23,917 / 474	972	251 173 265 77 15
WESTERN COMMAND	33,932	43 3 5 48 6,275 124 285 22 81 386 1,000 846 268 4 77 137 334 694 20,441 / 224	786	273 156 266 411 163
EASTERN COMMAND	21,823	48 4 14 3,479 734 222 10 62 497 505 802 233 3 16 13 201 425 12,472 / 140	497	169 133 123 81 16
SECUNDERABAD DIVISION	11,794	3 3 13 948 21 596 31 14 65 186 302 26 3 9 122 310 5,497 / 68	225	145 41 133 21 35
BURMA DIVISION	4,793	5 1 804 35 76 1 15 15 109 183 23 4 2 103 100 2,044 / 31	123	28 25 44 3 2
CHINA GARRISON	1,753	77 ... ... 53 ... 3 17 20 78 38 6 ... ... 6 35 594 / Si	29	6 7 22 ... ...
SOMALILAND EXPEDITIONARY FORCE	555	... 3 ... 347 ... ... 5 7 36 12 ... 1 ... 1 59 / 31	12	... ... 1 ... ...

* Remaining + admitted = total treated. Remaining + admitted + died out of hospital = total cases.
† Excluding troops in China and Somaliland only.
‡ Excluding troops in Extra India not in the Indian Command, Sikkim-Tibet Mission Force and Aden Column Field Force, from the Army of India.
§ As far as returns have been received.

Groups and Commands.	1. Average Strength. January.	February.	March.	April.	May.	June.	July.	August.	September.	October.	November.	December.	Total.
I.—Burma Coast and Bay Islands.	1,352	1,426	1,199	1,037	955	909	1,014	1,165	1,145	1,205	1,370	1,574	14,332
	32	25	21	10	17	38	51	41	31	25	23	32	346
II.—Burma Inland	2,611	2,710	2,755	2,470	2,612	2,469	2,799	2,943	3,101	3,335	3,925	3,254	34,984
	81	77	51	31	33	50	75	68	59	69	104	74	772
III.—Assam	845	960	988	915	901	877	852	868	869	1,017	1,215	1,084	11,391
	25	29	22	10	23	27	29	22	17	19	35	35	293
IV.—Bengal and Orissa	2,511	2,764	2,845	2,350	2,352	2,733	2,359	2,358	2,419	2,542	3,344	2,629	30,706
	78	76	61	49	53	52	66	78	75	70	102	60	819
V.—Gangetic Plain and Chutia Nagpur.	8,973	8,536	7,670	6,090	6,140	6,316	6,319	6,761	6,611	8,075	6,970	6,083	84,544
	136	124	103	94	108	111	118	120	149	170	139	127	1,507
VI.—Upper Sub-Himalaya	15,638	17,994	19,868	16,502	15,953	15,629	15,884	15,914	16,257	18,076	14,996	17,071	199,782
	324	383	383	386	404	352	352	372	134	388	410	464	4,552
VII.—North-Western Frontier, Indus Valley, and North-Western Rajputana.	19,474	20,747	20,909	16,370	16,376	15,787	17,524	18,236	18,521	19,226	17,396	14,933	215,354
	675	665	516	302	299	314	357	375	366	458	562	561	5,450
VIII.—South-Eastern Rajputana, Central India, and Gujarat.	13,230	15,920	14,440	11,866	11,937	12,062	12,353	12,610	13,203	14,836	15,310	15,851	163,768
	331	345	241	238	213	197	202	238	302	354	422	418	3,501
IX.—Deccan	14,971	16,351	16,795	14,948	14,662	14,222	14,558	15,385	15,867	17,284	18,418	18,286	191,717
	256	314	279	265	249	242	267	289	273	291	366	340	3,461
X.—Western Coast	2,221	2,090	1,951	1,732	1,731	2,403	1,800	1,835	1,802	1,689	2,156	2,123	23,733
	48	84	61	58	60	59	41	51	43	48	50	47	690
XI.—Southern India	6,487	6,436	6,489	5,743	5,511	5,659	5,097	6,006	5,870	5,839	5,408	6,155	71,660
	151	143	136	119	114	139	145	121	110	96	105	126	1,505
XII.—Hill Stations	19,185	19,133	19,937	21,326	20,945	20,499	21,030	21,616	21,865	21,040	18,254	17,887	242,717
	602	620	553	728	666	672	613	683	668	547	555	575	7,482
ARMY OF INDIA *	136,168	136,173	139,226	111,993	109,550	109,329	110,624	113,318	117,629	130,765	138,614	137,785	1,481,174
	3,248	3,303	3,783	2,652	2,626	2,604	2,597	2,668	2,618	2,762	3,240	3,268	34,369
INDIA †	132,845	133,082	126,119	108,922	106,540	106,381	107,764	110,591	114,577	126,930	134,678	133,966	1,442,395
	3,155	3,150	2,586	2,429	2,346	2,381	2,109	2,544	2,503	2,657	3,118	3,148	32,435
NORTHERN COMMAND	35,306	37,629	40,002	36,211	35,102	33,549	35,469	36,953	37,610	38,421	34,940	38,563	440,355
	1,068	1,152	953	953	895	837	855	935	872	913	1,085	1,151	11,669
WESTERN COMMAND	32,890	35,954	35,733	30,309	30,058	30,519	30,734	31,857	33,533	37,731	38,838	39,010	407,166
	890	895	771	718	647	646	619	715	772	842	968	931	9,414
EASTERN COMMAND	22,859	23,646	23,667	20,636	20,319	20,427	20,320	20,976	20,733	22,817	23,175	22,309	261,875
	522	511	444	417	470	495	489	503	498	510	536	549	5,944
SECUNDERABAD DIVISION	12,403	12,991	12,389	11,103	10,949	11,139	11,588	11,759	11,671	11,863	11,593	12,077	141,524
	270	263	227	206	209	230	252	228	202	189	204	223	2,703
BURMA DIVISION	4,752	4,981	4,689	4,068	4,247	4,044	4,476	4,770	4,917	5,241	5,949	5,365	57,499
	138	133	98	62	78	113	157	140	110	125	167	144	1,474

* Excluding troops in China and Somaliland only.   † Excluding troops in Extra India not in the Indian Command, Sikkim-Tibet Mission Force and Aden Column Field Force, from the Army of India.

TABLE XXX.

ABSTRACT of the CANTONMENT SANITARY REPORTS of the most UNHEALTHY STATIONS, SANITARY DEFECTS, IMPROVEMENTS, SUGGESTIONS, etc.

The ratios of sickness and mortality will be found in Table XXVIII.

NORTHERN COMMAND.

DERA ISMAIL KHAN.
JATTA.
DRAZAND.
KILA DROSH.
} No sanitary reports.

DHARMSALA.—The following defects have been brought to notice :—The cubic space has hitherto been defective in the lines. This as well as lighting and ventilation will be attended to in rebuilding the barracks which were practically all destroyed by the earthquake on the 4th April 1905.

The General Officer Commanding the Brigade and the Lieutenant-General Commanding the Forces, Northern Command, have no remarks to offer.

MALAKAND.—The hospital requires improved ventilation and lighting.

The General Officer Commanding the Division states that the matter is being enquired into locally.

WESTERN COMMAND.

RAJKOT.
DEESA.
SIRDARPORE.
KHERWARA.
SUTNA.
ADEN.
} No defects reported.

KOTRA.
TODGARH.
BEAWAR.
} No sanitary reports.

NASIRABAD.—A great number of minor defects have been brought before the Cantonment Committee and a Sub-Committee has been sitting to consider remedies. Their report will be considered when completed ; but the only difficulty will be provision of funds which must be spread over a term of years.

The Lieutenant-General Commanding the Forces remarks that these are local questions.

MHOW.—The following defects have been brought to notice :—The want of pucca paving in the last section of the main drain. The necessity for the demolition of insanitary houses in the bazaar, and the general correction of the drainage.

The General Officer Commanding the Division says that funds have been applied for.

The Lieutenant-General Commanding the Forces remarks that no action is necessary.

BOMBAY.—The following defects have been brought to notice :—(1) The necessity for converting the present intermittent water-supply to a continuous one. (2) Providing a water system of drianage. (3) Removing and rebuilding Pestonji's chawls for followers. (4) The provision of an operation room.

The General Officer Commanding the Brigade says in regard to the defects numbered above.—(1) An estimate for R1,43,903 is now before Government. (2) Is not practicable until (1) has been carried out. (3) An estimate for R1,67,711 is before Government for acquiring the ground, demolishing the present buildings and erecting new ones, etc. (4) The matter will be brought before the Standing Barrack Committee as a major work. The approximate cost has already been reported on as amounting to R4,000.

The General Officer Commanding the Division remarks as regards (4) of the defects brought to notice that the requisite action has been taken by the General Officer Commanding the Brigade.

The Lieutenant-General Commanding the Forces remarks on the above defects.—(1) The question was referred to Government but the Commander-in-Chief has ordered the postponement of work. (2) Until a decision has been arrived at in connection with the water-supply scheme, no action seems possible regarding the drainage system. (3) In 1903 proposals were submitted to Government, who replied that the project had been noted for consideration when funds were available. (4) The proposals have been noted for entry in the Military Works schedules for construction during 1907-08.

SANTA CRUZ.
MIR ALI KHEL.
GUMBAZ.
BUSHIRE.
KHORMAKSAR.
PERIM.
MAURITIUS.
} No sanitary reports.

EASTERN COMMAND.

SADIYA.
FATEHGARH.
} No defects reported.

FORT WILLIAM.—The following defects have been brought to notice.—(1) The want of a system of water carriage for the removal of excreta in the Fort. (2) The reconstruction of the Ordnance lines at Hastings. (3) The filling in of the lunette round the Fort. (4) The necessity which exists for the provision of an Officers' hospital. (5) The want of arrangements for the complete segregation of Enteric Fever cases.

The Officer Commanding the Brigade says as regards the above defects.—(1) is under consideration. (2) Has been entered in the 10 years programme. (3) This work has been discussed for many years past but it is far too costly to be likely to be carried out. (4) Estimates have been submitted. (5) This is now being carried out.

The Lieutenant-General Commanding the Forces remarks on (4) of the defects noted above that the estimates have been sanctioned by Government.

ALLAHABAD.—The following defects have been brought to notice.—(1) Bad drainage in the Fort. (2) The overcrowding in barracks of the Native Infantry Guard.

The General Officer Commanding the Brigade remarks as follows.—(1) is under consideration. (2) The question of reducing the Native Infantry at the Fort will be considered.

DEHRA DUN.—The following defects have been brought to notice :—(1) The very insanitary state of the native village of Birpur which is too close to and should be removed from the vicinity of the 9th Gurkha Rifles lines. (2) Water by the present mode of distribution to the 9th Gurkha Rifles lines is liable to contamination. Piped water-supply should be laid on as soon as possible. (3) Rice is largely

TABLE XXX—*continued.*

ABSTRACT of the CANTONMENT SANITARY REPORTS of the most UNHEALTHY STATIONS,
SANITARY DEFECTS, IMPROVEMENTS, SUGGESTIONS, etc.

The ratios of sickness and mortality will be found in Table XXVIII.

cultivated in too close proximity to Cantonments. The cultivation should be restricted as much as possible, especially in localities to windward of these lines during the rains. (4) Anopheles mosquitoes abound in many localities in the Cantonment. It is recommended that a sufficient sum of money be set aside for the entertainment of a mosquito brigade during the whole of the hot weather and rains under the personal supervision of a medical officer to drain the breeding places of those insects and where that is impossible, by continued treatment with kerosine oil and tar, of pools, disused wells, etc., to try to exterminate their larvæ.

The General Officer Commanding the Brigade says as regards the above defects.—(2) Pipe laying is in progress and will soon be completed. (3) All rice cultivation has been stopped in Cantonments but there is a lot in Garhi village which adjoins Cantonments but is not under Cantonment control. (4) No important breeding grounds are left. There are no funds for a Cantonment Brigade, but the matter will be dealt with regimentally. As regards the village of Garhi which is doubtless a breeding ground for mosquitoes nothing can be done by the Cantonment authorities.

The General Officer Commanding the Division says as regards the above defects.—(1) The Commanding Royal Engineer, Meerut Division, reported on 4th February 1906 that this would be completed by about the end of March 1906. (2) It is now under local consideration to include this village in Cantonments, or endeavour to restrict its extending further towards Cantonments. The heavy expenditure necessary for sanitary measures therein, if taken over, precludes the first-mentioned especially in view of the heavy State-aid already given to this Cantonment. (3) As regards the village of Garhi, *vide* remarks on (2).

The Lieutenant-General Commanding the Forces has no remarks to offer.

NAINI TAL.—The water obtained from the Municipal Water Works is distributed by hand to the various Government buildings.

The General Officer Commanding the Division says that when funds are available, water will be laid on to all buildings.

The Lieutenant-General Commanding the Forces has no remarks to offer.

SECUNDERABAD DIVISION.

TRIVANDRUM.
ST. THOMAS' MOUNT. } No defects reported.
OOTACAMUND.

MADRAS.—The following defects have been brought to notice.—(1) Nothing has been done to remedy the defects in the shore system of drainage, a scheme for the improvement of which was accepted by Government but deferred for want of funds. (2) A contagious diseases block is urgently required as no proper accommodation exists for infectious diseases.

The Colonel Commanding the Madras Brigade states that the defects have been brought to the notice of the Assistant Commanding Royal Engineer for action as soon as funds permit.

The Lieutenant-General Commanding the Division remarks that all these points have been brought forward in the List of Services for 1907-08.

BURMA DIVISION.

MAYMYO.—A need exists for the drainage of the swamps in and around the Native Infantry Cantonments.

The Lieutenant-General Commanding the Forces says that nothing is known about any swamps in and around the Native Infantry Cantonment.

TABLE XXXI.
INFLUENZA by months, stations, groups, and commands.

ADMISSIONS FROM INFLUENZA IN EACH MONTH.

Stations and Groups.	January.	February.	March.	April.	May.	June.	July.	August.	September.	October.	November.	December.	TOTAL.
Port Blair	...	...	...	...	...	...	2	3	..	...	...	...	5
Group I.—Burma Coast and Bay Islands	...	...	...	...	...	...	2	3	...	...	...	...	5
Fort William	...	...	...	...	...	...	...	...	25	...	...	...	25
Alipore	..	...	...	...	...	...	...	1	3	...	...	...	4
Group IV.—Bengal and Orissa	...	...	...	...	...	...	...	...	26	3	...	...	29
A													
Doranda	...	1	...	...	...	...	...	..	...	...	...	...	1
B													
Dinapore	...	...	...	...	...	...	...	...	...	...	...	...	...
Allahabad	1	...	...	...	...	...	...	...	...	...	...	...	1
Cawnpore	...	...	5	1	...	...	...	...	...	...	...	...	6
Group V.—Gangetic Plain and Chutia Nagpur	1	1	5	1	...	...	...	...	...	...	...	...	8
A													
Bareilly	1	...	...	...	...	...	...	...	...	...	...	...	1
Dehra Dun	...	...	...	...	...	2	...	...	...	...	...	8	10
Ambala	3	1	...	...	...	...	...	...	...	...	...	...	4
B													
Ferozepore	...	7	6	4	3	...	...	...	...	...	...	...	20
Jhelum	8	...	...	...	...	...	...	...	...	...	...	...	8
Rawalpindi	...	1	...	...	...	...	...	...	...	...	...	...	1
Group VI.—Upper Sub-Himalaya	12	9	6	4	3	2	...	...	...	...	...	8	44
A													
Mardan	19	7	1	3	8	...	...	...	...	...	...	...	38
Peshawar	...	...	...	...	...	5	...	...	...	...	...	...	5
Kohat	...	...	...	...	...	...	...	...	...	...	...	1	1
Edwardesabad	...	...	...	...	...	...	...	...	...	...	...	1	1
C													
Kurrachee	...	...	...	...	...	2	2	...	4	5	8	5	26
Group VII.—North Western Frontier, Indus Valley, and North-Western Rajputana	19	7	1	3	8	7	2	...	4	5	8	7	71
B													
Ajmer	...	...	..	3	...	1	...	...	...	...	...	...	4
Jhansi	...	...	...	...	...	2	...	...	...	...	...	...	2
Nowgong	...	7	1	...	...	...	...	...	...	...	...	...	8
Group VIII.—South Rajputana Central India, and Gujarat	...	7	1	3	...	3	...	...	...	...	...	...	14
A													
Saugor	...	...	1	1	...	...	...	...	...	...	...	...	2
Poona	...	...	1	...	...	...	...	...	...	...	...	...	1
Group IX.—Deccan	...	...	2	1	...	...	...	...	...	...	...	...	3

TABLE XXXII.
CHOLERA by months, stations, groups, and commands.

ADMISSIONS FROM CHOLERA IN EACH MONTH.

Stations and Groups.	January.	February.	March.	April.	May.	June.	July.	August.	September.	October.	November.	December.	TOTAL.
Port Blair	..	...	...	...	...	...	...	...	...	...	...	...	...
Group I.—Burma Coast and Bay Islands	...	...	...	...	...	...	...	...	...	...	...	...	...
Fort William	...	...	...	...	...	...	...	...	...	...	...	...	...
Alipore	...	...	...	...	...	...	...	...	...	...	...	..	...
Group IV.—Bengal and Orissa	...	...	...	...	...	...	...	...	...	...	...	...	...
A													
Doranda	...	...	...	...	...	...	...	...	...	...	...	...	...
B													
Dinapore	...	...	...	...	...	...	...	...	...	...	1	1	2
Allahabad	...	...	...	...	...	...	...	...	1	1	...	...	2
Cawnpore	...	...	...	...	...	...	...	...	...	...	...	...	...
Group V.—Gangetic Plain and Chutia Nagpur	...	...	...	...	...	...	...	...	1	1	1	1	4
A													
Bareilly	...	...	...	...	...	...	...	...	...	...	...	...	...
Dehra Dun	...	...	...	...	...	...	...	...	...	...	...	...	...
Ambala	...	...	...	...	...	...	...	...	...	...	...	...	...
B													
Ferozepore	...	...	...	...	...	...	...	...	...	...	...	...	...
Jhelum	...	...	...	...	...	...	...	...	...	...	...	...	...
Rawalpindi	...	...	...	...	...	...	...	...	...	...	...	...	...
Group VI.—Upper Sub-Himalaya	...	...	...	...	...	...	...	...	...	...	...	...	...
A													
Mardan	...	...	...	...	...	...	...	...	...	...	...	...	...
Peshawar	...	...	...	...	...	...	...	...	...	...	...	...	...
Kohat	...	...	...	...	...	...	...	...	...	...	...	...	...
Edwardesabad	...	...	...	...	...	...	...	...	...	...	...	...	...
C													
Kurrachee	...	...	...	...	...	...	...	...	...	...	...	...	...
Group VII.—North Western Frontier, Indus Valley, and North-Western Rajputana	...	...	...	...	...	...	...	...	...	...	...	...	...
B													
Ajmer	...	...	...	...	...	...	...	...	...	...	...	...	...
Jhansi	...	...	...	...	...	3	...	...	...	...	...	...	3
Nowgong	...	...	...	...	...	...	...	...	...	...	...	...	...
Group VIII.—South Rajputana Central India, and Gujarat	...	...	...	...	...	3	...	...	...	...	...	...	3
A													
Saugor	...	...	...	...	...	...	...	...	...	...	...	...	...
Poona	...	...	...	...	...	...	...	...	...	...	...	...	...
Group IX.—Deccan	...	...	...	...	...	...	...	...	...	...	...	...	...

* Stations where neither Influenza nor Cholera occurred are not shown in these tables. For the annual ratios see Table XXVIII.

Stations Groups, and Commands.	Admissions from Influenza in each month.													Admissions from Cholera in each month.												
	January.	February.	March.	April.	May.	June.	July.	August.	September.	October.	November.	December.	Total.	January.	February.	March.	April.	May.	June.	July.	August.	September.	October.	November.	December.	Total.
A																										
Bangalore	...	...	...	3	...	...	...	...	...	...	...	...	3	...	...	...	...	...	...	...	...	...	...	...	...	...
B																										
Madras	...	...	...	...	...	...	...	...	...	...	...	...	...	...	...	...	...	...	...	...	3	...	...	...	...	3
Group XI.—Southern India	...	...	...	3	...	...	...	...	...	...	...	...	3	...	...	...	...	...	...	...	3	...	...	...	...	3
Baklob	...	...	...	...	...	...	...	...	5	...	...	...	5	...	...	...	...	...	...	...	...	...	...	...	...	...
Abbottabad	...	...	...	...	...	...	...	...	...	...	...	...	...	...	...	...	...	...	...	...	...	1	...	...	...	1
Fort Sandeman	...	...	...	...	1	...	...	...	...	...	...	...	1	...	...	...	...	...	...	...	...	...	...	...	...	...
Quetta	...	1	...	...	...	...	...	...	...	...	...	...	1	...	...	...	...	...	...	...	...	...	...	...	...	...
Group XII.—Hill Stations	...	1	...	...	1	...	...	...	5	...	...	...	7	...	...	...	...	...	...	...	...	1	...	...	...	1
Rawalpindi Manoeuvres	...	...	...	...	...	...	...	...	...	...	...	3	3	...	...	...	...	...	...	...	...	...	...	...	...	...
Marching India	...	...	...	...	...	...	...	...	...	...	1	2	3	...	...	...	...	...	...	...	...	...	...	...	...	...
ARMY OF INDIA	32	25	15	15	12	12	4	3	35	8	9	20	190	...	...	...	...	...	3	...	4	2	1	1	...	11
Northern Command	30	16	7	7	11	5	...	...	5	...	...	5	86	...	...	...	...	...	...	...	...	1	...	...	...	1
Western „	...	8	3	4	1	5	2	...	4	5	8	5	45	...	...	...	...	...	3	...	...	...	...	...	...	3
Eastern „	2	1	5	1	...	2	...	...	26	3	...	8	48	...	...	...	...	...	...	...	1	1	1	1	...	4
Secunderabad Division	...	...	...	3	...	...	...	...	...	...	...	...	3	...	...	...	...	...	...	...	3	...	...	...	...	5
Burma Division	...	...	...	...	...	...	2	3	...	...	...	...	5	...	...	...	...	...	...	...	...	...	...	...	...	...

TABLE XXXIII.

ENTERIC FEVER by months, stations, groups, and commands.

TABLE XXXIV.

SIMPLE CONTINUED FEVER by months, stations, groups, and commands.

Stations* and Groups.	ADMISSIONS FROM ENTERIC FEVER IN EACH MONTH.														ADMISSIONS FROM SIMPLE CONTINUED FEVER IN EACH MONTH.													
	January.	February.	March.	April.	May.	June.	July.	August.	September.	October.	November.	December.	TOTAL.	January.	February.	March.	April.	May.	June.	July.	August.	September.	October.	November.	December.	TOTAL.
Port Blair	...	...	...	...	...	...	...	...	...	...	...	...	...	1	...	...	...	...	1	...	3	...	...	...	...	4
Rangoon	...	...	...	...	...	...	...	...	...	...	...	...	...	10	1	...	...	8	8	7	2	5	4	6	8	59
Group I.—Burma Coast and Bay Islands	...	...	...	...	...	...	...	...	...	...	...	...	...	11	1	...	...	8	9	7	4	5	4	6	8	63
Meiktila	...	...	...	...	...	...	...	...	...	...	...	...	...	1	...	...	...	...	...	...	...	...	...	...	...	1
Fort Dufferin	...	...	...	...	...	...	...	...	...	...	...	...	...	...	...	...	...	...	...	...	1	...	8	...	2	3
Bhamo	...	...	...	...	...	...	...	...	...	...	...	...	...	...	...	...	...	...	...	...	...	...	...	1	1	9
Group II.—Burma Inland	...	...	...	...	...	...	...	...	...	...	...	...	...	1	...	...	...	...	...	...	...	...	1	8	3	13
Fort William	...	...	...	...	...	...	...	...	...	...	...	...	...	...	...	...	...	...	4	...	8	...	...	...	...	12
Alipore	...	...	...	...	...	...	...	...	...	...	...	...	...	...	...	...	...	...	...	...	29	13	...	...	...	40
Barrackpore	...	...	...	...	...	...	...	...	...	...	...	...	...	...	...	...	...	...	...	...	...	...	...	1	...	1
Cuttack	...	...	...	...	...	...	...	...	...	...	...	...	...	...	...	1	...	...	...	...	...	...	...	...	...	1
Group IV.—Bengal and Orissa	...	...	...	...	...	...	...	...	...	...	...	...	...	...	...	1	...	...	4	...	8	28	13	1	...	54
B.																										
Dinapore	1	...	...	...	...	...	...	1	...	...	...	...	2	...	...	...	...	...	...	...	...	...	...	...	28	28
Benares	...	...	...	...	...	...	...	...	...	...	...	...	...	...	...	...	1	...	5	4	7	...	...	...	1	18
Fyzabad	...	...	...	...	...	...	...	...	...	...	...	...	...	8	...	3	7	8	15	0	...	4	...	...	1	45
Cawnpore	...	...	...	...	...	...	...	...	...	...	...	...	...	...	...	...	...	1	...	...	...	...	...	...	...	1
Group V.—Gangetic Plain and Chutia Nagpur.	1	...	...	...	...	...	...	1	...	...	...	...	2	2	...	3	8	8	15	11	5	7	4	...	23	92
A.																										
Dehra Dun	...	...	...	...	2	18	9	7	1	...	2	...	46	...	...	...	6	3	1	3	...	...	...	...	...	14
Meerut	...	...	1	...	...	...	...	...	...	...	...	...	1	...	...	...	...	...	...	...	...	...	...	...	...	...
Ambala	...	...	...	...	1	...	...	...	...	...	...	...	1	...	...	...	...	...	...	...	...	...	...	...	...	...
B.																										
Jullundur	...	...	...	...	...	...	...	...	...	...	...	...	...	...	...	1	...	...	...	...	...	...	...	...	...	1
Ferozepore	...	...	...	...	...	...	...	1	1	...	...	...	2	...	...	...	...	...	...	...	...	...	...	...	...	...
Mian Mir	...	...	...	1	1	...	...	...	...	...	...	...	2	1	...	...	3	...	4	6	...	...	...	...	...	3
Sialkot	...	...	1	...	...	...	...	...	...	...	...	...	1	...	...	...	...	5	1	1	...	...	23	2	...	34
Jhelum	...	...	...	...	...	...	...	...	...	...	...	...	...	2	...	...	5	4	...	...	1	4	3	3	...	25
Rawalpindi	1	...	...	...	...	...	...	...	...	...	...	...	2	1	1	1	2	5	4	...	...	...	...	...	...	...
Group VI.—Upper Sub-Himalaya.	1	...	2	1	3	19	10	7	2	1	2	1	49	1	1	1	12	13	11	10	1	...	4	36	5	88
A.																										
Nowshera	...	...	...	1	...	...	...	...	...	...	...	...	1	...	1	...	...	...	...	...	...	...	...	...	...	1
Peshawar	...	...	...	...	2	...	3	...	...	1	1	...	8	...	...	12	11	1	5	...	1	13	...	...	...	43
Kohat	...	...	...	...	1	...	...	...	...	...	1	...	2	...	...	1	3	...	...	...	...	2	...	...	...	6
Edwardesabad	...	...	...	...	...	...	...	...	1	...	1	...	2	...	...	2	4	...	...	7	21	8	7	...	...	58
Dera Ismail Khan	...	...	...	...	...	...	...	...	...	...	...	...	...	...	...	13	2	...	1	...	...	...	...	...	...	18
Group VII.—North Western Frontier, Indus Valley, and North-Western Raj-putana.	...	...	...	1	...	...	3	...	1	5	3	...	13	...	...	13	4	13	20	2	5	37	43	7	...	126
A.																										
Rajkot	...	...	...	...	...	...	...	...	...	...	...	...	...	...	...	...	2	1	...	...	...	...	...	3	...	9
Deesa	...	...	...	...	...	...	...	...	...	...	...	...	...	...	...	...	...	1	...	...	...	...	...	...	...	1
Ahmedabad	...	...	...	...	...	...	1	...	...	...	...	...	1	...	...	...	...	...	...	...	...	...	...	...	...	...
B.																										
Neemuch	...	1	...	...	...	...	1	...	...	...	...	...	2	...	...	...	...	3	...	...	...	...	...	...	...	...
Deoli	...	...	...	...	...	...	...	...	...	...	...	...	...	...	...	...	1	...	...	...	...	...	...	...	...	2
Nasirabad	...	...	...	...	...	1	...	...	...	...	...	...	...	...	...	...	1	...	...	...	...	...	...	...	...	1
Jhansi	...	...	...	...	...	...	...	1	1	...	1	...	3	16	5	3	2	...	1	...	...	3	3	...	...	43
Nowgong	...	...	...	...	...	...	...	...	...	...	...	...	...	1	...	...	2	...	...	4	0	1	1	...	...	15
Indore	...	...	...	...	...	...	...	...	...	...	...	...	...	...	...	...	1	...	...	1	...	...	...	...	...	2
Mhow	...	...	2	1	...	...	...	...	...	...	...	...	4	...	...	4	7	1	...	...	...	...	...	...	...	17
Group VIII.—South-Eastern Rajputana, Central India and Gujarat.	1	1	2	1	...	1	1	...	1	1	1	...	10	27	5	10	9	7	6	4	1	5	10	4	5	...

* Stations where neither Enteric Fever nor Simple Continued Fever occurred are not shown in these tables. For the annual ratio see Table XXVIII.

Stations, Groups, and Commands.	\multicolumn{13}{c}{Admissions from Enteric Fever in each month.}												\multicolumn{13}{c}{Admissions from Simple Continued Fever in each month.}											
	January.	February.	March.	April.	May.	June.	July.	August.	September.	October.	November.	December.	TOTAL.	January.	February.	March.	April.	May.	June.	July.	August.	September.	October.	November.	December.	TOTAL.
A Saugor	...	...	...	...	...	...	...	...	...	1	...	...	1	...	...	...	...	...	...	...	...	2	9	20	8	39
B																										
Aurangabad	...	1	...	...	...	...	...	...	...	...	...	...	1	...	2	...	...	...	...	...	...	...	...	...	...	2
Ahmednagar	...	...	...	...	...	...	...	...	...	...	...	...	...	1	...	...	...	...	...	...	...	...	...	...	...	1
Belgaum	1	...	...	...	...	...	...	...	1	2	2	2	8	9	10	5	6	9	11	12	9	24	14	7	16	132
Secunderabad	...	1	...	...	...	...	1	...	...	1	1	...	4	1	1	...	2	9	3	2	5	2	2	4	1	32
Belgaum	...	...	...	...	...	...	...	1	2	...	...	...	3	5	6	3	...	...	...	2	1	...	1	1	1	20
Satara	...	...	...	...	...	...	...	...	...	...	...	...	1	...	...	...	...	...	...	...	...	...	...	...	...	
Poona	...	...	...	...	...	...	...	...	...	...	...	...	...	5	3	1	1	...	...	...	2	1	10	7	6	37
Kirkee	...	...	...	...	...	...	...	...	...	...	...	...	...	...	...	2	4	9	11	15	9	5	6	2	4	67
Group IX.—Deccan	1	2	...	...	...	...	1	1	4	4	3	2	18	21	22	11	13	28	25	33	21	34	42	41	36	330
Bombay	...	...	...	...	...	...	...	...	...	1	...	...	1	...	1	...	...	8	3	1	4	2	1	...	...	20
Santa Cruz	...	...	...	...	...	...	...	...	...	...	...	...	...	1	...	...	...	...	...	...	...	...	...	...	...	1
Group X.—Western Coast	...	...	...	...	...	...	...	...	...	1	...	...	1	1	1	...	...	8	3	1	4	2	1	...	...	21
A Bellary	...	...	...	...	...	...	...	...	...	1	...	...	1	...	2	1	...	...	...	5	1	4	6	9	...	28
Bangalore	...	...	...	...	...	...	...	...	...	...	...	...	...	9	4	6	13	38	61	34	18	10	25	21	6	250
B																										
Trichinopoly	...	...	...	...	...	...	...	...	...	...	...	...	...	3	5	3	...	...	...	...	...	...	...	4	...	15
St. Thomas' Mount	...	...	...	...	...	...	...	...	...	...	...	...	...	...	...	...	...	...	...	...	...	2	2	2	1	7
Madras	...	...	...	...	...	...	...	...	...	...	...	...	...	...	2	...	5	2	...	4	4	6	1	4	4	32
C Visianagram	...	...	...	...	...	...	...	...	...	...	...	...	...	...	1	...	...	1	1	1	2	2	2	...	10
Group XI.—Southern India	...	...	...	1	...	...	...	...	...	...	...	...	1	12	14	10	18	40	62	44	24	24	39	41	11	342
Maymyo	...	...	1	...	...	...	...	...	...	...	...	...	1	...	...	...	...	...	...	...	...	...	...	...	...	...
Gantak	...	...	...	...	...	...	...	1	...	...	...	...	1	...	...	1	...	...	...	...	...	...	...	...	...	...
Almora	...	...	...	...	...	...	...	...	...	...	...	...	...	8	3	7	5	1	12	7	4	3	1	1	...	53
Lansdowne	...	...	...	...	...	...	...	...	...	...	...	...	...	1	1	1	...	3	2	1	...	...	...	...	...	9
Simla	...	...	...	...	...	...	...	...	...	...	...	...	...	...	1	...	...	8	16	2	...	5	...	...	...	31
Chitral	...	...	...	...	...	...	...	...	...	...	...	...	...	...	...	...	10	83	47	4	20	2	...	...	...	165
Kila Drosh	...	...	...	1	...	...	...	...	...	...	...	...	1	...	...	...	...	...	...	...	...	...	2	...	...	2
Malakand	...	...	...	...	...	...	...	...	...	...	...	...	...	...	...	...	...	...	...	...	...	...	2	1	...	1
Dargai	1	...	...	...	...	...	...	...	...	...	...	...	1	...	...	...	...	...	...	...	...	...	2	...	...	2
Abbottabad	...	...	...	...	1	1	...	11	5	6	4	1	29	...	...	...	9	...	43	51	30	7	7	11	...	203
Fort Lockhatr	...	...	...	...	...	...	...	...	...	...	...	...	...	...	1	...	...	...	...	...	...	...	...	...	...	1
Mir Ali Khel	...	...	...	...	...	...	...	...	...	...	...	...	...	...	...	...	...	...	1	...	...	1	...	...	...	2
Hindubagh	...	...	...	...	...	...	...	...	...	...	1	...	1	...	...	...	...	...	2	...	...	1	...	...	...	3
Quetta	...	...	...	...	...	...	...	...	...	...	...	...	...	...	...	...	...	...	...	...	...	...	...	...	...	2
Group XII.—Hill Stations	...	...	1	1	1	1	...	12	5	6	5	1	33	9	5	8	7	43	113	102	100	58	6	8	14	473
Rawalpindi Manœuvres	...	...	...	...	...	...	...	...	...	...	1	...	1	...	...	...	...	...	...	...	...	...	...	3	3	6
Marching, India	...	1	...	...	1	...	...	...	...	...	1	...	2	3	7	...	8	...	5	...	5	...	...	...	5	34
Extra India. (a) In the Indian Command. Sikkim-Tibet Mission Force	...	...	...	...	...	...	...	...	...	...	...	...	...	...	...	1	...	1	...	...	4	2	...	8
(b) Not in the Indian Command. Singapore	...	...	...	...	...	...	...	...	...	...	...	...	...	...	5	2	...	5	4	5	...	...	...	3	...	25
ARMY OF INDIA	4	4	5	5	8	20	15	22	12	13	14	8	130	38	63	53	68	175	175	225	183	131	173	181	127	1,768
Northern Command	1	...	1	5	5	2	4	12	6	8	6	4	52	1	4	14	6	39	119	103	107	13	38	30	29	612
Western "	1	2	2	1	...	...	1	2	3	2	3	1	18	39	17	16	14	26	20	26	16	15	38	34	24	285
Eastern "	1	...	1	...	2	18	9	8	2	...	2	1	44	11	4	12	14	18	32	25	12	18	33	13	31	223
Secunderabad Division	1	1	...	1	...	...	1	...	1	3	3	2	13	27	25	15	26	53	71	58	38	50	55	55	28	505
Burma	...	...	1	...	...	...	...	...	...	...	...	1	12	1	...	...	8	9	7	4	5	5	14	11	76

TABLE XXXV.

INTERMITTENT FEVER by months, stations, groups, and commands.

TABLE XXXVI.

REMITTENT FEVER by months, stations, groups, and commands.

ADMISSIONS FROM INTERMITTENT FEVER IN EACH MONTH.

ADMISSIONS FROM REMITTENT FEVER IN EACH MONTH.

STATIONS AND GROUPS.	Jan.	Feb.	Mar.	Apr.	May	June	July	Aug.	Sept.	Oct.	Nov.	Dec.	TOTAL	Jan.	Feb.	Mar.	Apr.	May	June	July	Aug.	Sept.	Oct.	Nov.	Dec.	TOTAL
Port Blair	6	2	2	8	8	65	44	26	15	2	3	3	181	...	...	...	...	...	...	...	...	1	...	...	...	1
Rangoon	3	5	..	1	...	1	5	2	...	...	...	...	20	...	2	1	...	...	2	...	...	...	...	...	...	5
GROUP I.—BURMA COAST AND BAY ISLANDS	9	7	2	9	8	66	49	28	15	2	3	3	201	...	2	1	...	...	2	...	...	1	...	...	...	6
Thayetmyo	...	1	...	...	...	...	...	...	...	...	...	...	1	...	...	...	...	...	...	...	...	...	...	...	...	...
Keng Tung	2	1	1	...	...	...	...	...	...	...	...	...	4	...	...	...	...	...	...	...	...	...	1	...	...	1
Fort Stedman	7	5	1	...	...	...	...	...	...	...	...	...	13	...	...	...	...	...	...	...	...	...	...	...	...	...
Meiktila	10	2	4	2	1	2	5	1	...	8	4	15	54	1	...	...	...	...	...	1	1	3	3	2	...	11
Fort D fferin	1	3	5	5	8	3	9	16	16	59	55	43	222	...	...	...	...	...	5	3	1	...	1	...	...	11
Bhamo	3	4	2	1	3	14	99	22	10	2	11	...	171	...	1	1	...	...	...	...	...	...	...	...	...	2
GROUP II.—BURMA INLAND	23	16	13	8	12	18	113	39	26	69	70	58	465	1	1	2	...	...	5	3	2	2	3	4	2	25
Manipur	5	5	3	1	4	12	21	19	14	35	32	17	168	1	...	...	...	...	...	...	...	...	...	...	...	1
Sadiya	1	1	...	...	1	4	...	...	1	1	...	...	9	...	...	...	...	...	...	...	...	...	...	...	...	...
Dibrugarh	4	5	1	...	1	...	1	1	4	6	2	1	26	...	...	...	...	...	...	...	...	...	...	1	...	1
GROUP III.—ASSAM	10	11	4	1	6	16	22	20	19	42	34	18	203	1	...	...	...	...	...	...	...	...	...	1	...	2
Fort William	9	9	16	17	16	26	77	111	50	25	38	15	409	...	...	1	2	2	7	22	19	...	...	...	...	54
Alipore	48	13	15	8	10	12	16	17	26	17	44	...	226	...	...	...	...	...	...	...	...	...	...	...	1	1
Barrackpore	26	21	8	7	7	3	4	16	10	10	40	28	180	3	1	1	...	...	...	...	...	1	...	...	...	6
Buxa	1	6	1	...	5	7	2	3	10	12	10	1	58	...	...	...	...	...	...	...	1	3	...	...	...	2
Cuttack	11	1	...	5	1	...	2	1	4	2	4	1	32	...	...	...	...	...	1	...	...	...	...	...	...	...
GROUP IV.—BENGAL AND ORISSA	95	50	40	37	39	48	101	148	100	66	136	45	905	4	1	2	2	2	8	22	19	1	1	1	...	63
A — Doranda	2	1	1	1	6	7	12	27	21	17	...	...	95	...	...	...	...	...	...	...	...	...	...	...	...	...
B — Dinapore	2	4	4	1	2	7	2	7	9	6	8	4	52	1	1	...	1	3	1	...	...	...	1	...	...	7
Benares	9	3	8	1	2	3	6	5	18	6	6	4	49	...	...	...	...	1	...	...	...	...	1	...	...	4
Allahabad	9	8	7	3	11	14	6	6	23	51	27	26	191	...	...	...	1	2	...	...	...	...	...	...	...	3
Fyzabad	18	9	8	6	24	19	18	25	19	22	4	3	175	11	3	1	5	5	5	5	...	...	5	5	...	43
Lucknow	8	8	2	5	5	6	9	12	22	29	23	13	142	...	...	1	...	1	...	1	...	...	3	1	...	6
Cawnpore	1	...	4	5	14	4	27	14	16	11	12	9	118	...	...	...	...	...	1	...	...	...	1	...	...	2
Fatehgarh	...	...	...	3	9	2	4	2	2	...	...	...	22	...	...	...	...	...	...	...	...	...	...	...	...	...
GROUP V.—GANGETIC PLAIN AND CHUTIA NAGPUR.	41	34	27	25	74	61	81	93	117	154	76	61	844	12	3	2	10	13	4	6	1	...	9	7	...	65
A — Bareilly	...	...	1	6	3	6	5	4	5	10	11	3	54	...	...	...	...	...	...	...	...	...	1	...	...	...
Roorkee	1	2	2	...	...	1	7	1	3	1	...	...	18	2	1	...	...	...	...	...	...	...	...	...	...	3
Dehra Dun	12	35	47	54	45	68	65	82	89	72	58	67	694	1	3	2	1	4	9	7	2	3	7	1	1	46
Meerut	3	6	4	3	3	3	2	5	4	3	1	3	40	1	...	1	5	1	2	1	...	...	...	...	...	11
Delhi	9	16	10	10	18	14	11	13	3	17	17	7	145	1	...	2	1	3	...	...	...	...	1	...	...	9
Ambala	...	1	2	3	1	1	25	10	7	4	...	...	54	...	...	...	...	...	...	...	...	...	...	...	...	...
B — Jullundur	5	6	4	5	3	3	10	8	4	12	6	6	72	1	1	...	3	...	...	...	1	...	...	...	...	7
Ferozepore	15	6	6	12	11	9	14	6	4	15	10	8	116	5	...	...	1	...	5	6	3	...	...	...	3	24
Mian Mir	7	16	6	11	18	7	11	7	9	22	11	4	129	...	...	...	...	...	2	...	1	2	4	1	1	11
Amritsar	...	...	...	1	1	2	2	...	1	...	...	...	7	...	...	...	...	...	1	...	...	...	1	...	...	1
Sialkot	25	17	19	16	14	26	36	27	22	19	13	7	241	...	...	...	...	...	2	3	...	...	...	...	1	7
Jhelum	8	11	11	5	6	13	5	7	4	20	27	37	154	...	2	1	1	...	...	...	...	...	...	2	...	6
Rawalpindi	22	4	8	4	6	6	14	15	17	3	10	8	117	...	...	...	...	...	11	3	3	2	3	...	1	20
Attock	2	1	...	...	...	...	...	...	...	4	4	3	14	...	...	...	...	...	...	...	...	...	...	...	...	...
GROUP VI.—UPPER SUB-HIMALAYA.	109	121	120	130	129	152	207	185	172	202	168	153	1,855	11	7	8	6	15	19	28	15	8	17	7	6	147

* Stations where neither Intermittent Fever nor Remittent Fever occurred are not shown in these tables. For the annual ratios, see Table XXVIII.

Stations and Groups.	ADMISSIONS FROM INTERMITTENT FEVER IN EACH MONTH.													ADMISSIONS FROM REMITTENT FEVER IN EACH MONTH.												
	January.	February.	March.	April.	May.	June.	July.	August.	September.	October.	November.	December.	TOTAL.	January.	February.	March.	April.	May.	June.	July.	August.	September.	October.	November.	December.	TOTAL.
A																										
Mardan	9	1	1	2	1	7	15	10	10	3	19	4	82	...	...	...	...	...	...	...	...	...	...	...	...	1
Newshera	49	16	23	12	26	25	15	20	13	14	40	48	321	3	8	6	2	6	4	3	1	1	1	3	3	41
Peshawar	32	23	11	14	34	25	24	15	15	19	36	19	267	1	2	2	...	...	1	...	...	...	...	1	...	7
Fort Jamrud	3	1	3	...	7	2	3	2	1	3	...	4	29	...	...	1	...	...	...	...	...	...	...	1	...	2
Kohat	44	26	31	33	50	84	53	31	21	54	9.	31	548	3	1	...	...	2	...	4	14	8	5.	2	...	2	42
Thal	6	8	2	5	6	4	3	15	3	7	19	21	99	...	...	...	...	...	...	...	...	...	...	...	...	...
Edwardesabad	34	38	13	3	11	14	12	22	10	47	70	30	304	10	3	2	...	1	2	3	2	3	2	1	1	30
Dera Ismail Khan	35	29	36	16	37	46	41	35	39	98	139	96	650	3	...	...	...	4	2	3	7	1	3	6	...	29
Jatta	2	...	...	...	...	...	2	1	...	3	2	3	13	...	...	...	...	...	...	...	...	...	...	...	...	...
Drasand	8	1	1	4	3	2	2	2	5	17	15	10	70	...	...	...	...	...	2	...	...	...	...	1	...	3
Fort Zam	4	4	3	...	1	...	...	...	...	5	5	8	30	2	...	1	...	...	...	...	...	...	...	1	...	4
Dera Ghazi Khan	10	4	4	2	1	2	4	5	11	9	14	1	67	...	...	...	1	...	1	...	1	...	1	...	...	3
Multan	13	25	10	6	13	5	8	1	5	43	24	8	161	...	1	...	2	3	...	...	2	...	1	1	...	10
B																										
Jandola	1	...	...	...	...	1	...	2	2	1	5	...	12	...	...	...	...	...	...	...	...	...	...	...	...	...
Sibi	32	10	9	1	1	...	...	...	...	3	8	133	197	...	...	...	...	...	...	...	...	...	...	...	5	5
C																										
Jacobabad	6	...	3	5	8	5	6	6	5	7	17	6	74	...	...	...	...	...	...	...	...	...	1	...	...	1
Hyderabad	13	15	2	1	6	3	2	...	4	2	4	7	59	...	...	...	...	...	...	...	...	...	...	...	...	...
Kurrachee	18	15	18	1	...	4	10	1	4	...	5	0	82	...	...	...	...	...	...	...	...	...	...	...	...	...
Group VII.— NORTH-WESTERN FRONTIER, INDUS VALLEY, AND NORTH-WESTERN RAJPUTANA.	322	236	170	105	205	229	200	168	148	335	512	435	3,065	22	15	12	7	16	13	25	20	12	11	14	11	178
A																										
Bhuj	13	6	...	...	...	...	...	...	...	...	...	...	19	...	...	...	...	...	...	...	...	...	...	...	...	...
Rajkot	7	15	16	15	3	4	6	13	5	7	9	7	105	...	...	...	...	...	...	...	...	...	...	...	...	...
Deesa	14	3	4	5	13	3	9	9	67	47	52	38	264	...	...	...	...	...	...	...	...	...	...	...	1	1
Ahmedabad	6	35	13	9	5	3	13	20	51	94	114	65	427	...	...	...	...	...	...	...	...	...	...	...	...	...
Baroda	18	9	3	2	3	1	3	3	43	139	396	314	934	...	...	...	...	...	...	...	1	2	4	1	2	11
B																										
Sirdarpore	...	3	2	2	5	1	...	...	5	6	2	5	31	...	...	...	...	...	...	...	...	...	...	...	...	...
Jhabwa	1	...	4	2	...	...	...	...	...	...	...	...	5	...	...	...	...	...	...	...	...	...	...	...	...	...
Kherwara	...	2	4	2	1	...	...	1	4	4	9	1	28	...	...	...	...	...	...	...	...	...	...	...	...	...
Kotra	2	...	3	...	...	...	...	1	3	9	4	2	23	...	...	...	...	...	...	...	...	...	...	...	...	...
Udaipur	...	...	...	...	...	...	...	...	...	2	...	...	2	...	...	...	...	...	...	...	...	...	...	...	...	...
Todgerh	1	...	...	...	...	1	...	...	...	...	...	...	2	...	...	...	...	...	...	...	...	...	...	...	...	...
Erinpura	1	1	1	3	7	...	3	3	7	8	21	6	61	2	...	1	...	1	...	...	...	...	...	1	...	5
Neemuch	1	...	...	...	...	1	...	...	1	1	1	...	10	...	...	...	...	...	...	...	...	...	...	...	...	...
Deoli	...	1	3	6	1	...	2	...	3	3	4	2	20	5	...	...	...	1	...	1	...	...	...	...	...	7
Nasirabad	8	2	...	1	1	...	...	1	9	5	6	13	46	...	...	...	...	1	...	...	...	...	...	...	...	2
Ajmer	...	5	4	1	1	1	2	2	...	1	3	1	22	1	...	...	...	...	...	...	...	...	...	...	...	1
Jaipur	...	...	...	2	...	...	...	...	...	...	1	...	3	...	...	...	...	...	...	...	...	...	...	...	...	...
Agra	5	1	3	4	3	2	...	2	11	15	3	...	50	1	...	...	...	...	...	...	...	...	...	...	...	1
Gwalior	...	...	...	...	...	...	...	...	...	...	...	...	...	...	...	...	...	...	...	...	...	...	...	...	...	...
Jhansi	19	7	4	6	3	5	8	8	58	54	48	51	271	1	...	...	...	...	...	...	...	...	...	...	5	5
Nowgong	10	2	6	3	2	2	3	6	13	33	23	15	118	1	...	2	...	...	...	...	1	...	...	...	3	3
Goona	3	2	1	...	1	...	...	2	1	1	...	...	11	...	...	...	...	...	...	...	...	...	...	...	...	...
Agar	...	...	...	...	...	1	...	...	1	...	2	6	11	...	...	...	...	...	1	...	...	...	...	...	...	...
Sehore	...	8	2	6	12	4	...	11	29	36	33	19	160	1	...	...	1	...	1	...	1	2	...	2	1	5
Indore	1	...	...	...	...	...	...	3	4	1	3	3	17	...	...	...	...	...	...	...	...	...	...	...	...	...
Mhow	10	20	8	15	8	10	12	10	38	78	17	15	241	...	1	...	...	...	...	...	...	...	...	...	...	1
Group VIII.— SOUTH-EASTERN RAJPUTANA, CENTRAL INDIA, AND GUJARAT.	120	123	85	80	71	41	61	94	345	539	765	567	2,891	10	...	4	2	2	...	3	3	8	3	1	10	46

TABLE XXXV—*continued.*
INTERMITTENT FEVER by months, stations, groups, and commands.

TABLE XXXVI—*continued.*
REMITTENT FEVER by months, stations, groups, and commands.

Stations and Groups.	___ ADMISSIONS FROM INTERMITTENT FEVER IN EACH MONTH.													ADMISSIONS FROM REMITTENT FEVER IN EACH MONTH.												
	January.	February.	March.	April.	May.	June.	July.	August.	September.	October.	November.	December.	TOTAL.	January.	February.	March.	April.	May.	June.	July.	August.	September.	October.	November.	December.	TOTAL.
A																										
Saugor	0	2	3	12	8	7	2	14	10	16	22	5	113	...	...	...	...	...	...	1	3	3	2	4	13
Sutna	...	...	...	...	1	...	...	...	2	3	...	...	6	...	...	...	...	...	...	...	...	1	...	...	1
Jubbulpore	2	...	3	1	2	...	...	5	30	24	13	89	...	...	...	...	...	...	1	...	...	...	1	3
Kamptee	5	2	7	14	4	4	6	4	12	10	1	4	79	...	...	...	...	...	...	...	...	...	...	1	1
Sitabaldi	0	6	2	...	...	2	4	3	2	2	...	2	29	...	...	...	...	...	...	...	...	...	...	...	...
B																										
Ellichpur	1	3	...	...	...	...	...	...	...	...	...	...	4	...	...	...	...	...	...	...	...	...	...	...	...
Aurangabad	9	26	16	21	20	15	13	30	30	38	42	15	274	...	...	...	...	1	1	...	...	...	...	...	2
Ahmednagar	...	1	3	3	1	1	1	5	1	3	3	1	23	...	...	...	...	...	...	...	...	...	...	...	...
Holarum	...	3	3	1	...	1	3	1	1	3	6	...	23	...	...	...	...	...	...	...	...	...	...	...	...
Secunderabad	9	0	7	...	10	0	12	4	5	11	23	10	121	1	2	...	1	...	1	2	4	2	1	1	15
Belgam	3	1	3	10	0	4	7	8	3	4	7	4	63	...	...	...	...	...	...	...	...	...	...	...	...
Satara	1	4	...	4	3	3	3	6	...	2	13	6	45	...	...	...	1	...	...	...	...	...	...	...	1
Poona	8	4	6	4	3	6	15	13	8	8	24	5	163	2	...	...	...	1	...	...	...	...	...	...	3
Kirkee	0	3	6	4	5	7	53	10	7	14	35	8	161	...	...	...	...	...	...	...	...	...	...	...	...
Sirur	...	1	3	3	2	...	5	1	2	...	6	8	33	...	...	...	...	...	...	...	...	...	...	...	...
GROUP IX.—DECCAN	67	63	62	82	72	63	129	99	88	150	214	87	1,170	3	2	1	1	2	2	6	6	5	2	6	38
Bombay	22	19	15	12	41	30	12	40	22	7	17	31	271	1	...	...	1	...	...	...	...	...	...	...	2
Santa Cruz	3	6	6	3	4	4	2	3	6	11	15	1	64	...	...	...	...	...	...	...	...	...	...	...	...
Cannanore	3	4	2	3	3	4	1	2	2	3	5	4	36	...	...	...	...	...	...	...	...	...	1	...	...
Trivandrum	...	...	1	...	2	1	2	...	1	...	1	2	10	...	...	...	...	...	...	...	...	...	...	...	...
GROUP X.—WESTERN COAST	28	29	24	18	53	39	17	45	31	21	38	38	381	1	...	...	1	...	...	...	...	1	...	...	3
A																										
Bellary	36	41	58	19	6	9	6	7	9	16	8	5	210	2	1	1	...	...	...	...	...	...	1	...	5
Bangalore	16	11	39	52	37	67	41	15	27	27	29	13	374	...	...	...	...	...	...	...	...	...	...	...	...
B																										
Trichinopoly	...	...	...	...	2	...	2	2	2	...	...	1	9	...	...	...	...	...	...	...	...	...	...	...	...
St. Thomas' Mount	13	7	3	15	19	9	8	10	5	...	...	...	89	...	...	...	...	...	...	...	...	...	...	...	...
Madras	1	2	1	...	3	6	3	...	...	...	2	30	48	...	...	...	...	...	...	...	...	...	...	...	...
GROUP XI.—SOUTHERN INDIA	6	61	106	86	60	91	60	34	37	40	37	49	729	2	1	1	...	...	...	...	...	...	...	...	5
Maymyo	9	14	7	26	11	8	19	9	6	7	13	6	135	...	...	...	1	...	1	...	1	...	1	...	4
Kohima	5	...	1	2	2	5	12	7	6	11	4	1	58	...	...	...	...	...	...	...	...	...	...	...	...
Shillong	18	5	4	2	16	7	11	13	11	28	18	53	189	1	...	...	...	...	...	...	...	...	...	...	1
Gantak	1	1	...	...	1	...	3	...	...	...	1	...	6	...	...	...	...	...	1	1	...	...	...	...	2
Almora	6	...	...	1	...	4	3	5	8	...	2	10	72	1	...	...	...	...	4	1	1	...	1	...	8
Naini Tal	...	...	...	...	1	4	5	5	8	2	...	...	36	...	...	...	...	...	...	...	...	...	...	...	...
Lansdowne	14	17	5	9	10	12	13	25	24	33	29	28	219	...	1	1	1	1	5	4	4	1	2	...	21
Siula	2	...	...	...	...	...	...	1	...	...	...	14	3	...	...	...	...	...	...	...	...	...	...	...	...
Dharmsala	5	2	9	5	5	5	7	4	7	...	14	65	...	...	...	...	...	...	...	...	...	...	...	...
Dakhil	7	2	5	5	5	10	26	51	30	7	7	35	193	1	1	5	7	2	6	7	6	5	1	...	4
Murree	...	...	...	...	1	...	1	2	...	...	...	...	3	...	...	...	...	...	...	...	...	...	...	...	...
Khyragully	...	...	...	1	...	...	...	1	...	...	...	...	3	...	...	...	...	...	...	...	...	...	...	...	...
Banagully	...	...	...	...	...	...	1	2	...	...	...	...	3	...	...	...	...	...	...	...	...	...	...	...	...
Kababgh	...	1	...	...	...	...	12	15	1	6	2	3	39	...	...	...	...	...	1	...	...	...	...	...	...
Chitral	2	1	2	0	7	3	1	13	27	18	6	3	91	...	2	2	1	...	2	...	3	...	2	2	15
Kila Drosh	2	2	1	3	7	6	5	3	4	6	4	1	44	...	...	2	1	...	4	...	...	...	1	...	8
Malakand	1	1	1	1	2	5	4	...	9	4	4	1	30	...	...	...	...	...	...	...	...	...	1	...	5
Dargai	...	...	...	...	...	3	13	11	15	11	18	1	79	...	...	...	...	...	...	...	...	...	...	...	...
Chakdara	66	37	27	41	38	198	123	68	57	30	33	22	800	3	...	...	...	...	6	21	14	9	1	2	67
Abbotabad	...	...	...	...	...	...	...	...	...	...	...	...	7	...	...	...	...	...	...	...	...	...	...	...	...
Cherat	2	10	4	1	1	1	9	...	33	26	25	9	123	...	...	...	...	...	...	...	...	...	...	...	...
Fort Lockhart	...	...	2	...	...	1	...	2	2	15	12	...	34	...	...	...	...	...	...	...	...	...	...	...	...
Hangu	...	...	...	...	...	...	...	...	...	...	...	...	...	...	...	...	...	...	...	...	...	...	...	...	...

Stations, Groups, and Commands.	Admissions from Intermittent Fever in each month.													Admissions from Remittent Fever in each month.												
	January.	February.	March.	April.	May.	June.	July.	August.	September.	October.	November.	December.	TOTAL.	January.	February.	March.	April.	May.	June.	July.	August.	September.	October.	November.	December.	TOTAL.
Mir Ali Khel	1	1	...	2	3	6	3	6	5	17	12	4	60	...	...	1	...	...	...	...	...	...	...	...	...	1
Fort Sandeman	7	3	7	12	7	5	21	34	34	18	9	11	165	...	1	3	...	1	...	9	3	1	2	...	20
Hindubagh	...	...	...	1	1	...	...	1	...	...	...	...	2	...	...	...	...	...	...	...	...	...	...	...	1
Musa Khel	...	...	...	1	1	...	5	4	6	..	...	1	18	...	...	...	...	...	1	...	...	...	...	...	1
Khan Mohamed Kot	1	1	...	...	...	2	1	...	...	4	...	...	9	...	...	...	...	...	...	...	...	...	...	...	1
Killa Saifulla	...	...	...	...	...	...	...	2	5	3	1	...	11	...	...	...	...	...	...	...	...	...	...	...	
Murgha	1	...	...	...	...	...	...	1	1	...	...	...	3	...	...	...	...	...	...	...	...	...	...	...	
Loralai	13	10	13	8	14	16	13	19	16	15	20	11	161	...	...	...	...	...	...	...	1	...	...	...	1
Gumbaz	1	...	...	...	...	...	1	2	2	...	...	1	7	...	...	...	...	1	...	...	...	...	...	...	1
Quetta	21	19	25	38	53	54	52	113	127	93	62	48	705	...	...	...	...	1	...	1	1	1	1	...	5
Peshin	...	...	...	...	1	1	...	...	1	1	1	...	5	...	...	...	1	...	...	...	...	...	...	...	
Shelabagh	2	...	...	2	...	1	1	...	...	1	1	2	10	...	...	...	...	1	...	...	...	...	...	...	
Spinnana	...	...	...	3	...	...	...	...	1	...	...	...	5	...	...	...	...	...	...	...	...	...	...	...	
Chaman	4	1	1	1	3	1	3	8	20	5	7	8	62	1	...	...	2	4	2	9	...	...	...	...	13
Mount Abu	1	...	1	...	...	...	...	...	...	3	1	1	7	...	...	...	...	...	...	...	...	...	...	...	
Ootacamund	...	...	...	...	...	5	8	2	3	4	2	1	25	...	...	...	...	...	...	...	...	...	...	...	
GROUP XII.—HILL STATIONS	194	131	116	176	252	375	372	433	459	379	281	276	3,444	7	3	10	19	12	25	61	39	27	12	10	8	233
Rawalpindi Manœuvres	...	...	...	..	...	...	...	...	...	...	31	122	153	...	...	...	...	...	...	...	...	...	...	3	5	8
Marching, India	209	129	41	65	36	69	30	24	48	35	261	222	1,169	7	4	...	1	3	9	6	1	2	3	20	25	81
EXTRA INDIA. (a) IN THE INDIAN COMMAND:—																									
Sikkim-Tibet Mission Force	...	...	1	2	1	2	3	5	9	26	23	6	78	...	...	...	...	...	...	1	6	12	13	2	...	33
Chabbar	1	...	...	1	...	1	4	12	28	7	2	3	58	...	...	...	...	...	...	...	...	...	...	...	
Jask	...	...	1	1	...	...	...	...	...	...	...	1	6	...	...	...	...	1	...	...	...	...	...	...	1
Muscat	1	...	...	3	1	...	1	1	...	...	...	1	6	...	...	...	...	...	...	...	...	...	...	...	
Bushire	...	...	1	...	1	...	1	1	...	...	...	...	5	...	...	...	...	...	...	...	...	...	...	...	
Aden	59	23	22	38	39	40	26	15	19	8	14	22	325	...	...	...	...	...	...	...	...	...	...	...	
Khormaksar	8	2	1	1	...	3	1	...	2	2	2	5	27	...	...	...	...	...	...	...	...	...	...	...	
Sheikh Othman	...	...	...	...	...	...	...	...	...	3	2	5	...	...	...	...	...	...	...	...	...	...	...	
Perim	1	...	...	...	...	...	...	...	...	...	1	2	...	...	...	...	...	...	...	...	...	...	...	
Aden Column Field Force	41	34	37	41	100	64	44	28	21	29	21	18	478	...	2	...	...	...	...	...	...	...	...	...	2
(b) NOT IN THE INDIAN COMMAND:—																									
Mauritius	46	309	411	461	521	229	161	64	63	55	45	78	2,533	...	...	26	22	26	7	2	5	3	...	...	91
Singapore	10	10	23	17	13	11	11	6	9	8	20	6	144	...	...	...	...	...	...	...	...	...	...	...	
ARMY OF INDIA	1,457	1,479	1,307	1,382	1,695	1,626	1,693	1,541	1,757	2,169	2,756	2,276	21,138	81	41	69	71	89	95	159	117	82	83	74	73	1,034
NORTHERN COMMAND	425	317	246	221	379	532	499	418	385	557	635	568	5,232	32	20	23	22	23	35	86	49	37	25	29	26	407
WESTERN COMMAND	383	298	259	278	317	269	339	450	734	867	1,137	945	6,276	13	...	7	8	11	8	13	15	16	10	7	16	124
EASTERN COMMAND	221	178	148	156	221	248	337	423	392	453	401	299	3,479	23	9	9	16	26	27	46	34	7	21	11	3	234
SECUNDERABAD DIVISION	81	77	119	98	76	110	86	47	49	61	77	67	948	3	3	1	1	...	1	2	4	2	2	1	1	21
BURMA DIVISION	41	37	2	43	31	92	181	76	47	78	86	67	801	1	3	3	1	...	8	3	3	3	4	4	2	35

M

TABLE XXXVII.
PNEUMONIA by months, stations, groups, and commands.

ADMISSIONS FROM PNEUMONIA IN EACH MONTH.

Stations* and Groups.	January.	February.	March.	April.	May.	June.	July.	August.	September.	October.	November.	December.	TOTAL.
Port Blair	...	...	...	...	...	...	...	...	...	...	...	1	1
Rangoon	1	1	...	...	...	...	...	...	...	...	...	...	2
Group I.—Burma Coast and Bay Islands	1	1	...	...	...	...	...	...	...	...	...	1	3
Thayetmyo	...	...	...	...	...	...	...	...	...	...	...	...	1
Keng Tung	...	...	1	...	...	...	...	...	...	...	...	...	
Meiktila	...	...	...	1	...	...	...	...	1	...	...	...	
Fort Dufferin	1	...	...	...	1	...	...	...	1	...	...	...	3
Bhamo	...	...	...	...	...	...	1	...	...	...	...	...	
Group II.—Burma Inland	1	...	1	...	...	...	...	...	1	1	...	...	5
Manipur	...	...	...	1	...	...	...	...	...	...	1	...	1
Sadiya	...	...	...	...	...	...	...	...	1	...	...	...	
Dibrugarh	...	3	...	...	...	...	...	...	...	...	...	...	3
Group III.—Assam	...	3	...	...	1	...	...	...	1	...	...	...	5
Fort William	1	4	1	...	1	...	...	...	1	2	...	...	6
Alipore	...	1	...	1	...	1	...	1	2	...	...	6	11
Barrackpore	...	...	1	...	1	1	...	...	...	2	6	...	7
Buxa	...	1	...	...	...	...	...	...	...	...	...	...	4
Cuttack	...	...	...	...	...	...	...	...	...	...	...	...	
Group IV.—Bengal and Orissa	2	6	2	...	2	1	1	...	1	2	4	7	28
Doranda	...	1	...	...	...	1	...	...	...	...	...	...	2
Dinapore	3	...	...	...	...	1	...	2	1	...			7
Benares	...	...	...	...	...	...	...	1	2	...	...	...	3
Allahabad	...	3	1	1	1	...	...	...	...	1	2	...	10
Fyzabad	...	...	...	1	...	...	...	...	...	...	...	1	
Lucknow	1	2	1	...	2	...	...	...	1	...	...	1	8
Cavnpore	...	...	1	...	...	1	...	...	...	1	1	...	5
Fatehgarh	...	...	...	...	...	...	...	...	...	...	...	...	
Group V.—Gangetic Plain and Chutia Nagpur	4	6	3	2	3	2	1	2	1	4	3	6	37
Bareilly	3	...	2	1	...	1	...	...	...	...	...	1	9
Roorkee	3	4	1	...	...	1	...	...	...	...	...	3	13
Dehra Dun	1	1	5	2	3	...	...	2	1	2	2	6	26
Meerut	2	1	...	...	...	1	...	...	1	1	...	3	9
Delhi	4	6	1	...	...	1	...	...	1	...	2	2	18
Ambala	...	2	2	2	...	...	...	...	...	...	...	...	6
Jullundur	2	1	...	2	...	...	...	1	...	11			18
Ferozepore	1	...	4	3	3	...	...	3	1	3			19
Mian Mir	6	9	3	...	2	...	...	3	3	4			30
Amritsar	...	...	...	...	...	8	...	...	1	2			7
Sialkot	5	...	1	...	...	...	...	2	...	3			12
Jhelum	3	1	3	2	1	...	2	...	...	2	2	21	35
Rawalpindi	9	4	8	5	1	...	1	...	1	2	2	20	54
Group VI.—Upper Sub-Himalaya	38	31	31	19	12	4	4	5	3	16	16	77	256

TABLE XXXVIII.
DYSENTERY by months, stations, groups, and commands.

ADMISSIONS FROM DYSENTERY IN EACH MONTH

Stations* and Groups.	January.	February.	March.	April.	May.	June.	July.	August.	September.	October.	November.	December.	TOTAL.
Port Blair	1	2	1	2	4	1	6	1	...	...	...	...	18
Rangoon	3	1	1	1	3	4	7	6	4	2	5	8	45
Group I.—Burma Coast and Bay Islands	4	3	2	3	7	5	13	7	4	2	5	8	63
Thayetmyo	1	...	...	...	...	...	...	...	...	...	...	...	1
Keng Tung	...	...	...	...	...	...	...	...	...	...	...	...	6
Meiktila	...	...	1	...	1	2	1	...	...	...	...	...	6
Fort Dufferin	...	...	2	...	3	6	3	5	6	8	10	5	46
Bhamo	...	...	...	...	3	3	5	...	2	...	1	...	16
Group II.—Burma Inland	1	...	3	...	7	11	9	6	8	8	11	5	69
Manipur	...	4	1	1	1	1	2	1	...	...	2	...	11
Sadiya	...	...	1	...	1	...	...	...	...	...	2	...	5
Dibrugarh	2	...	2	...	4	5	2	2	...	1	4	...	22
Group III.—Assam	2	4	4	2	6	6	4	3	...	1	6	...	38
Fort William	8	10	9	6	9	5	3	8	8	4	3	8	81
Alipore	4	2	1	6	2	2	1	...	3	6	12	...	45
Barrackpore	4	1	5	6	2	1	2	2	9	7	41	24	104
Buxa	2	1	1	...	...	...	1	1	...	1	...	...	7
Cuttack	...	1	...	...	...	1	...	1	...	1	3	...	7
Group IV.—Bengal and Orissa	18	15	16	18	14	8	7	15	27	23	51	32	244
Doranda	...	...	...	2	4	6	...	2	1	1	...	...	16
Dinapore	2	...	3	1	...	1	2	10	9	2	8	1	39
Benares	1	...	...	...	...	...	...	4	1	...	1	1	9
Allahabad	7	...	...	6	15	5	3	9	12	13	7	4	81
Fyzabad	3	...	1	2	1	1	...	3	6	...	3	2	24
Lucknow	3	2	1	...	1	4	4	5	4	5	6	5	33
Cavnpore	...	...	...	3	4	2	5	3	7	13	4	...	42
Fatehgarh	...	...	2	...	...	1	...	...	...	...	...	...	3
Group V.—Gangetic Plain and Chutia Nagpur	14	2	4	14	28	17	8	40	36	28	38	18	247
Bareilly	...	...	...	...	6	1	2	3	2	4	3	1	23
Roorkee	...	...	...	...	...	...	...	7	8	...	1	4	3
Dehra Dun	1	1	3	11	11	7	8	6	1	4	...	3	60
Meerut	...	...	...	...	...	5	3	...	...	...	3	2	15
Delhi	2	...	...	2	3	1	...	3	3	3	...	1	18
Ambala	2	...	2	...	3	1	...	2	4	2	9	2	27
Jullundur	1	...	1	1	8	2	1	9	2	6	1	1	33
Ferozepore	3	1	1	1	13	...	4	3	2	6	1	1	36
Mian Mir	4	6	2	1	8	7	4	4	3	7	3	4	50
Amritsar	...	...	...	...	...	...	...	...	...	...	...	2	2
Sialkot	3	...	3	1	7	3	1	7	5	6	2	...	39
Jhelum	3	...	3	1	5	2	...	5	2	5	8	9	39
Rawalpindi	5	1	1	...	11	9	3	2	5	3	8	4	53
Group VI.—Upper Sub-Himalaya	25	12	13	20	83	38	25	32	38	41	47	30	404

* Stations where neither Pneumonia nor Dysentery occurred are not shown in these tables. For the annual ratios see Table XXVIII.

Stations and Groups	Admissions from Pneumonia in each month.													Admissions from Dysentery in each month.												
	January.	February.	March.	April.	May.	June.	July.	August.	September.	October.	November.	December.	TOTAL.	January.	February.	March.	April.	May.	June.	July.	August.	September.	October.	November.	December.	TOTAL.
A																										
Mardan	10	3	2	...	...	...	...	...	...	...	2	...	17	1	2	...	...	2	1	1	3	6	1	3	1	21
Nowshera	13	6	5	4	1	1	1	...	...	...	1	5	37	7	2	2	1	28	18	3	5	5	12	8	9	100
Peshawar	22	8	3	...	...	...	...	1	3	2	1	8	51	6	10	1	...	7	15	2	5	13	13	20	11	103
Fort Jamrud	...	...	1	...	...	...	...	...	...	...	...	...	1	3	...	...	1	6	2	...	2	3	4	1	1	24
Kohat	13	14	8	7	3	1	1	6	1	2	4	23	88	11	3	3	5	4	6	18	4	6	7	9	82
Thal	...	1	1	...	...	...	...	...	...	...	...	...	2	...	...	...	1	3	...	4	2	1	...	1	13
Edwardesabad	5	15	10	2	...	...	...	1	3	5	3	44	6	1	3	2	5	7	4	5	7	17	11	14	82
Dera Ismail Khan	10	10	15	8	2	2	...	2	1	3	5	10	71	13	8	4	3	14	...	3	11	30	33	13	9	141
Jata	...	...	...	...	...	...	...	...	...	...	...	...	...	...	...	...	...	...	...	...	2	...	...	...	2
Drasand	1	...	...	1	...	...	...	...	...	...	...	...	2	...	...	...	...	...	2	1	1	2	5	1	12
Fort Zam	...	...	...	...	...	...	...	...	...	...	...	...	...	1	...	...	...	1	...	...	1	2	1	...	6
Dera Ghazi Khan	2	2	...	...	...	1	1	1	...	...	...	...	7	1	3	1	...	2	4	1	1	3	6	4	...	26
Multan	7	...	2	...	...	...	...	...	...	...	...	5	19	1	2	...	3	...	1	1	...	...	7	2	4	21
B																										
Jandola	...	...	...	...	...	...	...	...	...	...	...	...	...	1	...	1	1	...	1	1	3	...	...	1	9
Sibi	...	3	2	...	...	...	...	...	...	...	1	2	8	2	2	1	1	...	...	1	...	...	1	5	11	23
C																										
Jacobabad	1	...	...	...	1	1	...	1	1	...	...	2	7	1	...	...	...	1	2	1	4	3	5	...	1	18
Hyderabad	2	1	1	...	...	...	...	...	...	...	1	...	5	2	1	1	...	1	...	1	2	...	...	...	3	11
Kurrachee	1	1	3	...	...	2	1	...	...	...	...	1	9	4	...	3	3	...	...	1	1	1	...	1	14
GROUP VII.—N.-W. FRONTIER, INDUS VALLEY, AND NORTH-WESTERN RAJPUTANA	87	69	53	22	7	8	5	11	7	9	26	64	368	60	34	20	21	74	57	30	63	79	110	83	77	708
A																										
Bhuj	...	...	...	...	...	...	...	...	...	...	...	...	...	1	...	...	...	...	...	...	...	...	...	...	1
Rajkot	...	2	1	...	...	...	...	1	...	...	...	...	3	2	1	...	3	...	1	1	1	1	2	2	...	14
Deesa	3	3	2	3	1	...	...	1	1	1	2	2	19	3	1	...	...	1	2	2	3	5	...	2	...	15
Ahmedabad	...	4	...	...	...	...	1	1	...	1	2	1	10	...	2	...	...	4	1	2	9	4	2	2	6	32
Baroda	4	...	2	...	...	...	1	...	...	...	...	...	7	2	3	...	1	1	1	...	6	...	...	2	6	22
B																										
Sirdarpore	1	3	1	3	1	2	...	...	1	...	1	13	...	...	...	1	...	...	...	...	...	...	...	1
Kherwara	2	10	3	3	...	...	...	1	...	1	4	31	...	...	1	...	...	1	...	2	1	...	...	...	6
Kotra	1	4	2	...	...	...	1	1	2	4	7	...	...	1	...	1	...	...	...	...	...	...	6
Todgarh	...	...	...	...	...	...	...	...	...	...	...	1	...	...	1	...	...	...	...	...	...	1
Erinpura	2	1	4	2	...	1	...	1	1	...	...	12	...	...	...	2	...	1	2	1	2	...	1	2	11
Neemuch	1	...	...	...	...	1	1	...	...	...	3	...	...	...	...	...	...	...	1	...	...	...	...	...
Deoli	...	...	2	1	...	...	...	...	...	...	3	...	...	...	...	...	...	1	...	1	2	3	7
Beawar	...	...	2	...	1	...	...	...	...	2	...	...	...	...	...	...	...	1	...	...	...	...	...
Nasirabad	2	...	1	...	...	1	...	1	...	1	5	11	1	...	...	...	...	...	...	1	...	...	...	2
Ajmer	1	...	7	1	...	1	...	1	...	1	12	...	...	...	...	...	1	...	...	...	...	...	1
Jaipur	...	...	...	1	...	1	...	1	...	1	4	...	...	...	...	1	...	...	...	...	...	...	...
Agra	2	...	...	2	...	1	...	1	...	1	13	...	...	...	...	...	1	...	...	...	...	...	4
Jhansi	1	2	1	1	...	...	1	1	2	2	11	3	2	1	2	2	...	1	6	12	5	7	17	58
Nowgong	2	2	...	...	...	1	...	1	...	2	9	...	...	...	1	...	...	1	6	4	...	1	7	9
Goona	...	1	1	...	1	...	...	...	...	4	...	...	...	...	...	...	...	...	1	...	...	...	...
Agar	...	...	1	2	...	...	...	...	1	4	...	...	2	...	...	1	1	1	1	1	...	8
Sehore	3	1	...	...	...	...	...	...	...	4	...	...	...	2	...	...	1	5	1	1	1	1	15
Indore	...	...	...	...	...	...	...	...	2	...	3	...	1	...	2	2	...	1	3	1	1	1	3
Mhow	1	5	...	1	...	...	...	...	4	1	...	12	3	7	1	1	1	2	1	5	13	1	1	37
GROUP VIII.—SOUTH-EASTERN RAJPUTANA, CENTRAL INDIA, AND GUJARAT	27	44	29	19	4	4	4	6	6	14	18	25	200	17	17	3	18	13	6	12	45	30	25	22	39	247
A																										
Saugor	...	...	2	1	...	...	1	...	3	2	9	...	...	...	...	1	1	...	2	...	2	1	1	8
Sutna	...	...	...	...	...	...	...	...	...	...	2	...	...	...	...	...	...	1	...	...	...	...	1
Jubbulpore	...	...	...	...	...	2	...	1	3	6	1	...	1	2	...	6	2	4	7	10	1	6	40
Kamptee	...	...	1	1	...	...	...	...	1	3	...	...	1	1	2	1	2	2	1	...	1	...	13
Sitabaldi	...	...	...	...	...	...	...	...	...	...	...	...	1	1	...	2	...	1	...	...	...	6

TABLE XXXVII—continued.

PNEUMONIA by months, stations, groups, and commands.

TABLE XX[...]

DYSENTERY b[...]

STATIONS AND GROUPS	ADMISSIONS FROM PNEUMONIA IN EACH MONTH.													ADMISSIONS FRO[M]				
	January	February	March	April	May	June	July	August	September	October	November	December	TOTAL	January	February	March	April	May
B																		
Ellichpur	1	...	...	...	...	...	...	...	...	...	...	...	1	...	...	...	...	...
Aurangabad	6	...	1	...	2	...	2	...	...	...	2	...	13	1	2	2	1	...
Ahmednagar	...	1	...	...	...	...	...	...	...	1	1	...	3	...	...	...	...	...
Bolarum	...	1	...	...	...	1	...	...	...	...	...	...	6	4	3	2	...	4
Secunderabad	11	2	...	...	3	1	...	...	1	1	...	2	15	5	5	12	5	10
Belgam	3	4	1	2	1	...	1	...	2	...	...	2	13	1	5	...	2	1
Satara	...	...	...	...	...	...	...	...	...	1	...	...	1	2	1	...	1	1
Poona	2	1	2	2	1	...	1	...	1	...	1	...	11	4	...	1	2	5
Kirkee	2	...	3	1	...	...	...	...	1	2	3		12	1	...	2	2	...
Sirur	...	...	...	...	...	...	...	1	...	...	...		1	...	...	...	...	...
GROUP IX.—DECCAN	25	9	10	14	5	3	2	...	8	4	11	14	105	19	21	22	16	26
Bombay	5	5	1	1	...	1	...	2	...	...	...	1	17	2	2	1	1	8
Santa Cruz	4	...	1	2	1	...	...	...	1	...	...	...	9	1	...	...	1	1
Cannanore	...	...	...	...	...	...	1	...	...	...	...			...	...	...	...	...
GROUP X.—WESTERN COAST	9	5	2	3	1	1	1	2	1	1	...	1	27	3	2	1	2	9
A																		
Bellary	...	2	...	1	...	...	1	...	2	3	2	1	6	...	...	3	...	...
Bangalore	1	3	3	3	...	1	...	2	3	2	2	1	21	13	2	10	1	7
B																		
Trichinopoly	...	...	...	...	...	...	...	...	...	...	...		...	7	7	2	1	...
St. Thomas' Mount	...	...	...	...	...	1	...	1	...	...	...		1	...	...	1	1	1
Madras	...	...	1	...	...	...	...	...	...	...	...		2	...	...	1	...	...
C																		
Vizianagram	...	...	...	...	...	...	...	...	...	...	...			1	2	1	...	1
GROUP XI.—SOUTHERN INDIA	1	6	3	4	1	2	1	2	3	2	4	1	30	22	11	18	3	9
Maymyo	...	2	1	4	...	1	...	...	...	...	...		7	1	3	30	3	3
Kohima	1	1	...	...	...	1	...	...	...	...	1		4	...	2	...	...	...
Shillong	2	1	2	2	1	1	1	1	1	3	...		16	5	...	1	...	2
Gantak	...	...	...	...	...	...	1	...	...	2	1		7	...	1	...	...	...
Aimora	1	...	...	1	1	...	...	...	...	...	...			...	...	...	...	1
Naini Tal	...	...	1	...	...	...	1	2	...	1	1	1	11	2	...	1	1	19
Lansdowne	1	...	...	...	4	...	1	...	...	...	...		11	...	...	...	...	...
Simla	...	...	...	...	1	...	...	...	...	...	...		1	...	...	...	...	...
Jutogh	...	...	...	...	1	...	...	2	1	1	6		13	...	3	2	...	5
Dharmsala	...	1	2	...	...	1	4	6	...	3	...	4	22	...	...	...	...	1
Baklob	2	1	...	...	1	...	...	...	...	...	...			...	...	...	...	1
Khyraguily	...	...	...	...	...	...	1	...	...	...	...			...	1	1	1	...
Kalabagh	...	...	...	...	1	...	...	...	...	1	...		2	1	...	1	...	...
Chitral	...	...	...	...	1	...	...	...	...	...	1		2	1	...	...	...	...
Kila Drosh	1	...	...	...	...	...	1	1	...	2	7		38	...	...	1	...	2
Malakand	7	7	2	3	5	2	...	1	...	...	...		6	...	...	...	...	...
Dargai	3	...	3	...	...	1	...	...	...	...	1		4	...	...	...	...	1
Chakdara	1	...	...	1	...	...	...	...	...	1	...		6²	1	1	3	1	13
Abbottabad	7	9	5	11	8	2	4	6	2	...	2	6	62	...	...	...	...	...
Cherat	...	...	1	...	...	...	...	...	3	1	...		11	...	...	...	...	1
Fort Lockhart	...	5	...	...	...	2	1	...	...	...	...			...	...	...	...	...
Hangu	...	...	...	...	...	...	...	...	1	...	...		3	1	1	...	...	...
Mir Ali Khel	...	...	...	...	...	...	...	...	...	1	...		8	1	...	2	3	1
Fort Sandeman	1	1	...	...	2	1	...	...	...	...	3			...	...	1	...	...
Hindubagh	...	2	...	...	...	...	...	...	...	...	...		2	...	...	...	1	...
Musa Khel	...	...	...	...	...	...	...	...	...	...	...			...	...	...	...	...
Khan Mohamed Kot	...	...	...	...	...	...	...	...	...	...	...			...	...	...	...	...
Killa Saifulla	...	2	...	...	...	...	...	...	...	...	...		2	...	...	...	...	...

STATIONS, GROUPS, AND COMMANDS.	ADMISSIONS FROM PNEUMONIA IN EACH MONTH.													ADMISSIONS FROM DYSENTERY IN EACH MONTH.												
	January	February	March	April	May	June	July	August	September	October	November	December	TOTAL	January	February	March	April	May	June	July	August	September	October	November	December	TOTAL
Murgha	...	1	...	...	...	...	...	...	...	1	1	1	2	...	1	...	...	3	1	1	...	3	1	2	3	4
Loralai	3	...	...	...	...	...	...	...	1	1	...	1	6	1	1	...	...	3	1	1	3	1	2	3	...	16
Gumbaz	...	...	1	...	...	...	...	...	...	...	...	...	1	...	...	...	9	...	...	2	...	...	1	...	...	4
Quetta	3	4	...	4	2	1	2	1	2	5	...	...	24	...	...	...	9	3	10	4	6	7	21	13	9	82
Shelabagh	...	...	1	...	...	...	...	...	...	...	...	...	1	...	...	...	...	...	...	...	...	...	...	...	...	4
Spinwana	...	1	1	...	...	...	...	...	...	...	...	...	3	...	...	...	...	...	...	...	...	...	...	...	...	...
Chaman	4	4	1	1	...	...	...	...	1	...	3	...	14	...	1	...	...	3	3	...	...	1	2	4	1	15
Mount Abu	...	...	...	...	...	...	...	...	...	...	...	...	...	...	...	...	...	...	...	1	...	...	...	...	...	...
Ootacamund	...	...	...	...	...	...	...	...	1	1	...	...	2	...	...	...	...	...	...	1	...	...	...	...	...	1
GROUP XII.—HILL STATIONS	36	42	22	27	25	12	14	19	19	19	20	31	277	14	18	40	19	59	107	85	83	66	65	64	34	654
Rawalpindi Manœuvres	...	...	...	...	...	...	...	...	...	...	13	51	61	...	...	...	...	...	...	...	...	...	...	9	41	50
marching, India	24	21	8	12	2	...	3	2	2	2	21	35	132	67	53	46	7	21	29	7	7	5	15	45	52	354
EXTRA INDIA.																										
(a) In the Indian Command:—																										
Sikkim Tibet Mission Force	...	...	...	...	...	...	...	...	1	...	...	...	1	...	...	...	...	2	1	...	...	22	4	3	...	33
Chabbar	1	...	...	...	...	...	...	...	...	...	...	...	...	...	1	...	...	...	1	...	1	...	...	...	...	2
Jask	...	...	...	...	...	...	...	...	...	...	...	...	...	...	...	...	...	...	...	1	...	1	...	1	...	3
Muscat	...	...	...	...	...	...	1	...	...	...	...	...	...	...	...	...	...	...	...	...	...	2	1	...	...	3
Busbire	...	...	...	...	...	...	...	1	...	...	...	...	1	...	...	...	...	...	...	...	...	...	3	...	...	3
Aden	...	2	...	...	...	...	...	...	1	...	...	...	3	3	6	11	14	11	3	10	9	7	4	7	5	90
Khormaksar	...	...	...	...	...	...	...	...	...	...	...	...	...	2	...	1	...	...	...	2	...	...	...	1	...	6
Sheikh Othman	...	...	...	...	...	...	...	...	...	...	...	...	...	...	...	...	...	...	...	...	...	...	...	1	...	1
Perim	...	1	...	...	...	...	...	...	...	...	...	...	1	...	...	...	...	...	...	...	4	...	2	...	...	13
Aden Column Field Force	...	...	1	1	...	...	...	...	...	...	...	...	2	19	2	10	7	9	6	2	5	4	2	3	4	75
(b) Not in the Indian Command:—																										
Mauritius	...	...	...	...	...	...	...	...	...	...	...	...	1	1	4	13	7	4	3	4	5	8	2	...	...	57
Singapore	...	...	...	...	...	...	1	...	...	...	...	...	1	2	6	...	2	14	16	2	1	2	2	31	16	97
ARMY OF INDIA	256	245	165	124	64	37	37	49	44	76	137	313	1,548	295	211	230	176	397	360	389	423	372	398	460	399	4,017
Northern Command	128	106	83	51	30	14	16	27	12	30	53	198	748	75	46	32	25	130	123	91	125	141	153	156	144	1,271
Western Command	63	75	50	34	13	13	9	10	20	24	34	42	386	50	43	31	63	59	51	68	127	93	94	76	91	846
Eastern Command	26	29	17	12	15	8	6	8	5	15	23	33	197	45	28	28	51	98	89	54	89	82	67	112	59	802
Secunderabad Division	13	12	4	9	3	2	2	2	4	4	6	4	65	31	23	32	8	23	26	38	52	23	18	12	16	302
Burma Division	2	3	2	4	1	...	...	...	1	1	...	1	15	6	6	35	6	17	17	22	15	14	13	16	13	180

III.—PRISONERS, 1905.

JAILS.	Height above the sea-level in feet.*	Authority for height.†
ANDAMANS:—		
Port Blair Convict Settlement	85	S. G.
BURMA:—		
Mergui	14	S. G.
Tavoy	69	"
Moulmein	288	"
Shwegyin	128	"
Toungoo	156	"
Rangoon, Central, Europeans } Natives }	14	...
Maubin	...	...
Myaungmyo	...	"
Bassein, Central	40	S. G.
Insein	34	"
Henzada	44	"
Myanaung	74	"
Sandoway	...	...
Kyaukpyu	...	"
Akyab	32	S. G.
Paungdi	...	...
Prome	...	...
Thayetmyo, Central	145	S. G.
Taungdwingyi	492	"
Magwe	...	"
Minbu	...	...
Yamethin	653	S. G.
Meiktila	298	"
Pagan	...	...
Pakôkku	...	...
Myingyan, Central	243	S. G.
Mandalay "	249	"
Monywa	...	...
Shwebo	600	M. O.
Bhamo	351	S. G.
Katha	329	"
Kindat	361	"
EASTERN BENGAL AND ASSAM —		
Cachar (Silchar)	104	M. D.
Sibsagar	318	S. G.
Dibrugarh	342	"
Tezpur	292	"
Gauhati	134	I. B.
Silhet	257	M. D.
Mymensingh	59	M. D.
Dacca, Central	20	"
Tippera (Comilla)	36	"
Chittagong	87	"
Noakhali	43	"
Backergunge (Barisal)	13	"
Faridpur	46	M. D.
Pabna	...	"
Rajshahi Central (Rampur Bualia)	70	M. D.
Bogra	61	"
Malda	72	"
Dinajpur	123	"
Rangpur	123	"
Jalpaiguri	284	"
Shillong	4,987	S. G.
BENGAL:—		
Khulna	...	...
Jessore	33	M. D.
Baraset	...	...
Presidency, Central, European } Natives }	17	S. G.
Alipore	21	I. B.
Hooghly	34	S. G.
Bordwan	97	"
Krishnagar (Nadia)	72	"
Murshidabad (Berhampore)	67	M. D.
Purneah	120	S. G.
Naya Dumka	489	M. D.
Suri (Birbhum)	...	"
Hankura	298	M. D.
Midnapore, Central	149	"
Balasore	59	S. G.
Cuttack	74	"
Puri	17	"
Angul	...	...
Chaibassa (Singhbhum)	745	S. G.
Purulia (Manbhum)	...	...
Ranchi (Lohardaga)	2,128	S. G.
Palamau (Daltonganj)	...	"
Hazaribagh, Central	1,997	S. G.
Gaya	378	M. D.
Bhagalpur, Central	147	S. G.
Monghyr	148	"
Darbhanga	167	"
Champaran (Motihari)	217	"
Muzaffarpur	179	"
Patna (Bankipore)	177	"
BENGAL:—contd.		
Arrah (Shahabad)	191	S. G.
Chapra (Saran)	181	M. D.
Buxar, Central	204	S. G.
Sambalpur	490	"
Darjeeling	7,168	"
UNITED PROVINCES OF AGRA AND OUDH(a):—		
Korantadih (Ballia)	...	...
Ghazipur	227	S. G.
Azamgarh	256	"
Kasia	...	...
Gorakhpur	255	S. G.
Basti	292	"
Fyzabad	336	I. B.
Sultanpur	305	
Rai Bareli	351	S. G.
Partabgarh	311	"
Jaunpur	263	"
Benares, Central } District	256	"
Mirzapur	283	"
Allahabad, Central } District	298	"
Karwi	...	...
Banda	415	S. G.
Fatehpur	373	"
Hamirpur	367	"
Orai (Jalaun)	...	...
Cawnpore	417	S. G.
Unao	412	"
Lucknow, Central } District	400	"
Barabanki	378	"
Gonda	...	...
Bahraich	398	S. G.
Kheri	471	"
Sitapur	449	"
Hardoi	462	"
Etawah	498	"
Mainpuri	511	"
Etah	550	"
Fatehgarh, Central } District	444	I. B.
Shahjahanpur	507	S. G.
Pilibhit	...	"
Bareilly, Central } District	560	S. G.
Budaon	544	"
Aligarh	610	"
Bulandshahr	727	"
Moradabad	655	"
Bijnor	772	"
Dehra Dun	2,229	"
Saharanpur	903	"
Muzaffarnagar	790	"
Meerut	739	"
Muttra	576	"
Agra, Central } District	554	"
Jhansi	860	"
Lalitpur	...	...
Almora	5,494	S. G.
Pauri	...	...
Naini Tal	6,400	M. D.
PUNJAB:—		
Delhi	715	S. G.
Rohtak	712	
Hissar	689	I. B.
Karnal	800	S. G.
Umballa	902	"
Ludhiana	806	"
Hoshiarpur	1,058	"
Jullundur	960	"
Ferozepore	645	"
Amritsar	750	"
Lahore, Central } District } Female	706	"
Gurdaspur	...	...
Gujranwala	...	...
Sialkot	829	S. G.
Gujrat	...	"
Jhelum	827	S. G.
Rawalpindi	1,707	"
Shahpur	644	"
Mianwali	...	...
Lyallpur	...	...
Jhang	...	...
Montgomery, Central	600	I. B.
Moultan, Central } District	402	S. G.
Dera Ghazi Khan	395	"
Simla	7,230	"
N.-W. F. PROVINCE:—		
Peshawar	1,165	S. G.
Kohat	1,761	"
Bannu	1,279	"
Dera Ismail Khan	571	"
Abbottabad	4,152	"
BALUCHISTAN:—		
Quetta	5,511	S. G.
BOMBAY:—		
Shikarpur	194	S. G.
Sind Gang	...	"
Hyderabad, Central	134	I. B.
Kurrachee	38	S. G.
Rajkot	417	"
Ahmedabad, Central	170	"
Dhulia	842	"
Yerowda, Central (Poona)	1,951	I. B.
Bijapur	1,998	S. G.
Deccan Gang	...	"
Dharwar	2,385	S. G.
Thana	24	"
Bombay, Common } House of Correction }	20	"
Ratnagiri	110	M. D.
Karwar	12	S. G.
Aden	26	"
RAJPUTANA:—		
Ajmer	1,627	S. G.
CENTRAL PROVINCES:—		
Damoh	1,236	S. G.
Saugor	1,753	"
Jubbulpore, Central	1,306	"
Narsinghpur	1,305	I. B.
Mandla	1,487	S. G
Bilaspur	884	"
Raipur, Central	975	"
Balaghat (Burha)	...	...
Seoni	2,043	S. G.
Chhindwara	2,236	"
Hoshangabad	1,030	"
Nimar (Khandwa)	1,042	I. B.
Betul	2,189	S. G.
Nagpur, Central	1,025	"
Bhandara	851	"
Wardha	935	"
Chanda	658	"
Yeotmahl	1,476	"
Amraoti, Central	1,194	"
Akola, Central	920	"
Buldana	2,132	M. D.
HYDERABAD RESIDENCY JAIL:—		
Secunderabad	1,732	S. G.
MADRAS:—		
Mangalore	42	S. G.
Cannanore, Central	47	"
Bellary	1,483	"
Salem, Central	310	"
Coimbatore,	1,348	M. D.
Palamcottah	129	S. G.
Madura	438	"
Trichinopoly, Central	274	"
Tanjore	193	"
Cuddalore	60	"
Vellore, Central	698	"
Madras, Civil } Penitentiary, Central }	15	"
Nellore	57	"
Rajamundry, Central	112	M. D.
Viragapatam	14	S. G.
Berhampur	60	"
Russellkonda	...	...
COORG:—		
Mercara	3,832	S. G.

* These are not the exact heights of the jails themselves above sea-level, but usually those of the survey-marks or of the mercury-surface in barometer cisterns in the stations in which the jails are situated.

† S. G. = Surveyor-General of India; I. B. = Intelligence Branch of the Quarter-Master-General's Department; M. D. = Meteorological Department; M. O. = Medical Officers in charge of Station Hospitals in their Sanitary Reports.

(a) Late North-Western Provinces and Oudh.

TABLE XL.

RATIOS of ADMINISTRATIONS.

The ratios of admissions and deaths to strength are taken from Table XLII. The actuals will be found in Table XLIII.

RATIOS PER 1,000 OF THE AVERAGE STRENGTH.

	Burma	Eastern Bengal and Assam	Bengal	United Provinces	Punjab	N.-W.F. Province	Bombay	Central Provinces	Madras	India*	Andaman	India†
I.—AVERAGE ANNUAL STRENGTH	12,639	6,491	14,172	23,680	11,512	1,314	7,851	3,547	10,147	91,917	14,348	106,265
II.—CONSTANTLY-SICK-RATE OF EACH MONTH—												
January	14·4	38·6	38·2	25·6	27·9	31·2	23·1	18·1	19·6	26·1	62·0	31·1
February	14·9	41·7	45·3	25·0	27·7	34·1	23·9	18·9	19·2	27·7	66·1	33·0
March	14·6	39·0	44·9	24·3	25·8	36·1	22·2	16·3	20·8	29·7	65·0	32·0
April	16·1	43·3	43·2	26·9	24·6	45·6	24·8	18·1	19·7	27·7	65·1	32·9
May	16·2	42·5	38·1	26·7	20·9	37·0	25·9	18·5	19·1	27·1	67·8	32·7
June	16·2	42·5	36·7	25·7	27·8	31·0	25·9	13·7	17·9	26·3	76·6	31·1
July	18·3	46·2	37·3	28·1	27·6	38·1	27·9	22·3	19·1	28·3	89·2	36·5
August	18·7	47·3	41·7	28·4	27·9	33·4	32·5	25·4	19·0	27·5	77·6	35·8
September	17·7	47·6	40·0	31·0	24·1	30·3	31·6	29·9	19·9	29·6	63·8	34·4
October	17·2	42·9	38·2	27·7	21·7	31·0	30·9	26·2	22·6	27·8	60·3	32·1
November	17·3	64·3	...	25·2	21·9	27·6	30·4	20·5	19·2	27·1	57·9	31·2
December	15·9	55·5	32·7	25·7	25·3	25·7	28·9	20·3	18·0	26·2	58·2	30·5
Of the year	16·6	46·2	39·0	26·9	25·5	33·5	27·4	20·9	19·3	27·5	67·7	33·0
INCLUDING SUBSIDIARY JAILS AND LOCK-UPS	...	43·1	37·8	...	25·4	33·2	25·4	20·8	18·9	26·9	...	32·3
III.—ADMISSION-RATE OF THE YEAR—												
Influenza	1·1	16·6	9·6	20·3	7·0	14·16	...	9·0	·6	11·5	13·0	1·7
Cholera	·2	1·9	·6	·9	·3	...	·3	3·6	·8	...	·7	
Small-pox	·2	·2	·1	·2	·3	...	1·8	...	·3	...	·3	
Enteric Fever	·6	·9	·2	·4	·3	...	...	2·8	·7	...	·6	
Intermittent Fever	51·0	260·1	200·7	181·8	246·2	244·0	134·4	190·6	85·6	181·7	1,246·2	325·5
Remittent Fever	·8	2·2	1·5	·9	·5	1·5	1·1	1·7	1·0	1·2	6·1	1·9
Simple Continued Fever	19·0	1·2	3·3	7·3	5·0	...	1·1	4·2	34·0	9·7	...	8·4
Tubercle of the lungs	7·3	11·4	11·5	7·1	8·2	3·8	6·1	5·6	10·5	8·7	11·4	9·1
Pneumonia	3·1	8·0	6·2	12·1	14·2	13·7	13·5	5·9	8·0	9·7	11·0	9·9
Other Respiratory Diseases	10·7	38·9	34·1	22·0	41·2	12·2	29·0	16·6	24·8	26·4	57·4	30·6
Dysentery	28·3	207·0	196·4	36·5	28·4	45·7	41·7	41·4	64·0	81·6	164·4	92·7
Diarrhœa	17·6	90·0	90·9	21·2	42·9	30·4	39·5	36·1	3·5	37·2	55·3	41·4
Spleen Diseases	...	...	·5	1·5	1·1	·3	·8	2·5	...	·7	·6	·5
Scurvy	·2	6·2	1·1	·3	·7	·5	2·8	2·8	...	1·3	1·5	1·3
Anæmia and Debility	3·1	32·2	10·7	10·0	17·9	15·2	9·2	11·0	4·3	11·0	...	9·5
Abscess, Ulcer, and Boil	55·1	38·0	67·1	81·8	102·8	188·0	82·0	74·7	32·3	71·2	79·3	72·3
ALL CAUSES	320·3	1,081·4	953·5	568·8	707·7	560·4	602·7	629·5	460·3	646·8	1,898·3	815·8
INCLUDING SUBSIDIARY JAILS AND LOCK-UPS	...	1,031·7	947·2	...	704·2	1,008·1	606·6	630·1	555·0	655·5	...	815·5
IV.—DEATH-RATE OF THE YEAR—												
Cholera	...	1·87	·35	·17	·35	...	...	·28	1·38	·44	...	·38
Small-pox	·08	...	...	·04	·01	...	...	...	·04	...	·04	
Enteric Fever	·47	...	·14	·08	·09	...	·13	...	·39	·16	...	·14
Intermittent Fever	·47	1·87	2·75	·51	·52	...	·13	...	·79	·91	·84	·90
Remittent Fever	·08	·47	·21	·21	·09	·76	·51	...	·20	·22	2·6	·57
Simple Continued Fever	...	...	...	·03	...	...	...	...	·01	...	·01	
Tubercle of the lungs	3·80	3·14	4·02	2·66	4·43	1·52	2·29	1·97	2·37	3·19	6·48	3·63
Pneumonia	1·03	2·50	2·12	2·10	2·32	1·52	2·50	1·41	1·48	2·28	4·81	2·63
Other Respiratory Diseases	·63	·04	1·06	·10	·69	·76	1·53	1·13	·39	·92	·91	·92
Dysentery	2·61	7·50	6·56	1·94	·52	4·57	1·15	1·41	3·06	3·01	13·03	4·37
Diarrhœa	·93	·78	1·6	·84	·17	·76	1·15	...	·30	·74	·49	·71
Hepatic Abscess	·24	·16	·07	·04	...	...	...	...	·10	·08	·14	·08
Anæmia and Debility	·34	2·66	·49	·21	·69	...	·76	6	·20	·54	...	·47
Phagedæna, Slough, and Gangrene	...	...	...	...	·09	...	...	...	·10	·03	...	·03
ALL CAUSES	17·01	32·96	24·91	17·31	16·33	19·79	17·45	15·79	15·87	19·23	38·96	21·90
INCLUDING SUBSIDIARY JAILS AND LOCK-UPS	...	31·10	25·38	...	16·74	19·19	17·13	·56	15·00	19·11	...	21·66
V.—PERCENTAGE IN 100 ADMISSIONS—												
Influenza	·35	1·53	1·01	3·68	·99	14·74	...	1·43	·13	1·78	·68	1·43
Cholera	·07	·17	·06	·04	·07	...	...	·04	·78	·12	...	·8
Small-pox	·05	·01	·01	·03	·04	...	·30	...	·08	·05	...	·04
Enteric Fever	·20	·12	·02	·07	·05	...	...	...	·59	·10	...	·07
Intermittent Fever	15·93	24·05	30·49	31·95	34·70	35·52	22·30	30·27	18·25	28·10	65·65	39·00
Remittent Fever	·25	·20	·16	·16	·07	·16	·19	·27	·20	·18	·32	·23
Simple Continued Fever	5·93	·12	·33	1·28	·70	...	·19	·67	7·24	1·50	...	1·03
Tubercle of the lungs	2·27	1·05	1·42	1·25	1·15	·40	1·01	·90	2·25	1·35	·60	1·11
Pneumonia	1·04	·80	·67	2·12	2·01	1·43	2·2	1·57	1·89	1·51	·58	1·22
Other Respiratory Diseases	3·33	3·60	3·57	3·88	5·82	1·77	4·82	2·51	5·29	4·08	3·0	3·75
Dysentery	8·84	28·39	20·59	6·41	4·01	4·75	6·91	6·58	13·63	12·61	8·66	11·37
Diarrhœa	5·46	8·32	9·53	3·72	6·0	3·17	6·55	5·73	·76	6·16	2·91	5·07
Spleen Diseases	...	...	·05	·19	·04	·08	·42	...	·02	·10	·00	·07
Scurvy	·05	·58	·11	·14	·10	·16	·46	·45	...	·10	·08	·16
Anæmia and Debility	·96	2·58	1·12	1·77	2·53	1·58	1·32	1·75	·92	1·71	...	1·17
Abscess, Ulcer, and Boil	17·19	3·60	7·04	14·37	14·53	19·57	13·61	11·87	6·89	11·01	4·18	8·87
VI.—PERCENTAGE IN 100 DEATHS—												
Cholera	...	5·7	1·4	1·0	2·1	...	...	1·8	8·7	2·3	...	1·7
Small-pox	...	...	...	·6	·5	...	·7	...	...	·2	...	·2
Enteric Fever	2·8	...	·6	·5	·5	...	...	...	2·5	·8	...	·6
Intermittent Fever	2·8	5·7	11·0	2·9	3·2	...	·7	...	5·0	4·8	2·1	4·1
Remittent Fever	·5	1·4	·8	1·2	·5	3·8	2·9	...	1·2	1·1	7·3	2·6
Simple Continued Fever	...	...	...	...	·5	...	...	...	·0			
Tubercle of the lungs	22·3	10·4	16·1	15·4	27·1	7·7	13·1	12·5	14·9	16·6	16·6	16·6
Pneumonia	6·0	7·6	8·5	17·3	15·4	7·7	17·5	8·9	9·3	11·9	12·3	12·0
Other Respiratory Diseases	3·7	2·8	6·3	4·3	3·8	8·8	7·1	2·5	2·8	2·3		4·2
Dysentery	15·3	12·7	26·3	11·2	3·2	23·1	6·6	8·9	19·3	15·7	33·5	19·9
Diarrhœa	5·6	2·4	4·2	4·9	1·1	3·8	6·6	1·8	1·9	3·8	1·3	3·3
Hepatic Abscess	1·4	·5	·3	·2	...	...	...	...	·6	·4	...	·4
Anæmia and Debility	1·4	8·1	2·0	1·2	4·3	...	4·4	3·6	1·2	2·3	...	2·1
Phagedæna, Slough, and Gangrene	...	...	...	...	·5	...	...	...	1·8	·6	·2	·1

* Including Ajmer, Secunderabad, Quetta, Mercara and excluding Andamans. For complete detail of diseases, see Table LIII.
† Including Ajmer, Secunderabad, Quetta, Mercara and Andamans.

TABLE XLI.

RATIOS of GEOGRAPHICAL GROUPS.

The ratios of admissions and deaths to strength are taken from Table XLII. The actuals will be found in Table XLIII

	I	II	III	IV	V	VI	VII	VIII	IX	X	XI	XII
							N.-W. Frontier, Indus Valley, and N.-W. Rajputana.	S.-E. Rajputana, Central India, and Gujarat.				
	Burma Coast and Bay Islands.	Burma Inland.	Assam.	Bengal the Plain and Orissa.	Gange-tic Plain and Chutia Nagpur.	Upper Sub-Hima-laya.			Dec-can.	West-ern Coast.	South-ern India.	Hills.	India.*
I.—AVERAGE ANNUAL STRENGTH	3,808	3,831	1,371	11,901	22,083	11,330	8,407	4,346	7,017	2,318	9,240	575	91,917
II.—CONSTANTLY SICK-RATE OF EACH MONTH—													
January	12·7	18·2	32·1	46·1	24·0	27·2	28·1	32·0	21·5	19·4	20·1	17·3	26·1
February	12·9	19·3	30·8	54·5	26·6	25·2	27·9	32·1	21·8	16·9	20·4	26·5	27·7
March	11·8	20·7	27·0	52·1	26·6	24·3	25·8	25·9	21·2	16·0	21·7	20·8	26·7
April	13·3	22·4	25·6	46·6	31·9	24·3	25·8	27·7	23·9	18·9	81·1	19·4	27·7
May	15·0	18·8	31·8	40·8	30·3	25·7	26·0	28·0	26·6	18·1	20·3	24·9	27·1
June	15·0	18·7	41·3	41·1	28·5	24·9	26·6	26·0	23·6	20·5	18·5	22·5	26·3
July	16·1	23·2	49·3	40·8	30·7	24·9	27·1	34·1	29·8	24·2	19·5	30·3	28·3
August	15·9	25·1	60·7	40·6	32·1	27·9	26·5	32·1	31·9	23·7	19·1	35·6	29·5
September	14·4	25·4	61·5	38·6	35·6	23·8	24·9	31·1	37·2	20·7	20·1	37·5	29·6
October	13·8	25·9	71·3	34·8	32·0	25·1	23·5	33·3	34·2	19·3	23·1	39·7	27·8
November	16·5	19·5	113·1	39·3	27·8	21·9	23·7	29·1	23·0	23·0	19·6	29·2	27·1
December	14·4	19·8	72·5	42·1	26·2	23·9	24·4	28·1	26·8	22·6	18·3	29·2	26·2
Of the year	14·3	21·9	51·8	43·0	29·5	24·8	25·9	30·1	27·5	19·8	20·0	24·3	27·5
III.—ADMISSION-RATE OF THE YEAR—													
Influenza	...	3·7	3·6	19·9	12·7	18·7	30·9	...	4·6	...	·6	...	11·5
Cholera	...	·8	1·4	1·3	·4	·3	·4	...	·1	...	4·0	...	·8
Small-pox	·2	...	...	·2	·2	·6	1·0	·9	...	1·3	4	...	·3
Enteric Fever	·6	·8	...	1·2	·3	·6	...	...	...	3·0	2·3	...	·7
Intermittent Fever	32·1	94·5	228·6	248·0	230·0	236·6	212·0	160·4	158·0	141·1	87·7	252·2	181·7
Remittent Fever	·8	·8	4·3	1·2	1·1	1·4	1·1	·2	1·0	1·3	1·6	3·5	1·3
Simple Continued Fever	24·6	6·0	5·8	2·7	8·0	4·8	...	...	3·4	11·6	34·4	17·4	9·7
Tubercle of the lungs	6·4	9·4	3·6	15·9	7·9	7·5	8·7	5·5	3·0	1·2	10·8	1·7	8·7
Pneumonia	3·3	3·4	1·4	9·0	9·1	17·3	17·4	23·3	6·1	9·9	8·4	19·1	9·7
Other Respiratory Diseases	8·3	16·2	14·4	36·9	21·5	37·8	32·9	21·4	18·4	31·5	24·4	73·0	26·4
Dysentery	16·3	55·9	259·5	236·6	87·2	33·8	28·4	16·1	47·2	42·3	65·9	80·6	81·6
Diarrhœa	11·5	31·6	56·1	92·8	44·4	35·8	22·8	16·3	38·5	42·7	3·2	69·6	39·2
Spleen Diseases	...	...	·7	3·6	·6	·8	1·8	1·4	1·4	1·3	·1	...	1·3
Scurvy	·2	...	...	3·6	1·4	·3	2·4	·2	2·1	·1	...	...	·1
Anæmia and Debility	2·4	4·7	27·3	20·8	9·7	15·4	16·3	5·5	8·1	10·4	4·7	10·4	11·0
Abscess, Ulcer, and Boil	46·8	74·1	32·4	56·1	76·3	90·9	94·2	84·3	107·9	35·4	33·8	6·6	71·2
All Causes	261·9	454·5	817·7	1,036·1	685·9	660·0	651·1	523·5	672·1	531·5	485·5	878·3	646·8
IV.—DEATH-RATE OF THE YEAR—													
Cholera	...	...	1·44	1·09	·27	·35	·12	...	·14	...	1·52	...	·44
Small-pox	·11	...	...	...	·02	·17	...	...	...	·43	...	...	·04
Enteric Fever	·34	·78	...	·05	·14	·08	...	...	·14	·86	·22	...	·16
Intermittent Fever	·23	1·04	3·59	2·44	1·04	·25	·48	·23	·14	·43	·76	6·96	·91
Remittent Fever	...	·26	...	·42	·18	·25	·48	...	...	·43	·11	1·74	·22
Simple Continued Fever	...	...	...	...	...	·05	...	...	...	...	...	...	·01
Tubercle of the lungs	3·75	3·92	2·16	4·30	3·26	2·35	4·40	2·99	1·43	4·31	2·38	...	3·19
Pneumonia	·91	1·31	·72	2·94	2·49	2·43	3·81	3·45	1·14	1·73	1·52	6·96	2·28
Other Respiratory Diseases	·57	·78	·72	1·43	1·00	·84	·95	·92	1·00	2·16	·33	1·74	·92
Dysentery	1·36	5·43	14·38	6·13	3·58	1·09	1·31	·23	4·31	2·60	5·22	3·01	
Diarrhœa	·34	2·35	...	1·15	·95	·59	·24	...	·86	1·29	·32	...	·74
Hepatic Abscess	·23	·26	...	·17	·05	·05	...	...	...	...	·11	...	·08
Anæmia and Debility	·23	·26	3·75	·92	·41	·59	·48	...	·29	1·73	·22	...	·54
Phagedæna, Slough, and Gangrene	...	...	...	...	...	·08	...	...	·14	...	·11	...	·03
All Causes	13·85	24·28	51·76	28·15	20·02	15·09	19·39	12·43	12·83	25·45	15·37	24·35	19·23
V.—PERCENTAGE IN 100 ADMISSIONS—													
Influenza	...	·80	·41	1·92	1·85	2·79	4·75	...	·68	...	·13	...	1·78
Cholera	...	·17	·17	·12	·06	·04	·05	...	·02	...	·82	...	·12
Small-pox	·09	...	...	·01	·03	·05	·15	·18	...	·24	·09	...	·05
Enteric Fever	·22	·17	...	·11	·05	·09	...	...	...	·57	·47	...	·10
Intermittent Fever	12·27	20·79	26·35	23·93	33·54	35·37	32·55	30·61	23·52	26·54	18·06	28·71	28·10
Remittent Fever	·30	·17	·50	·11	·17	·21	·16	·04	·15	·24	·33	·40	·18
Simple Continued Fever	9·41	1·32	·66	1·16	1·16	·71	...	...	·51	2·19	7·09	1·98	1·50
Tubercle of the lungs	2·43	2·07	·41	1·53	1·16	1·12	1·33	1·05	·57	2·11	2·23	·20	1·35
Pneumonia	1·26	·75	·17	·87	1·32	1·84	2·67	4·20	·91	1·87	1·52	2·18	1·51
Other Respiratory Diseases	2·43	3·56	1·66	3·56	3·58	5·65	5·06	4·00	2·74	5·93	5·02	8·32	4·08
Dysentery	6·24	12·29	29·94	24·77	12·73	5·05	4·37	3·68	7·04	7·95	13·42	9·11	12·61
Diarrhœa	4·38	6·95	6·46	8·96	6·48	5·35	5·04	3·12	5·73	8·02	·65	7·92	6·06
Spleen Diseases	...	...	·06	...	·09	·11	·27	·26	...	...	·34	·02	·10
Scurvy	·09	...	·08	·35	·20	·06	·37	·04	·32	·08	...	...	·20
Anæmia and Debility	·01	1·03	3·15	2·01	1·41	2·31	2·50	1·05	1·21	1·95	·56	1·19	1·71
Abscess, Ulcer, and Boil	17·86	16·31	3·73	5·42	11·13	13·58	14·47	17·05	16·05	6·66	6·95	7·13	11·02
VI.—PERCENTAGE IN 100 DEATHS—													
Cholera	...	...	2·8	3·9	1·4	1·7	·6	...	1·1	...	9·9	...	2·3
Small-pox	·8	...	...	...	·3	1·1	·6	...	...	1·7	...	...	·2
Enteric Fever	2·5	3·3	...	·3	·7	...	...	...	...	3·4	1·4	...	·8
Intermittent Fever	1·6	4·3	6·0	8·7	5·2	1·7	2·5	1·9	1·1	1·7	4·9	28·6	4·8
Remittent Fever	...	1·1	...	1·5	·9	1·7	2·5	...	...	1·7	·7	7·1	1·1
Simple Continued Fever	...	...	...	...	...	·6	...	...	...	...	...	...	·1
Tubercle of the lungs	27·0	16·1	4·2	14·9	10·3	15·6	22·7	24·1	11·1	16·9	15·5	...	16·6
Pneumonia	6·6	5·4	1·4	10·4	12·4	16·1	19·0	27·8	8·9	6·8	9·8	28·6	11·9
Other Respiratory Diseases	4·1	3·2	1·4	5·1	5·0	5·6	4·9	7·4	7·8	8·5	1·4	7·1	4·8
Dysentery	9·8	21·6	27·8	21·8	17·0	7·2	6·7	1·9	...	16·9	16·9	21·4	15·7
Diarrhœa	2·5	9·7	...	4·2	4·8	3·9	1·2	...	6·7	5·1	2·1	...	3·8
Hepatic Abscess	1·6	1·1	...	·6	·2	·3	...	...	...	...	·7	...	·4
Anæmia and Debility	1·6	1·1	11·1	3·3	2·0	3·9	2·5	...	2·2	6·8	1·4	...	2·8
Phagedæna, Slough, and Gangrene	...	...	...	...	...	·6	...	...	1·1	...	·7	...	·2

* Including Aden and excluding Andamans. For complete detail of diseases, see Table LIII.

TABLE XLII.

RATIOS of JAILS, GROUPS, and ADMINISTRATIONS.

For actuals see Table XLIII.

JAILS AND GROUPS.	Average annual strength.	1. ADMISSION-RATE.						2. DEATH-RATE PER 1,000 OF STRENGTH.													Average number constantly sick per 1,000 of strength.
		Influenza.	Cholera.	Small-pox.	Enteric Fever.	Intermittent Fever.	Remittent Fever.	Simple Continued Fever.	Tubercle of the lungs.	Pneumonia.	Other Respiratory Diseases.	Dysentery.	Diarrhœa.	Hepatic Abscess.	Spleen Diseases.	Scurvy.	Anæmia and Debility.	Abscess, Ulcer, and Boil.	Phagedæna, Slough, and Gangrene.	ALL CAUSES.	
Mergui	63	...	...	...	...	...	...	...	...	15·9	15·9	...	...	...	...	...	15·9	...	254·0	15·9
Tavoy	103	...	...	...	...	126·2 9·71	...	...	...	...	9·7	9·7	...	...	...	...	9·7	...	378·6 9·71	19·4
Moulmein	636	...	...	...	...	95·9	...	...	3·1 1·57	3·1 1·57	12·6 6·29	67·6	14·2	...	...	...	75·5	...	427·7 15·72	20·4
Shwegyin	169	...	...	...	...	59·2	...	...	...	...	5·9	53·3 5·92	53·3 5·92	...	...	...	76·9	...	384·6 17·75	17·8
Toungoo	405	...	...	...	...	66·5	...	...	7·4 4·93	2·5	27·1	49·3	91·1	...	2·5	9·9 4·93	64·0	...	692·1 17·24	36·9
Rangoon, Central (Europeans).	20	...	...	...	...	...	...	50·0	...	...	...	200·0	...	...	...	...	50·0	...	600·0	161·0
Rangoon, Central (Natives).	2,143	...	...	...	...	8·4	...	57·4 5·43	7·5 ·93	3·3	5·6	13·5 ·47	1·4 ·47	·3 ·7	...	...	1·4	8·9	...	187·6 14·47	13·1
Maubin	339	...	...	...	...	13·9	...	...	...	...	12·1 2·79	11·1	8·4	...	...	...	2·8	30·6	...	141·8 11·14	8·4
Myaungmyo	627	...	...	...	1·6 1·59	19·1	6·4	...	3·2 3·19	...	6·4	11·2 3·19	367	...	...	...	162·7	...	374·8 14·35	19·1
Bassein, Central	1,091	...	...	...	3·7 1·83	30·2	1·8	49·5	15·6 3·67	11·0 3·67	19·2	11·9 ·92	1·8	...	...	1·8	63·2	...	303·4 11·92	18·3
Insein, Central	2,061	...	...	1·0 49	...	26·2 ·49	...	18·0	6·3 5·34	·5	1·9	1·0	...	...	...	·5	49·3	...	161·1 12·62	8·7
Henzada	398	...	...	...	...	62·8	2·5	...	...	...	...	10·1 2·5	2·5	...	...	17·6	67·5	...	278·9 7·54	10·1
Myanaung	79	...	...	...	...	25·3	...	...	...	...	...	88·5	...	...	...	25·3	38·0	...	265·8	12·7
Sandoway	69	...	...	...	...	130·4	...	...	...	...	...	87·0 14·49	14·5	...	...	...	87·0	1	550·7 14·49	43·5
Kyaukpyu	141	...	...	...	...	49·6	...	14·2	14·2 7·09	...	7·1 7·09	...	7·1 7·09	...	...	...	...	...	248·2 42·55	7·1
Akyab	443	...	...	...	...	15·8	...	...	2·3 2·26	13·5 2·26	13·5 6·77	11·3 2·26	2·3	...	2·3	2·3	4·5	...	146·7 18·05	9·0
GROUP I.—BURMA COAST AND BAY ISLANDS	8,808	...	...	·2 ·11	·6 34	32·1 ·23	·8	24·6	6·4 3·75	2·3 ·91	3·3 ·57	16·3 1·36	11·5 ·34	·7 ·2	...	2·4 ·23	46·8	...	261·9 13·85	14·3
Paungdi	184	...	...	...	...	·2	5·4	...	...	...	5·4 5·43	35·0 5·43	...	...	...	...	27	...	195·7 5·43	10·3
Prome	323	...	...	...	...	120·7	...	6·2 3·10	3·1 3·10	3·1	15·5	61·9 9·29	49·5 3·10	...	...	24·8	226·0	...	993·8 21·67	49·5
Thayetmyo, Central.	956	...	1·0	...	...	38·7	...	1·0	16·7 8·37	4·2 3·14	18·8 1·05	47·1 10·46	25·1 2·09	1·0 1·05	...	2·1	54·4	...	338·9 37·06	14·6
Taungdwingyi	44	...	...	...	...	22·7	...	...	22·7 22·73	...	...	...	...	...	...	...	22·7 22·73	...	22·7 22·73	5·4
Magwe	153	...	...	...	...	26·1	...	...	...	...	...	58·8 6·54	45·8	...	...	...	26·1	...	209·2 13·07	13·1
Minbu	1	...	...	...	...	...	...	...	...	...	...	...	...	...	...	...	...	...	...	
Yamethin	95	...	...	...	...	...	...	...	...	...	...	...	...	...	...	...	...	...	63·2	10·5
Meiktila	73	...	...	...	...	109·6 27·40	...	...	13·7 13·70	...	...	27·4	13·7	...	...	...	41·2	...	271·0 41·10	13·7
Pagan	48	...	...	...	...	...	...	...	41·7 20·83	...	...	...	...	...	...	...	125·0	...	312·5 62·50	41·7
Pakôkku	30	...	...	...	...	33·3	...	...	...	...	...	...	...	...	...	...	...	...	66·7 33·33	4·3
Myingyan, Central.	771	18·2	...	...	...	75·2	...	6·5	6·5 3·59	...	13·0	103·8 3·89	55·8 5·19	...	...	6·5	65·7	...	470·0 24·6	24·6
Mandalay, Central.	751	...	1·3	...	40 39	233·0 ·66	1·3 1·33	18·6	10·7 1·33	1·3	24·0 1·33	51·9 2·66	17·3	...	...	...	71·9	...	580·0 17·31	24·0
Monywa	82	...	...	...	...	12·2	...	...	24·4	24·4	12·2	...	...	...	...	...	...	...	134·1	12·9
Shwebo	147	...	...	...	...	40·8	...	...	...	...	13·6	34·0 13·61	54·2	...	...	...	54·4	...	272·1 13·61	13·6
Bhamo	63	...	...	...	...	285·7	...	15·9	31·7 5·87	47·6 15·87	79·4	31·7	95·2	...	...	...	317·5	...	1,539·7 47·02	63·5
Katha	65	...	...	...	...	15·4	...	...	15·4 38	5·4	15·4	46·2	15·4	...	30·8 15·38	46·2	...	203·3 30·7	15·4
Kindat	45	...	22·2	...	...	4·44	...	...	...	...	...	22·2 22	22·2	...	...	22·2	66·7	...	311·1 44·44	22·3
GROUP II.—BURMA INLAND.	3,831	3·7	·8	...	·8 ·78	94·5 1·04	·8 ·26	6·0	9·4 3·92	3·4 1·31	16·2 ·74	55·9 5·48	31·6 2·35	·3 ·26	...	4·7 ·26	74·1	...	454·5 24·38	21·9

Worked on the aggregates

TABLE XLII—continued.

RATIOS of JAILS, GROUPS, and ADMINISTRATIONS.   For actuals see Table XLIII.

JAILS AND GROUPS.	Average annual strength.	1. ADMISSION-RATE.								2. DEATH-RATE PER 1,000 OF STRENGTH.											Average number constantly sick per 1,000 of the sth.
		Influenza.	Cholera.	Small-pox.	Enteric Fever.	Intermittent Fever.	Remittent Fever.	Simple Continued Fever.	Tubercle of the lungs.	Pneumonia.	Other Respiratory Diseases.	Dysentery.	Diarrhœa.	Hepatic Abscess.	Spleen Diseases.	Scurvy.	Anæmia and Debility.	Abscess, Ulcer, and Boil.	Phagedæna, Slough, and Gangrene.	All Causes.	
Cachar	58	... ...	... ...	... ...	...	86·2 17·74		... ...	... ...	17·2 ...	103·4 ...	... ...	... ...	...	... ...	...	51·7 ...	34·5 ...	... ...	379·3 34·48	34·5
Sibsagar	71	... ...	28·2 28·17	... ...	...	98·6 ...		... ...	14·1 ...	42·3 ...	126·8 ...	112·7 ...	...	... ...	...	14·1 ...	84·5 ...	... ...	676·1 42·25	28·2
Dibrugarh	104	19·2 ...	... ...	... ...	...	538·5 ...		19·2 9·62	... ...	28·8 28·85	567·1 ...	153·8 ...	...	... ...	...	67·3 ...	28·8 ...	..	1,548·1 57·69	57·7
Tezpur	265	... ...	... ...	... ...	...	124·5 ...		7·5 3·77	... ...	90·6 7·35	79·2 ...	3·8 ...	...	... ...	...	41·5 2·64	34·0 ...	...	366·0 45·28	37·7
Gauhati	309	9·7 ...	... ...	... ...	...	339·8 3·24		25·9 ...	... ...	25·9 3·24	530·7 48·54	48·5 ...	...	... ...	...	16·2 3·24	35·6 ...	...	1,153·6 58·25	93·9
Sylhet	584	... ...	... ...	... ...	...	191·8 5·14	10·3	... ...	1·7 1·71	1·7 1·71	8·6 ...	169·5 ...	30·8	... ...	...	1·7 1·71	18·8 1·71	24·0 ...	801·4 54·79	39·4
GROUP III.— ASSAM.	1,391	3·6 ...	1·4 1·44	... ...	...	228·6 3·59	4·3	5·8 ...	3·6 2·16	1·4 ·71	14·4 ·72	259·5 14·38	56·1	·7 ...	...	·7 ·72	27·3 5·75	32·4 ...	867·7 51·76	51·8
Mymensingh	562	... ...	1·8 1·78	... ...	17·8	329·2 ...	1·8 1·78	... ...	44·5 5·34	10·7 1·78	33·8 ...	471·5 1·7	... ...	...	... ...	19·6	53·4 ...	48·0 ...	1,601·4 13·13	78·3
Dacca, Central	1,147	88·1 ...	1·7 1·74	... ...	...	96·8 ·87		... ...	6·1 1·74	4·4 1·74	82·0 ...	116·8 3·49	24·4 ·87	...	... ...	...	52·3 4·36	28·9 ...	...	654·8 21·86	21·8
Tippera	282	... ...	... ...	3·5	...	70·9 ...		... ...	28·4 ...	14·2 3·55	7·1 ...	21·3 3·55	202·1 ...	...	81·6 3·55	7·1 ...	7·1 ...	...	698·6 14·18	39·0
Chittagong	138	... ...	... ...	... ...	...	565·2 ...		... ...	14·4 7·25	7·2 ...	210·1 ...	311·6 ...	...	... ...	...	...	65·2 ...	...	1,434·8 14·49	36·2
Noakhali	145	... ...	... ...	... ...	...	151·7 ...		... ...	6·9 ...	13·8 13·79	20·7 ...	744·8 ...	27·6	...	... ...	...	...	13·8 ...	1,595·2 20·69	41·4
Backergunge	509	... ...	5·9 5·89	... ...	...	151·3 3·93		... ...	5·9 7·86	21·6 ·89	57·0 ...	776·0 170·8	320·2 ...	...	... ...	...	96·3 5·89	47·2 ...	1,656·6 62·87	78·6
Khulna	37	... ...	... ...	... ...	...	270·3 ...		... ...	... ...	... ...	...	486·5 54·05	27·0	...	... ...	...	...	...	1,000·0 54·05	27·0
Jessore	372	... ...	... ...	... ...	...	341·4 2·60	2·7	... ...	18·8 5·38	8·7 ...	18·8 ...	881·7 16·13	51·1	...	... ...	...	18·8 ...	48·4 ...	1,500·0 26·88	53·8
Baraset	119	... ...	... ...	... ...	...	529·4 ...		... ...	... ...	33·6 ...	344·5 8·40	58·8 ...	...	8·4 ...	...	...	84·0 ...	1,411·8 25·21	167·2
Presidency, Central (Europeans).	37	135·1 ...	... ...	... ...	...	... ...		... ...	... ...	27·0 ...	162·2 ...	...	...	... ...	...	...	27·0 ...	513·5 27·03	27·0
Presidency, Central (Natives).	1,175	50·2 ...	1·7 1·70	... ...	...	38·3 5·96		... ...	12·8 4·26	10·2 5·96	14·5 1·70	158·3 6·81	20·4 1·70	·9	·9 ...	11·1 ...	21·3 ...	595·1 33·19	28·1
Alipore, Central	1,817	39·6 2·75	·6 ·55	... ...	...	257·6 6·05		... ...	33·0 8·81	5·5 2·20	18·7 1·65	114·5 2·75	44·0 ...	...	... ...	7·7	53·9 ...	...	785·9 31·92	39·1
Hooghly	401	... ...	... ...	... ...	...	209·5 ...		... ...	2·5 ...	12·5 2·49	20·0 ...	152·1 4·99	226·9 2·49	2·5 2·49	...	15·0 ...	79·8 ...	965·1 14·96	34·9
Burdwan	253	... ...	... ...	... ...	...	735·2 3·95	7·9 7·91	... ...	19·8 3·95	19·8 15·81	102·8 3·95	131·4 3·95	143·2 ...	277·7	... ...	7·9 ...	67·2 ...	1,703·6 47·43	59·3
Krishnagar	156	... ...	... ...	... ...	...	230·8 ...		... ...	12·8 ...	19·2 ...	115·4 ...	128·2 1·723	25·6 ...	6·4	19·2 ...	96·2 ...	1,025·6 19·23	44·9
Faridpur	343	... ...	... ...	... ...	...	449·0 ...		... ...	32·1 8·75	20·4 2·92	17·5 5·83	346·9 8·75	29·2 ...	...	... ...	2·9	75·8 ...	1,530·6 29·15	61·2
Pabna	195	... ...	... ...	... ...	...	646·2 ...	25·6 5·13	... ...	15·4 10·26	46·2 5·13	...	256·4 ...	112·8	...	... ...	...	61·5 ...	1,369·2 20·51	41·1
Murshidabad	215	... ...	9·3 ...	... ...	...	151·4 ...	4·7	4·7 ...	4·7 ...	55·8 ...	88·4 ...	69·8 ...	...	... ...	...	37·2 ...	9·3	809·3 ...	23·3
Rajshahi, Central.	784	... ...	2·6 2·55	... ...	...	90·6 2·55		... ...	10·2 7·65	7·7 3·83	21·7 1·28	178·6 7·65	25·4 1·28	...	... ...	7·7	34·4 ...	556·1 30·61	30·6
Bogra	186	... ...	... ...	... ...	...	279·6 ...		... ...	... ...	26·9 5·38	...	150·5 ...	21·3	...	... ...	26·9	48·4 ...	661·3 5·38	21·5
Malda	95	... ...	... ...	... ...	...	1,400·0 ...	10·5	... ...	... ...	10·5 10·5	...	63·2 136·8	... ...	...	... ...	73·7 10·53	42·1 ...	2,000·0 10·53	63·2
Dinajpur	217	... ...	9·2 9·22	... ...	...	502·3 4·61	4·6 4·61	... ...	18·4 4·61	18·4 4·61	41·5 ...	506·9 ...	138·2	...	... ...	9·2	41·5 ...	1,465·4 27·65	50·9
Rangpur	248	... ...	... ...	... ...	8·1	697·6 ...		... ...	17·1 ...	112·9 ...	...	762·1 4·03	137·1 4·03	4·0 4·03	... ...	20·2	8·1 ...	48·4 ...	2,104·8 40·32	44·4

Jails and Groups	Average annual strength	Influenza	Cholera	Small-pox	Enteric Fever	Intermittent Fever	Remittent Fever	Simple Continued Fever	Tubercle of the lungs	Pneumonia	Other Respiratory Diseases	Dysentery	Diarrhœa	Hepatic Abscess	Spleen Diseases	Scurvy	Anæmia and Debility	Abscess, Ulcer, and Boil	Phagedæna, Slough, and Gangrene	All Causes	Average number constantly sick per 1,000 of strength
Jalpaiguri	109	...	...	...	...	266·1 / 9·17	...	...	9·2	...	45·9	229·4 / 13·35	100·9	...	...	...	36·7	73·4	9·2	1,073·4 / 27·52	55·0
Purneah	223	...	...	...	...	228·7 / 4·48	...	...	13·5	...	4·5	103·1	40·4	...	...	4·5	41·8	93·7	...	762·3 / 13·45	35·9
Naya Dumka	144	...	...	...	...	125·0	41·7	...	6·9	34·7	118·1 / 6·94	104·2 / 6·91	...	...	...	...	13·9	76·4	...	694·4 / 13·89	20·8
Suri	212	...	...	...	...	160·4	...	...	4·7 / 4·72	18·9	136·8	23·6	...	...	...	...	...	70·8	...	627·4 / 18·27	9·4
Bankura	223	...	...	9·0 / 4·48	...	103·1	4·5	...	13·5 / 4·48	9·0	71·7	130·0	17·9	...	...	4·5	...	85·2	...	637·8 / 13·45	31·4
Midnapore, Central	894	...	...	...	...	287·7	...	...	21·6 / 5·5	8·9 / 3·36	51·5 / 3·36	337·8 / 21·25	257·3 / 4·47	...	...	...	20·1 / 2·24	119·7	...	1,637·6 / 45·06	82·8
Balasore	144	...	...	...	...	256·9	...	180·6	6·9	13·4	34·7	437·5 / 6·94	41·7	6·9	...	...	...	20·8	...	1,222·2 / 27·78	27·8
Cuttack	336	...	...	...	...	297·6	...	...	2·98	3·0	11·9	139·9	145·8	...	...	...	11·9	130·9	...	931·5 / 11·90	20·8
Puri	125	...	...	8·0	...	152·0	...	...	8·0	8·0	...	112·0	...	...	...	...	...	8·0	...	392·0	16·0
Angul	58	...	...	...	...	206·9 / 17·24	...	...	...	...	34·5	86·2 / 6·13	124·1	...	...	...	...	258·6	...	1,019·0 / 34·48	51·7
Group IV.—Bengal and Orissa	11,921	19·0 / ·42	1·3 / 1·09	·1	1·2 / ·08	248·0 / 2·44	1·2 / ·42	2·7	15·9 / 4·20	9·0 / 2·91	36·9 / 1·43	256·6 / 6·13	92·8 / 1·18	·3 / ·17	·6	3·6 / ·03	20·8 / ·92	56·1	·3	1,036·1 / 28·15	43·0
A. Chaibassa	167	...	...	...	...	243·5	6·0 / 5·9	...	12·0	12·0 / 5·90	185·6	275·4 / 5·9	347·3	...	...	...	...	53·9	12·0	1,616·8 / 35·90	41·9
Purulia	263	...	...	...	...	361·2	...	...	41·8 / 3·80	26·6 / 7·60	102·7 / 3·80	216·7 / 26·02	262·4 / 26·02	...	...	...	26·6	68·4 / 3·80	...	1,433·5 / 45·03	41·8
Ranchi	209	...	...	...	...	86·1	...	...	43·1 / 19·14	9·6	191·4 / 4·78	...	...	...	...	...	...	4·8	...	401·9 / 28·71	19·1
Palamau	100	...	...	...	...	430·0	...	10·0	10·0	30·0	...	240·0 / 10·00	120·0 / 10·00	...	...	...	10·0	140·0	...	1,200·0 / 10·00	30·0
Hazaribagh, Central	923	...	2·1 / 1·08	...	...	252·4 / 1·08	...	...	15·2 / 5·41	5·4 / 3·25	69·3 / 2·17	253·5 / 15·17	17·7	...	10·8	...	19·5	124·6	...	1,055·3 / 35·73	85·6
B. Gaya	395	...	...	5·1	...	159·5 / 2·53	...	20·3	7·6 / 5·06	...	20·3	88·6 / 7·59	10·1	...	...	...	...	75·9	...	481·5 / 25·32	17·7
Bhagalpur Central	1,803	...	...	...	...	196·3 / 2·22	...	...	2·2 / 1·66	5·0 / 1·66	20·5	142·5 / 1·06	20·5	...	...	...	5·0	75·4	...	632·8 / 9·95	20·5
Monghyr	300	...	...	...	...	156·7 / 3·33	...	...	6·7 / 3·33	3·3	45·7	153·3 / 3·33	16·7 / 6·67	...	3·3	...	...	36·7	...	626·7 / 20·0	20·0
Darbhanga	221	...	...	...	...	140·3	...	...	4·5 / 4·52	9·0	...	217·2	72·4 / 9·05	...	...	...	40·7	54·3	...	669·7 / 13·57	36·2
Champarun	278	...	...	...	46·8	244·6 / 14·39	14·4	...	7·19	18·0	...	161·9 / 3·60	61·2	...	...	...	25·2 / 7·19	28·8	...	838·1 / 35·91	39·6
Muzaffarpur	320	...	...	...	...	246·9 / 6·25	...	...	12·5	12·5	...	165·6 / 9·18	56·2	...	...	...	9·4 / 6·25	40·6	...	631·2 / 25·00	21·9
Patna	297	...	...	...	...	178·5	...	...	10·1	...	30·3	67·3 / 3·37	26·9	...	...	...	6·7	74·1	...	673·4 / 20·20	30·3
Arrab	250	...	...	...	4·0 / 4·00	216·0	4·0	...	4·0	36·0	...	128·0 / 12·00	80·0	...	...	...	4·0	24·0	...	644·0 / 28·00	24·0
Chapra	267	...	3·7 / 3·75	...	...	247·2	...	...	26·2 / 11·24	3·7 / 3·75	7·5	363·3 / 3·75	41·2 / 5·75	...	...	...	11·2	44·9	...	850·1 / 33·71	22·5
Buxar, Central	1,228	...	...	...	...	982·0 / 1·63	...	...	4·1 / 2·44	3·3	32·6	221·5 / 4·89	274·4	...	...	...	8·1	58·6	...	1,757·3 / 10·59	45·6
Korantadih	43	...	...	...	...	488·4	...	...	...	...	...	23·3	46·5	...	...	...	...	46·5	...	651·2	11·38
Ghazipur	349	...	...	...	...	300·9	...	...	14·3 / 14·33	17·2 / 2·87	17·19	60·2 / 17·19	17·2 / 2·87	...	...	...	...	100·3	...	598·9 / 45·85	28·7
Azamgarh	289	3·5	...	...	...	200·7 / 3·46	...	...	6·9 / 3·46	17·3	38·1	27·7	20·8	...	6·9	...	13·8	6·9	...	597·5 / 2·22	17·3
Kasia	22	...	...	...	...	181·8	...	...	...	...	...	545·5 / 45·45	45·5 / 45·45	...	...	...	45·5	...	...	863·6 / 45·45	45·5
Gorakhpur	486	...	...	10·3 / 8·23	...	463·0	...	...	2·1 / 6·17	10·3 / 2·06	25·3	288·1 / 2·06	32·9	...	...	2·1	61·7	53·8	...	1,397·1 / 22·63	70·0
Basti	291	...	...	...	...	96·2	...	...	3·4 / 6·87	10·3	6·87	20·6	3·4	...	...	...	...	34·4	...	219·9 / 17·18	6·9
Fyzabad	305	...	3·3	...	...	180·3	...	...	10·4	...	...	26·2 / 3·28	29·5 / 6·56	...	...	...	36·1 / 29·5	...	...	708·2 / 22·51	2·5
Sultanpur	266	...	...	...	...	251·9	...	...	18·8 / 3·76	7·5 / 3·76	11·3	22·6	11·3	...	...	...	...	67·7	...	571·4 / 22·51	18·8
Rai Bareli	406	2·3	...	...	...	83·7	...	17·2	7·4	4·9	14·3	19·7 / 2·46	17·2 / 2·46	...	...	...	4·9	56·7	...	438·4 / 19·70	27·1

* Worked on the aggregates.

95

TABLE XLII—*continued.*

RATIOS *of JAILS, GROUPS, and ADMINISTRATIONS.* For actuals see Table XLIII.

JAILS AND GROUPS.	Average annual strength.	1. ADMISSION-RATE.										2. DEATH-RATE PER 1,000 OF STRENGTH.									Average number constantly sick per 1,000 of strength.
		Influenza.	Cholera.	Small-pox.	Enteric Fever.	Intermittent Fever.	Remittent Fever.	Simple Continued Fever.	Tubercle of the lungs.	Pneumonia.	Other Respiratory Diseases.	Dysentery.	Diarrhœa.	Hepatic Abscess.	Spleen Diseases.	Scurvy.	Anæmia and Debility.	Abscess, Ulcer, and Boil.	Phagedæna, Slough, and Gangrene.	All Causes.	
Partabgarh	176	...	...	...	...	108·0 / 5·68	...	...	...	...	28·4	31·1 / ...	34·1 / 5·68	...	...	...	39·8	73·9	5·7	483·0 / 11·36	32·7
Jaunpur	286	...	...	...	...	146·9	...	...	3·5	7·0 / 3·50	49·0	59·4	31·5	...	...	...	17·5	66·4	...	500·0 / 13·99	17·5
Banares, Central	1,250	46·4	...	...	...	240·0	...	...	4·8 / 4·50	10·4 / 3·20	6·4 / ·80	22·4 / ·20	23·2 / 2·40	...	2·4	3·2 / 2·40	92·8	...	...	572·8 / 1·60	27·2
Benares, District	324	...	...	...	...	95·7	...	...	9·3 / 6·17	6·2 / 6·17	6·2	37·0 / 6·17	15·4 / 3·09	...	...	...	13·7	...	...	487·7 / 37·04	24·7
Mirzapur	208	...	...	...	...	379·8	9·6 / 4·81	...	...	24·0 / 4·81	33·7	67·3	19·2	...	...	...	57·7	...	...	759·6 / 9·62	14·4
Allahabad, Central	1,337	...	...	...	...	96·5 / 2·34	110·7	18·0 / 6·73	3·7 / ·73	39·6 / 1·50	41·1	36·6 / ·75	...	...	...	115·9	...	...	653·7 / 17·99	39·6
Allahabad, District	357	...	...	...	...	70·0	...	...	7·2 / 1·20	23·3 / 5·39	37·7 / 1·80	37·7 / 3·5	1·8	...	...	...	19·7	80·8 / 1·80	...	474·0 / 16·16	26·9
Karwi	...	...	...	...	...	117·6	...	...	...	...	...	29·4	...	...	...	...	...	88·2	...	794·1	29·4
Baoda	225	...	...	...	...	184·4	...	...	13·3	13·3	57·8	62·2 / 8·89	97·8 / 4·44	35·6	...	...	22·2	222·2	...	1,760·0 / 35·56	48·9
Fatehpur	218	...	...	...	...	605·5 / 4·59	...	...	18·3 / 9·17	27·5	...	45·9	...	...	...	...	45·9	...	862·4 / 13·76	22·9
Hamirpur	110	...	...	...	...	490·9	9·1	...	36·4 / 9·09	63·6	36·4	...	...	...	...	...	163·6	...	1,263·6 / 45·45	36·4
Orai	151	...	...	...	...	430·5 / 6·62	...	33·1	6·6	19·9	53·0	19·9	...	...	...	125·8	6·6	1,072·8 / 13·25	39·7
Cawnpore	412	...	...	...	...	169·5 / 2·39	...	...	...	19·1	11·9 / 2·39	11·9	4·8	...	4·8	40·6	...	439·6 / 9·53	21·5
Unao	322	...	...	...	...	155·3	3·1	0·3 / 6·21	9·3 / 3·11	12·4	18·6	...	...	...	...	43·5	...	344·7 / 12·42	15·5
Lucknow, Central	1,450	...	...	1·0 / ·69	52·4	1·0 / ·69	...	4·8 / 3·45	6·9 / 1·38	6·0 / 2·07	11·7 / 1·38	9·0 / ·69	...	...	...	44·1	...	204·8 / 11·72	11·7
Lucknow, District	482	...	...	...	56·0	...	...	...	10·4	2·1	13·4 / 4·15	10·4	...	...	...	80·9	...	307·1 / 0·37	16·6
Barabanki	381	2·6	...	...	147·0	...	2·6	...	10·5	21·0	26·2	36·7	...	...	...	15·7	105·0	...	543·3 / 2·62	28·9
Gonda	442	13·	...	6·	...	49·8	...	9·0	13·6 / 4·52	43·0 / 4·52	38·5 / 4·52	6·8	4·5	...	2·3	24·9	18·1	...	359·4 / 18·10	27·1
Bahraich	260	...	...	...	280·8	...	11·5 / 3·85	...	7·7 / 3·85	15·4	15·4	15·4	...	...	...	11·5	115·4	...	647·3 / 11·54	23·1
Kheri	305	...	...	...	157·4 / 3·28	...	...	...	6·6 / 3·28	23·0	9·8	19·7	...	...	...	124·6	...	541·3 / 13·11	19·7
Sitapur	494	120·6 / 10·12	...	...	85·0	...	2·0	...	20·2 / 2·02	10·1 / 6·07	18·2 / 4·05	4·0	...	2·0	...	26·3	...	447·4 / 24·29	16·2
Hardoi	384	7·8	...	...	187·5	...	...	7·8 / 5·21	2·6 / 2·60	5·2	10·4	5·2	...	...	2·6	65·1	...	356·8 / 7·81	10·4
Etawah	300	...	...	...	140·0	...	...	·7 / 3·33	6·7	13·3	36·7 / 3·33	...	...	...	10·0	...	363·3 / 23·33	10·0
Mainpuri	333	...	...	3·0	192·2	...	...	6·0 / 6·01	15·0	18·0	78·1	45·0	...	9·0	12·0	121·1	...	657·-- / 12·01	45·0
Etah	328	...	...	3·0	289·6	...	...	9·1 / 6·10	51·8 / 12·20	42·7 / 3·05	64·0 / 3·05	9·1	...	105·7	198·2	...	953·4 / 30·63	45·7
Fatehgarh, Central	1,522	6·7	...	...	2·6 / ·65	135·1	2·0	...	12·4 / 7·18	9·1 / 4·57	6·5	36·8 / 3·26	33·3 / ·05	·7	·7	...	2·0	7·1	...	517·0 / 18·02	28·1
Fatehgarh, District	297	16·8	...	...	127·9	...	...	6·7 / 3·37	...	23·6 / 3·37	60·6 / 3·37	23·6	...	...	...	50·5	...	474·7 / 16·84	20·2
GROUP V.— GANGETIC PLAIN AND CHUTIA NAGPUR. A.	22,083	12·7 / ·23	·4 / ·27	·2	·1 / ·14	230·0 / 1·04	1·1 / ·18	8·0	7·9 / 3·26	9·1 / 2·49	24·3 / 1·00	87·2 / 3·58	44·4 / ·95	·0 / ·05	·6	1·4 / ·41	9·7	76·3 / ·09	·2	685·9 / 20·02	19·5
Shahjahanpur	314	...	...	...	258·0	...	...	6·4 / 6·37	22·3 / 3·18	19·1	25·5	47·8	...	...	...	98·7	...	694·3 / 12·74	22·3
Pilibhit	45	...	...	...	177·8	...	...	...	22·2	22·2	111·1	44·4	...	...	...	66·7	...	666·7	22·2
Bareilly, Central	1,771	...	...	·6	300·4	...	...	16·9 / 1·69	1·7	16·4 / ·56	4·5	16·9 / ·56	...	...	...	11·9 / ·56	15·6 / ·56	...	439·9 / 6·78	26·0

JAILS AND GROUPS.	Average annual strength	1. ADMISSION-RATE.									2. DEATH-RATE PER 1,000 OF STRENGTH.										Average number constantly sick per 1,000 of strength.
		Influenza.	Cholera.	Small-pox.	Enteric Fever.	Intermittent Fever.	Remittent Fever.	Simple Continued Fever.	Tubercle of the lungs.	Pneumonia.	Other Respiratory Disease.	Dysentery.	Diarrhœa.	Hepatic Abscess.	Spleen Diseases.	Scurvy.	Anæmia and Debility.	Abscess, Ulcer, and Boil.	Phagedæna, Slough, and Gangrene.	All Causes.	
Bareilly, District	734	287.5 / 1.36	...	...	...	167.6	...	...	12 3 / 2.72	21.8 / 1.31	38.6	)5.4	15.0	...	...	...	15.0	54.5	1.4	837.9 / 6.81	27.2
Budaon	370	...	...	...	...	64.9	...	...	2.7	...	...	5.4	16.2 / 5.41	...	...	...	2.7	170.3	...	408.7 / 10.81	18.9
Aligarh	386	...	...	...	5.2	111.4	13.0 / 3.59	...	5.2 / 2.59	23.3 / 12.95	25.9	23.3 / 2.59	23.3	2.6	...	...	15.5	116.6	...	590.7 / 23.94	23.3
Bulandshahr	242	...	...	4.1 / 4.13	...	86.8	...	...	12.4 / 4.13	20.7 / 4.13	...	57.9	41.3 / 4.13	8.3	...	...	4.1	70.2	...	413.2 / 28.93	20.7
Moradabad	368	...	...	...	...	326.1 / 2.72	...	...	...	5.4 / 2.72	13.6	46.2	13.6	...	...	...	...	103.3	...	657.6 / 8.15	21.7
Bijnor	233	...	...	...	...	98.7	8.6 / 4.29	...	...	...	34.3	12.9	34.3	...	4.3	...	8.6	128.8	...	523.6 / 4.29	30.0
Dehra Dun	80	...	...	...	...	162.5	50.0	...	12.5	...	...	...	12.5	...	...	...	50.0	25.0	...	525.0 / 25.00	25.0
Saharanpur	326	...	...	...	...	171.8	...	...	3.1	13.3 / 3.07	39.9	144.2 / 15.34	27.6 / 3.07	...	...	...	39.9	76.7	3.1	852.8 / 27.61	39.9
Muzaffarnagar	157	...	...	...	...	210.2	...	...	12.7	51.0 / 19.11	19.1	25.5 / 6.37	6.4	...	...	...	6.4	57.3	...	11.5 / 50.96	31.8
Meerut	551	9.1	...	...	...	143.4	...	...	3.6 / 1.81	9.1 / 3.63	41.7 / 3.63	7.3 / 5.44	10.9	3.6	...	...	32.7	87.1	...	529.9 / 25.41	25.4
Delhi	461	...	...	...	...	119.3	...	...	8.7 / 6.51	36.9 / 8.68	8.7	41.2	15.2	...	...	...	8.7 / 2.17	84.6	2.17	418.7 / 26.03	19.5
Rohtak	94	...	...	...	...	744.7	...	...	...	21.3	42.6	106.4	42.6	...	...	...	10.6	127.7	...	1,340.4 / 10.64	21.3
Hissar	168	6.0	...	...	...	452.4	6.0	...	6.0	17.9	77.4	29.8	125.0	...	...	6.0	41.7	285.7	...	1,577.4 / 23.81	41.7
Karnal	85	...	...	...	...	282.4	...	...	...	23.5	141.2 / 23.53	...	35.3	...	...	...	...	35.3	...	588.2 / 23.53	11.8
Umballa	560	...	...	...	...	108.9	...	46.4	7.1 / 3.57	14.3 / 1.79	25.0 / 3.57	46.4	64.3	...	...	1.8	5.4 / 1.79	117.9	...	678.6 / 19.64	28.6
B																					
Ludhiana	207	...	...	...	9.7	96.6	4.8	...	4.8 / 4.83	9.7	14.3	...	5.	...	...	...	...	87.0	...	458.9 / 4.83	19.3
Hoshiarpur	47	...	...	...	...	106.4	...	...	21.3	...	21.3	...	42.	...	...	...	...	85.1	...	617.0 / ...	10.2*
Jullundur	267	...	...	...	...	149.8	...	3.75	7.5 / 7.49	15.0	41.2	48.7	18.7	...	...	...	...	134.8	...	531.8 / 18.73	15.0
Ferozepore	424	...	...	2.4 / 2.36	2.4	4.7	2.4 / 2.36	73.1	4.7 / 2.36	9.4	89.6	2.4	65.0	7.1	...	...	16.	54.2	...	456.9 / 7.08	25.9
Amritsar	188	...	...	...	...	398.9	...	...	21.3	10.6	74.5	53.2	...	...	...	31.9	122.3	5.	1,000.0 / 5.32	21.3
Lahore, Central	1,466	...	2.0 / 2.05	...	...	530.0	...	...	3.4 / 2.73	10.2 / 2.05	66.8	42.3 / 2.05	67.5 / .68	...	...	...	.7	146.7	...	1,065.5 / 17.05	23.3
" District	461	...	...	...	4.3 / 2.17	253.8	...	...	6.5 / 2.17	15.2	49.9	41.2	71.6	...	...	...	6.5	75.0	...	676.8 / 10.85	19.5
" Female	150	...	...	...	6.7	273.3	...	20.0 / 20.00	6.7	73.3	11.3	...	26.7	...	...	...	60.0	...	733.3 / 26.07	26.7
Gurdaspur	221	13.6	...	...	...	72.4	13.6	...	4.5 / 4.5	13.6	72.6	6.2	36.2	...	...	4.5	9.0	81.4	...	402.7 / 4.52	13.6
Gujranwala	317	...	...	...	...	189.3	...	...	6.3	...	110.4	3.2	3.2	...	...	...	34.7	167.2	...	974.8 / 9.46	28.4
Sialkot	305	...	...	...	...	157.4	...	...	16.4 / 3.28	9.8	68.9	65.6	39.3	...	6.6	...	23.0	88.5	...	609.8 / 9.84	29.5
Gujrat	95	...	...	...	...	326.3	...	...	...	10.5 / 10.53	10.5	63.2	136.8	...	...	...	10.5	94.7	...	863.2 / 10.53	21.1
Jhelum	161	18.6	...	...	...	136.3	...	...	18.6 / 6.21	12.4	37.3	24.8	12.4	...	...	...	...	105.6	...	528.0 / 12.42	24.8
Rawalpindi	676	...	...	...	...	176.0 / 2.96	...	...	4.4 / 4.44	13.3 / 2.96	4.4 / 1.48	5.0	22.2 / 1.48	...	...	...	78.4 / 5.92	66.6 / 1.48	...	570.9 / 25.13	29.6
GROUP VI.—UPPER SOB-HIMALAYA.	11,930	18.7 / .08	.3 / .25	.3 / .17	.6 / .08	236.6 / .25	1.4 / .25	4.8 / .08	7.5 / 2.35	12.3 / 2.43	37.8 / .84	33.3 / 1.07	35.8 / .59	...	.8	.3	15.4 / .59	90.9 / .17	.3 / .08	669.0 / 15.09	24.8

* Worked on the aggregates.

TABLE XLII—*continued.*

RATIOS of JAILS, GROUPS, and ADMINISTRATIONS.   For actuals see Table XLIII.

JAILS AND GROUPS.	Average annual strength.	1. ADMISSION-RATE							2. DEATH-RATE PER 1,000 OF STRENGTH												Average number constantly sick per 1,000 of strength.
		Influenza.	Cholera.	Smallpox.	Enteric Fever.	Intermittent Fever.	Remittent Fever.	Simple Continued Fever.	Tubercle of the lungs.	Pneumonia.	Other Respiratory Diseases.	Dysentery.	Diarrhœa.	Hepatic Abscess.	Spleen Diseases.	Scurvy.	Anæmia and Debility.	Abscess, Ulcer, and Boil.	Phagedœna, Slough, and Gangrene.	All Causes.	
A Peshawar	538 {	3457 5·53	... ...	...	... ...	297·4 ...	...	...	7·4 3·72	16·7 1·86	14·9 ...	52·0 5·58	24·2 ...	...	...	1·9 ...	9·3 ...	180·3 ...	...	1,009·3 24·16	40·9
Kohat .	137 {	... ...	... ...	... ...	... ...	365·0 ...	...	...	...	21·9 7·30	14·6 ...	14·6 ...	...	...	7·3	14·6 ...	270·1 ...	...	948·9 14·60	29·2
Bannu .	134 {	... ...	... ...	... ...	... ...	231·3 ...	·5	...	7·5 ...	14·9 ...	74·6 7·46	52·2 ...	...	...	...	37·3 ...	119·4 ...	...	686·6 14·93	29·9
Shahpur .	221 {	... ...	... ...	... ...	... ...	176·5 ...	...	...	...	4·5 ...	9·0 ...	36·2 ...	...	...	...	4·5 ...	67·9 ...	...	438·9 27·15	
Mianwali	259 {	... ...	... ...	... ...	... ...	281·9 3·86	...	...	7·7 3·86	42·5 ...	54·1 ...	108·1 ...	150·6 ...	...	...	...	301·2 ...	...	1,490·3 15·44	38·6
Lyallpur	400 {	... ...	7·5 2·50	... ...	... ...	200·0 ...	...	...	2·5 2·50	25·0 7·50	17·5 ...	20·0 ...	22·5 ...	...	...	2·5 ...	2·5 ...	105·0 ...	...	712·5 20·00	25·0
Jhang .	177 {	... ...	... ...	... ...	... ...	175·1 ...	...	...	...	11·3 ...	45·2 ...	16·9 ...	16·9 ...	...	...	16·9 ...	152·5 ...	...	576·3 ...	16·9
Montgomery, Central.	1,564 {	33·6 ...	... ...	... ...	... ...	100·8 ·51	...	...	11·2 9·16	13·7 3·56	25·5 ·51	6·1 ...	8·1 ...	...	1·0	10·2 ...	78·9 ·51	...	423·6 18·33	22·4
Mooltan, Central.	1,171 {	... ...	... ...	... ...	... ...	375·7 ...	...	...	19·6 6·83	15·4 3·42	26·5 ...	26·5 1·71	31·6 ...	...	...	62·3 1·71	74·3 ...	...	749·8 17·93	41·0
" District.	609 {	13·1 ...	... ...	... ...	... ...	139·6 ...	...	...	9·9 3·28	16·4 3·28	29·6 1·64	16·4 1·64	14·8 ...	...	...	3·3	29·6 ...	...	382·6 13·14	21·3
Dera Ismail Khan.	415 {	... ...	... ...	... ...	... ...	455·4 ...	2·4 2·41	...	2·4 2·41	7·2 2·41	7·2 2·41	38·6 2·41	33·7 ...	2·4	...	12·0 ...	224·1 ...	2·4	1,050·6 19·28	31·3
Dera Ghazi Khan.	342 {	... ...	... ...	... ...	... ...	643·3 5·85	...	...	...	119·9 2·92	61·4 ...	172·5 ...	...	...	...	...	181·3 ...	...	1,587·7 11·70	26·3
C Shikarpur	542 {	... ...	... ...	7·4 ...	94·1 ...	5·5 5·54	...	5·5 3·69	18·5 1·85	27·7 5·54	...	9·2 ...	3·7 3·69	1·8 3·69	5·5 ...	14·8 ...	...	260·1 29·52	12·9		
Sind Gang	380 {	... ...	... ...	2·6 ...	44·7 ...	...	...	57·9 15·79	18·4 ...	10·5 2·63	2·6 ...	...	5·3 ...	...	...	...	163·4 18·42	5·3			
Hyderabad, Central.	871 {	... ...	... ...	3·4 ...	73·5 ...	...	6·9 2·30	21·8 8·04	60·8 1·15	48·2 2·30	41·3 1·15	...	8·0 2·30	12·6 ...	11·8 ...	20·7 ...	...	481·1 27·55	18·4		
Kurrachee	247 {	... ...	... ...	... ...	218·6 ...	16·2 ...	...	20·2 4·05	8·1 ...	72·9 ...	89·1 ...	72·9 ...	...	20·2 ...	4·0 ...	20·2 ...	1·79 ...	...	1,186·2 16·19	32·4	
GROUP VII.— N.-W. FRONTIER, INDUS VALLEY, AND N.-W. RAJPUTANA.	8,107 {	30·9 ·36	·4 ·12	1·0 ...	212·0 ·48	1·1 ·48	...	8·7 4·40	17·4 3·81	32·9 ·95	28·4 1·31	32·8 ·24	...	1·8 ·24	2·4 ·24	16·3 ·48	94·2 ·12	·1	651·1 19·39	25·9
A Rajkot .	107 {	... ...	... ...	... ...	743·0 ...	9·3 ...	...	...	9·3 ...	18·7 ...	28·0 ...	28·0 ...	...	...	...	...	84·1 ...	...	495·5 ...	18·7
Ahmedabad, Central.	863 {	... ...	... ...	3·5 ...	171·5 ...	...	...	8·1 3·48	38·2 4·64	9·3 1·16	23·2 ...	15·1 ...	...	1·2	13·9 ...	16·2 ...	...	390·5 15·00	27·8
B Ajmer .	412 {	... ...	... ...	2·4 ...	60·7 ...	...	...	4·9 2·43	14·6 4·85	7·3 2·43	7·3 ...	...	...	...	...	63·1 ...	...	390·8 16·99	21·8
Muttra .	238 {	... ...	... ...	... ...	33·6 ...	...	...	4·2 ...	12·6 ...	21·0 ...	...	...	8·4 ...	...	...	109·2 ...	...	331·9 8·40	21·0
Agra, Central	2,070 {	... ...	... ...	... ...	182·1 ...	...	...	5·3 3·86	22·2 3·38	25·6 ·48	11·6 ·48	14·5 ...	1·4 ...	...	4·3 ...	92·8 ...	...	552·2 11·59	35·3
" District	412 {	... ...	... ...	... ...	114·1 ...	...	...	7·3 2·43	19·4 4·85	31·6 ...	21·8 ...	6·7 ...	...	...	2·4	143·2 ...	...	553·4 9·71	21·8
Jhansi .	180 {	... ...	... ...	... ...	194·4 5·56	...	...	...	50·0 ...	50·0 ...	50·0 ...	...	5·6 ...	...	11·1	300·0 ...	...	1,116·7 22·22	33·9
Lalitpur	64 {	... ...	... ...	... ...	4·4 ...	...	...	...	...	31·3 ...	8·1 ...	...	...	...	...	125·0 ...	...	140·6 ...	31·3
GROUP VIII.— S. E. RAJPUTANA, CENTRAL INDIA, AND GUJARAT.	4,346 {	... ...	... ...	·9 ...	160·4 ·13	·2	...	5·5 2·9/	22·3 3·45	21·4 ·92	16·1 ·23	16·3 ...	1·4	·2	5·5	89·3 ...	...	523·5 12·43	30·1

311·1	44·4	...	...	22·2	22·2	155·6	44·4	...	...	...	22·2
...	...	...	...	...	...	...		...	...	...	22·2
114·9	...	...	11·5	...	...	114·9	...	...	...	...	11·5
...	...	...	11·49	...	...	...	...	...	...	...	...
9·5	1·2	...	7·1	4·8	2·4	...	3·6	...	...	...	11·9
...	...	...	2·38	...	...	...	...	...	...	...	...
140·6	...	...	...	15·6	15·6	15·6	...	...	...	...	31·3
...	...	...	...	...	...	...	...	...	...	...	...
295·1	...	...	...	16·4	32·8	32·8	98·4	...	...	...	...
...	...	...	...	...	...	...	...	...	...	...	...
120·7	25·9	...	8·6	...	25·9	69·0	43·1	...	...	...	8·6
...	...	...	8·62	...	...	...	...	...	...	...	...
109·1	...	...	9·1	...	0·1	54·5	27·3	...	...	...	...
...	...	...	9·09	...	...	18·18	...	...	...	...	...
240·5	...	...	2·2	13·4	40·1	69·0	17·8	...	...	...	2·2
...	...	...	2·23	...	2·23	4·45	...	...	...	...	...
236·4	...	...	...	36·4	36·4	72·7	36·4	...	...	...	...
...	...	...	...	18·18	...	...	...	...	...	...	...
117·6	...	...	...	...	...	19·6	39·2	...	...	...	98·0
...	...	...	...	...	...	...	...	...	...	...	...
100·0	...	...	...	25·0	...	25·0	...	...	...	...	25·0
...	...	...	...	...	...	...	...	...	...	...	...
473·7	...	...	...	...	26·3	118·4	131·6	...	...	...	13·2
...	...	...	...	...	...	13·16	...	...	...	...	...
573·0	...	11·2	...	22·5	22·5	80·9	123·6	...	...	...	11·2
...	...	...	...	...	...	...	11·24	...	...	...	...
320·0	...	...	...	60·0	80·0	120·0	20·0	...	...	...	80·0
...	...	...	...	...	...	...	...	...	...	...	...
350·8	...	20·0	13·8	9·2	13·8	23·1	70·8	...	...	1·5	4·6
...	...	...	3·08	3·08	...	1·54	...	...	...	...	...
246·2	...	...	...	15·4	...	30·8	15·4	...	...	...	...
...	...	...	...	...	...	...	...	...	...	...	...
130·4	...	...	...	21·7	65·2	...	21·7	...	...	21·7	...
...	...	...	...	...	...	...	...	...	...	21·74	...
137·3	...	...	...	19·6	58·8	39·2	39·2	...	...	...	19·6
...	...	...	...	19·61	...	...	...	...	...	...	19·6
129·4	11·8	...	...	...	...	47·1	23·5	...	...	...	11·8
...	...	...	...	...	...	...	...	...	...	...	...
156·9	...	...	...	...	19·6	58·8	39·2	...	...	...	...
...	...	...	...	...	...	19·61	...	...	...	...	...
119·2	...	3·8	...	11·5	11·5	65·4	11·5	...	...	30·8	7·7
...	...	...	...	3·85	3·85	...	...	...	...	3·85	...
207·2	...	...	3·0	3·0	9·0	54·1	69·1	...	...	...	6·0
...	...	...	...	...	6·01	...	...	...	...	...	...
58·8	...	...	14·7	14·7	...	29·4	...	...	...	...	44·1
...	...	...	...	...	...	...	...	...	...	...	...
421·9	...	...	...	8·2	21·9	183·6	60·3	...	...	...	16·4
...	...	...	...	2·74	2·74	...	2·74	...	...	...	...
95·0	...	...	·7	1·4	29·3	35·6	66·3	...	2·1	·	3·5
...	...	...	...	·70	...	...	·70	...	...	...	...
70·7	...	...	7·1	...	14·1	21·2	...	...	...	14·1	3·5
...	...	...	3·53	...	...	...	...	...	...	3·53	...
91·3	...	10·3	1·1	3·4	14·8	26·3	14·8	...	...	...	3·4
1·14	...	...	...	1·14	2·28	2·28	3·42	...	...	...	...
62·7	...	...	6·3	...	6·3	84·6	21·9	...	...	...	6·3
...	...	...	3·13	...	...	3·13	...	...	...	...	...
158·0	1·0	3·4	3·8	6·1	18·4	47·2	38·5	...	·4	2·1	8·1
·14	...	··	1·43	1·12	1·00	1·43	·86	...	...	·43	·2

* Worked on the aggregates.

99

TABLE XLII—continued.

RATIOS of JAILS. GROUPS, and ADMINISTRATIONS.

For actuals see Table XLIII

JAILS, AND GROUPS.	Average annual strength.	1. ADMISSION RATE.										2. DEATH-RATE PER 1,000 OF STRENGTH.									Average number constantly sick per 1,000 of strength
		Influenza.	Cholera.	Small-pox.	Enteric Fever.	Intermittent Fever.	Remittent Fever.	Simple Continued Fever.	Tubercle of the lungs.	Pneumonia.	Other Respiratory Diseases.	Dysentery.	Diarrhœa.	Hepatic Abscess.	Spleen Diseases.	Scurvy.	Anæmia and Debility.	Abscess, Ulcer, and Boil.	Phagædæna, Slough, and Gangrene.	All Causes.	
Thana	607	... ...	... ...	1·6 ...	— ...	89·0 ...	... ...	... ...	11·5 6·5	3·3 ...	56·0 ·1·65	51·1 3·29	46·1 ...	... ...	... ...	1·6 1·6	4·0 1·6	59·3 ...	... ...	454·7 14·83	} 26·.
Bombay, Common.	138	... ...	... ...	4·6 2·28	... ...	355·4 ...	2·3 2·28	... ...	13·7 6·85	13·7 4·57	22·8 2·28	38·8 2·28	123·3 4·57	6·8 ...	... ...	27·4 6·85	63·9 ...	... ...	1,180·4 52·31	} 32·0
Bombay, House of Correction.	215	... ...	... ...	... ...	... ...	160·0 ...	... ...	... ...	17·8 ...	4·4 4·44	41·4 8·89	8·9 ...	44·4 ...	... ...	... ...	26·7 ...	... ...	... ...	497·8 31·11	} 13·3
Ratnagiri	107	... ...	... ...	... ...	... ...	140·2 ...	... ...	... ...	9·3 9·35	... ...	... ...	18·7 9·35	... ...	... ...	... ...	18·7 ...	18·7 ...	... ...	289·7 18·69	} 9·3
Karwar	119	... ...	... ...	... ...	... ...	50·4 ...	... ...	... ...	8·4 ...	16·8 ...	... ...	16·8 ...	... ...	... ...	... ...	... ...	50·4 ...	... ...	428·6 8·40	} 16·8
Mangalore	144	... ...	... ...	... ...	6·9 ...	55·6 ...	... ...	41·7 ...	6·9 6·94	... ...	20·8 ...	20·8 2·00	13·9 ...	... ...	... ...	... ...	... ...	... ...	236·1 13·89	} 6·9
Cannanore, Central	678	... ...	... ...	... ...	8·8 2·95	75·2 1·47	2·9	31·0	8·8 1·47	17·7 1·47	23·6 1·47	60·5 8·85	7·4 ...	... ...	... ...	1·5 ...	14·7 ...	... ...	311·2 22·12	} 13·3
GROUP X,—WESTERN COAST.	2,318	... ...	... ...	1·3 ·43	3·0 ·85	141·1 ·43	1·3 ·43	11·6	11·2 4·31	9·0 1·73	51·5 2·16	42·3 4·31	42·7 1·29	... ...	13 ...	·4 ...	10·4 1·73	35·4 ...	... ...	531·5 25·45	} 19
A Bellary	568	... ...	... ...	3·5 ...	... ...	440·1 7·04	... ...	12·5	3·17 3·52	14·1 1·76	44·0 1·76	24·6 ...	... ...	... ...	... ...	3·5 ...	45·8 ...	... ...	906·7 15·85	} 28·2
Salem, Central	750	8·0 1·33	... ...	... ...	... ...	104·0 1·33	... ...	12·0	5·3 1·33	16·0 1·33	13·3 ...	56·0 6·67	... ...	... ...	... ...	... ...	22·7 ...	... ...	332·0 16·00	} 14·7
Coimbatore, Central	1,185	... ...	... ...	... ...	13·2 1·69	9·3 ...	... ...	54·8	10·1 2·53	13·5 1·69	16·0 ·84	103·7 1·69	... ...	·8 ·84	... ...	2·5 ·84	16·0 ...	... ...	438·4 12·65	} 25·3
B Palamcottah	386	... ...	... ...	... ...	... ...	25·9 ...	...		2·6	2·6	20·7	72·5 7·77	... ...	... ...	... ...	5·2 ...	85·5 ...	... ...	430·1 7·77	} 23·3
Madura	456	... ...	... ...	... ...	2·2 ...	... ...	2·2 ...		2·2 2·19	13·2 6·58	17·5 ...	28·5 ...	... ...	... ...	... ...	4·4 ...	13·2 ...	... ...	234·6 15·35	} 17·5
Trichinopoly, Central	985	... ...	... ...	... ...	... ...	29·4 2·03	... ...		18·3 2·03	2·0 ...	40·6 ...	27·4 ...	... ...	... ...	... ...	1·0 ...	39·6 ...	... ...	347·9 12·17	} 17·2
Tanjore	333	... ...	... ...	... ...	... ...	48·0 ...	36·0 3·00	15·0	15·0 6·01	3·0 ...	15·0 ...	54·1 3·00	12·0 ...	... ...	... ...	9·0 3·00	18·0 ...	... ...	462·5 30·03	} 21·0
Cuddalore	365	... ...	... ...	... ...	... ...	73·8 ...	... ...		2·7 2·73	... ...	16·4 ...	10·9 ...	... ...	... ...	... ...	5·5 ...	32·8 ...	... ...	795·1 5·46	} 19·1
Vellore, Central	1,220	... ...	... ...	... ...	... ...	29·5 ...	... ...	58·2	8·2 3·28	9·8 1·64	21·3 ...	20·5 1·64	... ...	... ...	... ...	... ...	60·7 ...	·8 ·82	430·3 12·30	} 21·3
Madras, Civil	34	... ...	... ...	... ...	... ...	... ...	... ...		... ...	... ...	... ...	... ...	... ...	... ...	... ...	... ...	... ...	... ...	... ...	}
Madras Penitentiary, Central.	1,072	3·7 2·80	1·9 ...	1·9 ...	57·8 ...	·9 ...	10·3	13·1 2·80	4·7 ...	37·3 ...	44·8 ...	·9 ·93	... ...	... ...	19·6 ...	28·9 ...	... ...	527·1 10·26	} 21·5
Nellore	83	10·6 3·53	... ...	... ...	7·1 ...	3·5 ...	10·6	7·1 7·07	3·5 3·53	... ...	28·3 7·07	7·1 ...	... ...	... ...	14·1 ...	14·1 ...	... ...	166·1 24·73	} 10·6
C Rajamundry, Central.		30·0 1·0	... ...	... ...	248·2 ...	... ...	146·1	11·0 1·00	6·0 3·00	35·0 ...	214·2 6·01	... ...	... ...	... ...	1·0 ...	41·0 ...	... ...	1,055·1 29·03	} 23·0
Vizagapatam	472	... ...	... ...	... ...	25·4 ...	... ...	2·1	6·4 4·24	12·7 ...	·2 ...	72·0 2·12	40·3 6·36	... ...	... ...	... ...	4·2 ...	... ...	214·0 19·07	} 8·5
Berhampur	127	... ...	... ...	... ...	69·8 ...	... ...	...	... ...	15·3 ...	7·8 ...	31·0 ...	31·0 ...	... ...	... ...	... ...	15·5 ...	... ...	248·1 7·75	} 7·8
GROUP XI.—SOUTHERN INDIA.	9,240	·6 ·11	4·0 1·52	·4 ...	2·3 ·32	37·7 ·76	1·6 ·11	34·4	10·8 2·38	8·4 1·52	24·4 ·22	65·2 3·60	3·2 ·32	·1 ·11	·1 ·11	... ...	4·7 ·22	33·8 ...	·1 ·11	485·5 15·37	} 20·0

Jails, Groups and Administrations	Average annual strength	Influenza	Cholera	Small-pox	Enteric Fever	Intermittent Fever	Remittent Fever	Simple Continued Fever	Tubercle of the lungs	Pneumonia	Other Respiratory Diseases	Dysentery	Diarrhœa	Hepatic Abscess	Spleen Diseases	Scurvy	Anæmia and Debility	Abscess, Ulcer, and Boil	Phagedæna, Slough, and Gangrene	All Causes	Average number constantly sick per 1,000 of strength
		1. Admission Rate.							**2. Death-rate per 1,000 of strength.**												
Shillong	50	...	...	...	...	140·0	...	...	...	20.0	20·0	...	80·0	...	...	...	...	...	...	40·0	7 2
Darjeeling	100	...	...	...	...	450·0 / 30·00	...	...	...	10·0	230·0	210·0 / 10·0	...	...	...	...	...	10·0 / 50·0	...	1640·0 / 40·00	70·0
Almora	67	...	...	...	...	313·4 / 14·93	...	149·3	...	29·9 / 14·93	29·0	59·7	101·5	...	...	...	29·9	?·??	...	1209·0 / 29·85	2·99
Pauri	9	...	...	...	...	...	...	...	...	111·1	...	555·5 / 111·11	111·1	...	...	...	...	111·1	...	1000·0 / 111·11	40·5?
Naini Tal	1	...	...	...	...	1,000·0	...	...	...	...	...	...	...	...	...	...	...	...	...	10·0·0	2·0?
Simla	16	...	...	...	...	62·5	...	...	...	...	...	...	...	...	...	...	...	...	...	62·5	1·7?
Abbottabad	90	...	...	...	...	244·4	...	...	...	22·2	11·1	44·4 / 11·11	44·4	...	...	...	33·3	44·4	...	677·8 / 11·11	11·1
Quetta	70	...	...	...	...	671·4	...	...	...	14·3	71·4	128·6	71·4	...	...	...	...	157·1	...	1,728·5	28·6
Mercara	87	...	...	...	...	11·5	...	...	11·5	34·5 / 34·48	23·0	...	11·5	...	...	...	...	34·5	...	218·4 / 45·98	11·5
Russellkonda	85	...	...	...	...	23·5 / 11·76	...	...	...	94·1 / 11·76	35·3	...	...	...	...	...	...	70·6	...	364·7 / 23·53	11·8
GROUP XII.—HILLS.	575	...	...	...	...	252·2 / 6·96	3·5 / 1·74	17·4	1·7	1·1 / 6·96	73·0 / 1·74	80·0 / 5·22	6·6	...	...	...	10·4	62·6	...	878·3 / 24·35	24·3
EXTRA INDIA—Aden	70	...	...	...	...	242·9	...	...	28·6	...	28·6	114·3	71·4	...	...	...	...	42·9	...	714·3 / 14·29	28·6
INDIA (a)	91,917	11·5 / ·16	·8 / ·44	·3 / ·04	·7 / ·16	181·7 / ·91	1·2 / ·22	9·7 / ·01	8·7 / 3·19	0·7 / 2·28	26·4 / ·92	51·6 / 3·01	39·2 / ·74	·1 / ·08	·6 / ·03	1·3 / ·08	11·0 / ·54	71·2 / ·07	·1 / ·03	646·8 / 19·23	27·5
BURMA	12,639	1·1	·2 / 0 8	·2 / ·47	·6 / ·47	51·0 / ·47	·8 / ·08	19·0	7·3 / 3·80	3·3 / 1·03	10·7 / ·63	28·3 / 2·61	17·6 / ·95	·2 / ·24	...	·2	3·1 / ·24	55·1	...	330·3 / 17·01	16·6
EASTERN BENGAL AND ASSAM	6,401	16·6	1·9 / 1·87	·2	1·9 / 1·87	260·1 / 1·87	2·2 / ·47	1·2	11·4 / 3·44	5·9 / 2·50	38·9 / ·94	307·0 / 7·50	1·0·0 / ·78	·3 / ·16	...	6·2 / ·31	32·2 / 2·06	28·9	-·2	1,081·4 / 32·96	46·2
BENGAL	14,172	9·6	·6 / ·35	·1	·2 / ·14	290·7 / 2·75	1·5 / ·21	3·2	13·5 / 4·02	6·2 / 2·12	34·1 / 1·06	196·4 / 6·56	90·9 / 1·06	·2 / ·07	·5	1·1	10·7 / ·49	67·1 / ·07	·3	953·5 / 24·91	39·0
UNITED PROVINCES	23,680	20·9 / ·25	·3 / ·17	·2 / ·04	·4 / ·08	181·8 / ·51	·9 / ·21	7 3	7·1 / 2·66	12·1 / 3·00	22·0 / 1·10	36·5 / 1·04	21·2 / ·84	·0 / ·04	1·1	·8	10·0 / ·21	81·8 / ·08	·2	566·8 / 17·31	26·9
PUNJAB	11,512	7·0	·5 / ·35	·3 / ·09	·3 / ·09	246·2 / ·52	·5 / ·09	5·0 / ·09	8·2 / 4·43	14·2 / 2·52	41·2 / ·69	23·4 / ·52	42·9 / ·17	·3	·7	17·9 / ·69	107·8 / ·17	·1 / ·0?	...	707·7 / 16·33	25·5
N.W. FRONTIER PROVINCE	1,314	141·6 / 2·28	...	...	...	344·0	1·5 / ·76	...	3·8 / 1·52	13·7 / 1·52	12·2 / ·76	45·7 / 4·57	30·4 / ·76	·8	1·5	15·2	128·0	·8	...	950·4 / 19·79	33·5
BOMBAY	7,851	...	...	1·8 / ·13	...	134·4 / ·13	1·1 / ·51	1·1	6·1 / 2·29	13·5 / 3·00	29·0 / 1·53	41·7 / 1·15	39·5 / 1·15	...	2·5 / ·25	2·8 / ·33	9·2 / ·76	82·0 / ·13	...	602·7 / 17·45	27·4
CENTRAL PROVINCES	3,547	9·0	·3 / ·28	...	...	190·6	1·7	4·2	5·6 / 1·97	9·9 / 1·41	16·6 / 1·13	41·4 / 1·41	36·1 / ·28	...	·56	...	11··? / ·56	74·7	·3 / ·28	629·5 / 15·79	26·9
MADRAS	10,147	·6 / ·10	3·6 / 1·38	·4	2·8 / ·39	85·6 / ·79	1·9 / ·20	34·0	10·5 / 2·27	8·9 / ·39	24·8 / ·39	64·0 / 3·06	3·5 / ·30	·1 / ·10	·1 / ·10	...	4·5 / ·20	37·3	·1 / ·0	469·3 / 15·87	19·3
ANDAMANS	14,348	13·0	...	...	...	1246·2 / ·84	6·1 / 2·86	...	11·0 / 6·48	11·0 / 4·81	57·4 / ·91	164·1 / 13·03	53·3 / ·4?	·2 / ·14	·07	1·5	...	79·3 / ·07	...	1,898·3 / ·07	67·7
INDIA (b)	106,265	11·7 / ·14	·7 / ·38	·3 / ·04	·6 / ·14	325·5 / ·90	1·9 / ·57	8·4 / ·01	9·1 / 3·63	9·9 / 2·63	30·6 / ·92	92·7 / 4·37	41·4 / ·71	·1 / ·05	·6 / ·04	1·3 / ·03	9·5 / ·47	72·3 / ·07	·1 / ·03	815·8 / 21·20	33·0

* Worked on the aggregates.
(a) Excluding Andamans.
(b) Including Andamans.

101

TABLE XLIII.

ACTUALS of JAILS, GROUPS, and ADMINISTRATIONS on which the ratios in Tables XL—XLII have been calculated.

JAILS AND GROUPS.	Average annual strength.	Influenza.	Cholera.	Small-pox.	Enteric Fever.	Intermittent Fever.	Remittent Fever.	Simple Continued Fever.	Tubercle of the lungs.	Pneumonia.	Other Respiratory Diseases.	Dysentery.	Diarrhœa.	Hepatic Abscess.	Spleen Diseases.	Scurvy.	Anæmia and Debility.	Abscess, Ulcer, and Boil.	Phagedæna, Slough, and Gangrene.	ALL CAUSES.	Tænia.	Ascaris lumbricoides.	Dracunculus Medinensis.	Strongylus duodenalis.	Other Entozoa.	Average number constantly sick.
Mergui . .	63	...	...	...	...	...	...	...	...	1	1	...	...	...	...	...	...	1	...	16	...	...	...	...	...	1
Tavoy . .	103	...	...	...	13	...	...	...	...	...	1	1	...	...	...	...	...	1	...	39	...	...	...	...	...	2
Moulmein .	636	...	...	...	61	...	...	2	2	8	43	9	...	...	...	...	...	4	...	272	1	...	...	...	...	11
Shwegyin .	160	...	...	...	10	...	...	...	...	1	9	9	...	...	...	...	...	13	...	65	...	...	...	...	...	3
Toungoo .	406	...	...	...	27	...	...	3	1	11	20	37	...	...	1	4	26	...	281	...	...	...	...	...	15
Rangoon, Central (Europeans)	20	...	...	...	...	...	1	...	...	...	4	...	...	...	...	...	1	...	12	...	...	...	...	...	...
Rangoon Central (Natives)	2,143	...	...	...	18	...	123	16	7	12	29	3	1	...	...	3	19	...	402	...	...	...	...	...	26
Maubin . .	359	...	...	...	5	...	...	...	...	4	4	3	...	...	...	1	11	...	52	...	...	...	...	...	3
Myaungmyo .	627	...	...	...	12	4	...	2	...	4	7	23	...	...	...	...	102	...	235	1	1	...	...	...	12
Bassein, Central	1,091	...	...	4	33	2	54	17	12	21	13	2	...	...	...	2	69	...	331	...	...	...	7	4	20
Insein, Central .	2,061	...	...	2	54	...	37	13	1	4	2	...	...	...	...	1	83	...	332	1	...	...	...	...	18
Henzada . .	398	...	...	...	25	1	...	...	...	...	4	1	...	...	7	27	...	111	...	...	...	...	...	4
Myanaung .	79	...	...	...	2	...	...	...	...	...	7	...	...	...	2	3	...	21	...	...	...	...	...	1
Sandoway .	69	...	...	...	9	...	...	...	...	...	6	1	...	...	...	6	...	33	...	...	...	...	...	3
Kyaukpyu .	141	...	...	...	7	...	2	2	...	1	...	1	...	...	...	...	...	35	...	...	...	...	...	1
Akyab . .	443	...	...	...	7	...	...	1	6	6	5	1	...	1	1	2	...	65	...	...	...	...	...	4
GROUP I.— BURMA COAST AND BAY ISLANDS.	8,808	...	...	2	283	7	217	56	29	73	144	101	2	...	2	21	412	...	2,307	3	1	...	7	4	126
Paungdi .	184	...	...	...	12	1	...	...	1	7	...	...	...	5	...	36	...	...	...	...	...	2
Prome . .	323	...	...	...	39	...	2	1	1	5	20	16	...	...	8	73	...	321	...	...	...	...	...	16
Thayetmyo, Central.	956	...	1	...	37	...	1	16	4	18	45	24	1	...	2	52	...	324	...	...	...	...	...	14
Taungdwingyi	44	...	...	...	...	...	...	1	...	...	...	...	...	...	1	...	...	...	...	...	...	...
Magwe . .	153	...	...	...	4	...	...	...	...	9	7	...	...	4	...	32	...	...	...	...	...	2
Minbu . .	1	...	...	...	...	...	...	...	...	...	...	...	...	...	...	...	...	...	...	...	...	...
Yamethin .	95	...	...	...	...	...	...	...	...	...	...	...	...	...	6	...	...	...	...	...	...	1
Meiktila .	73	...	...	...	8	...	...	1	...	2	1	...	...	3	...	20	...	...	...	...	...	1
Pagan . .	48	...	...	...	...	...	2	...	...	1	...	...	...	6	...	15	1	...	...	...	...	2
Pakôkku .	30	...	...	...	1	...	...	1	...	...	...	...	...	...	...	2	...	...	...	...	...	...
Myingyan, Central.	771	14	...	...	58	1	5	5	...	10	80	43	...	5	53	...	367	...	...	...	...	...	19
Mandalay, Central.	751	...	1	3	175	1	14	8	1	18	30	13	...	...	54	...	436	...	...	...	...	...	18
Monywa .	82	...	...	...	2	...	...	...	2	...	1	...	...	...	...	11	...	...	...	...	...	1
Shwebo .	147	...	...	...	6	...	...	...	...	2	5	9	...	...	8	...	40	...	...	...	...	...	2
Bhamo . .	63	...	...	...	18	...	1	2	3	5	2	6	...	...	20	...	97	1	...	...	...	...	4
Katha . .	65	...	...	...	1	...	...	1	1	...	3	1	...	...	2	3	...	19	...	...	...	...	...	1
Kindat .	45	...	...	...	2	...	...	...	1	...	...	...	...	...	3	...	14	...	...	...	...	...	1
GROUP II.— BURMA INLAND .	3,831	14	3	7	362	3	73	36	13	62	214	121	1	...	18	281	...	1,741	2	1	...	...	...	84

Jails and Groups	Average annual strength	Influenza	Cholera	Small-pox	Enteric Fever	Intermittent Fever	Remittent Fever	Simple Continued Fever	Tubercle of the lungs	Pneumonia	Other Respiratory Diseases	Dysentery	Diarrhœa	Hepatic Abscess	Spleen Diseases	Scurvy	Anæmia and Debility	Abscess, Ulcer and Boil	Phagedæna, Slough, and Gangrene	All Causes	Tænia	Ascaris lumbricoides	Dracunculus Medinensis	Strongylus duodenalis	Other Entozoa	Average number constantly sick
Cachar	58	...	...	... ...	...	5	...	...	...	..	1	6	..	...	...	...	3	2	...	22	..	...	...	...	...	2
		...	...	... ...	...	1	...	...	...	...	...	...	...	...	...	...	...	...	...	2	..	...	...	...	...	
Sibsagar	71	...	2	... ...	...	7	...	...	...	1	3	9	8	...	..	...	1	6	...	45	...	...	...	...	...	2
		...	2	... ...	...	...	...	...	...	...	...	...	...	...	...	...	...	...	...	3	...	...	...	...	...	
Dibrugarh	104	2	...	... ...	...	56	...	...	2	...	3	59	16	...	...	...	7	3	...	161	...	...	...	...	...	6
		...	...	... ...	...	...	...	...	1	...	...	3	6..	...	...	...	...	...	...	6	...	...	...	...	...	
Tezpur	265	...	...	... ...	...	13	...	...	2	...	...	24	21	1	...	...	11	9	...	150	...	...	...	...	...	10
		...	...	... ...	...	...	...	...	1	...	...	2	...	...	...	...	6	...	...	12	..	...	...	...	...	
Gauhati	309	3	...	... ...	...	105	...	8	...	...	8	154	15	...	...	...	5	11	...	353	...	1	...	...	...	29
		...	...	... ...	...	1	...	...	...	...	...	15	...	...	...	...	1	...	...	18	..	...	...	...	...	
Sylhet	584	...	...	... ...	...	112	6	...	1	1	5	99	18	...	1	11	14	...	...	468	...	...	...	8	...	23
		...	...	.. ...	...	3	...	...	1	1	...	...	...	...	1	1	...	...	...	32	...	...	...	2	...	
Group III.—Assam	1,391	5	2	... ...	...	318	6	8	5	2	20	361	78	1	...	1	35	45	...	1,207	...	1	...	8	...	72
		...	2	... ...	...	5	...	...	3	1	1	20	...	...	...	1	8	...	...	72	...	...	...	2	...	
Mymensingh	562	...	1	... ...	10	185	1	...	25	6	19	265	53	...	...	11	30	27	...	500	...	...	...	...	...	44
		...	1	... ...	...	...	1	...	3	1	...	1	...	...	...	...	...	...	...	13	...	...	...	...	...	
Dacca, Central	1,147	101	2	... ...	...	111	...	...	7	5	94	134	28	...	...	...	60	33	...	751	1	...	...	...	...	25
		...	2	... ...	...	1	...	...	2	2	4	4	1	...	...	...	5	...	...	25	...	...	...	...	...	
Tippera	282	...	...	1 ...	...	20	...	...	8	4	2	6	57	...	...	23	2	2	...	197	...	...	...	...	...	11
		...	...	.. ...	...	...	...	...	2	1	...	...	1	...	...	1	...	...	...	4	...	...	...	...	...	
Chittagong	138	...	...	.. ...	...	78	...	...	...	2	1	29	43	...	...	...	...	9	...	195	...	...	...	...	...	5
		...	...	.. ...	...	...	...	...	1	...	...	...	...	...	...	...	...	...	...	2	...	...	...	...	...	
Noakhali	145	...	...	.. ...	...	22	...	...	1	2	3	108	4	...	...	...	...	2	...	153	1	...	...	...	...	6
		...	...	.. ...	...	...	...	...	...	...	...	2	...	...	...	...	...	...	...	3	...	...	...	...	...	
Backergunge	509	...	3	... ...	...	77	...	...	3	11	20	395	163	...	...	...	49	24	...	1,001	...	1	...	...	1	40
		...	3	... ...	...	2	...	...	4	3	...	9	...	...	...	...	3	...	...	32	...	...	...	...	...	
Khulna	37	...	..	... ...	...	10	...	...	...	...	...	18	1	...	...	...	...	...	...	37	...	...	...	...	...	1
		...	..	... ...	...	...	...	...	...	...	...	2	...	...	...	...	...	...	...	2	...	...	...	...	...	
Jessore	372	...	...	... ...	...	127	1	...	7	1	7	328	19	...	...	...	7	18	...	555	...	...	...	...	...	20
		...	...	... ...	...	1	...	...	2	...	...	6	...	...	...	...	...	...	...	10	...	...	...	...	...	
Baraset	119	...	...	.. ...	...	63	...	...	...	...	4	41	7	...	...	1	...	10	...	169	...	...	...	...	...	8
		...	...	.. ...	...	...	...	...	...	...	...	1	...	...	...	...	...	...	...	3	...	...	...	...	...	
Presidency, Central (Europeans)	37	5	...	... ...	...	...	...	...	...	...	1	6	...	...	...	...	...	1	...	19	...	...	...	...	...	1
		...	...	... ...	...	...	...	...	...	...	...	1	...	...	...	...	...	...	...	1	...	...	...	...	...	
Presidency, Central (Natives)	1,175	59	2	... ...	...	45	...	...	15	13	17	186	24	1	...	1	13	25	...	61	1	...	...	1	...	33
		...	2	... ...	...	7	...	...	5	7	2	8	2	...	...	...	1	...	...	39	...	...	...	..	...	
Alipore, Central	1,817	72	1	... ...	...	468	...	...	60	10	34	208	80	...	...	...	14	96	...	1,425	3	...	...	3	...	71
		5	1	... ...	...	11	...	...	16	4	3	5	...	...	...	...	...	...	...	58	...	...	...	...	...	
Hooghly	401	...	...	... ...	...	84	...	...	1	5	8	61	91	1	...	...	6	32	...	387	...	1	...	...	...	14
		...	...	... ...	...	...	...	...	1	...	...	...	2	1	...	...	...	...	...	6	...	...	...	...	...	
Burdwan	253	...	...	... ...	...	186	2	...	5	5	26	34	39	7	...	...	2	17	...	431	...	...	...	...	...	15
		...	...	... ...	...	1	2	...	1	4	1	1	...	...	...	...	...	...	...	12	...	...	...	...	...	
Krishnagar	156	...	...	... ...	...	38	...	...	2	3	18	20	4	...	1	3	15	...	160	...	...	...	...	...		7
		...	...	... ...	...	...	...	...	...	...	3	...	...	...	...	...	...	...	...	3	...	...	...	...	...	
Faridpur	343	...	...	.. ...	...	154	...	...	11	7	6	119	10	...	...	1	26	...	525	...	...	...	...	...		21
		...	...	.. ...	...	...	...	...	3	1	2	3	...	...	...	...	...	...	...	10	..	...	...	...	...	
Pubna	195	...	...	... ...	...	126	5	...	3	9	50	22	..	...	12	...	267	...	...	...	...	8
		...	...	... ...	...	1	...	...	2	1	...	...	...	...	...	...	...	...	...	4	...	...	...	...	...	
Murshidabad	215	...	2	... ...	...	39	1	...	1	1	12	19	15	...	...	...	8	2	174	...	...	...	...	.	5
		...	...	... ...	...	...	...	...	...	...	...	...	...	...	...	...	...	...	...	...	...	...	...	...	...	
Rajshahi, Central	784	...	2	... ...	...	71	...	...	8	6	17	140	22	...	...	6	27	...	436	...	...	...	...	1	24
		...	2	... ...	...	2	...	...	6	3	1	6	1	...	...	...	...	...	...	24	...	...	...	...	...	
Bogra	186	...	...	.. ...	...	52	...	...	...	5	28	4	...	...	5	9	...	123	...	...	...	...	.	4
		...	...	.. ...	...	...	...	...	...	1	...	...	...	...	...	...	...	...	...	1	...	...	...	...	...	
Malda	95	...	..	... ...	...	133	1	...	1	1	6	13	...	...	7	4	...	190	1	...	..	...	...	6
		...	..	... ...	...	...	...	...	...	...	...	...	...	...	...	...	...	...	...	1	...	...	...	...	...	
Dinajpur	217	...	2	.. ..	...	109	1	...	4	4	9	110	30	...	...	2	9	...	318	...	...	...	...	.	13
		...	2	... ...	...	1	1	...	1	1	...	...	...	...	...	...	...	...	...	6	...	...	...	...	...	

TABLE XLIII—continued.

ACTUALS of JAILS, GROUPS, and ADMINISTRATIONS on which the ratios in Tables XL.—XLII have been calculated.

Jails and Groups	Average annual strength	Influenza	Cholera	Small-pox	Enteric Fever	Intermittent Fever	Remittent Fever	Simple Continued Fever	Tubercle of the lungs	Pneumonia	Other Respiratory Diseases	Dysentery	Diarrhœa	Hepatic Abscess	Spleen Diseases	Scurvy	Anæmia and Debility	Abscess, Ulcer, and Boil	Phagedæna, Slough, and Gangrene	All Causes	Tænia	Ascaris lumbricoides	Dracunculus	Strongylus duo-denalis	Other Entozoa	Average number constantly sick
Rangpur	248	...	...	...	2	173	...	...	...	3	28	189	34	1	...	5	2	12	...	522	...	2	...	...	...	11
		...	...	...	...	...	...	...	...	...	1	1	1	...	...	...	...	...	...	10	...	...	...	...	...	
Jalpaiguri	109	...	...	...	...	29	...	...	1	...	5	25	11	...	...	...	4	8	1	117	...	...	...	...	...	6
		...	...	...	...	1	...	...	...	...	...	2	...	...	...	...	...	...	...	3	...	...	...	...	...	
Purneah	223	...	...	...	...	51	...	...	3	...	1	23	9	...	...	1	10	22	...	170	...	...	...	...	...	8
		...	...	...	...	1	...	...	...	...	...	...	...	...	...	...	...	...	...	3	...	...	...	...	...	
Naya Dumka	144	...	...	...	...	18	...	6	...	1	5	17	15	...	...	...	2	11	...	100	...	...	...	...	...	3
		...	...	...	...	...	...	...	...	...	...	1	1	...	...	...	...	...	...	2	...	...	...	...	...	
Suri	212	...	...	...	...	34	...	...	...	1	4	29	5	...	...	...	...	15	...	133	...	...	...	...	1	2
		...	...	...	...	...	...	...	...	1	...	1	...	...	...	...	...	...	...	4	...	...	...	...	...	
Bankura	223	...	...	...	2	23	1	...	3	2	16	29	4	...	...	...	1	10	...	140	...	...	...	...	...	7
		...	...	...	...	...	...	...	1	...	...	...	...	...	...	...	...	...	...	3	...	...	...	...	...	
Midnapore, Central	894	...	...	...	...	259	...	...	22	8	46	302	730	1	...	4	18	107	...	1,464	...	1	...	...	...	74
		...	...	...	...	...	...	...	...	3	3	19	4	...	...	...	2	...	...	41	...	...	...	...	...	
Balasore	144	...	...	...	...	37	...	26	1	2	5	63	6	1	...	...	...	3	...	176	...	...	...	...	...	4
		...	...	...	...	...	...	...	...	...	...	1	...	...	...	...	...	...	...	4	...	...	...	...	...	
Cuttack	326	...	...	...	...	100	...	...	...	1	4	47	49	...	...	...	4	47	...	313	1	...	...	...	...	10
		...	...	...	...	...	...	...	1	...	...	1	...	...	...	...	...	...	...	4	...	...	...	...	...	
Puri	125	...	...	...	...	19	1	...	1	1	1	14	...	...	...	...	...	1	...	43	...	...	...	...	...	2
		...	...	...	...	...	...	...	...	...	...	...	...	...	...	...	...	...	...	...	...	...	...	...	...	
Angul	58	...	...	...	...	12	...	...	...	...	2	5	13	...	...	...	...	15	...	62	...	...	...	...	...	3
		...	...	...	...	1	...	...	...	...	...	...	...	...	...	...	...	...	...	2	1	...	...	...	...	
GROUP IV.—BENGAL AND ORISSA.	11,901	237	15	1	14	2,951	14	32	189	107	439	3,054	1,105	4	7	43	248	668	3	12,331	6	4	1	5	3	512
		5	13	...	1	29	5	...	50	35	17	73	14	2	...	1	11	...	...	335	1	...	...	...	...	
A Chaibassa	167	...	...	...	...	41	1	...	2	2	31	46	58	...	...	...	...	9	2	270	...	...	...	...	1	7
		...	...	...	...	...	...	...	...	1	...	1	...	...	...	...	...	...	...	6	...	...	...	...	...	
Perulia	263	...	...	...	...	95	...	11	7	7	27	57	69	...	...	...	7	18	...	377	...	...	...	...	...	11
		...	...	...	...	...	...	2	1	2	...	7	...	...	...	...	...	...	...	12	...	...	...	...	...	
Ranchi	209	...	...	...	...	18	...	...	9	...	2	40	...	...	...	...	...	1	...	84	...	...	...	...	...	4
		...	...	...	...	...	...	...	4	...	...	1	...	...	...	...	...	...	...	6	...	...	...	...	...	
Palamau	100	...	...	...	...	43	...	1	1	...	3	24	12	...	...	...	1	14	...	120	...	...	...	...	...	3
		...	...	...	...	...	...	...	...	...	...	...	1	...	...	...	...	...	...	1	...	...	...	...	...	
Hazaribagh, Central	923	...	2	...	...	233	...	14	5	5	64	234	44	...	...	10	18	115	...	974	...	...	...	...	...	79
		...	...	...	...	1	...	5	3	3	2	14	...	...	...	...	1	...	...	33	...	...	...	...	...	
B Gaya	393	...	...	2	...	63	...	8	3	...	8	35	4	...	...	...	...	30	...	191	...	...	...	...	...	7
		...	...	...	...	1	...	...	2	...	...	3	...	...	...	...	...	...	...	10	...	...	...	...	...	
Bhagalpur, Central	1,803	...	...	...	...	354	...	4	9	37	257	37	...	...	...	9	136	...	1,141	...	2	...	...	...	37
		...	...	...	...	4	...	...	3	...	...	3	...	...	...	...	...	...	...	18	...	...	...	...	...	
Monghyr	300	...	...	...	...	47	...	2	1	1	5	46	5	...	1	...	...	11	...	188	3	...	...	...	...	6
		...	...	...	...	1	...	...	1	...	...	2	...	...	...	...	...	...	...	6	...	...	...	...	...	
Darbhanga	221	...	...	...	...	31	...	2	2	48	16	...	...	...	...	9	12	...	148	...	...	...	...	...	8
		...	...	...	...	...	...	...	...	...	2	...	...	...	...	...	...	...	...	3	...	...	...	...	...	
Champarun	278	...	...	...	...	68	13	4	8	5	45	17	...	...	...	7	8	...	233	...	...	...	...	...	11
		...	...	...	...	4	...	2	...	...	1	...	...	...	...	...	2	...	...	10	...	...	...	...	...	
Muzaffarpur	320	...	...	...	...	73	...	4	4	53	18	...	...	...	3	13	...	202	...	...	...	...	...	7
		...	...	...	...	2	...	...	...	...	3	...	...	...	...	...	2	...	...	8	...	...	...	...	...	
Patna	292	...	...	...	...	53	...	3	9	20	8	...	...	...	2	22	...	200	...	...	...	...	...	9
		...	...	...	...	1	...	...	...	1	...	...	...	...	...	...	...	...	...	6	...	...	...	...	...	
Arrah	250	...	...	...	1	54	1	...	1	9	32	20	...	...	...	1	6	...	161	...	...	...	...	...	6
		...	...	...	...	1	...	...	...	...	3	...	...	...	...	...	...	...	...	7	...	...	...	...	...	
Chapra	267	...	...	1	...	66	...	7	1	2	97	11	...	...	...	3	12	...	235	...	...	1	...	...	6
		...	...	...	...	1	...	3	1	...	1	...	...	...	...	...	...	...	...	11	...	...	...	...	...	
Buxar, Central	1,228	...	...	...	...	1,207	...	5	4	40	272	337	...	...	...	10	72	...	2,158	...	1	...	...	3	56
		...	...	...	...	2	...	3	...	...	6	...	...	...	...	...	...	...	...	18	...	...	...	...	...	
Korantadih	43	...	...	...	...	21	...	...	...	...	1	2	...	...	...	...	2	...	28	...	...	...	...	...	...
Ghazipur	347	...	...	...	...	105	...	...	5	6	21	6	...	...	...	...	35	...	209	...	2	...	...	...	10
		...	...	...	...	...	...	...	5	...	6	1	...	...	...	...	...	...	...	16	...	...	...	...	...	
Azamgarh	289	...	1	...	...	58	...	2	5	11	8	6	...	...	2	4	20	...	164	...	...	2	...	...	5
		...	...	...	...	...	...	...	...	...	...	...	...	...	...	...	...	...	...	7	...	...	...	...	...	
Kasia	22	...	...	...	...	4	...	...	...	...	12	1	...	...	...	...	1	...	19	...	...	...	...	...	1
		...	...	...	...	...	...	...	...	...	...	...	...	...	...	...	...	...	...	1	...	...	...	...	...	
Gorakhpur	486	...	5	...	...	225	...	1	5	14	140	16	...	1	...	30	48	...	679	1	...	...	...	...	34
		...	...	4	...	...	...	...	3	...	1	1	...	...	...	...	...	...	...	11	...	...	...	...	...	
Basti	291	...	...	...	...	78	...	1	3	3	6	1	...	...	...	...	10	...	64	...	...	...	...	...	5
		...	...	...	...	1	...	...	2	...	...	...	...	...	...	...	...	...	...	5	...	...	...	...	...	
Fyzabad	305	...	1	...	...	55	...	...	5	8	9	...	1	1	...	9	26	...	216	1	...	...	...	...	9
		...	...	...	...	2	...	...	...	...	2	...	...	...	...	...	...	...	...	9	...	...	...	...	...	
Sultanpur	266	...	...	...	...	67	...	5	2	3	6	3	...	...	...	...	18	...	152	...	...	...	...	...	5
		...	...	...	...	1	...	...	...	...	...	...	...	...	...	...	...	...	...	5	...	...	...	...	...	

JAILS AND GROUPS.	Average annual strength.	Influenza.	Cholera.	Small-pox.	Enteric Fever.	Intermittent Fever.	Remittent Fever.	Simple Continued Fever.	Tubercle of the lungs.	Pneumonia.	Other Respiratory Diseases.	Dysentery.	Diarrhœa.	Hepatic Abscess.	Spleen Diseases.	Scurvy.	Anæmia and Debility.	Abscess, Ulcer, and Boil.	Phagedæna, Slough, and Gangrene.	All Causes.	Tænia.	Ascaris lumbricoides.	Dracunculus Medinensis.	Strongylus duodenalis.	Other Entozoa.	Average number constantly sick.
Rai Bareli	416 {	1	...	...	34	...	7	3	2	6	8	7	...	...	2	23		178	...	...	3	...	...	} 11
		...	...	...	...	...	...	...	1	1	...	...	...	...	...	8	...	...	...	...	
Partabgarh	176 {	...	...	...	19	...	...	...	...	5	6	6	...	...	7	13	1	85	...	...	...	...	} 4
		...	...	...	1	...	...	...	...	...	1	...	...	...	...	2	...	...	...	
Jaunpur	286 {	...	...	...	42	...	1	2	14	17	9	...	...	5	19	...	143	...	...	...	...	} 5
		...	...	...	...	...	...	1	...	...	...	...	...	...	4	...	...	...	
Benares, Central.	1,250 {	58	...	...	300	...	6	13	8	28	29	...	3	4	116	716	...	...	...	} 31
		...	...	...	...	...	6	4	1	4	3	...	3	...	42	...	...	...	
„ District	324 {	...	...	...	31	...	...	2	2	12	5	...	1	43	159	...	...	...	} 8
		...	...	...	...	...	3	2	2	...	2	1	...	...	12	...	...	...	
Mirzapur	208 {	...	...	...	79	2	...	5	7	14	4	...	12	158	...	...	3	} 3
		...	...	...	...	1	...	1	...	...	...	...	...	2	...	...	
Allahabad, Central.	1,337 {	...	...	...	129	...	148	24	5	53	55	49	...	155	874	1	1	} 53
		...	...	...	3	...	9	1	2	...	...	...	24	...	...	
„ District	557 {	...	...	...	39	...	4	13	21	31	1	...	11	45	264	...	} 15
		...	...	...	...	...	1	3	1	2	...	...	...	9	...	...	
Karwi	34 {	...	...	...	4	...	...	...	1	...	...	3	27	...	} 1
		...	...	...	...	...	...	...	...	...	...	...	
Banda	225 {	...	...	...	154	...	3	3	13	14	22	8	5	50	366	1	} 11
		...	...	...	...	...	...	2	1	...	...	...	8	...	
Fatehpur	218 {	...	...	...	132	...	...	4	6	10	...	10	185	...	} 5
		...	...	...	1	...	...	2	...	...	...	...	3	...	
Hamirpur	110 {	...	...	...	51	1	...	4	7	4	...	18	139	...	} 4
		...	...	...	...	...	1	1	...	...	...	5	...	
Orai	151 {	...	...	...	65	...	5	1	3	8	3	...	19	1	162	...	} 6
		...	...	...	1	...	1	...	...	...	...	...	2	...	
Cawnpore	419 {	...	...	...	71	...	...	8	5	5	2	2	17	130	...	1	} 9
		...	...	...	1	...	...	1	...	...	...	...	4	...	
Unao	322 {	...	...	...	30	...	1	3	4	6	...	14	111	...	} 5
		...	...	...	...	...	2	1	...	...	...	...	4	...	
Lucknow, Central.	1,450 {	...	...	1	76	1	7	10	10	17	13	...	64	297	...	} 17
		...	...	...	...	1	5	2	3	2	1	...	17	...	
„ District	482 {	...	...	...	27	...	...	5	1	6	5	...	39	148	...	} 8
		...	...	...	...	...	...	...	2	...	...	...	5	...	
Barabanki	381 {	1	...	...	56	...	1	4	8	10	14	...	6	40	207	...	} 11
		...	...	...	...	...	...	...	...	...	...	1	...	
Gonda	442 {	6	...	3	22	...	4	6	19	17	3	2	1	11	5	234	...	} 12
		...	...	...	...	...	2	2	2	...	...	...	8	...	
Bahraich	260 {	...	...	...	73	3	...	2	4	4	...	3	30	167	...	} 6
		...	...	...	...	1	...	...	...	...	...	...	3	...	
Kheri	305 {	...	...	1	48	...	...	2	7	3	6	...	38	166	...	} 6
		...	...	...	1	...	...	1	...	...	...	...	4	...	
Sitapur	494 {	100	...	...	42	...	1	10	5	9	2	1	13	221	1	} 8
		5	...	...	...	...	1	3	2	...	...	...	12	...	
Hardoi	384 {	3	...	...	72	...	3	1	2	4	2	...	1	25	137	...	} 4
		...	...	...	...	...	2	1	...	...	...	...	3	...	
Etawah	300 {	...	...	...	42	...	2	2	4	11	7	...	3	109	...	} 3
		...	...	...	...	...	1	...	...	...	...	...	7	...	
Mainpuri	333 {	...	...	1	64	...	3	5	6	26	15	3	4	41	219	...	1	} 15
		...	...	...	...	...	2	...	...	...	...	...	4	...	
Etah	328 {	...	...	...	95	...	1	3	17	14	21	3	...	35	65	316	...	} 15
		...	...	...	...	...	2	4	1	...	...	...	13	...	
Fatehgarh, Central.	1,532 {	96	...	4	207	3	19	14	10	87	51	1	3	109	792	4	} 43
		...	...	1	...	11	7	1	...	...	...	29	...	
„ District	297 {	5	...	...	38	...	2	...	7	18	7	...	15	141	...	1	} 6
		...	...	...	...	...	1	...	1	...	...	...	5	...	
GROUP V.— GANGETIC PLAIN AND CHUTIA NAGPUR.	22,083 {	280	9	5	5,080	25	176	175	200	542	1,926	951	1	14	30	214	1,685	4	15,146	13	6	7	3	4	} 651
		5	6	...	23	4	...	72	55	22	79	21	1	...	...	9	2	...	442	1	
A. Shahjahanpur.	314 {	...	...	...	81	...	2	7	6	8	15	...	31	218	...	} 7
		...	...	...	...	...	...	2	1	...	...	...	4	...	
Pilibhit	45 {	...	...	...	8	...	...	1	1	5	2	...	3	30	...	} 1
		...	...	...	...	...	...	...	...	...	...	...	...	
Bareilly, Central.	1,771 {	...	...	1	532	...	30	3	29	8	30	...	21	33	779	...	} 46
		...	...	...	...	...	3	...	1	...	...	...	12	...	

TABLE XLIII—*continued.*

ACTUALS of JAILS, GROUPS, and ADMINISTRATIONS on which the ratios in Tables XL—XLII have been calculated.

JAILS AND GROUPS.	Average annual strength.	Influenza.	Cholera.	Small-pox.	Enteric Fever.	Intermittent Fever.	Remittent Fever.	Simple Continued Fever.	Tubercle of the lungs.	Pneumonia.	Other Respiratory Diseases.	Dysentery.	Diarrhœa.	Hepatic Abscess.	Spleen Diseases.	Scurvy.	Anæmia and Debility.	Abscess, Ulcer, and Boil.	Phagedæna, Slough, and Gangrene.	All Causes.	Tænia.	Ascaris lumbricoides.	Dracunculus medinensis.	Strongylus denalis.	Other Entozoa.	Average number constantly sick.
Bareilly District	734	211 / 1	...	...	...	123	...	9 / 3	16 / 1	43 / ...	70 / ...	11 / ...	...	...	...	...	11	40	1	615 / 5	...	...	...	...	...	20
Budaon	370	...	...	...	24	...	...	1	...	... / ...	6 / 2	...	...	...	...	...	1	63	...	149 / 4	...	...	...	...	...	7
Aligarh	386	...	...	2	43	5 / 1	...	2 / 1	9 / 5	10 / ...	9 / 9	9 / ...	1	...	...	...	6	43	...	228 / 10	...	...	...	...	...	9
Bulandshahr	242	...	...	1	21	...	...	...	3 / 1	5 / 1	14	10 / 1	2	...	...	...	1	17	...	100 / 7	...	...	...	...	...	5
Moradabad	368	...	...	...	120	...	...	1	2 / 1	5 / ...	17	5 / ...	...	...	...	...	...	38	...	242 / 3	...	...	...	...	...	8
Bijnor	233	...	...	...	23	2 / 1	...	...	...	8 / ...	3	8 / ...	1	...	...	2	30	...	122 / 1	...	...	...	...	...	7
Dehra Dun	80	...	...	...	13	4	...	...	1 / ...	...	...	1 / ...	...	...	...	4	2	...	42 / 2	...	...	...	...	...	2
Saharanpur	326	...	...	...	56	...	...	1	5 / 1	13 / ...	47 / 5	9 / ...	...	...	...	13	25	1	278 / 9	...	...	...	2	...	13
Muzaffarnagar	157	...	...	...	33	...	...	2	8 / 3	3 / ...	4 / 1	1 / ...	...	...	...	1	9	...	96 / 8	...	...	...	...	...	5
Meerut	551	5	...	...	79	...	...	2 / 1	5 / 2	23 / 2	4 / 3	6 / ...	2	...	...	18	48	...	292 / 14	...	...	...	...	...	14
Delhi	461	...	...	...	55	...	...	4 / 3	17 / 4	4 / ...	19 / ...	7 / ...	...	...	...	4	39	...	193 / 12	...	2	...	...	...	9
Rohtak	94	...	...	...	70	...	...	...	2 / ...	4 / ...	10	4 / ...	...	...	...	1	12	...	126 / 1	...	...	...	...	...	2
Hissar	168	1	...	...	76	1	...	1	3 / ...	13 / ...	5	21 / ...	1	...	...	7	48	...	265 / 4	...	11	...	...	...	7
Karnal	85	...	...	...	24	...	...	...	2 / ...	12 / 2	...	3 / ...	...	...	...	...	3	...	50 / 2	...	...	...	...	...	1
Umballa	560	...	...	...	61	...	26	4	8 / ...	11 / ...	26	36 / ...	...	...	...	1	66	...	380 / 11	...	1	...	...	...	16
B Ludhiana	207	...	...	2	20	1	...	1	2 / ...	3 / ...	...	11 / ...	...	...	...	...	18	...	95 / 1	...	...	...	...	...	4
Hoshiarpur	47	...	...	...	5	...	...	1	...	1 / ...	...	2 / ...	...	...	...	...	4	...	29 / ...	...	1	...	...	...	...
Jullundur	267	...	...	...	40	...	...	2	4 / ...	11 / ...	13	5 / ...	...	...	...	...	36	...	142 / 5	...	...	...	...	...	4
Ferozepore	474	...	...	1	2	...	31	1	4 / ...	35 / ...	1	28 / ...	3	...	...	7	23	...	195 / 3	...	...	...	2	...	11
Amritsar	188	...	...	...	75	...	...	4	2 / ...	14 / ...	10	...	...	...	...	6	23	1	188 / 1	...	...	...	...	...	4
Lahore, Central	1,466	...	3	...	777	...	...	5 / 4	15 / 3	95 / ...	62	99 / 3	...	...	...	1	215	...	1,562 / 25	2	...	2	...	...	34
„ District	461	...	...	1	117	...	...	3 / 1	7 / ...	23 / ...	19	33 / ...	...	...	3	35	...	312 / 5	...	...	...	...	...	4
„ Female	150	...	...	1	41	...	...	3 / 3	1 / ...	11 / ...	2	4 / ...	...	...	...	9	...	110 / 4	1	...	2	...	...	4
Gurdaspur	221	3	...	...	16	3	...	1	3 / 1	5 / ...	8	8 / ...	...	...	1	2	18	...	89 / 1	...	...	...	...	...	3
Gujranwala	317	...	...	...	60	...	...	2	...	35 / ...	1	1 / ...	...	...	...	11	53	...	309 / 3	...	1	...	...	...	9
Sialkot	305	...	...	...	48	...	...	5	3 / ...	21 / ...	20	12 / ...	...	...	2	7	27	...	186 / 3	...	...	...	...	...	9
Gujrat	95	...	...	...	31	...	...	1	...	1 / ...	6	13 / ...	...	...	...	1	9	...	82 / 1	...	...	...	...	...	2
Jhelum	161	3	...	...	30	...	...	3	2 / ...	6 / ...	4	2 / ...	...	...	...	...	17	...	85 / 2	...	...	2	...	...	3
Rawalpindi	676	...	...	...	119 / 2	...	...	3 / 2	0 / 1	1 / ...	2	15 / ...	...	...	...	53 / 4	45	...	392 / 17	...	7	...	...	...	20
GROUP VI.— UPPER SUB-HIMALAYA.	11,030	223 / 1	3 / 3	4 / 2	7 / 1	2,923 / 3	17 / 3	57 / 1	89 / 28	147 / 29	451 / 10	401 / 13	413 / 7	9	5 / ...	184 / 7	1,084	3	7,981 / 180	4	1	31	2	...	246

JAILS AND GROUPS.	Average annual strength	Influenza.	Cholera.	Small-pox.	Enteric Fever.	Intermittent Fever.	Remittent Fever.	Simple continued Fever.	Tubercle of the lungs.	Pneumonia.	Other Respiratory Diseases.	Dysentery.	Diarrhœa.	Hepatic Abscess.	Spleen Diseases.	Scurvy.	Anæmia and Debility.	Abscess, Ulcers, and Boil.	Phagedæna, Slough, and Gangrene.	All Causes.	Tænia.	Ascaris lumbricoides.	Dracunculus Medinensis.	Strongylus Duodenalis.	Other Entozoa.	Average number constantly sick.
A.																										
Peshawar	538	186 3	...	...	...	160 ...	...	...	4 2	9 1	8 ...	28 3	13	...	1	5	97	...	543 13	...	...	1	...	...	22
Kohat	137	...	...	...	...	50	...	...	...	3 1	2	2 ...	2	...	1	2	37	...	130 2	...	...	2	...	...	4
Bannu	134	...	...	...	31 ...	1	...	...	1	2 1	10	7	...	...	5	16	...	92 2	...	...	1	...	...	4
Shahpur	221	...	...	...	39 ...	...	...	1	...	2	8	...	...	1	15	...	97 6	2	...	5	...	...	5
Mianwali	259	...	...	...	73 1	...	...	1	2 ...	11 ...	14 ...	28 ...	39 ...	...	...	78	...	386 4	...	...	5	...	...	10
Lyallpur	400	...	3 1	...	80 ...	...	...	1	10 3	7	8	9	...	1 1	42	...	285 8	...	...	...	...	...	10
Jhang	177	...	...	...	31 ...	...	...	...	2	8	3	3	...	3	27	...	102 ...	1	...	...	...	...	3
Montgomery, Central.	1,964	66 ...	...	...	198 1	...	...	22 18	25 7	50 1	12 ...	16	...	2	20 1	155	...	832 36	6	...	7	...	1	44
Mooltan, Central.	1,171	...	...	...	440 ...	...	...	23 8	18 4	31 ...	31 2	37	...	...	73 2	87	...	378 21	3	...	...	...	...	48
„ District	609	8 ...	...	...	85 ...	...	...	6 2	10 2	18 1	10 1	9	...	2	18	...	233 5	3	...	3	...	...	13
Dera Ismail Khan.	415	...	...	...	189 ...	1 1	...	1	3 1	3	16 1	14 1	1	1	5	93	1	436 8	...	...	20	...	...	13
Dera Ghazi Khan.	342	...	...	...	220 2	...	...	...	41 ...	21	50 ...	...	...	...	62	...	543 4	...	...	1	...	...	9
C.																										
Shikarpur	542	...	...	4 ...	51 ...	3 3	...	3 2	10 1	15 3	...	5	...	2 2	1	3 2	8	...	141 16	...	...	...	...	...	7
Sind Gang	380	...	...	1 ...	17 ...	...	...	22 6	7 ...	4 1	1	...	2	...	...	...	64 7	...	...	...	...	...	2
Hyderabad, Central.	871	...	...	3 ...	64 ...	...	...	6 2	19 7	53 ...	42 2	36 1	7	11	12	18	...	419 24	1	...	2	...	...	16
Kurrachee	247	...	...	...	54 ...	4	...	5 1	2	18	22	18	...	5 1	5	39	...	203 4	...	...	...	...	...	8
GROUP VII.— N.-W. FRONTIER, INDUS VALLEY, AND N.-W. RAJPUTANA.	8,407	260 3	3 1	8 ...	1,782 4	9 4	...	73 37	146 32	277 8	230 11	276 2	15 ...	20 2	137 2	791 4	1	5,474 163	16	...	46	...	1	218
A.																										
Rajkot	107	...	...	...	26 ...	1	...	...	1	2	3	3	...	...	...	9	...	53 ...	...	...	2	...	...	2
Ahmedabad, Central.	863	...	...	3 ...	148 ...	...	...	7 3	33 4	8 1	20	13	...	1	12 ...	14	...	337 13	...	...	5	...	...	24
B																										
Ajmer	412	...	...	1 ...	25 ...	...	...	2 1	6 2	3 1	3	...	...	...	...	26	...	161 7	...	...	24	...	...	9
Muttra	238	...	...	1 ...	8 ...	...	...	1	3	5 ...	...	2	...	...	26	...	72 2	...	...	...	...	...	5
Agra, Central	7,070	...	...	...	577 ...	...	...	11 8	46 7	53 1	24 1	30 ...	3	...	9	192	...	1,143 24	3	...	2	...	...	73
„ District	412	...	...	1 ...	47 ...	...	...	3 1	8 2	13 1	9	11 ...	...	...	1	59	...	228 4	...	...	1	...	...	9
Jhansi	180	...	...	...	35 ...	...	...	...	...	9 1	9	9 ...	1	...	2	54	...	201 4	...	...	...	...	...	7
Lalitpur	64	...	...	...	31 ...	...	...	...	2	5 ...	...	...	...	8	...	73 ...	...	...	...	...	...	2
GROUP VIII.— S.-E. RAJPUTANA, CENTRAL INDIA, AND GUJARAT.	4,346	...	...	4 ...	697 1	...	...	24 13	97 15	93 4	70 1	71 ...	6	1	24	388	...	2,275 54	3	...	34	...	...	131

107

R 2

TABLE XLIII—continued.

ACTUALS of JAILS, GROUPS, and ADMINISTRATIONS on which the ratios in Tables XL—XLII have been calculated.

JAILS AND GROUPS.	Average annual strength.	Influenza.	Cholera.	Small-pox.	Enteric Fever.	Intermittent Fever.	Remittent Fever.	Simple Continued Fever.	Tubercle of the lungs.	Pneumonia.	Other Respiratory Diseases.	Dysentery.	Diarrhœa.	Hepatic Abscess.	Spleen Diseases.	Scurvy.	Anæmia and Debility.	Abscess, Ulcer, and boil.	Phagedæna, Slough, and Gangrene.	ALL CAUSES.	Tænia.	Ascaris lumbricoides.	Dracunculus Medinensis.	Strongylus duodenalis.	Other Entozoa.	Average number constantly sick.
A.																										
Damoh	45	...	...	...	...	14	2	...	...	1	1	7	2	...	...	..	1	11	...	65	...	...	5	...	...	2
Saugor	87	...	...	...	...	10	...	...	1	...	...	10	...	...	...	...	1	4	...	35	2	...	...	...	...	1
Jubbulpore, Central	840	...	...	...	...	8	1	...	6	4	2	...	3	...	...	...	10	42	...	185 8	...	...	2	...	...	8
Narsinghpur	64	...	...	...	...	9	...	...	...	1	1	1	...	...	...	...	2	2	...	31 1	...	...	...	...	...	1
Mandla	61	...	...	...	...	18	...	...	...	1	2	2	6	...	...	...	...	17	...	86	...	...	...	...	...	3
Bilaspur	116	1	...	...	...	14	3	...	1	...	3	8	5	...	...	...	1	2	...	47 3	1	...	...	...	...	2
Sambalpur	110	...	...	...	...	12	...	...	1	...	1	6	3	...	...	...	...	3	...	34 4	...	...	...	...	...	1
Raipur, Central	442	5	...	...	...	108	...	...	1	6	18	31 1	8	...	...	...	1	30	...	295 7	...	...	...	...	...	9
Balaghat	55	...	...	...	...	13	...	...	...	2	2	4	2	...	...	...	...	10	...	55 1	...	.	...	...	...	2
Seoni	51	...	...	...	...	6	...	...	...	...	1	...	2	...	...	...	5	2	...	43 1	2	...	...	...	...	2
Chhindwara	40	...	...	...	...	4	...	...	1	...	1	1	...	...	...	...	1	5	...	20	...	...	1	...	...	1
Hoshangabad	76	...	...	...	...	36	...	...	...	...	2	9	10	...	...	...	1	7	1	81 2	...	...	...	...	...	4
Nimar	89	...	...	...	...	51	...	1	...	2	2	8	11	...	...	...	1	17	...	125 3	3	...	1	...	...	4
Betul	50	...	...	...	...	16	...	...	...	3	4	6	1	...	...	...	4	13	...	71	...	...	...	...	...	1
Nagpur, Central	50	22	...	...	...	228	...	11	9	6	9	15	46	...	...	1	3	29	...	567 11	...	...	...	...	...	16
Bhandara	65	...	...	...	...	16	...	...	...	1	...	2	1	...	...	...	...	2	...	29	...	...	...	...	...	1
Wardha	46	...	...	...	...	6	...	...	1	3	...	1	...	...	...	1	...	2	...	20 1	...	...	...	...	...	...
Chanda	51	...	...	...	...	7	...	...	1	3	2	2	...	...	...	...	1	4	...	25 2	...	...	...	...	...	1
B.																										
Secunderabad	85	...	...	...	...	11	1	...	...	...	4	2	...	...	...	...	1	8	...	61	...	...	1	...	...	2
Yeotmahl	51	2	...	...	...	8	...	...	...	1	3	2	...	...	...	...	...	3	...	30 2	...	...	2	...	...	1
Amraoti, Central	260	2	1	...	...	31	...	1	...	3	3	17	3	...	8	1	2	10	...	135 6	...	...	...	...	...	5
Akola, Central	333	...	...	...	...	69	...	1	1	3	18	23	...	...	...	...	2	48	...	256 6	1	...	1	...	...	10
Buldana	68	...	...	...	...	4	...	...	1	1	...	2	...	...	...	...	3	5	...	29	...	...	3	...	...	2
Dhulia	365	...	...	...	...	154	...	...	...	3	8	67	22	...	...	...	6	31	...	384 3	1	...	18	...	...	13
Yerrowda, Central	1,432	...	...	...	...	136	...	...	1	2	42	51	95	3	...	1	5	267	...	1,280 3	...	1	37	...	...	71
Bijapur	283	...	...	...	...	2	...	...	2	4	...	6	...	...	4	1	1	1	...	101 5	...	...	20	...	...	5
Deccan Gang	876	...	...	...	...	80	...	9	1	3	13	23	13	...	...	...	3	175	...	504 15	1	...	23	...	1	22
Dharwar	310	...	...	...	...	20	...	...	1	...	2	27	7	...	...	...	2	4	...	119 4	...	...	17	...	...	7
GROUP IX.— DECCAN	7,017	32	1	...	...	1,109	7	24	27	43	129	331	270	...	3	15	57	757	1	4,716 90	9	2	125	...	1	195

| | | 1. ADMISSIONS. | | | | | | | | | | | | | 2. DEATHS. | | | | | | | | | | | | | | |
JAILS AND GROUPS.	Average annual strength.	Influenza.	Cholera.	Small-pox.	Enteric Fever.	Intermittent Fever.	Remittent Fever.	Single Continued Fever.	Tubercle of the lungs.	Pneumonia.	Other Respiratory Diseases.	Dysentery.	Diarrhœa.	Hepatic Abscess.	Spleen Diseases.	Scurvy.	Anæmia and Debility.	Abscess, Ulcer, and Boil.	Phagedæna, Slough, and Gangrene.	ALL Causes.	Tænia.	Ascaris lumbricoides.	Dracunculus Medinensis.	Strongylus duodenalis.	Other Entozoa.	Average number constantly sick.
Thana	607 {	...	... 1 ...	54		... 7 4	2	34 1	31 2	28 ...	...	1	3 1	36	...	276 9		...	10	...	} 16
Bombay, Common.	438 {	...	... 2 ... 1 ...	157	1 1	... 6 3	6 2	10 1	17 1	54 2	3	12 3	29	...	517 23		...	5	...	} 14
Bombay House of Correction.	225 {	...	... ...	36	...	... 4 1	1	10 2	2	10 ...	...	6	...	...	112 7		...	1	...	} 2
Ratnagiri	107 {	...	...	15		... 1 1	...	...	2	... 1	...	2	2	...	31 2		...	...	...	} 1
Karwar	119 {	...	...	6		... 1	2	...	2	... ...	...	...	...	6	...	51 1		1	...	...	} 2
Mangalore	144 {	...	... 1	8		6 1	1	...	3 1	2 ...	...	...	...	...	31 2		...	...	...	} 1
Cannanore, Central	678 {	...	6 2	51 1	2	21 1	6 1	12 1	16	41 6	5 ...	...	1	10	...	211 15		...	...	...	} 9
GROUP X.— WESTERN COAST.	2,318 {	...	3 7 1 2	327 1	3 1	27 10	23 5	71	93 10	99 3	3 ...	1	24 4	82	...	1 232 59		1	16	...	} 46
A. Bellary	568 {	...	2 ...	250 4	...	7	18	5 2	25 1	14 1	...	...	2	26	...	515 9		...	62	...	} 16
Salem, Central	750 {	6 1	...	78 1	...	9 1	4	12 1	10	42 5	...	...	...	17	...	249 12		...	9	...	} 11
Coimbatore, Central.	1,186 {	...	8 2	11	...	15	12 3	16 2	19 1	123 2	...	1	3 1	19	...	520 15	1	...	52	1	} 30
B. Palamcottah	386 {	...	...	10	...	...	1	1	8 3	28 ...	...	...	2	33	...	166 3		...	5	...	} 9
Madura	456 {	...	1	20	1	...	1	6 3	8	13 ...	...	...	2	6	...	107 7		...	7	1	} 8
Trichinopoly, Central.	986 {	...	...	29 2	...	15 2	...	2	40	27 1	...	...	1	39	...	243 12		2	12	3	} 17
Tanjore	333 {	...	...	16	12 1	5	5 2	1	5 1	18	4 ...	...	3	6	...	154 10		...	1	...	} 7
Cuddalore	366 {	...	...	27	...	...	1	...	6	4	...	...	2	12	...	108 2		...	4	...	} 7
Vellore, Central	1,220 {	...	...	36	...	71	10 4	12 2	26	25 2	...	...	...	74	1 1	525 15		5	10	1	} 26
Madras, Civil	34 {	...	...	...	...	...	...	...	...	...	...	...	...	...	...	...		...	...	...	} ...
Madras Penitentiary, Central.	1,072 {	...	4 2 2 3	62	1	11	14 3	5	40	48 ...	1 1	...	21	31	...	565 11		9	3	...	} 23
Nellore	283 {	...	3 1	2	1	3	2 1	1 1	...	8 2	2 ...	...	4	4	...	47 7		...	3	...	} 3
C. Rajamundry, Central.	999 {	...	30 10	243	...	146	11 1	6 3	35	214 6	...	...	3	41	...	1,051 29	1	4	48	1	} 23
Vizagapatam	471 {	...	...	12	...	1	3 2	6	2	34 1	19 3	...	...	2	...	101 9		...	...	...	} 4
Berhampur	129 {	...	...	9	...	...	...	2	1	4	4 ...	...	...	2	...	32 1		...	...	...	} 1
GROUP XI.— SOUTHERN INDIA.	9,240 {	6 1	37 21 14 2	811 7	15 1	318 22	100	78 14	225 2	602 24	39 3	1	1	43 2	312	1 1	4,486 142	2	20	216	6 2	} 185

109

TABLE XLIII—concluded.

ACTUALS of JAILS, GROUPS, and ADMINISTRATIONS on which the ratios in Tables XL—XLII have been calculated.

JAILS, GROUPS, AND ADMINISTRATION.	Average annual strength.	1. ADMISSIONS.										2. DEATHS.									ALL CAUSES.						Average number constantly sick.
		Influenza.	Cholera.	Small-pox.	Enteric Fever.	Intermittent Fever.	Remittent Fever.	Simple Continued Fever. Tubercle of the lungs.	Pneumonia.	Other Respiratory Diseases.	Dysentery.	Diarrhœa.	Hepatic Abscess.	Spleen Diseases.	Scurvy.	Anæmia and Debility.	Abscess, Ulcer, and Boil.	Phagedæna, Slough and Gangrene.			Tænia.	Ascaris lumbricoides.	Dracunculus Medinensis.	Strongylus duodenalis.	Other Entozoa.	
Shillong	50	...	...	...	...	7	...	...	...	1	1	...	4	...	...	...	...	...	...	17	...	...	...	...	...	...
Darjeeling	100	...	...	...	...	45 / 3	...	...	...	1	23	21	18	...	...	...	1	5	...	164 / 4	5	3	...	...	...	} 7
Almora	67	...	...	...	...	21 / 1	...	10	...	2	2	4	7	...	...	...	2	6	...	81 / 2	...	1	...	...	...	} 2
Pauri	9	...	...	...	...	...	...	...	...	1	...	5 / 1	1	...	...	...	...	1	...	9 / 1	...	...	...	...	...	} ...
Naini Tal	1	...	...	...	...	...	...	...	...	...	...	...	...	...	...	...	...	...	...	1	...	...	...	...	...	} ...
Simla	16	...	...	...	...	...	...	...	...	...	...	...	...	...	...	...	...	...	...	1	...	...	...	...	...	} ...
Abbottabad	90	...	...	...	...	22	...	...	...	2	1	4	4	...	...	...	3	4	...	61 / 1	...	...	...	...	1	} 1
Quetta	70	...	...	...	...	47	...	...	...	1	5	9	5	...	...	...	...	11	...	121	...	...	...	...	...	} 2
Mercara	87	...	...	...	...	1	...	...	...	3	2	...	1	...	...	...	...	3	...	19 / 4	...	...	...	...	...	} 1
Russellkonda	85	...	...	...	...	...	...	2	...	1	8	3	...	...	...	...	...	6	...	31 / 2	...	...	...	...	...	} 1
GROUP XII.—HILLS	575	...	...	...	...	145 / 4	2	10	1	11	42	45 / 3	40	...	...	...	6	36	...	505 / 14	5	4	...	...	1	14
EXTRA INDIA— Aden	70	...	...	...	...	17	...	...	...	2	2	8	5	...	...	...	...	3	...	50 / 1	...	...	...	...	...	2
INDIA (a)† { Remaining from 190 Admitted. / Died / Died out of hospital	91,917	13 / 1,057 / 15	1 / 73 / 40	... / 31 / 4	4 / 6 / 15	364 / 16,704 / 84	5 / 103 / 20	27 / 872 / 1	103 / 803 / 293	55 / 890 / 210	118 / 2,428 / 85	255 / 7,495 / 277	56 / 3,603 / 68	... / 10 / 7	5 / 58 / 3	1 / 118 / 7	88 / 1,014 / 50	225 / 6,543 / 6	2 / 13 / 3	2,051 / 59,451 / 1,763 / 25	2 / 63 / 1	2 / 41 / ...	4 / 476 / 1	2 / 29 / 4	17 / ...	2,532
BURMA	12,639	14 / ...	3 / ...	2 / 1	8 / 6	645 / 6	10 / 1	240 / 48	92 / 13	47 / ...	135 / 8	353 / 33	222 / 12	3 / 2	... / ...	2 / 3	39 / ...	696 / ...	...	4,043 / 215	5 / ...	2 / ...	... / ...	7 / ...	4 / ...	210
EASTERN BENGAL AND ASSAM	6,401	106 / ...	12 / 12	1 / ...	12 / 12	1,565 / 3	14 / ...	8 / 23	73 / 16	57 / 6	249 / 15	1,965 / 93	576 / 15	2 / 1	...	40 / 2	206 / 17	249 / ...	1	6,922 / 241	2 / ...	5 / ...	... / ...	8 / 1	2 / ...	296
BENGAL	14,172	136 / ...	5 / ...	2 / 1	3 / 2	4,120 / 39	21 / 3	45 / 57	192 / 30	83 / 15	483 / 15	2,783 / 93	1,288 / 15	3 / 1	7 / ...	15 / 7	151 / ...	91 / ...	4	13,513 / 353	12 / 1	6 / ...	1 / ...	6 / ...	5 / ...	553
UNITED PROVINCES.	23,680	496 / ...	6 / ...	9 / 2	4 / 12	4,304 / 5	21 / ...	173 / 63	169 / 71	286 / 26	522 / 6	864 / 20	501 / 1	1 / ...	6 / ...	19 / 5	236 / 2	1,936 / ...	4	13,470 / 410	14 / ...	4 / 1	10 / ...	2 / ...	2 / ...	636
PUNJAB	11,512	81 / ...	6 / 4	3 / 1	4 / 6	2,834 / 1	6 / 1	57 / 51	94 / 29	164 / ...	474 / 8	327 / 6	494 / 2	...	3 / ...	8 / 8	206 / 2	1,184 / ...	1	8,147 / 188	18 / ...	1 / ...	32 / ...	... / ...	1 / ...	294
N.-W. F. PROVINCE.	1,314	186 / 3	6 / ...	... / ...	... / ...	452 / ...	2 / ...	... / ...	5 / 2	18 / 2	16 / 1	60 / 6	40 / 1	...	1 / ...	2 / ...	20 / ...	247 / ...	1	1,262 / 26	... / ...	1 / ...	23 / ...	... / ...	1 / ...	44
BOMBAY	7,851	... / ...	... / ...	14 / 1	... / ...	1,055 / 1	9 / 1	9 / 18	48 / 24	106 / 1	228 / 12	327 / 9	310 / 2	...	20 / 2	24 / 3	72 / 6	644 / 1	...	4,732 / 137	3 / ...	3 / ...	135 / ...	1 / ...	1 / ...	215
CENTRAL PROVINCES.	3,547	32 / 1	1 / 1	... / ...	... / ...	676 / ...	6 / 1	15 / ...	20 / 7	35 / 5	57 / 4	147 / 5	128 / 1	...	... / ...	10 / 2	39 / 3	265 / 1	...	2,233 / 56	7 / ...	... / ...	14 / ...	... / ...	1 / ...	74
MADRAS	10,147	6 / 1	37 / 14	4 / 2	25 / 8	869 / 2	19 / ...	345 / 24	107 / 15	90 / 4	252 / 1	649 / 31	36 / 3	1 / 1	1 / ...	... / ...	44 / 2	328 / ...	1	4,762 / 161	2 / ...	20 / ...	216 / ...	6 / ...	1 / 2	196
ANDAMANS	14,348	186 / ...	... / ...	... / ...	... / 12	17,881 / 41	89 / 41	... / ...	163 / 93	156 / 69	824 / 11	2,359 / 187	793 / 7	3 / 2	1 / 1	21 / ...	... / ...	1,138 / ...	...	27,237 / 559	... / ...	... / ...	5 / ...	1 / ...	2 / ...	971
INDIA (b)‡‡ { Remaining from 190 Admitted. / Died / Died out of hospital	1,06,265	22 / 1,243 / 15	1 / 73 / 40	... / 4 / 15	4 / 34,545 / 16	595 / 107 / 61	8 / 892 / 1	27 / 266 / 380	300 / 1,034 / 279	61 / 3,252 / 98	159 / 9,555 / 464	387 / 443,613 / 75	90 / 5 / 9	... / 3 / 4	5 / 139 / 8	3 / 1,014 / 50	88 / 7,186 / 7	274 / 13 / 3	2	2,021 / 96,688 / 2,327 / 57	2 / 63 / 1	2 / 41 / ...	4 / 481 / 1	2 / 30 / ...	1 / 19 / 5	3,503

* Remaining + Admitted = total treated ; Remaining + admitted + died out of hospital = total cases.
† Including Ajmer, Secunderabad, Quetta, and Mercara, and excluding Andamans.
  (a) Including the subsidiary jails, the total figures are :—Average strength 97,078. Average constantly sick 2,614. Number of deaths 1,955. Number of admissions 63,630.
‡ Including Ajmer, Secunderabad, Quetta, Mercara, and Andamans.
  (b) Including subsidiary jails, the total figures are :—Average strength 1,11,424. Average constantly sick 3,535. Number of deaths 2,414. Number of admissions 90,807.

GEOGRAPHICAL GROUPS.	1. AVERAGE STRENGTH. 2. CONSTANTLY SICK.												
	January.	February.	March.	April.	May.	June.	July.	August.	September.	October.	November.	December.	Average for the year.
GROUP I.—BURMA COAST AND BAY ISLANDS.	8,197	8,300	8,217	8,377	8,577	8,718	8,819	9,063	9,176	9,357	9,161	9,468	8,803
	104	106	97	111	129	131	143	144	131	129	151	135	126
GROUP II.—BURMA INLAND .	3,849	3,784	3,760	3,754	3,825	3,956	3,929	3,902	3,854	3,823	3,791	3,718	3,831
	70	73	78	84	73	74	91	93	98	91	74	74	81
GROUP III.—ASSAM . .	1,276	1,297	1,371	1,404	1,448	1,462	1,440	1,451	1,425	1,401	1,377	1,352	1,398
	41	40	37	36	46	60	71	88	89	100	161	98	72
GROUP IV.—BENGAL AND ORISSA .	11,515	11,457	11,452	11,685	11,771	11,973	12,077	12,353	12,323	12,656	12,819	12,798	12,001
	531	621	600	541	480	491	493	496	470	412	465	497	512
GROUP V.—GANGETIC PLAIN AND CHUTIA NAGPUR.	21,170	21,801	22,081	21,238	21,784	22,333	22,781	23,979	23,030	22,666	22,411	21,375	22,083
	509	562	560	618	657	637	698	771	827	716	624	561	651
GROUP VI.—UPPER SUB-HIMALAYA .	11,500	11,347	11,417	11,600	11,653	11,843	12,031	12,230	12,413	12,504	12,340	12,017	12,020
	313	286	278	279	392	295	300	342	320	280	270	193	296
GROUP VII.—N.-W. FRONTIER, INDUS VALLEY, AND N.-W. RAJPUTANA.	8,215	8,245	8,141	8,301	8,336	8,313	8,397	8,493	8,660	8,501	8,541	8,190	8,407
	231	230	213	214	222	222	218	225	116	193	203	207	218
GROUP VIII.—S.-E. RAJPUTANA, CENTRAL INDIA, AND GUJARAT.	4,248	4,179	4,197	4,186	4,245	4,280	4,396	4,417	4,504	4,537	4,500	4,448	4,346
	136	134	107	116	119	114	150	143	140	140	131	125	131
GROUP IX.—DECCAN . .	7,065	7,054	6,932	6,832	6,814	7,012	7,181	7,325	7,116	7,057	6,915	6,935	7,017
	152	161	147	163	181	166	114	151	263	241	202	186	195
GROUP X.—WESTERN COAST .	2,471	2,478	2,381	2,295	2,243	2,245	2,372	2,370	3,319	2,283	2,300	2,200	2,318
	48	42	38	42	41	46	55	55	48	41	53	52	46
GROUP XI.—SOUTHERN INDIA .	8,540	8,565	8,566	8,693	8,863	9,051	9,331	9,649	9,740	9,856	9,909	9,884	9,240
	171	175	188	183	130	161	182	182	200	228	194	181	185
GROUP XII.—HILLS . .	520	529	530	567	602	621	595	596	613	573	583	533	57
	9	14	11	11	25	14	18	20	93	17	17	17	14
INDIA*	88,720	88,344	88,178	88,823	90,167	92,013	93,324	95,679	93,400	93,383	94,060	92,661	91,917
	2,330	2,448	2,356	2,461	2,443	2,110	2,643	2,519	2,825	2,616	2,552	2,429	2,532

ADMINISTRATIONS.	1. AVERAGE STRENGTH. 2. CONSTANTLY SICK.												
	January.	February.	March.	April.	May.	June.	July.	August.	September.	October.	November.	December.	Average for the year.
BURMA	12,045	11,974	11,937	12,131	12,401	12,674	12,747	12,960	13,030	13,180	13,259	13,196	12,639
	174	179	171	195	201	205	233	242	230	237	230	210	210
EASTERN BENGAL AND ASSAM .	6,133	6,138	6,148	6,215	6,352	6,228	6,336	6,061	6,591	7,032	6,285	6,307	6,101
	237	296	240	263	270	273	301	315	314	301	403	350	295
BENGAL	13,795	13,832	13,827	13,947	14,134	14,168	14,499	14,461	14,571	14,735	14,127	14,153	14,172
	527	626	631	602	537	531	540	596	552	514	478	463	552
UNITED PROVINCES .	22,475	27,217	22,153	22,575	23,157	23,755	24,374	25,773	24,035	24,851	24,431	22,150	23,650
	576	578	541	608	618	611	686	737	774	688	615	594	636
PUNJAB	11,521	11,259	11,370	11,421	11,761	11,425	11,415	11,614	11,695	11,632	11,623	11,652	11,512
	311	315	293	281	306	318	315	324	291	352	350	372	294
N.-W. F. PROVINCE . .	1,316	1,322	1,329	1,394	1,308	1,257	1,311	1,316	1,356	1,457	1,306	1,283	1,314
	41	45	48	50	48	39	50	44	42	42	36	33	44
BOMBAY	7,880	7,793	7,700	7,590	7,604	7,790	7,860	8,005	8,063	8,033	7,917	7,591	7,851
	181	189	171	188	197	202	219	270	255	248	241	228	215
CENTRAL PROVINCES . .	3,533	3,499	3,502	3,476	3,506	3,595	3,669	3,865	3,610	3,545	3,514	3,416	3,547
	64	66	57	63	65	49	81	93	108	93	72	70	74
MADRAS	9,425	9,452	9,435	9,555	9,728	9,959	10,218	10,539	10,924	10,761	10,535	10,918	10,147
	185	182	206	228	186	178	191	200	215	242	209	196	196
INDIA †	88,720	83,348	89,278	88,825	90,167	92,013	93,324	95,679	95,401	95,323	94,060	92,661	91,917
	2,330	2,448	2,356	2,461	2,443	2,410	2,519	2,519	2,835	2,456	2,552	2,428	2,532
ANDAMANS . . .	14,210	14,191	14,161	14,216	14,355	14,356	14,411	14,417	14,127	14,414	14,433	14,293	14,348
	882	938	910	972	673	1,102	1,285	1,320	950	865	836	844	971
INDIA (a) . . .	102,942	102,547	102,339	102,341	104,522	106,407	107,735	110,105	107,539	109,797	108,493	97,164	106,365
	3,202	3,386	3,376	3,392	3,416	3,520	3,918	3,930	3,775	3,525	3,588	3,272	3,503

* Including Aden and excluding Andamans. † Including Ajmer, Secunderabad, Quetta, and Mercara, and excluding Andamans.
(a) Including Ajmer, Secunderabad, Quetta, Mercara and Andamans.

TABLE XLIV.

ABSTRACT of the SANITARY SHEETS of the most UNHEALTHY JAILS, SANITARY DEFECTS, IMPROVEMENTS, SUGGESTIONS, etc.

BURMA.

Pagan.—Variations of temperature were predisposing causes of disease especially among the weak and infirm leper prisoners. The Inspector-General remarks :—" This jail will be abolished as soon as accommodation is provided for all leper prisoners of the province at Myingyan in a building which will form an *annexe* of the jail there."

Bhamo.—The sickness and mortality are due in great part to the swamp on the west side of the jail and to the opium-eating and smoking habits of a good many of the prisoners.

EASTERN BENGAL AND ASSAM.

Dibrugarh.—The overcrowding in the jail was avoided by using the old workshop as a sleeping barrack. The sickness and mortality were said to be due not to any particular local cause, but to the general unhealthiness of the town and district and also to the large number of prisoners admitted to the jail in indifferent or bad health.

Gauhati.—The under-trial ward was, except during July and October, overcrowded throughout the year. The whole of the jail site is water-logged during the rains, owing to defects in drainage, and this is considered the chief cause of the prevalence of dysentery. Three additional temporary isolation wards for the dysentery patients were erected. With the permission of His Honour the Lieutenant-Governor of the Province, an isolation ward to accommodate 14 patients was opened at Bharalo camp, about ½ a mile from the jail. This proved a veritable sanatorium and more than justified its existence by the excellent results obtained, all the worst cases having been removed thither and having with two exceptions resulted in recovery, partial or complete. The following sanitary improvements were effected during the year :— (1) From the 1st of April the system of trenching in the jail garden was discontinued, the night-soil being removed to a distance of about ¾ of a mile. (2) Exten-ion of the pipe water-supply. (3) An improved additional night latrine with *Bailey's* seats was provided for the female ward. (4) A new Larymore boiler was bought. (5) A new permanent barrack affording accommodation for 61 male convicts was erected. The Inspector-General remarks :—" Until the jail is better drained, it is likely to remain unhealthy. Its situation, unfortunately, does not lend itself to very satisfactory drainage. Arrangements are being made for the incineration of *all* night-soil and urine".

Backergunge.—Overcrowding existed during the latter half of the year. The cane shop is invariably used as a sleeping barrack at night. The ventilation is defective, free circulation of air being interfered with by high partition walls. The drainage is bad, and high-tide water comes into the jail through the drains ; the whole site is water-logged during the rainy season. The jail is surrounded on three sides by a very offensive *khal*, which is to all intents and purposes a sewer. The unhealthiness is said to be due to the defective site and its water-logged condition, which, this year, was aggravated by the excessive rainfall. The following sanitary improvements were effected during the year. The cow-shed below the three-storied barrack was abolished ; a new 22-seated latrine was put up ; six new cells were built ; the construction of a new female and a new civil ward was begun and is now in progress ; the jail mortuary has been provided with fly-proof nettings, stone table, sink, etc. ; two more aluminium night-soil carts were purchased and are in use ; the north enclosure was included in the jail compound ; the old garden ward was put into thorough repair and is now capable of accommodating 44 under-trials and 19 convicts and two latrines for this ward were purchased. The Inspector-General remarks :—"Although considerable improvements have been effected in this notoriously unhealthy jail, much remains to be done, but not without large expenditure. The jail site is bad in the extreme as it is situated in the centre of a *bazaar*, and is practically surrounded by a *khal* into which the latrines on the banks discharge their contents. At high tides the *Khal* water flows into the jail through the drains, to prevent which sluice gates will have to be provided. The water-supply brought, as it is, in a tank by boat from the middle of the river, is not above suspicion, as possibly, to save trouble, water is sometimes taken from less distant and more convenient sources. For this reason it has been suggested that the reserve tank water in the neighbourhood of the jail might be taken. From analysis it compares favourably with the river water and can be thoroughly supervised. To combat dysentery and dropsical complaints, which have been such a serious factor in the sickness of this jail, it has been decided that a camp at a distance from the jail and in open ground should be pitched ; it is hoped that this may do much to reduce the existing unhealthiness and afford a better chance of recovery to the sick."

Rangpur.—The convict and under-trial wards were overcrowded for 167 days of the year. All the dormitories except wards Nos. 1 and 2 are on the ground floor and are low, dark, damp, badly ventilated dungeons. Wards Nos. 4, 5, 6, 7 and 8 have good, cross ventilation, but as in these there are several rows of prisoners between the windows, good ventilation is imperative yet possibly not sufficient so long as the massive doors remain and are in spite of orders kept shut. The Public Works Department have, therefore been asked to cut off the upper 20 inches or 2 feet of each shutter on one side, *i. e.,* on the side on which the monsoon rains do not beat, so that free air cannot be excluded. There are *pucca* drains within and around the jail, but in spite of them the ground floor is damp. The surrounding country is water-logged during the rains. The sickness has to some extent been due to bad drainage around the jail while the mortality, apart from four deaths from malaria contracted outside the jail, was due to causes other than local. The Inspector-General remarks :—" An unhealthy jail in an unhealthy district. It is hoped that the general health will be much improved when wards 4 to 8 have been converted into workshops and godowns and, in wards built on the roofs of these, extra sleeping accommodation provided."

BENGAL.

Khulna.—The jail was overcrowded for a few days only, chiefly in August and September. The workshed is not well ventilated, and the same applies to the female ward. The sickness was mostly due to malaria and dysentery, which are prevalent in the town and district. The following sanitary improvements were effected during the year. An incinerator was obtained for burning the stools of dysentery cases, and a boiler and wringer were erected towards the end of the year for washing blankets. The In-pector-General remarks :—" Usually a very healthy jail. Is in reality a small subsidiary jail—accommodation for only 49 prisoners. Usually a smaller number in jail, hence one death which may be purely accidental means a mortality of 20 per mille. The construction of a larger jail has been sanctioned and money allotted."

Burdwan.—The jail was overcrowded from 16th May to 15th November. The ventilation of the sleeping wards is not good. The jail surroundings are very bad, there being several large tanks and a village in a very insanitary state just outside the enclosure wall. The jail site is low and the sleeping wards low and too wide. The year was a very wet one, and early in September the jail drains proved unsatisfactory. The sickness was principally due to the unhealthiness of the district generally. Many were suffering from bowel and lung complaints at the time of admission. Owing to persistent overcrowding from May to November, almost all the physically fit prisoners had to be transferred to other jails, leaving the old and feeble behind. The chief cause of mortality was chest diseases. The Inspector-General remarks :—" The jail is old-fashioned, but the removal of shutters has done much to improve ventilation. Ill. health this year is largely due to the enormous number of prisoners admitted from the local courts, and the consequent overcrowding, and necessary transfer of the more healthy prisoners to other jails."

Midnapore Central.—The accommodation is found to be hardly sufficient in the under-trial ward. The barracks are low and the roof ventilation in all the buildings is defective but is being rectified. The drainage is very defective, there being not a single surface drain in the compound. The water-supply is suspicious, though great care is exercised in its collection and distribution. The site of the jail is good, but the buildings are old and of a very inferior type. The sickness was largely due to malaria and bowel complaints, the cause of the latter is suspected to be the water-supply. The following sanitary improvements were effected during the year. Cage latrines were fixed in all the sleeping barracks ; a new surface drain was made, and ventilators fixed in the cookshed and female workshed. The Inspector-General remarks :—" I have submitted a special report to Government and Government has agreed to my proposals for radical improvement of the ventilation of the wards and workshops. It will be costly, but must be faced." Apart from this defect the general health of new prisoners is admittedly bad. The presence of such a high percentage of oral sepsis cases, at least shows the general bad health of the inhabitants of the district."

Chaibassa.—The under-trial ward was overcrowded for 133 days. It has been noticed that when a large number of flies are about, bowel complaints are specially prevalent. Malarial fever was less prevalent. The prophylactic issue of quinine was carefully carried out from July to the end of October. A large number of prisoners, admitted during the year, were in poor health.

112

TABLE XLIV—*continued.*

ABSTRACT of the SANITARY SHEETS of the most UNHEALTHY JAILS, SANITARY DEFECTS, IMPROVEMENTS, SUGGESTIONS, etc.

Purulia.—Accommodation is insufficient both in the workshed and sleeping wards. The superficial area of the jail enclosure and the hospital accommodation are insufficient and the main building itself is very poor. Fever and respiratory diseases increased in February and March due to unseasonable and excessive rainfall and overcrowding. Diarrhœa, dysentery and fever also occurred in an almost epidemic form in June and July, due to drought followed by excessive rain. There was also from June to December an epidemic of mumps which was imported from the town where the disease was epidemic. Out of seven deaths from dysentery, five were in prisoners who had suffered from the disease at home. The following sanitary improvements were effected during the year. Siphon ventilation was provided in Nos. 2, 3 and 4 *pucca* wards and cross ventilation in the hospital; weighing machines and boilers erected; segregation ward built for newly-admitted prisoners; cage latrines fixed in the hospital; and thorough white-washing carried out. The Inspector-General remarks :—"General health not as good as usual; weather very unseasonable; a separate hospital wanted. Scheme for new barracks is under consideration. Old barrack defective; much overcrowding."

Hazaribagh Central.—In hospital, in segregation and *hafat* wards, the number confined occasionally exceeded the accommodation. This was met by distributing the excess number in any particular ward amongst the other wards so that no actual overcrowding took place. No particular reasons for the sickness and mortality are given, except that this jail contains a large number of very old and decrepit men and convalescents from other jails. The following sanitary improvements were effected during the year. Iron latrines were put up in segregation ward. and wards Nos. 3 and 4 as well as one outside the jail for warders; bathing platforms were erected in almost all the wards; cookshed was provided with fly-proof netting; floor of cowshed made *pucca*; washing platform in under-trial ward completed; drainage in juvenile ward and in the main enclosure improved. and a *pucca* drain constructed outside the jail for removing waste water. The Inspector-General remarks:—" Large number of improvements have been and are being carried out. Jail more unhealthy than usual this year but for no very definite reasons. New water-supply completed; further improvements have been sanctioned."

Champarun.—The sickness and mortality cannot be attributed to any special local conditions except that the jail is situated on the bank of the stagnant Dhanouti river, which is full of weeds, frequently smells of rotting vegetation and acts as a breeding place for mosquitoes. The dysentery cases were probably due to badly cooked rice, as, since double husking and special attention to cooking has been insisted on, the disease has stopped. The following sanitary improvements were effected during the year. Six-seated Alipur pattern latrines were erected for the hospital as well as for warders outside the jail; blanket boiler and wringing machine were provided for cleaning the clothing and bedding of patients and ordinary prisoners; the female hospital, sanctioned in 1904, was completed and occupied; one glass frame was made for preserving the dysentery stools for observation. The Inspector-General remarks —" Large number of improvements carried out. This jail has usually been very healthy. This year was not a healthy one and many prisoners from (Nepal) *terai* came to jail."

Darjeeling.—Two sheds are necessary outside the jail with a view to the removal of patients suffering from cholera and other infectious diseases. The hill-men are dirty. and it is difficult to keep them as clean as plains-men. The Inspector-General remarks :—" A small jail. Usually very healthy. This year there were four deaths, one received in a moribund state. A scheme is in hand for providing a new segregation ward which is much needed."

UNITED PROVINCES.

Kasia.—The jail was overcrowded for 13 days during the year. The sickness and mortality were due to the admission into jail of men in a poor state of health.

Banda.—There was overcrowding for 80 days. Of the total number (1,116) of admissions, 412 were in bad or indifferent health. Many of those admitted especially from Mohoba, were saturated with malaria. The mortality was not due to any special local conditions, but entirely to accidental causes. The Inspector-General remarks :—" This is an excellent jail and its sanitary condition is very good. apart from the fact that the water-supply is very insufficient in quantity. A project for a new water-supply is being prepared, but the question is a difficult one as all the wells in the neighbourhood are gradually drying up. The high death-rate was entirely due to accidental causes and was in no way caused by any defects in the jail."

Hamirpur.—There was no overcrowding during the year. The tank at the back of the jail is a breeding place for *anopheles.* The Inspector-General remarks :—" A small jail and very old fashioned, but in good sanitary condition. Famine prevailed in the district and the prisoners who died were in bad health on admission to jail. "

Etah.—There was overcrowding for 127 days. The extra numbers were transferred to other jails. The dormitories and hospital are not provided with night latrines. The east and south sides of the jail remain water-logged for several months after the rainy season owing to the jail site being lower than its surroundings. Most of the deaths occurred from January to March, when the weather conditions were unseasonable and excessively cold. A large number of prisoners was received in jail suffering from privation and all, excepting two, who died were admitted in bad health. One prisoner was attacked by plague which prevailed all around the jail in a severe form, most of the cases being pneumonic.

Muzaffarnagar.—Wards Nos. 1, 2, 7, 9 and 10 were overcrowded for 71 days. The overcrowding was relieved by the transfer of prisoners and the use of one factory building. A segregation camp was also open for a part of the year. The dormitories and hospital have no night latrines. The ventilation is defective owing to the partition walls being too high. The land outside the jail is badly drained. The district is very malarious, subject to sudden variations of temperature and is extremely cold in winter. The tubercle cases suffered from the disease on admission to jail. The Inspector-General remarks:—" This jail is too small for the district, but there is plenty of room in jails near at hand and no difficulty about arranging transfers to prevent overcrowding. Only prisoners in good health are transferred, so the proportion of men in bad and indifferent health left in the jail is high; hence the high mortality. The sanitary condition of the jail which is old and built on an obsolete plan is good."

BOMBAY.

Bombay Common.—The jail was overcrowded throughout the year. The overcrowding was relieved by using for barrack accommodation two workshops throughout the year and a third for part of the year. Ventilation is defective. The Inspector-General remarks : —" The prison is structurally defective, the hospital accommodation insufficient and the surroundings insanitary. The question of the building of a new Bombay prison is under consideration. The ague cases which occurred mostly in October, November, December, January and February, were nearly all due to relapses. The ventilation of the hospital was greatly improved and more cubic space per patient provided."

TABLE XLV.

INFLUENZA by months, Jails, Groups, and Administrations.

TABLE XLVI.

CHOLERA by months, Jails, Groups, and Administrations.

JAILS* AND GROUPS.	ADMISSIONS FROM INFLUENZA IN EACH MONTH.													ADMISSIONS FROM CHOLERA IN EACH MONTH.												
	January.	February.	March.	April.	May.	June.	July.	August.	September.	October.	November.	December.	TOTAL.	January.	February.	March.	April.	May.	June.	July.	August.	September.	October.	November.	December.	TOTAL.
Thayetmyo, Central	...	...	...	...	...	...	...	...	...	...	...	...	...	...	...	...	...	...	...	...	...	1	...	...	...	1
Myingyan, Central	...	...	...	9	5	...	...	...	...	...	...	...	14	...	...	...	...	...	...	...	...	...	...	...	...	...
Mandalay, Central	...	...	...	...	...	...	...	...	...	...	...	...	...	...	...	...	...	...	...	...	...	...	...	...	1	1
Kindat	...	...	...	...	...	...	...	...	...	...	...	...	...	...	...	...	...	...	...	...	...	...	...	...	1	1
GROUP II.—BURMA INLAND	...	...	...	9	5	...	...	...	...	...	...	...	14	1	...	...	...	...	...	...	...	...	1	...	...	1	3
Sibsagar	...	...	...	...	...	...	...	...	...	...	...	...	...	...	...	...	...	...	...	...	...	...	...	...	2	2
Dibrugarh	...	...	1	1	...	...	...	...	...	...	...	...	2	...	...	...	...	...	...	...	...	...	...	...	...	...
Gauhati	...	3	...	...	...	...	...	...	...	...	...	...	3	...	...	...	...	...	...	...	...	...	...	...	...	...
GROUP III.—ASSAM	...	3	1	1	...	...	...	...	...	...	...	...	5	...	...	...	...	...	...	...	...	...	...	...	2	2
Mymensingh	...	...	...	...	...	...	...	...	...	...	...	...	...	...	...	...	...	...	...	...	...	...	...	1	...	1
Dacca, Central	...	...	...	...	...	...	...	93	8	...	...	...	101	...	...	...	...	...	...	...	...	2	...	...	...	2
Backergunge	...	...	...	...	...	...	...	...	...	...	...	...	...	...	...	...	2	...	...	...	...	...	...	1	...	3
Presidency, Central (Europeans)	...	...	...	...	...	2	...	...	...	2	1	...	5	...	...	...	...	...	...	...	...	...	...	...	...	...
Presidency, Central (Natives)	...	...	...	5	7	6	7	7	1	10	8	8	59	2	...	...	...	...	...	...	...	...	...	...	...	2
Alipore, Central	29	22	21	...	...	...	...	...	...	...	...	...	72	1	...	...	...	...	1	...	...	...	...	...	...	2
Murshidabad	...	...	...	...	...	...	...	...	...	...	...	...	...	...	...	...	1	1	...	...	...	...	...	...	...	2
Rajshahi Central	...	...	...	...	...	...	...	...	...	...	...	...	...	...	...	...	...	1	...	...	...	1	...	...	...	2
Dinajpur	...	...	...	...	...	...	...	...	...	...	...	...	...	...	...	...	...	...	...	...	...	2	...	...	...	2
GROUP IV.—BENGAL AND ORISSA	29	22	21	5	7	6	9	7	94	18	10	9	237	3	...	1	2	2	...	...	...	2	3	2	...	15
A																										
Hazaribagh Central	...	...	...	...	...	...	...	...	...	...	...	...	...	...	...	...	...	...	...	2	...	...	...	...	2
B																										
Chapra	...	...	...	...	...	...	...	...	1	...	...	...	1	...	...	...	...	...	1	...	...	...	...	...	...	1
Azamgarh	...	...	...	...	...	...	...	...	...	...	...	...	...	...	...	...	...	...	...	5	...	...	...	...	5
Gorakhpur	...	...	...	...	...	...	...	...	...	...	...	...	...	...	...	...	...	...	1	...	...	...	...	...	...	1
Fyzabad	...	...	...	...	...	...	...	1	...	...	...	...	1	...	...	...	...	...	...	...	...	...	...	...	...
Rai Bareli	...	...	...	57	1	...	...	...	...	...	...	...	58	...	...	...	...	...	...	...	...	...	...	...	...
Benares, Central	...	...	...	...	...	...	...	...	...	...	...	...	1	...	...	...	...	...	...	...	...	...	...	...	...
Barabanki	...	...	...	...	...	...	6	...	...	...	...	...	6	...	...	...	...	...	...	...	...	...	...	...	...
Gonda	...	...	5	5	5	1	10	47	30	3	2	1	109	...	...	...	...	...	...	...	...	...	...	...	...
Sitapur	...	...	3	...	...	...	...	...	...	...	...	...	3	...	...	...	...	...	...	...	...	...	...	...	...
Hardoi	...	71	24	...	...	...	1	...	...	...	...	...	96	...	...	...	...	...	...	...	...	...	...	...	...
Fatehgarh, Central	...	...	5	...	...	...	...	...	...	...	...	...	5	...	...	...	...	...	...	...	...	...	...	...	...
Fatehgarh District																										
GROUP V.—GANGETIC PLAIN AND CHUTIA NAGPUR	...	71	37	63	6	1	10	49	31	9	2	1	280	...	...	...	...	...	2	7	...	...	...	...	9
A																										
Bareilly District	1	...	16	30	11	...	13	56	33	31	9	11	211	...	...	...	...	...	...	...	...	...	...	...	...
Meerut	1	...	1	3	...	...	...	...	...	1	...	...	5	...	...	...	...	...	...	...	...	...	...	...	...
Hissar	...	...	...	1	...	...	...	...	...	...	...	...	1	...	...	...	...	...	...	...	...	...	...	...	...
B																										
Lahore, Central	...	...	...	...	...	...	...	...	...	...	...	...	...	...	...	...	...	...	...	...	1	2	...	...	3
Gurdaspur	3	...	...	...	...	...	...	...	...	...	...	...	3	...	...	...	...	...	...	...	...	...	...	...	...
Jhelum	...	...	3	...	...	...	...	...	...	...	...	...	3	...	...	...	...	...	...	...	...	...	...	...	...
GROUP VI.—UPPER SUB-HIMALAYA	5	...	20	33	11	...	13	56	33	32	9	11	223	...	...	...	...	...	...	...	1	2	...	...	3
A																										
Peshawar	...	...	...	87	22	23	16	18	20	...	...	...	186	...	...	...	...	...	...	...	...	...	...	...	...
Lyallpur	...	...	...	...	...	...	...	...	...	...	...	...	...	...	...	...	...	...	...	...	3	...	...	...	3
Montgomery, Central	32	14	7	13	1	...	...	...	...	...	...	...	66	...	...	...	...	...	...	...	...	...	...	...	...
Mooltan District	8	...	...	...	...	...	...	...	...	...	...	...	8	...	...	...	...	...	...	...	...	...	...	...	...
GROUP VII.—N. W. FRONTIER, INDUS VALLEY, AND N.-W. RAJPUTANA	40	14	7	99	23	23	16	18	20	...	...	...	260	...	...	...	...	...	...	...	...	3	...	...	3

* Jails where neither Influenza nor Cholera occurred are not shown in these tables.
For the annual ratios, see Table XLII.

JAILS, GROUPS AND ADMINISTRATIONS.	ADMISSIONS FROM INFLUENZA IN EACH MONTH.													ADMISSIONS FROM CHOLERA IN EACH MONTH.												
	January.	February.	March.	April.	May.	June.	July.	August.	September.	October.	November.	December.	TOTAL.	January.	February.	March.	April.	May.	June.	July.	August.	September.	October.	November.	December.	TOTAL.
A																										
Bilaspur	...	...	...	1	...	...	...	...	...	...	...	...	1	...	...	...	...	...	...	...	...	...	...	...	...	...
Raipur, Central	...	...	...	5	...	...	...	...	...	...	...	...	5	...	...	...	...	...	...	...	...	...	...	...	...	...
Nagpur, Central	3	2	2	1	...	...	...	...	14	...	...	...	22	...	...	...	...	...	...	...	...	...	...	...	...	...
B																										
Yeotmahl	...	...	2	...	...	...	...	...	...	...	...	...	2	...	...	...	...	...	...	...	...	...	...	...	...	...
Amraoti, Central	...	2	...	...	...	...	...	...	...	...	...	...	2	...	...	...	...	...	...	...	1	...	...	...	...	1
GROUP IX.—DECCAN	3	4	4	7	...	...	...	...	14	...	...	...	32	...	...	...	...	...	...	...	1	...	...	...	...	1
A																										
Salem, Central	...	...	...	6	...	...	...	...	...	...	...	...	6	...	...	...	...	...	...	...	...	...	...	...	...	...
B																										
Madras Penitentiary, Central	...	...	...	...	...	...	...	...	...	...	...	...	...	...	...	...	...	...	...	...	3	1	...	...	...	4
Nellore	...	...	...	...	...	...	...	...	...	...	...	...	...	...	...	...	...	...	...	...	...	...	2	1	...	3
C																										
Rajamundry, Central	...	...	...	...	...	...	...	...	...	...	...	...	...	...	...	...	...	...	...	...	...	30	...	...	...	30
GROUP XI.—SOUTHERN INDIA	...	...	...	6	...	...	...	...	...	...	...	...	6	...	...	...	...	...	...	...	3	31	2	1	...	37
INDIA*	77	114	90	221	52	30	48	130	192	59	21	21	1,057	4	...	1	2	2	...	2	12	39	5	3	3	73
Burma	...	...	...	9	5	...	...	...	...	...	...	...	14	1	...	...	...	...	...	...	...	1	...	...	1	3
Eastern Bengal and Assam	...	3	1	1	...	...	...	93	8	...	...	...	106	...	...	...	1	2	...	...	...	2	3	2	2	12
Bengal	29	22	21	5	7	6	9	7	1	10	10	9	136	3	...	1	1	...	...	...	2	...	...	...	...	8
United Provinces	2	71	54	93	17	1	23	105	64	41	11	12	496	...	...	...	...	...	...	1	5	...	...	...	...	6
Punjab	43	14	10	13	1	...	...	18	20	...	...	...	81	...	...	...	...	...	...	...	1	5	...	...	...	6
N.-W. Frontier Province	...	...	...	87	22	23	16	18	20	...	...	...	186	...	...	...	...	...	...	...	...	...	...	...	...	...
Bombay	...	...	...	...	...	...	...	...	...	...	...	...	...	...	...	...	...	...	...	...	...	1	...	...	...	1
Central Provinces	3	4	4	7	...	...	...	...	14	...	...	...	32	...	...	...	...	...	...	...	3	31	2	1	...	37
Madras	...	...	...	6	...	...	...	...	...	...	...	...	62	...	...	...	...	...	...	...	3	31	2	1	...	37
Andamans	69	27	14	4	2	...	...	...	43	3	13	11	186	...	...	...	...	...	...	...	...	...	...	...	...	...
India †	146	141	104	227	54	30	48	130	235	62	34	32	1,243	4	...	1	2	2	...	2	12	30	5	3	3	73

* Excluding Andamans.
† Including Andamans.

TABLE XLVII.

ENTERIC FEVER by months, Jails, Groups, and Administrations.

TABLE XLVIII.

SIMPLE CONTINUED FEVER by months, Jails, Groups, and Administrations.

JAILS* AND GROUPS.	ADMISSIONS FROM ENTERIC FEVER IN EACH MONTH.													ADMISSIONS FROM SIMPLE CONTINUED FEVER IN EACH MONTH.												
	January.	February.	March.	April.	May.	June.	July.	August.	September.	October.	November.	December.	TOTAL.	January.	February.	March.	April.	May.	June.	July.	August.	September.	October.	November.	December.	TOTAL.
Rangoon, Central (Europeans)	...	...	...	...	...	...	...	...	...	...	...	...	...	...	...	...	...	1	...	...	...	...	...	...	...	1
,, ,, (Natives)	...	...	...	...	...	...	...	...2	...	...	...	...	...	3	12	12	7	15	14	10	10	9	11	9	11	123
Myaungmyo	...	...	...	...	1	...	...	...	...	...	...	...	...	...	...	...	...	...	...	...	...	...	...	...	...	...
Bassein, Central	...	...	2	...	...	1	1	...	...	...	...	...	4	1	6	1	6	5	11	11	5	2	2	1	...	54
Insein, ,,	...	...	...	...	...	...	...	...	...	...	...	...	...	2	3	8	4	2	1	14	2	1	...	...	...	37
Kyaukpyu	...	...	...	...	...	...	...	...	...	...	...	...	...	...	...	2	...	...	...	...	...	...	...	...	...	2
GROUP I.—BURMA COAST AND BAY ISLANDS	...	...	2	...	1	1	1	...	...	...	...	...	5	6	21	31	17	25	26	38	17	12	13	10	11	217
Prome	...	...	...	...	...	...	...	...	...	...	...	...	...	...	...	...	...	...	...	...	2	...	...	...	...	2
Thayetmyo, Central	...	...	...	...	...	...	...	...	...	...	...	...	...	...	...	...	...	...	...	1	...	...	...	...	...	1
Myingyan, ,,	...	...	...	...	...	...	...	...	...	...	...	...	...	...	...	5	...	...	...	...	...	...	...	...	...	5
Mandalay, Central	...	...	...	1	1	...	...	1	...	...	...	...	3	...	...	...	...	...	...	1	3	1	5	3	1	14
Bhamo	...	...	...	...	...	...	...	...	...	...	...	...	...	...	...	...	...	...	...	...	...	...	...	...	1	1
GROUP II.—BURMA INLAND	...	...	...	1	1	...	...	1	...	...	...	...	3	...	...	5	...	...	...	2	5	1	5	3	2	23
Gauhati	...	...	...	...	...	...	...	...	...	...	...	...	...	2	5	...	1	...	...	...	...	...	...	...	...	8
GROUP III.—ASSAM	...	...	...	...	...	...	...	...	...	...	...	...	...	2	5	...	1	...	...	...	...	...	...	...	...	8
Mymensingh	...	...	...	...	...	2	3	...	...	...	1	4	10	...	...	...	...	...	...	...	...	...	...	...	...	...
Rangpur	...	...	...	...	2	...	...	...	...	...	...	...	2	...	...	...	...	...	...	...	...	...	...	...	...	...
Naya Dumka	...	...	...	...	...	...	...	...	...	...	...	...	...	...	...	...	...	...	...	...	...	...	...	2	4	6
Bankura	...	...	...	1	...	...	...	1	...	...	...	...	2	...	...	...	...	...	...	...	...	...	...	...	...	...
Balasore	...	...	...	...	...	...	...	...	...	...	...	...	...	...	...	...	...	...	15	3	2	3	3	...	26
GROUP IV.—BENGAL AND ORISSA	...	...	...	1	2	3	...	1	...	...	1	4	14	...	...	...	...	...	...	15	3	2	3	5	4	32
A																										
Palamau	...	...	...	...	...	...	...	...	...	...	...	...	...	1	...	...	...	...	...	...	...	...	...	...	...	1
B																										
Gaya	...	...	...	...	...	...	...	...	...	...	...	...	...	1	...	...	...	...	...	...	...	...	...	7	8
Champaran	...	...	...	...	...	...	...	...	...	...	...	...	...	...	...	...	...	...	...	...	2	2	...	...	4
Arrah	...	...	...	1	...	...	...	...	...	...	1	...	...	...	...	...	...	...	...	...	...	...	...	...	...
Rai Bareli	...	...	...	...	...	...	...	...	...	...	...	...	1	4	1	1	...	...	...	...	...	...	...	7
Allahabad, Central	...	...	...	...	...	...	...	...	...	...	1	...	3	3	9	16	14	16	20	42	21	3	1	148
Unao	...	...	...	...	...	...	...	...	...	...	...	...	...	...	...	...	...	...	...	...	1	1
Lucknow, Central	...	...	...	...	...	...	...	...	...	...	1	...	...	...	...	...	...	1	...	...	...	...	...
Barabanki	...	...	...	...	...	...	...	...	...	...	...	...	...	...	...	...	...	...	1	...	...	4
Gonda	...	...	...	...	...	...	...	...	...	...	...	...	...	...	...	2	1	...	1	...	...	...	4
Sitapur	...	...	...	...	...	...	...	...	...	...	...	...	...	...	...	1	...	...	...	...	...	...
Mainpuri	...	...	...	1	...	...	...	...	...	...	...	...	...	...	...	...	...	...	...	...	1	...
Etah	...	...	...	...	...	...	...	...	1	2	...	4	...	...	...	...	...	...	...	...	...	...
Fatehgarh, Central	...	...	...	1	...	...	1	...	...	...	...	...	...	...	...	...	...	1	...	...	...
GROUP V.—GANGETIC PLAIN AND CHUTIA NAGPUR	...	...	1	1	1	...	1	...	1	2	...	7	3	7	4	11	18	15	16	24	44	22	4	8	176
A																										
Bareilly, Central	...	...	...	...	...	...	1	...	...	...	...	...	...	...	...	...	...	...	...	...	...	...
Aligarh	...	...	...	1	1	...	...	...	...	...	...	2	...	...	...	...	...	...	...	...	...	...
Umballa	...	...	...	...	...	...	...	...	...	...	...	...	...	4	...	...	7	8	3	4	26
Ferozepore	...	...	...	...	...	1	...	...	...	...	1	3	2	1	2	3	7	3	1	2	2	1	2	4	31
Lahore District	...	...	1	...	...	...	...	...	...	...	2	...	...	...	...	...	...	...	...	...	...	...
,, Female	...	...	...	...	...	1	...	...	...	...	...	...	...	...	...	...	...	...	...	...	...
GROUP VI.—UPPER SUB-HIMALAYA	...	...	...	1	2	1	1	...	1	...	7	3	3	1	3	11	3	1	2	9	9	5	8	57

* Jails where neither Enteric Fever nor Simple Continued Fever occurred are not shown in these tables. For the annual ratios, see Table XLII.

116

Jails, Groups, and Administrations	Admissions from Enteric Fever in each month.													Admissions from Simple Continued Fever in each month.												
	Jan.	Feb.	Mar.	Apr.	May	Jun.	Jul.	Aug.	Sep.	Oct.	Nov.	Dec.	Total	Jan.	Feb.	Mar.	Apr.	May	Jun.	Jul.	Aug.	Sep.	Oct.	Nov.	Dec.	Total
A																										
Nimar	...	...	...	...	...	...	...	...	...	...	...	...	...	1	...	...	...	...	...	...	...	...	...	...	...	1
Nagpur, Central	...	...	...	...	...	...	...	...	...	...	...	...	...	5	1	3	...	1	1	...	...	2	...	...	...	13
B																										
Amroati, Central	...	...	...	...	...	...	...	...	...	...	...	...	...	3	...	1	...	2	1	...	...	...	...	...	...	1
Deccan Gang	...	...	...	...	...	...	...	...	...	...	...	...	...	3	...	3	...	2	1	...	...	...	...	...	...	9
Group IX. —Deccan	...	...	...	...	...	...	...	...	...	...	...	...	...	9	1	7	...	3	2	...	...	2	...	...	...	24
Mangalore	...	...	...	...	...	...	...	1	...	...	...	...	1	4	2	2	1	1	...	2	...	2	3	1	...	6
Cannanore, Central	...	...	...	...	...	...	4	2	...	...	...	...	6	...	...	...	...	...	...	...	...	1	3	4	1	21
Group X.—Western Coast	...	...	...	...	...	...	4	2	1	...	...	...	7	4	2	2	1	1	...	2	...	3	6	5	1	27
A																										
Bellary	...	...	...	...	...	...	...	...	...	...	...	...	...	...	...	7	...	...	...	1	...	...	...	3	2	7
Selem, Central	...	...	...	...	...	...	...	...	...	...	...	...	...	2	2	...	...	...	...	2	1	1	...	3	2	9
Coimbatore, Central	...	...	...	...	1	...	4	7	...	4	1	1	18	2	2	...	3	6	7	6	11	7	11	6	4	65
B																										
Madura	...	...	...	...	...	...	1	...	...	...	...	...	1	...	...	...	1	1	1	...	2	...	...	...	...	5
Tanjore	...	...	...	...	...	...	...	...	...	...	...	...	...	14	7	6	3	8	2	6	3	3	5	9	...	71
Vellore, Central	...	...	...	...	...	...	...	...	...	...	...	...	...	4	3	3	...	...	...	...	...	...	...	...	...	11
Madras Penitentiary, Central	...	...	...	...	...	...	...	...	...	...	...	2	2	...	...	...	...	...	1	...	...	...	...	...	...	...
Nellore	...	...	...	...	...	...	...	...	...	...	...	...	...	...	...	...	...	...	...	...	...	...	...	2	...	3
C																										
Rajamundry Central	...	...	...	...	...	...	...	...	...	...	...	...	...	...	...	...	38	23	8	7	11	10	29	14	6	146
Vizagapatam	...	...	...	...	...	...	...	...	...	...	...	...	...	...	...	...	...	...	...	...	...	1	...	...	1
Group XI.—Southern India	...	...	...	...	...	1	4	8	...	4	1	3	21	20	12	16	45	39	19	20	29	21	46	28	23	318
Almora	...	...	...	...	...	...	...	...	...	...	...	...	...	...	...	...	3	4	...	1	...	...	2	...	10
Group XII.—Hills	...	...	...	...	...	...	...	...	...	...	...	...	...	...	...	...	3	4	...	1	...	...	2	...	10
INDIA*	...	...	4	4	6	6	14	11	3	6	3	7	64	47	50	56	78	100	69	94	81	94	104	62	57	892
Burma	...	...	2	1	2	1	1	1	...	...	...	...	8	6	21	26	17	25	26	40	22	13	18	13	13	240
Eastern Bengal and Assam	...	...	...	...	2	2	3	...	...	...	1	4	12	2	5	...	1	...	...	...	...	...	...	...	...	8
Bengal	...	...	1	1	...	...	...	1	2	...	...	...	3	2	...	...	...	15	3	4	3	5	11	...	...	45
United Provinces	...	...	...	2	2	...	2	...	1	2	...	...	9	1	7	4	11	21	19	16	23	42	22	6	1	173
Punjab	...	...	1	...	...	2	...	...	...	...	1	...	4	3	2	1	3	11	3	1	2	9	9	5	8	57
North-West Frontier Province	...	...	...	...	...	...	...	...	...	...	...	...	...	...	...	...	...	...	...	...	...	...	...	...	...	...
Bombay	...	...	...	...	...	...	1	...	...	...	...	...	1	3	...	3	...	2	1	...	...	...	2	...	...	9
Central Provinces	...	...	...	...	...	...	...	...	...	...	...	...	...	3	1	4	...	2	1	...	...	2	...	...	...	15
Madras	...	...	...	...	...	1	8	10	1	4	1	3	28	24	14	18	46	40	19	22	29	24	32	33	24	345

* Excluding Andamans.

TABLE XLIX.

INTERMITTENT FEVER by months, Jails, Groups, and Administrations.

TABLE L.

REMITTENT FEVER by months, Jail Groups, and Administrations.

TABLE XLIX. — ADMISSIONS FROM INTERMITTENT FEVER IN EACH MONTH.

Jails and Groups.	January.	February.	March.	April.	May.	June.	July.	August.	September.	October.	November.	December.	TOTAL.
Tavoy	2	...	1	1	...	...	...	...	8	1	...	...	13
Moulmein	2	3	...	1	...	5	13	5	15	14	3	...	61
Shwegyin	...	1	1	1	...	3	2	2	...	...	...	...	10
Toungoo	1	3	1	2	...	3	1	4	1	1	4	6	27
Rangoon, Central, Natives	4	2	1	1	...	2	2	...	1	4	...	1	18
Maubin	...	...	...	...	1	1	...	...	1	...	...	...	5
Myaungmyo	1	...	2	...	1	1	...	2	1	...	...	...	12
Bassein, Central	1	3	3	1	1	2	4	3	2	6	6	...	33
Inscin	5	2	5	3	...	...	1	7	6	9	4	11	54
Henzada	1	1	1	2	2	...	3	4	4	1	...	5	25
Myanaung	...	...	...	1	...	...	...	...	...	1	...	...	2
Sandoway	...	...	2	...	1	1	...	1	2	4	1	...	9
Kyaukpyu	...	...	...	3	...	1	...	...	...	...	...	...	7
Akyab	1	...	...	1	...	...	...	...	2	...	...	3	7
GROUP I.—BURMA COAST AND BAY ISLANDS	18	16	17	17	7	19	28	27	44	43	19	28	283
Paungdi	...	1	...	...	1	...	3	2	...	1	1	3	12
Prome	...	2	4	9	1	2	4	4	1	3	1	8	39
Thayetmyo, Central	3	5	...	2	6	4	3	3	5	2	1	3	37
Hiagwe	...	...	...	...	2	1	...	...	...	1	...	...	4
Meiktila	...	...	...	1	2	1	2	1	...	1	...	...	8
Paiokku	...	...	...	...	...	...	1	...	...	...	...	...	1
Myingyan, Central	3	...	7	12	2	4	3	6	4	7	3	5	58
Mandalay	5	6	3	5	5	13	12	8	27	41	33	17	175
Monywa	...	...	...	1	...	...	...	...	...	...	...	...	1
Shwebo	...	1	...	...	...	...	4	...	...	...	...	...	6
Bhamo	2	...	...	2	1	3	1	5	1	2	1	...	18
Katha	...	...	...	...	...	...	1	...	...	...	...	...	1
Kindat	...	...	...	...	...	...	2	...	...	...	...	...	2
GROUP II.—BURMA INLAND	13	17	14	31	19	27	37	29	59	57	42	37	362
Cachar	...	...	...	...	...	...	...	...	...	1	...	4	5
Sibsagar	2	3	1	...	...	...	...	...	1	...	...	...	7
Dibrugarh	4	5	5	2	10	6	...	11	3	4	5	1	56
Tezpur	...	...	...	4	2	2	5	10	4	1	3	3	33
Gauhati	1	...	...	2	1	3	24	20	16	7	8	3	105
Sylhet	4	3	8	7	6	8	14	19	11	20	4	6	112
GROUP III.—ASSAM	11	12	15	11	24	40	39	63	35	32	19	17	318
Mymensingh	11	12	16	12	19	17	24	20	15	15	6	18	185
Dacca, Central	5	7	5	9	2	9	13	9	7	13	18	14	111
Tippera	1	1	2	2	1	3	1	3	...	3	2	1	20
Chittagong	3	1	5	8	3	6	10	14	8	4	11	5	78
Noakhali	...	...	1	3	3	5	5	1	...	...	3	1	22
Backergunge	6	4	8	11	11	7	2	5	6	6	4	1	77
Khulna	...	1	...	...	...	2	...	...	...	1	...	...	10
Jessore	9	8	12	7	5	8	8	9	23	12	15	11	127
Baraset	10	4	4	2	3	2	5	11	12	4	3	3	63
Presidency Central, Natives	6	7	3	7	4	2	3	1	6	6	...	...	45
Alipore, Central	16	26	34	30	27	37	35	27	36	79	83	38	468
Hooghly	8	1	4	9	11	6	8	5	10	9	8	8	84
Burdwan	6	3	16	8	4	8	16	12	30	30	26	27	186
Krishnagar	6	2	5	5	3	3	...	...	5	3	4	...	36
Faridpur	2	3	3	2	11	7	19	19	23	18	23	24	154
Pubna	1	11	5	12	9	7	3	17	16	7	11	10	126
Murshidabad	1	3	1	8	3	...	4	4	4	7	3	...	39
Rajshahi, Central	8	2	2	1	1	9	6	9	2	15	8	9	71
Bogra	5	3	3	9	3	3	2	13	2	1	4	4	52
Malda	3	2	5	4	1	3	3	16	17	23	37	19	133
Dinajpur	6	1	12	10	10	8	8	10	13	13	5	7	109
Rangpur	19	13	5	7	13	15	6	9	21	33	20	12	173
Jalpaiguri	2	1	...	2	4	4	5	5	...	...	3	3	79
Purneah	3	3	7	5	12	5	5	1	8	2	1	1	51
Naya Dumka	...	...	1	1	...	2	3	3	2	3	...	...	18
Suri	...	...	...	1	...	8	...	7	...	4	4	2	23
Bankura	3	6	2	1	...	...	...	1	6	2	2	2	34
Midnapore, Central	10	11	3	14	34	31	16	29	37	39	20	9	259
Balasore	12	1	4	2	8	...	1	...	...	...	4	5	37
Cuttack	5	...	3	2	8	10	13	18	23	5	1	2	110
Puri	2	...	...	4	2	3	1	...	...	3	...	4	19
Angul	1	...	...	...	...	...	4	3	1	...	...	2	12
GROUP IV.—BENGAL AND ORISSA	180	132	177	223	214	231	231	279	328	368	335	247	2,951

TABLE L. — ADMISSIONS FROM REMITTENT FEVER IN EACH MONTH.

Group.	TOTAL.
GROUP I.—BURMA COAST AND BAY ISLANDS	7
GROUP II.—BURMA INLAND	3
GROUP III.—ASSAM	6
GROUP IV.—BENGAL AND ORISSA	14

* Jails where neither Intermittent Fever nor Remittent Fever occurred are not shown in these tables. For the annual ratios see Table XLII.

Jails and Groups.	Admissions from Intermittent Fever in each month.													Admissions from Remittent Fever in each month.												
	January.	February.	March.	April.	May.	June.	July.	August.	September.	October.	November.	December.	Total.	January.	February.	March.	April.	May.	June.	July.	August.	September.	October.	November.	December.	Total.
A																										
Chaibassa	2	...	...	1	1	3	...	5	3	9	7	10	41	...	...	...	...	...	...	...	...	...	...	...	1	1
Purulia	4	2	8	4	5	7	23	8	12	4	9	9	95	...	...	...	...	...	...	...	...	...	...	...	...	...
Ranchi	1	...	3	2	1	2	2	1	1	3	1	1	18	...	...	...	...	...	...	...	...	...	...	...	...	...
Palamau	2	4	1	...	...	2	5	7	10	5	3	4	43	...	...	...	...	...	...	...	...	...	...	...	...	...
Hazaribagh, Central	6	10	8	19	20	10	15	28	66	24	14	13	233	...	...	...	...	...	...	...	...	...	...	...	...	...
B																										
Gaya	...	...	...	1	2	3	4	17	17	4	9	6	63	...	...	...	...	...	...	...	...	...	...	...	...	...
Bhagalpur, Central	11	16	31	29	17	24	49	23	27	41	48	38	354	...	...	...	...	...	...	...	...	...	...	...	...	...
Monghyr	1	5	1	13	3	4	1	5	2	4	4	4	47	...	...	...	...	...	...	...	...	...	...	...	...	...
Darbhanga	1	1	...	1	2	2	6	2	4	9	3	...	31	...	...	...	...	...	...	...	...	...	...	...	...	...
Champarun	10	6	8	12	3	13	5	2	...	1	5	3	68	...	...	...	6	7	...	...	...	...	...	...	...	13
Muzaffarpur	8	3	7	9	7	6	6	8	5	7	8	5	79	...	...	...	...	...	...	...	...	...	...	...	...	...
Patna	1	1	4	8	5	1	3	7	2	5	12	4	53	...	...	...	...	...	...	...	...	...	...	...	...	...
Arrah	1	2	4	7	10	6	1	6	3	5	3	6	54	...	...	1	...	...	...	...	...	...	...	...	...	1
Chapra	4	3	5	3	5	6	8	6	6	9	9	2	66	...	...	...	...	...	...	...	...	...	...	...	...	...
Buxar, Central	66	46	64	235	200	154	79	104	62	86	60	51	1,207	...	...	...	...	...	...	...	...	...	...	...	...	...
Korantadih	...	3	1	4	...	1	3	3	5	1	...	...	21	...	...	...	...	...	...	...	...	...	...	...	...	...
Ghazipur	3	8	3	13	16	4	6	8	7	20	11	6	105	...	...	...	...	...	...	...	...	...	...	...	...	...
Azamgarh	4	...	...	5	10	3	2	12	7	5	6	4	58	...	...	...	...	...	...	...	...	...	...	...	...	...
Kasia	1	...	...	1	...	...	...	...	...	2	...	...	4	...	...	...	...	...	...	...	...	...	...	...	...	...
Gorakhpur	3	3	14	8	16	9	30	35	54	24	18	11	225	...	...	...	...	...	...	...	...	...	...	...	...	...
Basti	1	1	2	3	6	...	1	5	3	3	3	...	28	...	...	...	...	...	...	...	...	...	...	...	...	...
Fyzabad	3	1	8	9	...	1	8	8	4	6	3	4	55	...	...	...	...	...	...	...	...	...	...	...	...	...
Sultanpur	2	3	4	9	...	2	...	6	27	8	3	3	67	...	...	...	...	...	...	...	...	...	...	...	...	...
Rai Bareli	2	...	1	2	3	4	1	4	9	7	...	1	31	...	...	...	...	...	...	...	...	...	...	...	...	...
Partabgarh	...	3	2	...	...	1	1	2	2	7	1	...	19	...	...	...	...	...	...	...	...	...	...	...	...	...
Jaunpur	7	4	7	10	1	1	1	3	3	3	1	1	42	...	...	...	...	...	...	...	...	...	...	...	...	...
Benares, Central	23	10	8	10	14	12	6	12	59	69	56	21	300	...	...	...	...	...	...	...	...	...	...	...	...	...
" District	5	2	1	2	1	1	2	4	6	3	4	...	31	...	...	...	...	...	...	...	...	...	...	...	...	...
Mirzapur	11	1	2	2	7	44	...	2	1	1	7	1	79	...	1	...	...	...	...	...	...	1	...	...	...	2
Allahabad Central	4	5	...	9	11	9	1	6	26	35	15	8	129	...	...	...	...	...	...	...	...	...	...	...	...	...
" District	4	1	1	...	6	4	2	4	5	9	3	...	39	...	...	...	...	...	...	...	...	...	...	...	...	...
Karwi	...	...	...	1	2	...	...	1	...	...	...	...	4	...	...	...	...	...	...	...	...	...	...	...	...	...
Banda	5	7	9	16	15	16	12	12	15	15	22	10	154	...	...	...	...	...	...	...	...	...	...	...	...	...
Fatehpur	10	12	4	15	9	3	3	21	24	16	8	7	132	...	...	...	...	...	...	...	...	...	...	...	...	...
Hamirpur	7	5	4	4	4	1	2	13	4	2	2	6	54	...	...	...	...	...	1	...	...	...	...	...	...	1
Orai	2	3	4	1	3	2	8	10	14	5	13	...	65	...	...	...	...	...	...	...	...	...	...	...	...	...
Cawnpore	1	4	6	1	1	...	3	6	14	14	12	9	71	...	...	...	...	...	...	...	...	...	...	...	...	...
Unao	...	...	...	1	3	5	3	11	17	7	2	1	50	...	...	...	...	...	...	...	...	...	...	...	...	...
Lucknow, Central	2	8	1	...	4	11	5	11	12	19	1	2	76	...	...	...	...	1	...	...	...	...	...	...	...	1
" District	3	2	...	3	...	2	1	4	9	1	2	...	27	...	...	...	...	...	...	...	...	...	...	...	...	...
Barabanki	2	1	2	4	4	3	7	12	11	1	5	4	56	...	...	...	...	...	...	...	...	...	...	...	...	...
Gonda	3	...	3	...	3	1	2	3	1	4	1	1	22	...	...	...	...	...	...	...	...	...	...	...	...	...
Bahraich	4	...	1	2	1	1	3	12	22	13	9	5	73	...	...	...	...	...	2	...	...	...	1	...	...	3
Kheri	6	5	1	...	2	3	...	13	2	3	12	1	48	...	...	...	...	...	...	...	...	...	...	...	...	...
Sitapur	1	...	...	2	1	1	1	2	25	3	3	3	42	...	...	...	...	...	...	...	...	...	...	...	...	...
Hardoi	6	6	...	5	7	7	2	3	14	6	6	10	72	...	...	...	...	...	...	...	...	...	...	...	...	...
Etawah	...	1	2	5	9	1	9	6	3	1	1	1	42	...	...	...	...	...	...	...	...	...	...	...	...	...
Mainpuri	1	13	4	6	11	13	4	3	3	3	1	2	64	...	...	...	...	...	...	...	...	...	...	...	...	...
Etah	3	22	9	16	13	5	5	5	7	6	3	1	95	...	...	...	...	...	...	...	...	...	...	...	...	...
Fatehgarh, Central	20	37	3	18	15	19	21	36	10	11	10	7	207	...	...	...	...	...	...	...	...	1	...	...	2	3
" District	7	...	4	3	2	6	3	5	1	2	4	1	38	...	...	...	...	...	...	...	...	...	...	...	...	...
Group V.—Gangetic Plain and Chutia Nagpur	274	270	255	534	481	439	365	532	646	551	443	290	5,080	...	1	1	6	8	3	...	...	2	1	...	3	25
A																										
Shahjahanpur	4	2	1	14	15	12	4	3	13	8	4	1	81	...	...	...	...	...	...	...	...	...	...	...	...	...
Pilibhit	...	...	...	...	1	...	2	3	1	...	...	1	8	...	...	...	...	...	...	...	...	...	...	...	...	...
Bareilly, Central	20	11	52	43	32	26	34	75	85	73	35	46	532	...	...	...	...	...	...	...	...	...	...	...	...	...
" District	31	11	5	...	8	25	25	...	...	...	8	10	123	...	...	...	...	...	...	...	...	...	...	...	...	...
Budaon	2	...	...	1	3	...	...	4	11	1	...	2	24	...	...	...	...	...	...	...	...	...	...	...	...	...
Aligarh	1	...	2	3	10	6	3	2	1	6	4	5	43	...	...	1	...	...	2	1	...	...	...	1	...	5
Bulandshahr	2	2	...	6	2	...	2	4	2	1	...	...	21	...	...	...	...	...	...	...	...	...	...	...	...	...
Moradabad	12	4	6	6	4	4	4	8	31	18	21	2	120	...	...	...	...	...	...	...	...	...	...	...	...	...
Bijnor	...	...	2	2	3	2	2	4	3	2	1	2	23	...	...	...	...	1	...	...	1	...	...	...	...	2
Dehra Dun	1	...	...	2	...	2	...	1	2	3	2	...	13	...	1	...	1	...	...	1	...	1	...	...	...	4
Saharanpur	4	2	5	6	2	4	2	5	13	4	7	2	56	...	...	...	...	...	...	...	...	...	...	...	...	...
Muzaffarnagar	...	1	...	1	1	1	2	7	11	5	3	1	33	...	...	...	...	...	...	...	...	...	...	...	...	...
Meerut	6	6	7	11	5	7	7	11	9	5	2	3	79	...	...	...	...	...	...	...	...	...	...	...	...	...
Delhi	6	3	3	6	3	3	7	5	1	14	4	...	55	...	...	...	...	...	...	...	...	...	...	...	...	...
Rohtak	6	4	3	6	13	8	4	4	5	5	5	7	70	...	...	...	...	...	...	...	...	...	...	...	...	...
Hissar	4	2	...	4	4	...	6	8	5	24	11	8	76	1	...	...	...	...	...	...	...	...	...	...	...	1
Karnal	...	...	1	1	2	3	8	3	1	1	4	...	24	...	...	...	...	...	...	...	...	...	...	...	...	...
Umballa	10	3	5	2	8	7	7	16	...	1	1	1	61	...	...	...	...	...	...	...	...	...	...	...	...	...
B																										
Ludhiana	...	...	1	5	...	2	3	...	...	1	3	5	20	...	...	...	...	...	...	...	...	1	...	...	...	1
Hoshiarpur	2	...	1	...	1	...	...	...	...	...	1	...	5	...	...	...	...	...	...	...	...	...	...	...	...	...
Jullundur	2	1	3	4	3	1	5	4	...	1	7	9	40	...	...	...	...	...	...	...	...	...	...	...	...	...
Ferozepore	...	...	...	...	1	...	...	...	...	1	...	...	2	1	...	...	...	...	...	...	...	...	...	...	...	1
Amritsar	3	4	7	4	2	4	5	2	3	14	13	14	75	...	...	...	...	...	...	...	...	...	...	...	...	...
Lahore, Central	34	42	87	57	70	61	64	76	35	30	107	114	777	...	...	...	...	...	...	...	...	...	...	...	...	...
" District	10	7	1	3	12	24	6	8	7	14	14	11	117	...	...	...	...	...	...	...	...	...	...	...	...	...
" Female	4	...	2	2	4	6	1	...	4	9	4	5	41	...	...	...	...	...	...	...	...	...	...	...	...	...
Gurdaspur	2	...	...	1	3	4	2	2	1	1	...	...	16	...	...	...	...	1	2	...	...	...	...	...	...	3
Gujranwala	6	4	3	8	9	7	13	1	3	...	5	1	60	...	...	...	...	...	...	...	...	...	...	...	...	...
Sialkot	14	...	2	3	3	6	1	5	2	7	1	4	48	...	...	...	...	...	...	...	...	...	...	...	...	...
Gujrat	2	4	4	1	2	4	2	1	2	4	3	2	31	...	...	...	...	...	...	...	...	...	...	...	...	...
Jhelum	2	1	2	...	2	1	3	3	2	2	9	3	30	...	...	...	...	...	...	...	...	...	...	...	...	...
Rawalpindi	14	13	3	3	11	22	13	2	5	5	22	6	119	...	...	...	...	...	...	...	...	...	...	...	...	...
Group VI.—Upper Sub-Himalaya	204	127	208	205	239	252	237	267	258	260	301	265	2,823	2	1	1	1	2	4	2	1	2	...	1	...	17

TABLE XLIX—*concluded.*

*INTERMITTENT FEVER by months, Jails, Groups, and
Administrations.*

TABLE L—*concluded.*

*REMITTENT FEVER by months, Jails,
Groups, and Administrations.*

JAILS AND GROUPS.	January.	February.	March.	April.	May.	June.	July.	August.	September.	October.	November.	December.	TOTAL.	January.	February.	March.	April.	May.	June.	July.	August.	September.	October.	November.	December.	TOTAL.
A																										
Peshawar	19	13	61	...	21	...	...	...	...	21	15	10	160													
Kohat	...	3	2	1	3	7	4	8	7	8	3	2	50													
Bannu	1	1	4	1	...	3	6	3	2	3	4	4	31									1				1
Shahpur	4	...	...	1	1	3	7	6	1	3	3	8	39													
Mianwali	3	9	9	4	2	3	10	5	4	9	5	6	73													
Lyallpur	20	10	7	4	3	1	2	2	2	3	11	6	80													
Jhang	6	3	8	2	1	3	2	...	...	1	4	1	31													
Montgomery, Central	16	6	7	13	70	27	6	4	9	21	49	49	198													
Mooltan, Central	24	23	27	32	102	98	47	41	33	21	49	13	440													
„ District	15	13	14	5	6	5	4	...	4	3	7	9	85													
Dera Ismail Khan	9	8	5	6	12	14	26	7	8	27	32	35	189													1
Dera Ghazi Khan	5	9	4	12	32	13	6	19	16	31	43	31	230													
C																										
Shikarpur	7	15	12	7	1	...	...	1	1	3	3	1	51								1		1		1	3
Sind Gang	1	2	4	2	1	1	1	...	2	...	2	...	17													
Hyderabad, Central	1	3	3	12	...	1	12	5	6	5	9	7	64													
Kurrachee	4	4	8	4	...	2	2	5	6	5	6	8	54								1		3	...	1	4
GROUP VII.—N.-W. FRONTIER, INDUS VALLEY, AND N.-W. RAJPUTANA	139	130	175	105	205	105	156	109	96	152	219	190	1,782	...	...	...	...	1	...	...	5	2	...	1	9
Rajkot	3	1	3	2	2	...	...	4	3	4	2	2	26													1
Ahmedabad, Central	3	3	2	6	3	1	4	7	21	46	35	17	148													
B																										
Ajmer	3	2	4	7	3	...	...	1	1	...	3	...	25													
Muttra	2	1	2	2	...	...	...	...	...	...	...	...	8													
Agra, Central	25	70	12	30	47	32	69	41	30	31	17	14	377													
„ District	7	4	3	5	3	4	5	4	...	5	2	1	47													
Jhansi	2	3	1	...	2	4	3	...	...	7	2	5	35													
Lalitpur	1	...	...	...	...	2	...	...	2	16	5	...	31													
GROUP VIII.—S.-E. RAJPUTANA, CENTRAL INDIA, AND GUJARAT	45	33	27	63	62	42	83	61	83	91	62	45	697								1					1
A																										
Damoh	...	...	...	...	2	...	...	2	5	...	...	2	14													2
Saugor	1	...	...	1	...	...	...	1	3	2	...	2	10													
Jubbulpore, Central	...	...	...	...	...	...	...	2	4	...	...	...	8													1
Narsinghpur	...	2	1	2	...	...	...	...	...	1	2	1	9													
Mandla	1	...	2	4	1	...	3	3	2	...	1	2	18													
Bilaspur	...	...	...	1	4	...	...	1	...	3	...	1	14								7		1			3
Sambalpur	...	...	...	...	...	...	...	...	...	...	2	5	12													
Raipur, Central	5	3	5	4	3	6	9	15	16	18	15	6	108													
Balaghat	...	...	...	...	2	...	2	2	3	...	1	...	13													
Seoni	1	...	...	2	...	2	...	...	1	...	1	...	6													
Chhindwara	...	...	...	...	...	...	...	1	...	...	...	...	6													
Hoshangabad	2	3	2	1	3	5	4	5	4	2	3	5	36													
Nimar	1	2	...	1	2	1	2	14	12	5	8	5	51													
Betul	3	2	...	1	2	1	...	1	1	3	2	1	16													
Nagpur, Central	...	2	...	...	1	4	8	16	61	73	46	17	228													
Bhandara	...	...	2	...	2	1	...	8	2	...	...	...	16													
Wardha	1	...	...	...	...	...	...	1	...	3	...	1	6													
Chanda	1	...	1	...	...	...	1	...	1	...	1	...	7													
B																										
Secunderabad	...	1	2	3	...	...	...	1	1	...	1	...	11													1
Yeotmahl	1	...	...	...	...	...	...	...	1	...	...	...	8													
Amraoti, Central	3	2	1	...	3	2	6	3	3	1	...	...	31													
Akola, Central	5	14	3	3	2	5	8	5	9	...	4	11	69													
Buldana	1	...	...	...	...	...	...	1	...	...	...	...	2													
Dhulia	25	7	2	1	4	2	...	25	36	24	18	9	154													
Yerrowda, Central	7	10	21	3	7	...	8	15	18	10	18	11	136													
Bijapur	1	1	...	3	...	...	2	2	2	...	5	3	20													
Deccan Gang	10	4	11	6	1	3	7	...	4	6	3	8	80													
Dharwar	1	2	1	3	1	1	...	1	2	3	2	2	20													
GROUP IX.—DECCAN	74	76	60	44	41	41	61	110	146	166	143	97	1,109	...	...	1	...	...	...	...	4	1	1	...		7
Thana	3	4	6	...	1	...	7	5	8	6	6	8	54											1		...
Bombay, Common	14	16	13	8	10	9	11	13	9	20	23	12	157								1				1	1
„ House of Correction	1	3	...	2	...	...	2	1	4	6	5	9	36													
Ratnagiri	...	1	1	...	...	...	1	...	3	3	3	3	15													
Karwar	4	...	...	...	...	...	...	1	1	...	...	...	6													
Mangalore	...	...	2	1	...	...	...	5	2	1	2	...	8													
Cannanore, Central	10	5	4	2	2	4	6	5	5	2	2	4	51								1				1	2
GROUP X.—WESTERN COAST	32	29	25	13	14	13	27	31	31	37	45	28	327								1				2	3

JAILS, GROUPS AND ADMINIS- TRATIONS.	ADMISSIONS FROM INTERMITTENT FEVER IN EACH MONTH.													ADMISSIONS FROM REMITTENT FEVER IN EACH MONTH.												
	January.	February.	March.	April.	May.	June.	July.	August.	September.	October.	November.	December.	TOTAL.	January.	February.	March.	April.	May.	June.	July.	August.	September.	October.	November.	December.	TOTAL.
A																										
Bellary	30	42	18	6	4	14	10	6	24	47	35	12	250	...	...	...	...	...	...	...	...	...	...	...	...	...
Salem, Central	3	6	9	3	8	6	2	5	8	12	9	7	78	...	...	...	...	...	...	...	...	...	...	...	...	...
Coimbatore, Central	3	...	4	1	2	...	...	...	...	...	...	1	11	...	...	...	...	...	...	...	...	...	...	...	...	...
B																										
Palamcottah	...	2	...	...	2	2	2	...	...	...	...	2	10	...	...	...	...	...	...	...	...	...	...	...	...	...
Madura	2	1	...	1	3	2	3	1	2	1	2	2	20	...	...	...	...	...	...	...	...	...	...	...	1	1
Trichinopoly, Central	7	6	4	1	1	1	1	2	2	1	1	2	29	...	...	...	...	...	...	...	...	...	...	...	...	...
Tanjore	1	...	...	1	...	...	1	2	2	3	3	3	16	...	...	1	2	...	1	1	...	1	3	...	3	12
Cuddalore	1	...	1	4	6	4	3	2	...	...	2	2	27	...	...	...	...	...	...	...	...	...	...	...	...	...
Vellore, Central	3	3	4	2	5	...	5	1	7	2	4	4	36	...	...	...	...	...	...	...	...	...	...	...	...	...
Madras Peniten- tiary, Central	4	6	7	7	4	4	4	10	7	3	5	1	62	...	...	...	...	...	...	...	...	...	...	1	...	1
Nellore	...	1	...	...	...	...	...	...	...	...	1	...	2	...	...	...	...	...	...	...	...	...	...	1	...	1
C																										
Rajamundry, Central	19	48	88	11	9	5	7	2	14	35	5	5	248	...	...	...	...	...	...	...	...	...	...	...	...	...
Vizagapatam	...	1	...	...	...	...	...	2	2	2	5	2	12	...	...	...	...	...	...	...	...	...	...	...	...	...
Berhampur	2	2	...	1	...	...	2	1	1	...	...	...	9	...	...	...	...	...	...	...	...	...	...	...	...	...
GROUP XI.— SOUTHERN INDIA	75	117	135	38	44	38	40	32	69	108	72	42	810	...	...	1	2	...	1	1	...	...	3	2	4	15
Shillong	...	...	...	...	...	...	...	1	3	...	2	1	7	...	...	...	...	...	...	...	...	...	...	...	...	...
Darjeeling	5	1	8	3	2	1	5	4	4	1	6	5	45	...	...	...	...	...	...	...	...	...	...	...	...	...
Almora	2	7	1	...	1	3	...	1	3	2	1	21	...	...	...	...	...	...	...	...	...	...	...	...	...
Naini Tal	1	...	...	...	...	...	...	...	...	...	...	1	...	...	...	...	...	...	...	...	...	...	...	...	...
Simla	...	...	...	...	...	...	...	...	...	...	...	1	...	...	...	...	...	...	...	...	...	...	...	...	...
Abbottabad	...	4	3	1	2	4	1	2	2	...	1	2	22	...	...	...	...	...	...	...	...	...	...	...	...	...
Quetta	4	2	2	2	1	5	4	4	7	5	9	2	47	...	...	...	...	...	...	...	...	...	...	...	...	...
Mercara	...	...	...	...	...	1	...	...	...	...	...	1	...	...	...	...	...	...	...	...	...	...	...	...	...
Russellkonda	...	...	...	...	...	...	...	...	...	...	...	...	...	...	...	...	...	...	...	...	...	...	...	1	1	2
GROUP XII.— HILLS.	12	14	14	6	5	11	14	11	17	9	21	11	145	...	...	...	...	...	...	...	...	...	...	1	1	2
EXTRA INDIA:— ADEN	...	1	1	1	1	4	1	...	1	3	4	...	17	...	...	...	...	...	...	...	...	...	...	...	...	...
INDIA*	1,083	974	1,123	1,292	1,356	1,264	1,319	1,551	1,843	1,877	1,725	1,207	16,704	4	3	5	10	3	16	8	4	15	10	10	11	109
Burma	31	33	31	48	26	46	65	56	83	100	61	65	645	...	1	...	1	3	1	...	1	...	...	...	...	10
Eastern Bengal and Assam	93	67	64	112	112	139	160	213	159	187	181	148	1,665	1	...	...	...	3	3	1	...	1	2	2	...	14
Bengal	227	177	250	469	400	376	322	363	431	430	383	283	4,120	1	...	1	7	8	...	1	...	1	...	...	1	21
United Provinces	277	244	210	333	340	325	327	480	667	547	360	216	4,304	...	2	1	1	2	5	2	1	3	1	1	2	21
Punjab	217	171	204	183	340	240	255	219	140	211	358	313	2,834	2	...	...	...	2	...	...	1	...	...	...	...	6
N.-W. F. Prov- ince	29	29	75	9	38	28	37	20	19	59	50	53	452	...	...	...	...	...	...	1	1	...	...	...	...	2
Bombay	87	92	88	59	32	35	60	91	127	144	149	91	1,055	...	...	...	...	2	...	...	...	4	1	1	1	9
Central Prov- inces	30	34	22	26	27	26	40	67	135	122	68	59	676	...	1	...	...	...	...	...	...	3	1	1	...	6
Madras	83	122	141	41	47	42	47	38	73	110	76	47	869	...	1	2	...	1	...	1	...	1	...	4	5	10
Andamans	1,024	897	980	1,110	1,511	2,221	3,330	2,173	1,301	1,118	1,017	1,070	17,881	5	12	1	4	10	10	13	6	6	12	4	5	88
India†	2,107	1,871	2,103	2,411	2,897	3,485	4,649	3,724	3,234	2,995	2,742	2,377	34,585	9	15	6	14	23	26	21	10	21	27	14	16	197

* Including Ajmer, Secunderabad, Quetta, and Mercara and excluding Andamans.
† Including Ajmer, Secunderabad, Quetta, Mercara and Andamans.

TABLE LI.

PNEUMONIA by months, Jails, Groups, and Administrations.

TABLE LII.

DYSENTERY by months, Jails, Groups, and Administrations.

JAILS* AND GROUPS.	ADMISSIONS FROM PNEUMONIA IN EACH MONTH.													ADMISSIONS FROM DYSENTERY IN EACH MONTH.												
	January.	February.	March.	April.	May.	June.	July.	August.	September.	October.	November.	December.	TOTAL.	January.	February.	March.	April.	May.	June.	July.	August.	September.	October.	November.	December.	TOTAL.
Mergui	...	...	...	...	...	...	...	...	...	...	...	...	...	...	...	...	...	...	...	1	...	...	...	...	...	1
Tavoy	...	...	...	...	...	...	...	...	...	...	...	...	...	...	...	...	...	...	1	...	...	...	...	...	...	
Moulmein	...	1	...	...	...	...	...	...	...	1	...	...	2	...	...	2	6	1	7	...	3	6	11	2	1	43
Shwegyin	...	...	...	...	...	...	...	...	...	...	...	...	...	2	...	...	2	1	...	2	1	1	...	...		9
Toungoo	...	...	...	...	...	1	...	...	...	...	...	...	1	...	...	1	...	1	...	3	6	2	3	3	1	20
Rangoon, Central (Europeans)	...	...	...	...	...	...	...	...	...	...	...	...	...	...	...	...	1	...	...	1	1	...	...	...	...	4
Rangoon, Central (Natives)	1	2	...	1	2	1	...	...	...	...	...	...	7	2	2	2	...	4	3	2	3	1	3	4	3	29
Maubin	...	1	...	...	...	...	...	...	...	...	...	...		...	...	1	...	...	...	...	1	1	...	1	...	4
Myaungmyo	...	...	...	...	...	...	...	...	...	...	...	...	...	2	1	...	...	...	1	...	2	...	1	...	...	7
Hessein, Central	3	1	1	3	1	...	1	1	...	1	...	...	12	1	...	2	...	1	2	3	...	2	1	...	...	13
Insein, Central	...	...	...	...	...	...	...	...	...	...	...	...		...	...	...	1	1	...	1	2	...	...	...	...	2
Sandoway	...	...	...	...	...	...	...	...	...	...	...	...	...	...	...	...	1	1	...	...	2	...	...	...	...	6
Akyab	...	...	...	...	...	...	...	1	1	4	...	...	6	...	...	2	2	...	...	...	...	...	...	1	...	5
GROUP I.—BURMA COAST AND BAY ISLANDS	4	4	1	4	3	2	1	2	3	5	...	...	29	7	3	8	9	12	14	18	19	16	20	13	5	144
Paungdi	...	...	...	...	...	...	...	...	...	...	...	1	1	1	2	...	1	1	1	5	4	1	2	1	...	7
Prome	...	...	...	...	...	...	...	...	...	...	...	1	1	1	2	...	1	2	1	2	6	3	9	5	11	20
Thayetmyo, Central	...	1	...	1	...	...	...	1	...	...	1	4	1	2	2	1	2	1	2	6	3	9	5	11	45
Taungdwingyi	...	...	1	...	...	...	...	...	...	...	...	...	1	...	...	...	...	1	...	1	...	...	...	...	9
Magwe	...	...	...	...	...	...	...	...	...	...	...	...	...	...	...	...	5	3	...	...	1	...	...	...	2
Meiktila	...	...	...	...	...	...	...	...	...	...	...	...		...	...	...	...	...	...	...	1	...	...	...	9
Myingyan, Central	...	...	...	...	...	...	...	...	...	...	...	...		6	3	6	10	6	3	7	8	4	7	11	9	80
Mandalay, Central	...	...	...	...	...	...	...	1	...	...	...	1	4	4	5	2	1	2	7	5	4	...	2	3	39
Monywa	...	2	...	...	...	...	...	...	...	...	...	...	2	...	1	...	...	...	...	...	...	...	...	...	1
Shwebu	...	...	...	...	...	...	...	...	...	...	...	...		2	...	...	...	...	...	1	...	2	...	...	5
Bhamo	1	...	...	...	...	...	...	...	...	2	...	3	...	...	...	...	...	...	...	1	...	1	...	...	2
Katha	...	...	...	...	...	...	...	...	...	...	...	...		...	...	1	1	1	...	...	...	...	...	...	3
Kindat	...	...	...	...	...	...	...	...	...	...	...	...		...	...	...	...	...	...	...	...	...	...	...	1	1
GROUP II.—BURMA INLAND	2	3	1	1	...	...	...	1	1	2	...	...	13	15	15	14	15	11	13	26	26	16	20	20	25	214
Cachar	...	...	...	...	...	...	...	...	...	1	...	1	...	...	...	2	1	...	2	...	1	...	...	...	6
Sibsagar	...	...	...	...	...	...	...	...	...	...	...	...		...	...	...	2	1	...	...	1	...	1	...	...	9
Dibrugarh	...	...	...	...	...	...	...	...	...	1	...	1	2	...	...	16	12	8	3	5	10	3	...	...	59
Tezpur	...	...	...	...	...	...	...	...	...	...	...	...		1	...	4	4	6	2	3	3	...	1	...	24
Gauhati	...	...	...	...	...	...	...	1	...	...	...	1	8	17	5	19	23	27	14	23	13	9	5	1	104
Sylhet	...	...	...	...	...	...	...	1	...	...	...	1	4	4	...	12	14	3	4	11	18	13	8	6	92
GROUP III.—ASSAM	...	...	...	...	...	...	...	1	1	...	...	2	14	21	6	30	59	51	31	41	40	33	18	8	361
Mymensingh	...	1	...	...	1	...	...	...	...	...	2	6	16	19	29	20	26	33	28	18	15	18	20	23	265
Dacca, Central	...	...	...	3	...	...	1	...	...	1	5	4	5	4	12	9	19	8	0	11	13	19	24	134
Tippera	...	...	...	...	...	...	...	...	...	...		4	...	1	...	4	4	...	...	...	...	...	6
Chittagong	...	...	1	...	...	1	...	...	...	3	5	2	1	2	4	3	3	...	1	3	2	29
Noakhali	...	...	1	...	1	...	...	...	1	2	8	8	5	9	7	6	10	18	9	2	18	5	108
Backergunge	2	1	2	...	3	1	...	1	...	11	25	7	30	31	33	41	39	40	16	35	63	35	395
Khulna	...	...	...	...	...	...	...	...	...		...	3	...	3	4	3	2	...	...	...	...	...	18
Jessore	...	...	...	...	...	...	...	1	1	37	47	29	37	15	7	35	6	27	32	26	30	328
Barasct	...	...	...	...	...	...	...	...		1	...	12	8	4	2	1	2	5	2	3	41
Presidency, Central, Europeans	...	...	...	...	...	...	...	...	...		...	...	1	1	2	...	1	6
Presidency, Central, Natives	4	...	...	3	2	...	1	...	1	12	12	5	7	6	5	7	7	23	54	30	14	16	156
Alipore, Central	...	...	...	2	...	3	2	...	2	1	10	15	17	19	17	12	17	16	20	27	11	5	32	202
Hooghly	2	...	...	1	...	...	1	...	2	4	3	2	4	4	3	3	8	8	4	5	10	61
Burdwan	1	...	...	1	...	1	1	1	5	3	...	1	3	2	6	2	4	2	3	3	34
Krishnagar	...	...	...	1	...	...	...	...	3	5	3	8	11	5	9	20	10	18	19	10	2	119
Faridpur	2	...	1	2	...	1	...	...	7	2	2	3	6	7	3	4	2	5	5	9	50
Pabna	2	...	...	...	...	...	1	...	3	2	3	1	4	3	2	1	...	2	1	...	19
Murshidabad	...	3	1	...	...	...	1	6	4	3	7	15	19	4	14	18	13	10	16	17	140
Rajshahi, Central	...	...	...	...	...	...	...		1	...	4	4	1	...	6	3	2	1	4	3	28
Bogra	...	...	...	...	...	...	...		3	...	...	2	...	...	1	...	...	6
Malda	1	...	...	1	...	1	...	1	...	3	1	22	18	12	12	9	6	5	2	110
Dinajpur	...	...	...	...	...	1	...	4	17	17	18	20	16	19	15	21	19	13	8	189
Rangpur	...	1	...	...	...	1	1	...	3	...	2	...	1	1	...	6	1	2	5	2	25
Jalpaiguri	...	...	...	...	...	...	...		1	...	1	1	...	2	3	1	5	2	4	23
Purneah	...	...	...	1	...	...	...		1	...	2	...	3	...	2	1	1	3	17
Naya Dumka	...	...	...	...	...	1	...	2	4	2	2	3	4	5	4	3	7	29
Suri	...	...	...	...	...	...	...		2	3	1	9	1	3	3	1	2	29
Bankura	...	...	1	...	...	1	...	2	19	17	9	11	38	31	52	41	24	20	23	18	304
Midnapore, Central	...	1	1	2	...	...	...	2	4	2	2	3	4	5	12	0	3	4	3	63
Balasore	1	...	1	...	...	...	1	2	2	2	3	8	3	3	47
Cuttack	...	...	1	...	...	...	...	1	1	5	...	1	2	...	1	14
Puri	...	...	1	...	...	...	...		...	2	1	...	1	5
Angul	...	...	...	...	...	...	...		...	...	...			
GROUP IV.—BENGAL AND ORISSA	10	11	9	7	20	9	8	5	8	6	15	117	199	93	210	273	730	744	333	297	291	263	279	256	3,654

JAILS AND GROUPS	January	February	March	April	May	June	July	August	September	October	November	December	TOTAL	January	February	March	April	May	June	July	August	September	October	November	December	TOTAL
A																										
Chaibassa	...	...	...	...	...	...	...	1	...	1	...	2	4	5	5	...	...	2	6	6	7	6	2	46
Purulia	...	1	...	1	2	...	...	1	1	...	...	7	...	4	3	2	3	10	13	9	6	6	1	57
Ranchi	...	...	...	...	...	...	1	...	...	...	...	5	4	7	5	4	3	0	2	2	2	1	40
Palamau	...	...	...	...	...	...	...	...	...	...	...	3	1	2	...	3	6	5	2	...	...	24
Hazaribagh,Central	...	...	1	...	1	...	...	1	...	2	...	5	17	9	7	29	30	23	24	36	34	17	8	10	234
B																										
Gaya	...	...	...	...	...	...	...	...	...	...	...	1	...	...	...	1	...	7	10	8	4	4	35
Bhagalpur,Central	1	1	...	...	...	1	...	1	...	6	...	9	11	7	14	23	11	18	20	52	32	39	15	10	237
Monghyr	...	...	1	...	...	...	...	...	...	...	1	2	2	4	5	2	3	5	6	10	3	2	2	46
Darbhanga	...	...	...	...	...	...	...	...	...	...	1	1	7	3	3	3	5	7	1	5	8	9	1	43
Champarun	...	...	...	...	...	...	...	...	...	...	...	1	1	4	3	2	13	17	1	1	1	45
Muzaffarpur	...	...	...	3	...	...	...	...	1	...	4	0	1	6	3	10	5	6	6	7	...	2	...	53
Patna	...	...	...	...	...	...	...	...	...	...	1	3	...	2	...	1	...	5	5	...	2	20
Arrah	...	...	...	...	...	...	...	...	1	...	1	...	3	6	2	2	1	7	5	2	1	3	32
Chapra	...	...	...	...	...	1	...	...	...	...	1	8	2	4	6	6	2	6	16	15	17	11	9	97
Buxar, Central	1	...	...	...	...	1	...	2	...	4	1	...	2	15	17	34	26	85	35	26	17	13	272
Kerantarih	...	...	...	...	...	...	...	...	...	...	...	...	...	1
Ghazipur	...	...	...	...	...	2	...	...	...	3	5	1	...	2	...	1	1	3	3	2	3	4	21
Azamgarh	...	...	...	...	...	...	...	2	1	2	3	...	...	...	1	...	1	2	...	1	1	2	8
Kasia	...	...	...	...	...	...	...	...	...	...	...	2	2	2	1	1	1	1	1	12
Gorakhpur	...	1	...	1	...	...	1	1	1	...	5	6	4	7	17	16	14	11	17	17	25	4	2	140
Basti	...	...	...	...	...	...	1	...	...	...	...	...	...	...	...	1	1	2	6
Fyzabad	...	...	...	...	...	...	...	...	...	...	...	...	1	1	8
Sultanpur	...	...	...	...	...	1	...	1	...	2	1	...	1	...	3	1	...	1	6
Rai Bareli	...	1	...	...	...	...	1	...	...	2	1	...	2	1	...	...	...	8
Partabgarh	...	...	...	...	...	...	...	...	...	...	...	1	1	2	2	8
Jaunpur	...	1	...	...	...	...	...	...	...	2	1	...	1	2	3	...	3	2	3	17
Benares, Central	...	2	2	4	...	1	1	...	...	3	13	1	...	...	1	3	7	3	8	5	28
" District	...	...	1	...	...	...	...	...	2	2	...	...	...	...	2	1	...	12
Mirzapur	...	...	...	1	...	...	3	...	5	3	...	1	5	1	1	1	...	14
Allahabad, Central	...	1	1	...	1	...	...	...	5	3	...	2	2	2	4	4	7	13	6	1	8	51
" District	1	2	3	...	...	2	...	1	2	2	13	2	...	1	1	2	2	5	2	2	3	1	31
Karwi	...	...	...	...	...	...	...	...	...	...	...	...	1	...	1
Banda	...	1	...	1	...	...	...	...	3	1	...	2	...	1	...	2	14
Fatehpur	1	1	...	...	...	...	...	...	4	...	...	...	...	1	...	8
Hamirpur	1	1	...	...	...	...	...	...	2	4	...	3	...	1	...	8
Orai	...	...	...	...	...	...	...	1	...	1	1	1	1	...	2	1	...	8
Cawnpore	...	...	...	...	...	...	...	...	...	1	...	1	...	1	1	5
Unao	...	1	...	...	1	...	1	...	...	3	1	1	...	1	...	1	1	...	7
Lucknow, Central	2	2	...	...	...	...	...	...	4	10	...	1	...	1	...	2	9	2	1	1	17
" District	1	1	...	...	...	...	1	...	2	5	...	...	...	1	...	1	2	1	6
Barabanki	...	1	...	...	...	...	...	...	1	4	...	1	1	1	...	3	10
Gonda	2	3	3	2	3	...	1	1	2	1	1	19	...	2	...	...	3
Bahraich	...	...	2	...	...	...	...	...	2	1	1	...	...	...	1	4
Kheri	...	...	...	1	...	...	...	1	2	1	...	1	1	...	3
Sitapur	1	...	...	1	1	1	...	...	3	10	1	...	1	2	1	3	1	9
Hardoi	...	...	...	...	...	1	...	...	2	...	1	2	2	...	4
Etawah	...	...	...	1	...	...	1	...	...	2	1	...	1	2	4	1	11
Mainpuri	...	2	...	1	...	...	...	1	5	3	...	1	3	3	2	26
Etah	1	5	6	2	...	...	1	...	1	3	17	4	2	3	2	3	...	21
Fatehgarh, Central	...	5	2	1	1	3	...	...	1	1	14	1	4	3	12	20	9	3	5	6	9	6	9	87
" District	...	...	...	...	...	...	...	...	2	2	3	1	3	1	4	3	...	18
GROUP V.—GAN-GETIC PLAIN AND CHUTIA NAGPUR	12	31	27	21	8	7	9	8	8	16	29	24	200	89	52	85	163	159	158	137	324	287	207	129	117	1,956
A																										
Shahjahanpur	...	...	...	...	1	...	2	...	2	2	...	7	...	1	1	...	...	1	...	3	2	8
Pilibhit	...	...	1	...	...	...	...	...	...	1	...	1	...	...	...	1	...	5
Bareilly, Central	...	1	...	...	...	...	...	...	...	1	3	...	1	...	3	1	...	1	5
" District	2	2	4	4	2	1	...	...	...	1	16	10	5	2	2	3	2	5	12	12	6	3	6	70
Budaon	...	...	...	2	3	1	...	...	2	...	9	...	...	...	2	...	1	...	2
Aligarh	...	1	...	...	...	...	1	...	...	3	...	...	...	1	3	...	1	14
Bulandshahr	...	...	...	1	...	...	1	...	1	3	1	...	1	3	...	4	2	2	1	17
Moradabad	...	...	...	...	...	...	1	...	...	1	6	1	...	4	...	2	...	3
Bijnor	...	...	...	...	...	...	...	...	...	...	1	1	...	...	3
Dehra Dun	...	...	...	...	...	...	...	1	...	1	...	...	...	1	...	1	2
Saharanpur	3	...	...	1	...	...	...	...	1	5	1	...	1	2	2	3	4	8	9	6	4	5	47
Muzaffarnagar	1	1	...	...	...	...	2	...	3	...	...	1	...	...	1	...	4
Meerut	...	...	1	...	...	1	...	...	...	5	...	...	1	1	...	...	4
Delhi	5	...	2	3	...	3	1	1	...	1	2	17	1	...	2	...	2	4	2	4	1	1	19
Rohtak	...	...	...	...	...	...	...	...	1	...	1	...	...	3	1	2	1	10
Hissar	1	2	...	...	...	...	...	...	...	3	...	1	...	...	...	5
Karnal	...	...	...	2	...	...	...	...	...	3	1	...	...	...	1
Umballa	...	1	1	...	...	1	...	...	2	...	8	1	2	...	...	1	2	3	4	36
B																										
Ludhiana	...	...	...	1	...	...	...	...	1	...	2	1	...	...	2	...	1	1	13
Jullundur	...	...	1	1	...	1	...	...	...	3	1	...	...	2	1	3	...	1
Ferozepore	...	...	1	...	1	...	1	1	...	4	2	...	...	2	...	14
Amritsar	...	...	...	...	...	...	...	2	1	4	2	...	1	2	1	...	3	14
Lahore, Central	1	4	1	1	1	1	...	...	1	4	15	5	1	4	2	2	7	2	8	14	5	7	5	62
" District	1	2	3	...	1	...	...	...	...	7	3	...	4	...	3	3	...	19
" Female	...	...	...	...	...	1	...	...	1	1	...	...	...	2
Gurdaspur	...	...	...	...	...	...	1	...	3	1	...	1	...	1	...	8
Gujranwala	...	...	...	...	...	...	...	...	...	1	1	1	...	1	1
Sialkot	...	...	...	...	1	...	...	...	2	3	4	2	1	2	3	1	2	1	20
Gujrat	...	...	...	...	...	...	...	...	2	1	1	...	...	2	1	6
Jhelum	...	...	1	...	...	...	...	...	1	1	...	...	...	1	4
Rawalpindi	4	...	2	...	...	...	...	2	1	9	...	2	...	...	...	2
GROUP VI.— UPPER SUB-HIMALAYA	19	16	20	14	11	5	11	3	5	7	14	22	147	34	21	14	20	32	29	23	58	56	45	37	34	403

TABLE LI—*concluded.*

PNEUMONIA by months, Jails, Groups, and Administrations.

TABLE LII—*concluded.*

DYSENTERY by months, Jails, Groups, and Administrations.

JAILS AND GROUPS.	January.	February.	March.	April.	May.	June.	July.	August.	September.	October.	November.	December.	TOTAL.	January.	February.	March.	April.	May.	June.	July.	August.	September.	October.	November.	December.	TOTAL.
A																										
Peshawar	...	...	...	5	1	...	1	...	1	1	...	...	9	2	2	...	...	3	3	1	4	4	2	5	2	28
Kohat	...	...	1	...	...	...	...	2	...	...	...	...	3	...	...	...	...	...	1	...	1	...	...	...	...	2
Bannu	1	...	...	...	...	...	...	...	...	...	...	...	1	...	...	...	...	2	...	...	2	4	...	2	...	10
Shahpur	...	...	...	...	...	...	...	...	...	...	...	...	...	...	...	...	...	...	...	...	1	1	...	1	...	3
Mianwali	...	1	2	7	1	...	...	...	...	...	...	...	11	2	1	...	3	1	3	2	3	4	3	2	4	28
Lyallpur	1	1	...	2	...	...	...	...	4	2	...	...	10	3	1	1	...	1	2	...	1	...	...	...	...	8
Jhang	...	...	...	...	...	...	1	...	...	...	1	...	2	...	...	...	...	...	...	...	1	...	...	1	...	3
Montgomery, Central	...	3	...	3	2	3	3	4	...	...	1	2	25	1	1	...	...	...	...	...	1	1	4	4	12
Mooltan, Central	...	3	3	4	3	1	2	2	...	...	...	...	18	3	...	1	1	1	...	1	5	5	4	5	5	3
,, District	2	...	...	1	...	...	...	...	...	4	3	...	10	2	...	...	...	...	...	...	...	1	1	2	3	10
Dera Ismail Khan	...	1	2	...	...	...	...	...	...	...	...	...	3	...	3	1	1	...	...	2	...	...	1	3	5	10
Dera Ghazi Khan	...	...	...	...	...	...	...	...	...	...	...	...	...	1	...	...	2	1	1	4	3	3	2	4	...	21
C																										
Shikarpur	5	3	1	1	...	...	...	...	...	...	...	...	10	...	...	...	...	...	...	...	...	...	...	...	...	
Sind Gang	1	1	2	1	...	2	1	3	...	3	...	9	22	...	...	...	...	...	...	...	...	1	2	...	...	4
Hyderabad, Central	1	9	1	1	...	...	1	...	1	...	2	3	19	5	1	2	3	1	4	17	7	2	...	...	...	42
Kurrachee	...	...	...	...	...	...	1	...	1	...	...	...	2	...	...	1	3	5	2	1	3	3	...	3	1	22
GROUP VII.—N.W. FRONTIER, INDUS VALLEY, AND N.W. RAJPUTANA	10	22	12	25	7	6	7	14	6	9	9	10	145	19	9	5	14	16	16	26	32	30	17	31	24	239
A																										
Rajkot	...	...	...	...	...	...	...	...	...	...	...	1	1	1	...	...	...	...	...	1	...	...	...	...	...	3
Ahmedabad, Central	3	8	9	4	3	1	1	...	...	...	...	1	33	1	1	...	4	1	...	...	5	4	...	3	...	20
B																										
Ajmer	...	4	2	...	...	...	...	...	...	...	...	...	6	1	...	1	...	...	...	...	...	...	...	...	...	3
Muttra	...	...	...	...	...	1	...	1	1	...	...	...	3	...	...	...	...	...	...	...	...	...	...	1	...	1
Agra, Central	8	5	4	1	2	3	5	3	5	4	2	2	45	...	...	...	...	2	2	4	6	2	...	7	...	24
,, District	...	...	...	4	1	...	...	...	1	1	...	1	8	...	2	...	...	...	...	2	...	...	3	...	2	9
Jhansi	...	...	...	...	...	...	...	...	...	...	...	...	...	1	...	1	1	...	...	...	...	1	1	...	1	9
Lalitpur	...	...	...	...	...	...	...	...	...	...	...	...	...	...	...	...	...	...	...	...	...	1	...	...	...	2
GROUP VIII.—S.E. RAJPUTANA, CENTRAL INDIA AND GUJARAT	11	17	15	9	6	5	7	5	7	5	3	7	97	4	4	2	5	5	3	7	11	9	5	10	5	70
A																										
Damoh	...	1	...	...	...	...	...	...	...	...	...	...	1	...	...	...	...	...	...	...	2	2	1	1	...	7
Saugor	...	...	...	...	...	...	...	...	...	...	...	...	...	...	...	...	...	...	...	5	...	3	1	1	...	10
Jabbulpur, Central	...	...	...	1	...	...	...	2	...	...	...	...	4	...	...	...	...	1	...	...	...	1	...	...	...	3
Narsinghpur	...	...	...	1	...	...	1	...	...	...	...	...	1	...	...	...	1	...	...	...	1	...	...	...	...	2
Mandla	...	...	...	...	...	...	...	...	...	...	...	...	...	...	1	...	...	...	2	...	2	2	...	1	...	8
Bilaspur	...	...	...	...	...	...	...	...	...	...	...	...	...	...	1	...	...	...	...	2	3	1	...	2	...	6
Sambalpur	...	...	...	...	...	...	...	...	...	...	...	...	...	1	...	...	...	...	...	...	...	...	...	...	...	1
Raipur, Central	...	...	...	...	...	1	...	2	1	...	...	...	6	2	3	2	3	1	6	8	6	2	...	...	...	31
Balaghat	...	...	1	...	1	...	...	...	...	...	...	...	2	...	1	...	...	...	...	...	...	...	...	...	...	4
Seoni	...	...	...	...	...	...	...	...	...	...	...	...	...	1	...	...	...	...	...	...	...	...	1	...	...	1
Chindwara	...	...	...	...	...	...	...	...	...	...	...	...	...	1	...	...	...	...	...	...	1	3	2	...	...	9
Hoshangabad	...	...	...	...	...	...	1	...	...	...	...	...	2	1	...	...	...	...	1	...	2	...	4	1	2	8
Nimar	...	...	...	1	...	...	...	...	...	...	...	...	2	...	...	...	...	...	...	2	2	...	...	1	...	6
Betul	...	...	...	...	2	...	...	...	...	...	...	...	6	...	...	...	...	...	1	8	3	2	...	1	...	15
Nagpur, Central	2	...	...	...	...	...	...	1	...	1	...	...	5	...	...	...	...	...	...	3	1	...	...	1	...	3
Bhandara	...	...	...	...	...	...	...	...	...	...	...	...	...	...	1	...	...	...	...	...	...	...	...	...	...	...
Wardha	1	...	...	...	...	...	...	...	...	...	...	...	1	...	...	...	...	...	...	...	...	...	...	...	...	2
Chanda	...	...	...	...	...	...	1	...	...	...	...	...	1	...	...	1	...	...	...	...	...	...	...	...	...	2
B																										
Secunderabad	...	...	...	...	...	...	...	...	...	...	...	...	...	...	...	...	1	...	...	...	...	...	...	1	1	4
Yeotmahl	...	...	...	...	...	...	...	...	...	...	...	...	...	...	...	...	1	...	...	2	...	...	...	...	1	12
Amrauti, Central	1	...	1	1	...	...	...	...	...	...	...	...	3	2	...	1	...	1	...	10	5	...	...	1	...	17
Akola, Central	...	1	...	...	...	...	...	...	...	...	...	...	1	...	...	...	...	...	3	6	3	...	4	2	2	18
Buldana	...	...	1	...	...	...	...	...	1	...	...	...	1	...	...	...	...	...	6	24	11	4	11	8	67
Dhulia	...	...	...	...	...	1	1	...	...	...	...	...	3	2	...	...	...	1	6	10	15	1	1	8	5	55
Yerrowda, Central	1	...	...	...	...	...	1	...	...	...	...	...	2	1	1	2	6	2	2	8	...	1	1	2	6
Bijapur	...	...	...	...	...	...	...	...	...	...	...	...	...	1	...	...	...	...	...	2	...	1	...	...	23
Deccan Gang	...	...	...	...	...	...	1	1	...	...	...	...	3	3	2	5	1	...	5	5	...	1	...	5	23
Dharwar	...	...	...	...	...	...	...	...	...	...	...	...	...	4	...	5	6	4	3	3	...	...	...	2	27
GROUP IX.—DECCAN	4	6	2	4	5	1	6	3	5	4	2	...	43	9	12	8	23	13	9	41	89	57	21	26	23	332
Thana	1	1	...	...	...	...	...	...	...	...	...	...	...	3	2	3	2	1	5	6	2	...	5	...	31
Bombay, Common House of Correction	2	2	...	...	1	...	1	1	...	...	...	...	6	1	1	...	2	1	...	...	3	4	2	4	17
,,	...	...	...	...	...	...	...	...	...	...	...	...	1	...	...	...	...	...	...	...	...	3	...	2
Ratnagiri	...	...	...	...	...	...	...	...	...	...	...	...	...	...	...	...	...	...	...	...	...	1	...	1	2
Karwar	...	...	...	...	...	...	...	...	...	...	...	...	2	...	...	...	...	...	...	...	...	...	...	...	2
Mangalore	...	...	...	...	...	...	...	...	...	...	...	...	...	1	...	...	...	...	...	...	...	2	...	...	3
Cannanore, Central	2	1	1	...	1	1	...	2	1	...	1	1	12	1	1	1	...	7	3	4	3	3	9	7	41
GROUP X.—WESTERN COAST	5	4	2	...	2	1	1	4	2	...	1	1	23	4	5	4	3	4	11	10	11	6	10	14	14	58

Jails, Groups, and Administrations	Admissions from Pneumonia in each month													Admissions from Dysentery in each month												
	Jan.	Feb.	Mar.	Apr.	May.	June.	July.	Aug.	Sept.	Oct.	Nov.	Dec.	TOTAL	Jan.	Feb.	Mar.	Apr.	May.	June.	July.	Aug.	Sept.	Oct.	Nov.	Dec.	TOTAL
A																										
Bellary	...	1	...	2	...	...	2	1	...	1	...	...	8	...	...	3	...	...	4	...	6	...	1	...	2	14
Salem, Central	1	...	1	...	...	...	1	1	5	...	...	4	12	2	1	2	3	7	3	6	6	5	2	2	1	42
Coimbatore, Central	1	...	...	...	1	5	2	3	2	1	...	...	16	3	9	3	4	5	7	35	16	6	13	12	10	123
B																										
Palamcottah	...	1	...	...	...	...	...	...	...	...	...	...	1	...	...	...	...	...	3	3	3	2	1	9	7	28
Madura	1	1	...	2	...	...	2	...	...	...	...	...	6	...	...	...	...	2	1	2	3	3	2	...	...	13
Trichinopoly, Central	...	...	...	...	1	1	...	...	...	...	...	...	2	...	...	3	4	3	...	4	...	...	...	5	5	27
Tanjore	...	...	1	...	...	...	...	...	...	...	...	...	1	1	3	...	...	...	1	...	2	...	...	3	3	18
Cuddalore	...	...	...	...	...	...	...	...	...	...	...	...	...	...	...	...	...	1	1	2	...	...	...	...	...	4
Vellore, Central	3	1	...	...	...	...	2	1	1	2	1	1	12	2	4	...	...	2	5	5	...	2	1	1	5	25
Madras Penitentiary, Central	1	1	...	1	...	...	...	...	...	...	...	1	5	1	1	4	1	6	3	3	9	5	4	8	1	48
Nellore	...	...	...	...	...	...	...	1	...	1	...	...	1	1	1	1	3	...	3	...	...	...	...	1	...	8
C																										
Rajamundry, Central	...	...	1	...	1	...	...	1	...	2	1	...	6	6	2	10	10	3	1	8	20	46	88	9	11	214
Vizagapatam	1	...	...	1	...	...	...	1	1	...	...	2	6	3	2	1	2	...	1	2	3	12	4	3	1	34
Berhampur	...	...	...	...	...	1	...	...	1	...	...	...	2	1	...	...	...	...	...	2	1	...	...	...	...	4
GROUP XI.— SOUTHERN INDIA	8	6	5	4	3	8	7	8	10	8	3	8	78	22	23	29	24	32	32	70	63	83	119	37	48	603
Shillong	...	1	...	...	...	...	...	...	...	...	...	...	1	4	...	...	...	...	2	3	6	...	4	2	...	21
Darjeeling	...	...	...	...	...	...	...	1	...	...	...	...	1	...	...	...	...	1	2	...	...	1	...	...	...	4
Almora	...	...	...	...	...	1	...	1	...	...	...	...	2	...	...	...	1	2	...	...	1	...	...	...	...	4
Pauri	1	...	...	...	1	...	...	...	...	...	...	...	2	...	...	...	2	2	...	...	1	...	...	...	...	5
Abbottabad	...	...	...	...	...	...	...	1	1	...	...	...	2	...	...	...	1	1	...	...	1	...	...	1	...	4
Quetta	...	...	...	...	...	...	1	...	...	1	...	...	3	...	...	2	...	...	2	...	3	1	...	...	1	9
Mercara	...	...	...	...	...	...	...	...	...	...	...	...	...	...	...	...	...	...	...	...	...	1	...	...	...	...
Russellkonda	...	...	...	...	...	...	...	...	...	...	...	...	...	...	...	...	...	...	1	...	1	...	...	1	...	3
GROUP XII.— HILLS	1	2	...	...	1	1	3	1	1	...	...	...	11	4	...	2	...	4	10	9	5	8	1	1	2	46
EXTRA INDIA.— ADEN	...	...	...	...	...	...	...	...	...	...	...	...	...	...	...	...	1	2	...	2	...	2	...	1	...	8
INDIA*	86	22	94	89	65	42	53	55	54	64	72	100	896	410	330	377	390	607	592	751	978	901	763	635	552	7,496
BURMA	6	7	2	3	3	2	1	2	4	6	2	2	42	22	16	22	24	23	27	44	45	32	40	33	30	353
EASTERN BENGAL AND ASSAM	4	4	8	2	12	2	3	2	3	1	5	11	57	100	89	121	193	217	203	201	191	155	154	193	133	1,065
BENGAL	8	10	3	9	11	4	8	7	4	8	12	4	88	162	135	142	237	197	200	280	410	370	260	100	198	2,783
UNITED PROVINCES	24	41	33	31	14	14	19	11	15	22	27	33	286	53	25	36	61	89	74	60	103	123	101	65	71	864
PUNJAB	16	17	19	22	11	7	10	10	8	6	15	23	164	30	17	8	18	17	22	18	41	45	39	39	33	327
N.-W. F. PROVINCE	2	2	3	5	1	...	1	2	1	1	...	...	18	2	5	1	1	6	5	1	9	9	3	11	7	60
BOMBAY	12	15	14	7	4	3	3	7	4	6	4	17	100	13	14	10	27	21	20	46	66	42	17	25	26	327
CENTRAL PROVINCES	4	2	2	4	5	1	...	5	2	4	3	1	35	4	4	4	9	3	1	19	45	30	14	9	5	147
MADRAS	10	8	6	4	4	9	8	8	12	9	3	9	90	23	25	30	24	33	39	75	68	86	124	65	56	649
ANDAMANS	10	13	5	8	9	21	22	12	17	11	19	11	138	227	202	287	257	254	234	236	172	150	117	101	120	2,359
INDIA	96	135	99	97	74	63	75	67	71	75	91	111	1,054	617	532	664	827	801	829	987	1,150	1,051	1,041	880	718	9,855

* Including Ajmer, Secunderabad Quetta, and Mercara and excluding Andamans.
Including Ajmer Secunderabad, Quetta, Mercara and Andamans.

TABLE LIII.

DETAIL of DISEASES.

DISEASES	EUROPEAN ARMY OF INDIA.								NATIVE ARMY OF INDIA. Present 123,234, Enrolled 145,012			JAIL POPULATION OF INDIA, 166,265	
	MEN, 70,994				WOMEN, 3,375		CHILDREN, 5,151						
	Admissions.	Constantly sick.	Deaths.	Invalids.	Admissions.	Deaths.	Admissions.	Deaths.	Admissions.	Deaths.	Invalids.	Admissions.	Deaths.
Small-pox	9?	9·07	1	...	16	3	9	4	77	1	...	31	4
Cow-pox	10	·39	...	...	...	...	1	...	62	...		5	...
Chicken-pox	18	·82	...	...	...	4?	...		135	...	..	805	
Measles	13	·63	1	...	2	...	12	2	317	1	...	59	2
Rubella	12	·77	...	...	...	...	2	...	14	...	...	4	...
Scarlet fever	8	·78	...	...	...	...	5	...	...	...	...	...	...
Typhus fever	...	...	...	...	...	...	...	...	1	1	...	...	...
Plague	13	1·41	4	...	...	...	1	...	79	45	...	66	46
Relapsing fever	...	...	...	...	...	...	...	...	11	...	...	...	...
Dengue	415	9·01	1	...	8	...	3	...	14	...	...	..	...
Influenza	1,014	25·70	2	...	53	...	8	...	100	...	...	1,243	15
Whooping cough	...	...	...	...	...	...	23	...	1	...	..	...	...
Mumps	5	·20	..	...	1	...	12	...	1,648	...	...	483	1
Diphtheria	1	·01	1	...	1	...	5	4	...	...	...	...	...
Cerebro-spinal fever	2	·05	2	...	...	...	...	...	1	1	...	11	10
Simple continued fever	3,415	120·57	...	5	63	...	93	...	1,760	...	...	892	1
Enteric fever	1,145	220·50	213	52	29	5	26	2	130	35	...	64	15
Mediterranean fever	4	·83	...	...	...	...	...	...	43	1	...	...	...
Cholera	8	·18	7	...	...	...	...	...	11	7	...	69	38
Choleraic diarrhoea	2	·01	2	...	...	...	...	...	1	...	...	4	2
Epidemic diarrhoea	1	·01	...	...	...	...	...	...	...	...	...	1	...
Dysentery	936	71·91	33	37	47	3	34	2	3,912	21	14	9,855	464
Beri-beri	46	9·09	5	25	...	...	...	...	15	2	2	177	28
Intermittent fever	7,786	307·63	8	124	163	1	257	5	20,582	45	35	34,585	96
Remittent fever	151	9·46	1	2	9	...	9	1	975	62	2	197	61
Phagedæna	...	...	...	...	...	...	...	...	1	...	...	...	1
Sloughing phagedæna	...	...	...	...	...	...	...	...	1	...	...	1	...
Erysipelas	31	1·84	3	...	3	...	2	...	24	3	...	69	14
Pyæmia	3	·50	2	...	...	...	1	1	4	...	...	7	6
Septicæmia	3	·89	1	1	...	...	...	...	5	4	...	5	5
,, puerperal	...	...	...	...	4	5	...	...	...	...	...	5	2
Tetanus	1	·01	1	...	...	...	...	...	5	4	...	10	7
Tubercle, not defined	...	...	...	...	...	...	...	...	6	...	1	...	...
,, general	1	·12	.	...	...	...	...	...	1	...	...	...	...
,, of the nose	1	·12	...	1	...	...	...	...	...	...	...	...	...
,, of the meninges	...	...	...	...	...	...	...	...	1	1	...	2	2
,, of the brain	...	...	...	...	...	...	1	1	...	...	...	...	...
,, of the membranes of the brain	...	...	...	...	...	...	2	2	...	...	...	...	...
,, of the larynx	...	...	...	...	...	...	...	...	...	...	...	1	...
,, of the lungs and larynx	...	...	...	...	...	...	...	...	...	...	...	...	1
,, of the lungs	144	29·87	16	116	18	3	6	2	383	62	167	915	363
,, of the lungs and general	...	...	...	...	...	...	...	...	...	...	...	9	7
,, of the lungs and pericardium	1	·03	1	...	...	...	...	...	...	...	...	...	...
,, of the lungs and intestines	2	·06	2	...	...	...	...	...	...	...	...	10	13
,, of the pleura and meninges	...	...	...	...	...	...	...	...	1	1	...	...	...
,, of the lungs and kidney	...	...	...	...	...	...	...	...	...	...	...	1	1
,, of the lungs and bladder	1	·13	1	...	...	...	...	...	...	...	...	...	...
,, of the lungs and pleura	...	...	...	...	...	...	...	...	1	...	...	1	1
,, of the pleura	...	...	...	...	...	...	...	...	...	...	...	1	1
,, of the intestines	...	...	...	...	...	...	2	1	2	1	...	14	16
,, of the bladder	...	...	...	...	...	...	...	...	...	...	...	1	...

DISEASES.	MEN.				WOMEN.		CHILDREN.		NATIVE ARMY OF INDIA.			JAIL POPULATION OF INDIA.	
	Admissions.	Constantly sick.	Deaths.	Invalids.	Admissions.	Deaths.	Admissions.	Deaths.	Admissions.	Deaths.	Invalids.	Admissions.	Deaths.
Tubercle of the peritonæum	...		...	...	...	...	1	1	4	1	1	3	3
„ of the scrotum	...	...	...	...	...	...	...		...			1	
„ of lymphatic-glands	...	...	...	...	...	...		...	17	1	3	36	2
„ of the kidneys	...	...	...	...	...	...	...	...	...	...	...	1	1
„ of the testicles	1	'.8	...	2	...	...	...	...	1	...	1	2	
„ of bones	2	'27	...	1	...	...	...	...	1	...	1	...	...
„ of joints	1	'07	...	...	...	...	1	...	1	1	...	4	...
„ of the spine	...	...	...	...	...	...	...	...	1	...	1	...	...
„ of the skin	...	...	...	...	...	...	...	...	3	...	1	...	...
„ of the mesenteric glands		...	...	...	...	...	...	...	...	...		5	3
Leprosy	...	...	...	...	...	...	...	...	15	...	1	143	22
Yaws	...	...	...	...	...	...	...	...	...		...	2	...
Syphilis	2,535	3'10'42	13	75	3	...	...	...	92,	6	67	1,732	15
„ inherited	...	...	...	...	...	...	2	...	...	...	...	...	...
Gonorrhœa	5 332	46'61	...	8	...	...	...	...	895	...	10	427	...
Hydrophobia	2	'02	2	...	...	...	...	...	6	6	...	...	...
Anthrax	1	...	1	...	...	...	...	...	...	...	...	3	
Actinomycosis	...	'18	...	1	...	...	...	...	...	...	...	...	...
Kala-azar	4	1'00	...	3	...	...	3	...	10	9	1	...	...
Bilharzia hæmatobia	22	3'91	...	23	...	...	...	...	...	...	...	6	...
Bothriocephalus latus	3	'24	...	...	...	...	...	...	...	...	...	4	...
„ liguloides	...	...	...	...	...	...	...	...	...	...	...	1	...
Tænia solium	147	3'68	...	...	4	...	6	...	18	...	...	37	1
„ mediocanellata	13	'26	...	...	...	...	...	...	3	...	...	4	...
Cysticercus of the Tænia solium	...	...	...	...	...	...	...	...	...	...	...	1	...
Cysticercus tenuicollis	6	'10	...	...	...	...	...	...	...	...	...	...	...
Echinococcus hominis	2	'.5	...	1	...	...	...	...	...	...	...	1	...
Ascaris lumbricoides	4	'05	...	...	...	...	10	...	27	...	...	41	1
„ Mystax	...	...	...	...	...	...	...	...	...	...	...	1	...
Guinea-worm	...	...	...	...	...	...	...	...	609	1	1	481	...
Strongylus duodenalis	...	...	...	...	...	...	...	...	1	1	...	30	5
Thread-worm	2	'03	...	...	...	...	1	...	13	...	...	7	...
Musca domestica	1	'01	...	...	...	...	...	...	...	...	...	...	...
Pediculus capitis	...	...	...	...	...	...	...	...	1	...	...	...	...
„ vestimenti	...	...	...	...	...	...	...	...	10	...	...	...	...
Phthirius inguinalis	22	'44	...	...	...	...	...	...	...	...	...	...	...
Acanthia lectularia	1	'01	...	...	...	...	...	...	...	...	...	...	...
Culex anxifer	3	'04	...	...	1	...	2	...	...	...	...	...	...
Scabies	519	20'61	...	...	3	...	6	...	1,610	...	...	940	...
Mycetoma	...	...	...	...	...	...	...	...	1	...	...	3	
Tinea favosa	...	...	...	...	...	...	...	...	1	...	...	13	...
Ringworm	265	11'71	...	...	...	...	3	...	306	...	...	147	...
Tinea versicolor	5	'28	...	...	...	...	...	...	1	...	...	3	
Thrush	...	...	...	...	...	...	1	...	...	...	...	...	
Parasite not defined	...	...	...	...	...	...	...	...	2	...	...	...	
Eruption by external use of drug		...	...	...	...	...	...	...	...	...	...	1	...
Surfeit	...	...	...	...	...	...	...	...	2	...	...	...	...
Scurvy	7	'90	...	...	...	...	...	...	211	4	10	130	8
Alcoholism	193	9 25	7	...	...	...	...	...	6	1	...	...	...
Delirium tremens	3	'04	1	...	...	...	...	...	—	...	...	...	...
Rheumatic fever	27	4'38	1	2	1	...	5	...	39	2	1	8	...
Rheumatism	894	62'37	...	43	18	...	1	...	1,421	4	91	952	7

TABLE III—*continued.*

DETAIL of DISEASES.

DISEASES	EUROPEAN ARMY OF INDIA								NATIVE ARMY OF INDIA			JAIL POPULATION OF INDIA	
	MEN				WOMEN		CHILDREN						
	Admissions.	Constantly sick.	Deaths.	Invalids.	Admissions.	Deaths.	Admissions.	Deaths.	Admissions.	Deaths.	Invalids.	Admissions.	Deaths.
Gout	13	·84	...	1	1	...	...	...	7	...	...	5	...
Osteoarthritis	...	...	...	...	...	...	...	...	12	...	2	...	...
Cyst	33	1·78	1	1	3	...	...	...	37	...	1	8	...
,, of kidney	...	...	...	...	...	...	...	...	...	...	...	1	1
Pterygium	1	·13	...	...	...	...	...	...	22	...	...	5	...
Chalazion	...	...	...	...	...	...	...	...	...	...	...	1	...
New growth, non-malignant, not defined	4	·14	..	...	...	...	1	...	9	...	...	24	...
Lipoma	6	·50	...	...	...	...	...	...	10	...	...	9	...
Fibroma	18	·98	...	1	5	...	...	...	8	...	1	8	...
Chondroma	4	·38	...	...	...	...	...	...	...	...	...	...	...
Osteoma	2	·29	...	1	...	...	...	...	...	...	...	1	1
Myxoma	2	·23	...	...	...	...	...	...	...	...	...	...	...
Mucous polypus	...	...	...	...	...	...	...	...	...	...	...	2	...
Myoma	...	...	...	...	...	...	...	...	...	...	...	1	...
Angioma	4	·31	...	...	...	...	...	...	...	...	...	...	...
Lymphadenoma	1	·04	...	...	...	...	...	...	2	2	...	...	...
Papilloma	33	2·35	...	1	...	...	...	...	2	...	...	2	...
Warts	2·81	17·14	...	...	...	...	...	...	3	...	...	5	...
Condyloma	...	...	...	...	...	...	...	...	...	...	...	13	...
Adenoma	...	...	...	...	1	...	8	...	...	...	...	...	...
Sarcoma	1	·02	...	...	...	...	...	...	1	...	1	...	...
New growth, malignant, not defined	...	...	...	...	...	...	...	...	...	...	...	10	1
Sarcoma	4	·81	3	...	...	...	...	...	2	...	1	9	7
Carcinoma	...	...	...	...	3	1	...	...	1	1	1	7	4
Epithelioma	1	·02	...	...	...	...	...	...	...	...	...	4	—
Columnar carcinoma	3	·30	2	1	...	...	...	...	...	...	...	...	...
Colloid cancer	2	·11	1	...	...	...	...	...	...	...	...	..	...
Rickets	...	...	...	...	...	...	4	...	...	...	...	...	...
Cretinism	...	...	...	...	...	...	1	...	...	...	...	...	...
Myxœdema	...	...	...	...	...	...	...	...	...	...	...	1	...
Anæmia	81	5·43	...	8	19	...	6	...	333	2	10	528	30
Idiopathic anæmia	...	...	...	...	...	...	...	...	1	1	...	1	...
Purpura	...	...	...	...	...	...	...	...	4	...	...	8	3
Leucocythæmia	1	·42	...	1	...	...	...	...	...	...	...	...	...
Hodgkin's disease	...	...	...	...	...	...	...	...	1	...	...	3	...
Hæmophilia	1	·02	...	...	...	...	...	...	...	...	...	2	...
Diabetes mellitus	7	·90	2	2	...	...	...	...	11	1	4	12	3
,, insipidus	1	·35	...	2	...	...	...	...	...	...	...	7	1
Immaturity at birth	...	...	...	...	...	...	24	24	...	...	...	1	...
Fissure of the skull	...	...	...	...	...	...	1	...	...	...	...	...	...
Fissure of foot	...	...	...	...	...	...	...	...	4	...	...	...	...
Teeth deficient	1	·07	...	...	...	...	...	...	...	—	...	...	—
Malposition of testes	1	·05	...	...	...	...	...	...	...	...	...	...	...
Testicle diminutive and absent	2	·03	...	1	...	...	...	...	...	...	...	...	...
Intestines impervious	...	...	...	...	...	...	1	1	...	...	...	...	...
Congenital phimosis	27	2·07	...	...	...	...	19	...	1	...	...	...	...
Vitelline duct persistent (Meckel's diverticulum)	...	...	...	...	...	...	...	...	1	...	...	...	...
Hymen impervious	...	...	...	...	1	...	...	...	...	...	...	...	...
Debility	1,629	102·91	1	152	502	...	2·6	14	1,110	8	214	424	12
Old age	...	...	...	...	...	...	...	...	1	...	1	62	8
Neuritis	34	3·51	...	5	...	...	...	...	41	...	4	3	...

DISEASES.	EUROPEAN ARMY OF INDIA.								NATIVE ARMY OF INDIA.			JAIL POPULATION OF INDIA.	
	MEN.				WOMEN.		CHILDREN.						
	Admissions.	Constantly sick.	Deaths.	Invalids.	Admissions.	Deaths.	Admissions.	Deaths.	Admissions.	Deaths.	Invalids.	Admissions.	Deaths.
Multiple neuritis	3¹	3'45	...	7	1	...	...	...	6	...	1	4	...
Degeneration of the nerves	2	'32	...	1	...	...	...	...	.	...	...	...	...
Spinal meningitis	...	'01	...	...	...	...	...	...	...	...	...	...	...
Myelitis	5	1'61	1	3	...	...	...	...	5	3	2	..	...
Anterior poliomyelitis	1	'49	...	1	...	...	...	...	1	...	...		...
Degeneration of spinal cord	1	...	...	1	...	...	...	...	...	...	...	...	...
Progressive muscular atrophy	2	'15	...	1	...	...	...	...	1	...	...	1	...
Primary lateral sclerosis	1	'09	...	2	...	...	...	...	2	...	2	7	...
Posterior sclerosis	3	'75	...	1	...	...	...	...	4	1	...	4	—
Postero-lateral sclerosis	1	'13	...	...	...	...	...	...	3	...	...	...	...
Disseminated ,,	1	'24	—	1	...	...	...	...	3	...	1	...	...
Cerebral meningitis	6	'79	5	...	2	1	3	3	4	4	...	19	16
Pachymeningitis	4	4 .	3	...	2	2	3	3	2	...	...	...	...
Hæmatoma of duramater	1	'03	...	...	...	...	...	...	...	...	...	...	...
Leptomeningitis	1	'06	2	...	...	...	...	...	...	...	...	...	...
Hæmorrhage into the membranes of the brain	...	...	...	...	...	...	...	...	...	...	...	6	7
Encephalitis	1	'02	1	...	...	...	...	...	1	...	...	1	1
Abscess of the brain	7	'14	6	...	...	...	...	...	...	...	...	2	2
Sclerosis ,, ,,	1	'01	...	...	...	...	...	...	...	...	...	2	...
Softening ,, ,,	...	...	...	...	...	...	...	...	...	...	...	1	2
Sanguineous apoplexy	6	1'05	2	1	...	...	...	...	2	...	...	4	3
Hyperæmia of the brain	...	...	...	...	...	...	...	...	...	...	...	2	...
Bulbar paralysis	...	...	...	...	...	...	...	...	1	1	...	2	...
Internal hydrocephalus	...	...	...	...	...	...	1	1	...	...	...	...	...
Apoplexy	...	...	...	...	...	...	...	...	2	2	...	6	5
Paralysis	5	1'11	...	...	...	...	...	...	47	...	5	7	...
Paraplegia	3	'35	...	...	...	...	...	...	8	...	2	16	4
Hemiplegia	11	1'61	...	3	...	...	...	...	16	...	7	20	2
Monoplegia	1	'07	...	...	...	...	...	...	4	—	...	...	...
Local paralysis	7	'80	...	2	...	...	...	...	11	...	...	10	...
Incomplete paralysis	4	'31	...	...	...	...	...	...	12	...	...	...	...
Tremor	...	...	...	...	...	...	...	...	4	...	1	...	...
Paralysis agitans	...	...	...	...	...	...	...	...	1	...	1	1	...
Chorea	5	'84	1	1	...	...	2	...	1	...	...	...	...
Wry-neck	1	'04	...	...	...	...	1	...	4	...	...	5	...
Facial spasm	...	...	...	...	...	...	...	...	...	...	...	4	...
Convulsions	...	...	...	...	...	...	...	...	1	1	...	...	...
,, Infantile	...	...	...	...	...	...	17	15	...	...	...	1...	1...
,, puerperal	...	...	...	...	2	...	...	...	...	...	...	1	1
Epilepsy	70	9'22	2	38	3	...	1	...	34	2	11	127	6
Laryngismus stridulus	...	...	...	...	...	...	4	...	...	...	...	1	...
Vertigo	24	1'03	...	1	...	...	...	...	11	...	...	2	...
Headache	47	1'45	1	1	2	...	1	...	31	...	...	25	...
Megrim	3	'11	...	...	...	...	...	...	29	...	...	11	...
Anæsthesia	...	...	...	...	...	...	...	...	5	...	...	2	...
Neuralgia	223	10'43	...	2	17	...	2	...	365	...	8	93	...
Facial hemiatrophy	...	...	...	...	...	...	...	...	3	...	...	...	...
Hysteria	10	'67	...	1	8	...	...	...	7	...	...	7	...
Somnambulism	1	'02	...	...	...	...	...	...	...	...	...	...	...
Aphasia		'05	...	...	...	...	...	...	...	...	...	...	...

TABLE LIII—*continued.*

DETAIL of DISEASES.

DISEASES	EUROPEAN ARMY OF INDIA.								NATIVE ARMY OF INDIA.			JAIL POPULATION OF INDIA.	
	MEN.				WOMEN.		CHILDREN.						
	Admissions.	Constantly sick.	Deaths.	Invalids.	Admissions.	Deaths.	Admissions.	Deaths.	Admissions.	Deaths.	Invalids.	Admissions.	Deaths.
Hiccough	...	...	...	...	...	...	...	...	1	...	...	2	...
Nervous weakness	30	2·80	...	6	3	...	1	...	10	...	1	3	...
Idiocy	...	·34	...	2	...	...	...	...	1	...	1	5	
Mania	10	2·0)	...	13	2	...	...	...	20	3	9	49	
,, puerperal	...	...	...	...	2		...	...	...		...		
Melancholia	28	5·29	...	27	1	...	...	...	17	...	10	9	1
Dementia	10	2·28	...	9	...	...	...	...	10	...	7	4	1
Mental stupor	2	·27	...	...		...	...	...	5	...	1	1	
General paralysis of the insane .	...	...	...	...	...	...	...	...	...		2		...
Delusional insanity . . .	21	4·38	...	14	...	...	...	...	4	...	2	4	
Conjunctivitis	370	18·49	...	5	11	...	92		2,061	...	5	1.050	
,, granular . . .	...	..	...					1	86	...	1	45	...
Ecchymosis of the conjunctiva .	...	...	...	...		...		...	...	...	...	3	...
Chemosis ,, ,, .	...	...	...	...			...	...	1	...			...
Chronic hyperæmia . . .	...	...	...	...			...	...	...	...	1	1	...
Keratitis	10	1·16	...	3	...	...	...	...	47	...	3	27	...
Ulcerative keratitis . . .	35	3·34	...	1	1	...	...	1	190	...	6	197	...
Degeneration of the cornea .	...	...	...	...	...	...	...	...	...	...	...	2	
Opacity ,, ,, . .	1	·03				...	...	...	14	...	6	4	...
Staphyloma	...	...	...		...	...	...	...	...	...	...	6	...
Scleritis	...	...	...		..	...	...	...	1	...	1	2	...
Iritis	46	3·72	...	2	1	...	...	...	45	...	3	36	...
Synechia	...	...	...		...	...	...	...	...			1	...
Mydriasis	2	·08	...	...	...	...	...	...	1	...	...	...	...
Choroiditis	2	·23	...	...	...	...	...	...	2	...	1	...	...
Glaucoma	...	...	...	...	...	...	...	...	12	...	6	6	...
Hypopyon	...	...	...	...	...	...	...	...	...	...	...	3	...
Optic neuritis	5	·70	...	5	...	...	...	...	5	...	3	...	...
Atrophy of optic nerve . .	1	·02	...	1	...	...	...	...	2	...	1	...	...
Retinitis	4	·30	...	2	...	...	...	...	3	...	1	...	...
Hæmorrhage into retina . .	2	·16	...	...	...	...	...	...	...	...	...	...	...
Degeneration and atrophy of retina .	...	...	...	1	...	...	...	...	1	...	1	...	...
Detachment of retina . .	1	·05	...	...	...	...	...	...	...	...	...	...	
Lenticular cataract . . .	2	2·22	...	4	...	...	...	...	10		7	38	
Capsular ,, . . .	1	·01			1	...	...	...	...		...		
Aphakia	1	·13	...	1	...	...	...				...		
Dislocation of lens . .	1	·02		...	...	...	...				...	2	
Panophthalmitis . .	...				...	...					...	2	
Amblyopia and amaurosis . .	7	·51		2	..		...		10		3	1	...
Functional night blindness . .	1	·05		1		...	...		11		1	3	...
Blinding from intense light .	...		...		...	...			1	...		...	
Ametropia	4	·22			...	...		...	8	...	8		...
Myopia	21	1·44		2	...	...			3	...	3	1	...
Hypermetropia . . .	16	1·02		9	...				2	...			...
Astigmatism . . .	15	1·03	...	6	...					...			...
Presbyopia . . .	1	·15			...				...	...			...
Asthenopia . . .	1	·15			...		...		4	...			...
Squint . . .	2	·06	...	...	...		1		2	...	...		...
Nystagmus . . .	...			1	...				1	...			
Stricture and obliteration of puncta and canaliculi . . .	1	·02	...		...	...	...		1	...			

DISEASES.	MEN.				WOMEN.		CHILDREN.		NATIVE ARMY OF INDIA.			JAIL POPULATION OF INDIA.	
	Admissions.	Constantly sick.	Deaths.	Invalids.	Admissions.	Deaths.	Admissions.	Deaths.	Admissions.	Deaths.	Invalids.	Admissions.	Deaths.
Chronic dacryo-cystitis	2	...	...	1	...	...	...	...	2	...	1	...	
Abscess of lacrymal sac	1	·06	...	...	...	...	...	...	1	...		5	...
Fistula of lacrymal sac	2	·07	...	...	...	...	...	...	1	...	...		...
Obstruction of nasal duct	...			...	...	...	...	...	1		...	1	
Epiphora	...	...		...	...	...	...	...	1	...	...	...	...
Blepharitis marginalis	16	1·06	...	1	...	...	...	...	7	...	...	12	...
Stye	17	·23	...	...	1	...	...	...	86	...	...	16	
Abscess of the eyelids	1	·08		...	...	...	...	...	...	...	...	...	...
Ecchymosis	2	·08	...	...	...	...	...	...	1	...	...	...	...
Trichiasis	...	...	...	...	...	...	...	...	...	...	...	5	...
Entropion	...	...	...	...	...	...	...	...	1	...	...	9	...
Œdema of the eyelids	1	·03	...	...	...	...	...	...	1	...	...	1	...
Ptosis	1	·01	...	...	...	...	...	...	2	...	...	...	...
Proptosis	1	·06	...	...	...	...	...	...	...	...	...	...	...
Inflammation of the external ear	599	24·23	...	4	7	...	11	...	255	...	1	156	...
Abscess ,, ,, ,,	3	·18	...	...	1	...	...	...	11	...	...	4	...
Hæmatoma of the auricle	1	·03	...	...	...	...	...	...	...	...	...	1	...
Accumulation in external meatus of wax and epidermis	7	·08	...	...	...	...	1	...	3	...	...	...	...
Inflammation of the middle ear	162	8·91	...	11	...	...	...	...	65	...	...	17	...
,, ,, ,, suppurative	26	2·41	1	4	1	...	...	...	31	1	...	47	...
Ulceration of the membrana tympani	2	·11	...	...	...	...	...	...	...	...	...	1	
Perforation ,, ,,	136	13·42	...	36	...	...	...	...	12	...	2	2	...
Tinnitus	1	·24	...	1	...	...	...	...	...	...	...	...	...
Deafness	23	1·02	...	8	...	...	...	...	16	...	6	2	...
Rhinitis	1	1·27	...	1	...	...	...	...	5	...	...	1	...
Coryza	58	1·14	...		1	...	...	...	55	...	...	48	...
Ozæna	6	·36	...	...	2	...	...	...	...	...	...	14	...
Abscess of the nose	...	...	...	...	...	...	...	...	...	...	...	3	...
Perichondritis	1	·01	...	...	...	...	...	...	...	...	...	...	...
Abscess of the septum	...	...	...	...	...	...	...	...	...	...	...	1	...
Perforation of the septum	1	·03	...	...	...	...	...	...	...	...	...	...	...
Epistaxis	8	·20	...	...	...	...	...	...	8	...	...	25	...
Inflammation of the accessory sinuses	1	·10	...	...	...	...	...	...	2	...	...	1	...
Empyema	2	·05	...	1	...	...	...	...	...	...	...	...	...
Inflammation of the naso-pharynx	2	·10	...	...	...	...	...	...	13	...	...	25	...
Hypertrophy of the pharyngeal tonsil	...	...	...	...	...	...	...	...	...	...	...	2	...
Adenoid vegetations	...	...	...	...	...	...	1	...	...	...	...	...	...
Pericarditis	4	·10	2	...	1	...	...	...	4	1	...	8	4
Hydropericardium	...	...	...	...	...	...	...	...	...	...	...	3	...
Endocarditis	2	·03	5	...	...	...	1	1	1	...	...	4	2
Valvular disease of the heart	167	24·18	18	95	7	1	2	...	41	11	10	83	33
Abscess of the muscular substance of the heart	...	...	...	...	...	...	...	...	...	...	...	1	...
Fatty degeneration of the muscular substance of the heart	3	·12	6	...	...	...	...	...	3	1	3	15	12
Hypertrophy of the heart	3	·55	...	2	...	...	...	...	1	...	...	3	2
Dilatation of the heart	7	·63	2	3	...	...	1	...	14	...	2	8	5
Excessive growth of fat in the heart	...	...	...	...	...	...	...	...	...	1	...	...	...
Aneurysm of the heart	1	·01	1	...	...	...	...	...	1	...	...	...	...
Thrombus	...	...	...	...	...	...	...	...	1	2	...	1	1
Angina pectoris	1	·07	1	2	...	...	...	...	...	...	...	...	...
Syncope	3	·10	2	...	1	...	...	...	8	13	...	9	10

TABLE LIII—*continued.*

DETAIL of DISEASES.

DISEASES.	EUROPEAN ARMY OF INDIA.								NATIVE ARMY OF INDIA.			JAIL POPULATION OF INDIA.	
	MEN.				WOMEN.		CHILDREN.						
	Admissions.	Constantly sick.	Deaths.	Invalids.	Admissions.	Deaths.	Admissions.	Deaths.	Admissions.	Deaths.	Invalids.	Admissions.	Deaths.
Disordered action of the heart	527	50·50	...	76	3	...	1	...	19	1	11	9	...
Dilatation of arteries	...	...	...	...	...	...	...	...	...	...	...	2	...
Aneurysm of arteries	* 6	·69	4	...	1	...	...	...	3	1	2	6	
Aneurysm by anastomosis	...	...	...	...	...	...	...	...	...	...	...	1	...
Traumatic aneurysm	1	·14	...	...	...	...	...	...	...	...	...	—	...
Rupture of artery	...	...	1	...	...	...	...	...	...	...	...	...	...
Thrombosis	3	·05	...	...	...	...	...	...	1	...	...	1	...
Embolism	...	...	...	...	1	●	...	...	...	...	...	1	2
Phlebitis	36	2·19	...	6	5	...	...	...	15	...	1	11	...
Obstruction of veins	...	...	...	...	...	...	...	...	1	...	...	2	...
Obliteration of the veins	1	·07	...	...	...	...	...	...	...	...	...	...	...
Thrombosis ,, ,,	19	·83	...	10	4	...	...	...	...	...	...	1	...
Phlegmasia dolens	...	...	...	...	...	...	...	...	...	...	...	2	...
Varix	96	7·35	...	5	4	...	...	...	18	...	8	5	...
Varicose aneurysm	2	·37	...	...	1	...	...	...	...	...	...	...	...
Nævus	1	·08	...	...	...	...	1	...	1	...	...	...	...
Hay asthma	5	·07	...	...	...	...	...	...	...	...	...	...	...
Laryngitis	43	2·46	...	2	2	...	2	1	129	1	...	24	3
Œdema of glottis	...	...	...	...	...	...	...	...	1	1	...	...	...
Perichondritis	...	...	...	...	...	...	...	...	...	...	...	1	...
Aphonia	...	...	...	...	...	...	...	...	2	...	...	...	...
Tracheitis	4	·10	...	...	...	...	...	...	7	...	...	...	...
Bronchitis	1,344	55·69	1	9	35	...	223	10	2,865	9	17	2,348	41
Dilatation of bronchi	3	·67	...	2	...	...	...	...	1	1	...	2	...
Spasmodic asthma	26	2·54	...	3	6	...	1	...	148	1	12	632	5
Congestion of the lungs	4	·45	...	...	...	...	...	...	5	...	...	4	...
Hæmorrhage from the lungs	1	·08	...	...	...	...	...	...	...	...	...	...	...
Hæmoptysis	8	·36	...	...	...	...	...	...	8	...	1	26	...
Pulmonary apoplexy	...	...	...	...	...	...	...	...	...	...	...	2	1
Œdema of the lungs	...	...	...	...	...	...	...	...	...	...	...	3	4
Pneumonia	294	23·41	44	1	3	...	17	8	1,545	234	4	1,054	279
Broncho-pneumonia	18	1·65	1	...	...	...	34	12	25	6	...	26	13
Abscess of the lungs	1	·22	...	1	...	...	...	...	...	...	...	1	1
Gangrene of the lungs		·09	...	...	...	...	...	...	...	...	...	16	10
Cirrhosis of the lungs	2	·23	1	1	...	...	...	...	1	...	1	...	...
Phthisis	3	·91	...	2	2	...	...	...	6	1	2	4	1
Emphysema	6	3	...	...	...	...	...	...	4	...	...	5	2
Atelectasis	...	...	...	...	...	...	1	1	...	...	...	...	...
Pleurisy	156	10·72	...	1	4	...	4	...	284	6	8	153	13
Empyema	10	2·15	1	...	...	...	...	...	2	2	2		2
Hydrothorax	...	...	...	...	...	...	...	...	1	...	...	...	...
Pneumothorax	...	...	...	...	...	...	...	...	1	...	...	...	...
Adhesions of pleura	1	·12	...	...	...	...	...	...	1	...	1	...	...
Inflammation of the lips	...	...	...	...	...	...	...	...	5	...	...	1	...
Fissure of the lips	...	...	...	...	...	...	...	...	2	...	...	...	...
Stomatitis	22	1·37	...	...	1	...	4	...	55	...	...	89	...
Ulceration of the mouth	2	·14	...	...	...	...	...	...	13	...	...	5	...
Gangrene ,, ,,		...	...	...	...	...	1	...	...	...	...	4	1
Disorders of dentition	...	...	...	...	...	...	43	10	1	...	...	...	...
,, ,, with convulsions		...	...	...	...	...	3	3	1	...	...	...	...
,, diarrhœa		...	...	...	...	...	26	2	...	...	...	...	...
Inflammation of the dental pulp	...	...	...	...	...	...	...	...	...	...	...	...	...

Diseases.	EUROPEAN ARMY OF INDIA.								NATIVE ARMY OF INDIA.			JAIL POPULATION OF INDIA.	
	MEN.				WOMEN.		CHILDREN.						
	Admissions.	Constantly sick.	Deaths.	Invalids	Admissions.	Deaths.	Admissions.	Deaths.	Admissions.	Deaths.	Invalids.	Admissions	Deaths.
Suppuration of the dental pulp	...	...	...	...	...	...	...	...	1	...	...	1	...
Caries of dentine	224	7'19	...	14	6	...	4	...	35	...	...	24	...
Necrosis of cementum	...	...	...	...	...	...	...	...	1	...	...	...	...
Inflammation of the dental periosteum	92	2'80	...	...	3	...	3	...	15	...	...	...	...
Gum-boil	192	5'50	...	...	...	...	...	...	142	...	1	218	...
Inflammation of the gums and periosteum	5	'15	...	...	...	...	2	...	10	...		26	
Suppuration of the periosteum, gums, and alveoli	5	'85	...	2	...		...		23	...	1	3	...
Ulceration of the gums and periosteum	1	'05	...	...	...	...	...	...	3	...	...	26	...
Caries of alveoli	11	'23	...	...	...	...	...	...	9	...	...	6	...
Necrosis of ,,	4	'55	...	...	...	...	...	...	2	...	...	1	...
Hypertrophy of gums	...	...	...	...	...	...	...	...	1	...	...	...	...
Impaction of teeth	1	...	...	...	...	...	...	...	1	...	...	6	...
Toothache	1	'01	...	...	...	...	...	...	3	...	...	1	...
Glossitis	2	'08	...	...	...	...	...	...	11	...	...	2	...
Abscess of the tongue	...	...	...	...	...	...	...	...	...	...	...	2	...
Ulceration of the tongue	2	'08	...	...	...	...	...	...	1	...	...	3	...
Sore throat	705	22'91	...	...	8	...	17	...	175	...	...	78	...
Ulceration of fauces	2	'05	...	...	...	...	...	...	3	...	...	1	...
Tonsillitis	120	2'60	...	...	4	...	7	...	...	...	...	...	...
Follicular tonsillitis	1,046	36'90	...	...	19	...	52	...	269	...	...	68	...
Quinsy	70	3'11	...	...	2	...	...	...	8	...	...	30	...
Hypertrophy of the tonsils	...	...	...	...	...	...	5	...	3	...	1	...	...
Elongated uvula	2	'05	...	...	...	...	...	...	2	...	...	5	...
Inflammation of the salivary glands	3	'14	...	...	...	...	...	...	10	1	...	6	...
Salivation	...		...	...	...	...	...	...	...	...	...	1	...
Inflammation of the pharynx and oesophagus	23	'76	...	...	...	...	...	...	49	...	...	7	...
Post-pharyngeal abscess	...	...	...	...	...	...	...	...	...	...	...	3	...
Ulceration of the pharynx	...	...	...	...	...	...	...	...	...	...	...	2	1
Gastritis	258	10'50	5	6	21	...	18	1	77	...	2	70	2
Ulceration of the stomach	...	...	...	...	4	2	1	...	4	...	1	5	1
,, ,, ,, perforating	...	...	...	...	...	...	...	...	...	...	...	6	5
Gangrene of the stomach	...	...	...	...	...	...	...	...	...	...	...	1	1
Hæmatemesis	...	...	...	...	...	...	...	...	...	...	...	7	7
Dilatation of the stomach	8	'39	...	1	...	...	...	...	3	...	...	...	...
Perforation of stomach	...	...	...	...	...	...	...	...	...	...	...	1	...
Indigestion	869	29'49	...	5	55	...	3	...	239	...	3	557	...
Vomiting	3	'08	...	1	4	...	...	...	3	...	...	...	...
Gastralgia	3	'04	...	...	...	...	...	...	2	...	...	4	...
Loss of appetite	...	...	...	...	...	...	...	...	1	...	...	4	...
Inflammation of the intestines	...	...	...	...	2	...	2	...	...	...	...	25	...
Enteritis	118	7'51	2	1	7	...	35	26	50	8	...	60	17
Typhlitis	139	11'95	4	7	4	...	2	...	58	2	...	18	6
Colitis	38	2'46	...	1	2	...	4	...	37	...	...	27	3
Catarrhal inflammation of the intestines	113	3'76	...	1	4	...	22	7	20	...	...	390	5
Ulceration of the intestines	3	'24	1	...	1	1	...	...	4	1	...	3	2
,, ,, ,, perforating	...	...	...	...	...	...	...	...	...	...	...	3	2
Degeneration of intestines	...	...	...	...	...	...	...	...	1	1	...	...	...
Gangrene of intestines	...	...	...	...	...	...	...	...	...	...	...	1	1
Hypertrophy of intestines	...	...	...	...	...	...	...	...	...	...	...	1	...
Appendicitis	...	...	...	...	...	...	...	...	...	...	...	4	1

TABLE LIII—*continued.*

DETAIL of DISEASES.

DISEASES.	EUROPEAN ARMY OF INDIA.								NATIVE ARMY OF INDIA.			JAIL POPULATION OF INDIA.	
	MEN.				WOMEN.		CHILDREN.						
	Admissions.	Constantly sick.	Deaths.	Invalids.	Admissions.	Deaths.	Admissions.	Deaths.	Admissions.	Deaths.	Invalids.	Admissions.	Deaths.
Hæmorrhage from the intestines	3	·22	...	...	...	...	...	...	4	1	...	14	1
Fæcal accumulation in the intestines	4	·13	...	...	...	...	1	...	3	...	...	8	...
Tympanites	...	...	...	...		...	...		1	...	...		...
Sprue	3	·29	...	3	...	...	...	...	1	...	...	4	3
Hernia	101	8·14	1	19	1	...	8	...	41	1	11	57	1
Intussusception	1	·01	1	...	...	...	...	...	3	1	...	5	5
Volvulus	1	·02	1	...	...	...	...	...	...	...	...	2	2
Internal strangulation of the intestines	...	...	...	...	...	...	...	...	...	...	...	3	3
Stricture of the intestines	2	·13	1	1	...	...	...	...	...	...	...	1	1
Obstruction ,,	3	·38	1	1	1	1	...	...	1	1	...	6	3
Compression ,,	5	·10	...	...	...	...	...	...	...	...	...	...	...
Perforation ,,	...	...	...	...	...	...	...	...	...	...	...	3	3
Intestinal dyspepsia	...	...	...	...	...	...	1	...	...	...	...	...	...
Rupture of the intestines	...	...	...	...	...	...	...	...	...	...	...	1	1
Constipation	118	2·86	...	...	10	...	9	...	40	...	...	126	...
Colic	415	9·74	...	...	18	...	4	...	225	...	...	331	...
Diarrhœa	1,083	37·96	...	2	60	...	207	25	877	7	1	4,395	75
Enteralgia	...	...	...	...	...	...	...	...	1	...	...	1	...
Proctitis	3	·24	...	1	...	...	...	...	...	...	...	8	...
Periproctitis	4	·45	...	1	...	...	...	...	6	1	...	7	...
Abscess of the rectum and anus	6	·58	...	...	...	...	...	...	6	...	...	7	...
Ulceration ,, ,, ,,	5	·44	...	1	...	...	...	...	5	...	1	17	1
Fissure of the anus	16	·90	...	...	...	...	...	...	17	...	1	13	...
Fistula in ano	26	2·60	...	...	1	...	...	...	48	...	3	63	...
Recto-vesical fistula	...	...	...	...	...	...	...	...	1	...	1	...	...
Recto-vaginal ,,	...	...	...	...	1	...	...	...	...	...	...	...	...
Hæmorrhage from the rectum	1	·04	...	...	...	...	...	...	3	...	...	...	...
Prolapse of the rectum	3	·22	...	2	...	...	1	...	8	...	3	16	...
Piles	375	18·25	...	1	8	...	...	...	145	...	8	479	...
Hepatitis	452	34·05	5	14	2	...	1	1	46	...	1	40	1
Abscess of the liver	152	25·45	83	49	...	...	...	...	19	1	...	13	9
Cirrhosis ,, ,,	8	·73	2	4	...	...	...	...	6	...	1	67	40
Perihepatitis	14	·98	...	1	...	...	...	...	...	...	...	1	...
Congestion of the liver	748	35·67	...	8	13	...	7	...	25	...	1	42	...
Acute yellow atrophy of the liver	1	...	1	...	...	...	...	...	...	...	...	1	1
Fatty degeneration of the liver	...	...	...	...	...	...	...	...	...	...	...	1	1
Atrophy of the liver	...	...	...	...	...	...	...	...	...	...	...	1	1
Hypertrophy of the liver	...	...	...	...		...	...	...	1	...	...	12	2
Jaundice	347	17·26	...	...	3	...	4	...	195	1	...	158	...
Cholecystitis	53	6·24	...	...	...	...	2	...	8	...	...	55	1
Gallstones	2	·27	...	...	1	...	...	...	...	...	...	2	1
Accumulation of bile	...	...	...	...	...	...	...	...	1	...	...	...	...
Biliary colic	3	·24	...	...	...	...	...	...	3	...	...	...	...
Perforation of gall bladder		...	...		...	...	...	...	...	...	...	1	1
Inflammation of the pancreas	...	...			...	...	...	...	1	1	...	4	3
Peritonitis	6	·42	4	...	2	...	1	1	16	6	...	21	14
Ascites	1	·09	...	...	...	...	...	...	5	1	1	19	4
Adhesions of peritoneum	...	·01	...	...	...	...	...	...	...	...	...	...	...
Splenitis	12	·81	...	...	...	...	...	...	76	...	6	24	1
Perisplenitis	1	·05	...	...	...	...	...	...	...	...	...	1	1
Abscess of the spleen	...	...	...	...	...	...	...	...	...	...	...	1	1
Congestion of the spleen	...	...	...	...	...	...	...	...	1	...	...	...	...

Diseases	EUROPEAN ARMY OF INDIA.								NATIVE ARMY OF INDIA.			JAIL POPULATION OF INDIA.	
	MEN.				WOMEN.		CHILDREN.						
	Admissions.	Constantly sick.	Deaths.	Invalids.	Admissions.	Deaths.	Admissions.	Deaths.	Admissions.	Deaths.	Invalids.	Admissions.	
Hypertrophy of the spleen	...	...	...	...	...	...	...	...	17	...	1	33	1
Inflammation of lymph glands	1,100	126'38	...	7	1	...	9	...	304	...	3	233	
Suppuration of lymph glands	52	5'84	...	2	...	...	...		18	...	...	89	
Hypertrophy of lymph-glands	1	'03	...	...	...	...	...		5	...	...	1	
Inflammation of lymphatics	26	'89	...	...	1		1		19	...	...	18	
Dilatation of lymphatics	...				...		...				...	1	
Elephantiasis	...		...	...	...	...	...	...	1		1	16	
Inflammation of the thyroid body	1	'02	...	...			...	...	1	...		1	
Hypertrophy ,, ,,	...	...	...				...		1	...			
Goître	13	'84		4	1	...		...	12	...	1	3	
Exophthalmic goître	...	'27		1			...		...	...	...		
Acute nephritis	23	1'85	3	3	1		...		14		2	16	2
Bright's disease	2	'77	...	...	...	...	...		6	2	1	29	8
Chronic nephritis	15	2'46	3	8	3		...		4		...	64	17
Granular kidney	1	'68	...	1	...				5	1	1	6	1
Abscess of kidney	...	...		...			...	...	2	...		1	
Disseminated suppurative nephritis	...	...	..	...			...	...	...	...		1	1
Pyonephrosis	1	'54		...	...			...		...		1	1
Perinephritic abscess	1	'11	..	...			...		...		...	1	1
Pyelitis	2	'56	1	1	1				1	...		1	
Congestion of kidney	1	'03	...	...	...	...	...		1	...	...	1	
Hydronephrosis	...		...		...			...	1	...		1	1
Movable kidney	2	'16	...	1	2		...		2		1	2	
Calculus in kidney	11	1'63		2	...		...	...	21	1	1	0	
,, in ureter	2	'66		1	...	...	...	...	5	...	...	...	...
Thrombosis	1	'01	...	...	...	...	...	...	...		...	...	...
Nephralgia	1	'25	...	...	...	...	...	...	4	...	...	1	
Hæmaturia	27	2'17	...	...	...	...	...	...	10	1	...	9	1
Hæmoglobinuria	2	'11	...	1	...				...		...	2	...
Chyluria	...	'03	...	...	...		...		...			2	
Albuminuria	5	'29	...		1	...		...	3	...		9	1
Lithuria	...		...	...	...				4	...		1	
Oxaluria	...	'05	...	...	...				...	...	...		
Phosphaturia	1	'17	...		...	...	...		...	...		...	
Inflammation of the bladder	38	2'58	...	1	3	...	...		12	2	...	17	2
Suppuration ,, ,,	...				...		...		2	...			
Calculus in the bladder	...		...	...	...		...	...	13	...		12	2
Irritability of the bladder	...		...		...		...		1	..	...	3	
Retention of urine	2	'04	...		1	...		...	6	...		3	
Incontinence of urine	30	2'13	...	1	1	...	...		1	...		2	
Urethritis	50	2'18			1		...		3	...		6	...
Gleet	1	'03					...		1	...		3	...
Abscess of the urethra	1	'03	...					...		...	...		
Ulcer of the urethra	1	'08						...	...		...		
Hæmorrhage from the urethra	2	'06	...	...	...		...		...			...	
Stricture of the urethra	55	5'47	...	4	...		...	...	18	...	2	38	2
Urethral fistula	3	'14	...		...	...			5		2	2	1
Extravasation of urine	...				...		...		1	1		3	1
Impacted calculus	...				...				...	...		3	
Inflammation of the prostate	1	'01		1			...		...			...	
Prostatorrhœa	...				...	...			1	...	...	...	
Abscess of the prostate	1	'07	...	...		...			1	...		1	..

135

TABLE LIII—*continued*.

DETAIL of DISEASES.

DISEASES	EUROPEAN ARMY OF INDIA.								NATIVE ARMY OF INDIA.			JAIL POPULATION OF INDIA.	
	MEN.				WOMEN.		CHILDREN.						
	Admissions.	Constantly sick.	Deaths.	Invalids.	Admissions.	Deaths.	Admissions.	Deaths.	Admissions.	Deaths.	Invalids.	Admissions.	Deaths.
Hypertrophy of the prostate	...	...	...	...	...	...	...	...	1	...	...	4	1
Calculus in the prostate	...	...	...	...	...	...	...	...	...	...	...	6	...
Posthitis	...	...	...	...	...	...	...	...	1	...	...	...	...
Œdema of the prepuce	1	'02	...	...	...	...	1	...	...	...	...	...	...
Phimosis	72	4'78	...	...	...	...	...	...	20	...	...	53	...
Paraphimosis	27	1'50	...	...	...	...	...	...	4	...	...	17	...
Balanitis	216	7'43	...	...	...	...	1	...	5	...	...	3	...
Abscess of the penis	...	...	...	...	...	...	—	...	1	...	...	...	...
Ulcer „ „	27	1'71	...	...	...	...	...	...	42	...	...	34	...
Gangrene „ „	1	'01	...	...	...	...	...	...	...	...	...	...	1
Œdema „ „	1	'01	...	...	...	...	...	...	...	...	...	...	...
Soft chancre of the penis	3,090	286'24	...	...	...	...	...	...	560	1	2	118	...
Inflammation of the scrotum	1	'03	...	...	...	...	...	...	3	...	...	5	...
Abscess of the scrotum	6	'59	...	...	...	...	...	...	3	...	...	12	...
Sloughing „ „	...	...	...	...	...	...	...	...	1	...	...	15	1
Œdema „ „	...	...	...	...	...	...	...	...	...	...	...	2	...
Soft chancre „ „	...	...	...	...	...	...	...	...	1	...	...	9	...
Pruritus	...	...	...	...	...	...	...	...	1	...	—	...	...
Inflammation of the spermatic cord	1	'06	...	...	...	...	...	...	2	...	...	1	...
Hydrocele „ „	6	'57	...	...	...	...	2	...	6	...	1	1	...
Hæmatocele „ „	2	'16	...	...	...	...	...	...	1	...	...	2	...
Varicocele	69	4'85	...	2	...	...	...	...	7	...	...	...	...
Hæmatocele of the tunica vaginalis	...	...	...	...	...	...	...	...	4	...	...	6	...
Hydrocele „ „	42	2'98	...	...	...	...	...	...	34	...	1	73	...
Inflammation of the testicle	...	...	...	...	...	...	...	...	13	...	...	44	...
Orchitis	495	26'79	...	1	...	...	...	...	227	1	2	89	...
Epididymitis	30	1'12	...	...	...	...	...	...	30	...	...	6	...
Abscess of the testicle	1	...	...	...	...	...	...	...	1	...	...	1	...
Atrophy of the testicle	...	...	...	...	...	...	...	...	...	...	...	1	...
Spermatorrhœa	3	'11	...	...	...	...	...	...	5	...	...	...	...
Inflammation of the ovary	...	...	...	...	4	...	...	...	...	...	...	1	...
Inflammation of the Fallopian tube	...	...	...	...	1	...	...	...	...	...	...	...	...
Hydrosalpinx	...	...	...	...	1	...	...	...	...	...	...	...	...
Inflammation of uterine ligaments	...	...	...	...	1	...	...	...	...	...	...	...	...
Perimetritis	...	...	...	...	1	...	...	...	...	...	...	...	...
Parametritis	...	...	...	...	3	1	...	...	...	...	...	...	...
Metritis	...	...	...	...	9	...	...	...	...	...	...	...	...
Endometritis „	...	...	...	...	18	...	...	...	...	...	...	3	...
Ulcer of the uterus	...	...	...	...	1	...	...	...	...	...	...	...	...
Hæmorrhage from the uterus	...	...	...	...	3	...	...	...	...	...	...	2	...
Abrasion of the uterus	...	...	...	...	3	...	...	...	...	...	...	...	...
Anteversion „ „	...	...	...	...	1	...	...	...	...	...	...	...	...
Retroversion „ „	...	...	...	...	1	...	...	...	...	...	...	...	...
Retroflexion „ „	...	...	...	...	2	...	...	...	...	...	...	...	...
Prolapsus „ „	...	...	...	...	2	...	...	...	...	...	...	...	...
Distension „ „	...	...	...	...	...	...	...	...	...	...	...	1	...
Procidentia	...	...	...	...	1	...	...	...	...	...	...	...	...
Stricture of the canal of the uterus	...	...	...	...	1	...	...	...	...	...	...	...	...
Inflammation of the vagina	...	...	...	...	1	...	...	...	...	...	...	1	...
„ „ vulva	...	...	...	...	...	...	1	...	...	...	...	...	...
Amenorrhœa	...	...	...	...	1	...	...	...	...	...	...	...	...
Dysmenorrhœa	...	...	...	...	10	...	1	...	...	...	...	2	...

DISEASES.	EUROPEAN ARMY OF INDIA.								NATIVE ARMY OF INDIA.			JAIL POPULATION OF INDIA.	
	MEN.				WOMEN.		CHILDREN.						
	Admissions.	Constantly sick.	Deaths.	Invalids.	Admissions.	Deaths.	Admissions.	Deaths.	Admissions.	Deaths.	Invalids.	Admissions.	Deaths.
Menorrhagia	...	...	...		16	...		...	...	...	...	10	...
Metrorrhagia	...		...	...	9	...	...	...	...	...	...	5	2
Leucorrhœa	...		...	...	5	...	1	...	...	...	...	3	...
Vaginismus	...		...	...	...	...	...	...	...	...	...	1	
Cramp and spurious labour pains	...	...	...	...	9	...	...	...	...	...	...	...	...
Catarrh of cervix uteri	...	...	...	...	1	...		...	...	...	...	...	...
Hæmorrhage from the uterus (699)	...	...	...	...	8	...	...	...	...	...	...	3	...
Abortion	...	...	...	...	107	...		...	...	...	...	9	2
Apoplectic ovum	...	...	...	...	1	...	...	...	...	...	...	...	...
Carneous mole	...	...	...	...	1	...	...	...	...	...	...	...	...
Atony of the uterus	...	...	...	...		...	...	...	...	...	...	1	...
Mechanical obstacle to the expulsion of the fœtus	...		...	...	3	1	...	...	...	...	...	4	...
Hæmorrhage, unavoidable, from placenta prævia	...		...	...	2	...	...	...	...	...	...	...	...
Hæmorrhage accidental, from detachment of placenta	...	...	...	...	4	...	...	...	...	...	...	3	...
Rupture of perinæum	...	...	...	...	6	...	...	...	...	...	...	...	...
Still-birth	...	...	...	...	2	...		...	...	...	...	...	...
Post-partum hæmorrhage	...	...	...	...	1	...	...	...	...	...	...	3	...
Retention of the placental fragments	...	...	...	...	...		...	...	...	...	...	1	...
Abscess of the areola	...	...	...	...	2	...	...	...	...	...	...	...	...
Mastitis	...	...	...	...	3	...	...	...	...	...	...	8	...
Mastitis, puerperal	...	...	...	...	2	...	...	...	...	...	...	...	...
Suppuration of mammary gland	...	...	...	...	4	...	...	...	...	...	...	3	...
,, ,, puerperal	...	...	...	...	8	...	...	...	...	...	...	...	...
Inflammation of the male breast	1	'05	...	...	...	...	...	...	3	...	...	...	...
Ostitis	13	1'07	...	1	...	...	...	...	9	...	...	4	...
Septic osteo-myelitis	...	...	...	...	...	...	...	...	...	...	...	2	...
Periostitis	53	5'19	...	4	2	...	...	...	69	...	4	23	...
Chronic abscess of bones	4	'46	...	...	...	...	...	...	1	...	...	...	...
Caries of bones	3	'71	...	2	...	...	...	...	2	...	2	13	1
Necrosis of bones	16	2'19	...	5	...	...	...	...	5	...	1	41	2
Spontaneous fracture		...	...	...	...	...	...	...	...	...	...	1	...
Arthritis	2	'12	...	1	...	...	...	...	6	...	...	...	...
Synovitis, Ankylosis of joints (including of the spine)	801	48'29	...	17	4	...	1	...	517	...	16	134	2
	10	'76	...	3	...	...	...	...	9	...	5	7	...
Dislocation of articular cartilage	9	'58	...	4	...	...	...	...	3	...	2	...	
Loose body	1	'01	...	...	...	...	...	...	2	...	1	...	...
Dislocation of joints	...	...	...	...	...	...	...	...	...	...	...	1	...
Inflammation of the spine	1	'12	...	...	...	...	...	...	...	...	...	...	...
Caries of the spine	...		...	...	...	...	...	...	...	...	...	1	...
Psoas, lumbar, post-pharyngeal, abscesses	...	...	...	...	...	...	...	...	1	...	...	4	1
Posterior curvature of the spine	2	'05	...	2	...	...	...	...	2	...	2	...	...
Angular ,, ,, ,,	...	...	...	...	...	...	...	...	...	...	...	1	...
Lateral ,, ,, ,,	1	'23	...	1	...	...	...	...	1	...	1	...	...
Ankylosis of the spine	...	...	...	2	...	...	...	...	...	...	...	...	...
Inflammation of muscles	1	'01	...	...	...	...	...	...	4	...	...	4	...
Suppuration ,, ,,	4	'64	...	...	...	...	...	...	...	...	...	3	...
Atrophy ,, ,,	...	...	...	...	...	...	...	...	1	...	...	2	...
Spontaneous rupture of muscles	1	'02	...	...	...	...	...	...	...	...	...	...	...
Contracture of muscles	...	...	...	...	...	...	...	...	1	...	...	...	...
Pseudo-hypertrophic paralysis	...	...	...	...	...	...	1	...	...	...	...	...	...

S

TABLE LIII—*continued.*

DETAIL of DISEASES.

DISEASES.	EUROPEAN ARMY OF INDIA.								NATIVE ARMY OF INDIA.			JAIL POPULATION OF INDIA.	
	MEN.				WOMEN.		CHILDREN.						
	Admissions.	Constantly sick.	Deaths.	Invalids.	Admissions.	Deaths.	Admissions.	Deaths.	Admissions.	Deaths.	Invalids.	Admissions.	Deaths.
Idiopathic muscular atrophy	...	...	...	...	...	...	...	...	...	...	2	...	...
Myalgia	180	6.89	...	...	5	...	...	...	408	...	14	133	...
Contracture of fasciæ	1	·11	...	1	...	...	...	...	7	...	1	1	...
Inflammation of tendons	1	·04	...	...	...	...	...	...	...	...	...	...	...
Gangrene ,, ,,	...	...	...	...	...	...	...	...	1	...	...	...	...
Adhesions ,, ,,	3	·02	...	...	...	...	...	...	...	...	...	...	...
Contraction ,, ,,	2	·18	...	...	...	...	...	...	...	...	...	...	...
Tenosynovitis	1	·04	...	...	...	...	...	...	11	...	...	1	...
Thecal abscess	...	...	...	...	...	...	...	...	1	...	...	8	...
Ganglion	9	·28	...	...	...	...	...	...	5	...	...	...	...
Inflammation of bursæ	40	2·00	...	1	...	...	...	...	28	...	1	...	...
Abscess of bursæ	1	·12	...	...	...	...	...	...	...	...	...	...	...
Bunion	6	·30	...	...	...	...	...	...	1	...	...	...	...
Bursal cyst	...	...	...	...	...	...	...	...	1	...	...	...	...
Bursal tumour	1	·01	...	...	...	...	...	...	2	...	...	...	...
Club-foot	2	·02	...	2	...	...	...	...	...	...	...	...	...
Flat-foot	11	·60	...	7	...	...	1	...	2	...	1	1	...
Deformities of the great toe	3	·28	...	...	...	...	...	...	1	...	...	...	...
Hallux valgus	10	·38	...	...	...	...	...	...	...	...	...	...	...
Hammer toe	50	4·68	...	2	...	...	...	...	...	...	...	...	...
Inflammation of the connective tissue	774	31·6,	...	1	5	...	3	...	382	2	5	573	23
Abscess ,, ,, ,,	939	44·81	2	5	14	...	22	...	1,650	1	4	3,493	6
Gangrene ,, ,, ,,	1	·07	1	1	...	...	...	...	2	...	1	10	2
Hæmorrhage from ,, ,,	...	...	...	...	...	...	...	...	1	...	...	...	...
Œdema of the ,, ,,	14	·33	...	...	...	...	...	...	7	...	...	39	7
Elephantiasis	...	...	...	...	...	...	...	...	...	...	...	15	...
Emphysema	...	...	...	...	...	...	...	...	1	...	...	...	...
Undue formation of fat	...	...	...	...	...	...	...	...	1	...	1	...	...
Erythema	16	·65	...	...	1	...	...	...	6	...	...	7	...
Roseola	...	...	...	...	...	...	...	...	2	...	...	...	...
Pityriasis rosea	5	·18	...	...	...	...	...	...	...	...	...	1	...
Urticaria	42	1·09	...	...	2	...	4	...	72	...	...	53	...
Prickly heat	21	·67	...	...	...	...	...	...	14	...	...	12	...
Eczema	489	25·03	...	4	5	...	19	...	416	...	2	341	...
Impetigo	63	2·77	...	...	1	...	5.	...	55	...	...	79	...
Pityriasis rubra	3	·11	...	...	...	...	...	...	3	...	...	3	...
Prurigo	...	...	...	...	...	...	...	...	4	...	...	3	...
Lichen	8	·46	...	...	...	...	...	...	7	...	...	7	...
Psoriasis	57	4·77	...	...	...	...	...	...	16	...	1	18	...
Miliaria	1	·02	...	...	...	...	...	...	2	...	...	2	...
Herpes	41	1·23	...	...	...	...	2	...	86	...	...	33	...
Zona	40	1·43	...	...	...	...	...	...	65	...	...	57	...
Pemphigus	209	7·70	1	...	1	...	19	...	9	...	...	11	1
Dermatitis herpetiformis	2	·07	...	...	...	...	...	...	16	...	...	4	...
Acne	17	·64	...	...	...	...	...	...	9	...	...	5	...
Gutta rosea	2	·16	...	...	...	...	2	...	...	...	...	...	...
Sycosis	33	1·90	...	2	...	...	...	...	13	...	...	5	...
Seborrhœa	6	·65	...	...	...	...	...	...	2	...	...	...	...
Ichthyosis	...	...	...	...	...	...	...	...	2	...	...	...	...
Leucodermia	1	·03	...	1	...	...	...	...	1	...	...	...	...
Melasma	1	·04	...	...	...	...	...	...	...	...	...	...	...
Alopecia	1	·13	...	...	...	...	...	...	...	...	...	...	...

DISEASES.	EUROPEAN ARMY OF INDIA.								NATIVE ARMY OF INDIA.			JAIL POPULATION OF INDIA.	
	MEN.				WOMEN.		CHILDREN.						
	Admissions.	Constantly sick.	Deaths.	Invalids.	Admissions.	Deaths.	Admissions.	Deaths.	Admissions.	Deaths.	Invalids.	Admissions.	Deaths.
Area	1	·12	...	...	...	...	...		...	...	...	1	...
Chilblain	...	...	...	...	...	...	...	...	20	...	...	3	...
Ulcer	516	28·71	...	...	14	...	15	...	3,200	...	5	2,861	...
Cicatrices	...	...	...	...	...	...	...	...	...	...		2	...
Boil	932	39·56	...	...	9	...	28	...	2,867	...	...	1,332	1
Carbuncle	16	·81	1	...	...	...	...	...	...	...	...	197	5
Gangrene	...	...	...	...	...	...	...	...	23	...	...	2	...
Whitlow	175	7·22	...	...	...	...	2	...	400	...	...	334	...
Onychia	210	9·52	...	...	1	...	...	...	27	...	...	23	...
Corn	30	·76	...	...	...	...	...	...	11	...	1	3	...
Cheloid	...	...	...	...	...	...	...	...	5	...	...	1	...
Wen	53	2·24	...	...	...	...	...	...	26	...	...	14	...
Molluscum contagiosum	...	...	...	...	...	...	...	...	3	...	...	...	...
Hyperidrosis	14	·37	...	1	...	...	...	...	1	...	...	...	...
Bromidrosis	3	·11	...	...	...	...	...	...	1	...	...	...	...
Pruritus	...	...	...	...	...	...	...	...	...	...	...	4	...
Lupus	3	·15	...	2	...	...	...	...	2	...	...	2	...
Delhi boil	2	·14	...	...	...	...	...	...	96	1	...	...	...
Mycosis fungoides	...	...	...	...	...	...	...	...	2	...	...	...	...
Rhinoscleroma	...	...	...	...	...	...	...	...	...	...	...	1	...
ACCIDENTAL :—													
Heat-stroke	304	14·02	29	12	3	1	2	...	48	3	...	85	13
Sun-stroke	99	3·95	3	3	1	1	2	...	10	2	...	33	12
Heat-apoplexy	66	2·84	23	1	1	1	...	...	9	7	...	34	14
Effects of cold	...	...	...	...	...	...	...	...	1	...	...	...	...
Effects of chemical irritants and corrosives	2	·08	...	...	...	...	...	...	...	...	...	26	...
Lightning stroke	...	...	...	...	...	...	...	...	1	...	...	...	...
Multiple injury	12	1·83	6	2	...	...	...	...	...	...	...	48	4
Suffocation from submersion	...	...	6	...	...	...	1	2	3	9		...	1
„ „ strangulation	...	...	...	...	...	...	...	...	...	...	...	4	...
Starvation	...	...	...	...	...	...	...	...	1	...	...	2	...
Exhaustion	...	...	...	...	...	...	...	...	...	1	...	...	...
Shock	...	...	...	...	...	...	...	...	2	3	...	1	...
Burns and scalds (general and local)	82	4·59	...	1	2	...	7	...	285	...	2	268	4
Frost-bite	...	...	...	...	...	...	...	...	9	...	...	2	...
Abrasions	643	22·57	...	...	4	...	...	...	2,065	...	...	129	...
Contusions	1,469	66·23	2	6	4	...	13	...	2,801	1	6	713	...
Wounds	1,683	65·87	1	1	11	...	31	...	3,457	2	12	3,010	6
„ gunshot	81	6·23	3	1	...	...	...	...	96	2	16	8	1
Strains and sprains	1,540	62·19	...	1	3	...	2	...	1,681	...	5	208	...
Dislocation of spine	6	·48	...1	...	...	...	...	...	1	1	...	...	...
Dislocation of other bones	100	6·97	...	3	...	...	...	...	91	...	7	21	...
Rupture of muscles, tendons, and ligaments	5	·49	...	1	...	...	...	...	10	...	...	...	...
Rupture of membrana tympani	1	·05	...	...	...	...	...	...	...	...	...	...	...
Fracture of the vault of the skull	...	...	...	...	...	...	...	...	2	1	1	1	...
Rupture of intestines	1	·01	1	...	...	...	...	...	...	...	...	...	...
„ kidney	1	·08	...	...	...	...	...	...	...	...	...	...	...
„ lung	...	...	...	...	...	...	...	...	...	...	...	1	...
Fracture of skull	...	...	...	1	...	...	...	...	...	1	...	4	4
„ of the base of the skull	13	·65	9	2	...	...	...	...	11	5	...	4	3
„ of spine	1	...	...	1	...	...	...	...	1	2	...	3	4

TABLE LIII—*continued.*

DETAIL of DISEASES.

DISEASES.	EUROPEAN ARMY OF INDIA.								NATIVE ARMY OF INDIA.			JAIL POPULATION OF INDIA.	
	MEN.				WOMEN.		CHILDREN.						
	Admissions.	Constantly sick.	Deaths.	Invalid-.	Admissions.	Deaths.	Admissions.	Deaths.	Admissions.	Deaths.	Invalids.	Admissions.	Deaths.
Fracture of spine with dislocation	...	...	2	...	...	...	...	...	3	2	1	...	...
„ of other bones	450	44·61	1	25	2	...	9	...	428	4	28	418	2
Foreign bodies in tissues and organs	15	·40	...	...	...	...	...	...	19	...	...	15	1
Rupture of spleen	...	...	...	...	...	...	...	...	1	...	...	...	...
Effects of irritants	...	·01	...	...	...	...	...	...	1	...	...	...	...
Compression of nerves	3	·29	...	...	...	...	...	...	...	...	...	...	...
Concussion of the brain	34	4·10	2	2	...	...	3	...	49	2	...	3	2
Compression of the brain	1	...	1	...	...	...	...	...	1	2	...	...	...
Subconjunctival hæmorrhage	1	·03	...	...	...	...	...	...	...	...	...	...	...
Compression of chest	...	...	...	...	...	...	...	...	10	...	...	...	...
Concussion of the spinal cord	2	·08	...	...	...	...	...	...	...	...	1	...	...
Contusion of abdomen with rupture of viscera	...	...	...	...	...	...	...	...	1	1	...	...	...
Rupture of viscera	...	...	...	...	...	...	...	...	7	1	...	1	1
„ of spleen	...	...	...	...	...	...	...	...	...	...	...	2	4
„ of aorta	...	...	...	...	...	...	...	...	...	...	...	...	2
„ of urethra	...	...	...	...	...	...	...	...	1	...	...	...	...
Foreign bodies in the alimentary canal	...	...	...	...	...	...	1	...	...	...	...	1	...
Separation of epiphyses	1	·14	...	...	...	...	1	...	...	...	...	...	...
Internal derangement of joints	4	·14	...	2	...	...	...	...	2	...	...	...	...
Killed by earthquake	...	...	...	...	...	...	...	...	13	185	...	...	...
POISONS:—													
Arsenic	...	...	...	...	...	...	...	...	2	...	...	1	1
Mercury	5	·88	4	...	...	...	...	...	7	...	...	1	...
Oxalic acid	3	·03	...	...	...	...	...	...	...	...	...	...	...
Lime	...	...	...	...	...	...	...	...	...	...	...	1	...
Alcohol	1	·02	3	...	...	...	...	...	...	...	...	...	...
Carbolic acid	1	·05	...	...	1	...	...	...	...	...	...	1	...
Izal	...	...	...	...	...	...	...	...	...	...	...	1	1
Carbonic acid gas	...	...	...	...	...	...	...	...	...	...	...	...	4
Indian hemp	...	...	...	...	...	...	...	...	2	...	...	...	...
Strychnine	1	·07	...	...	...	...	...	...	...	...	...	...	...
Opium	...	...	...	...	...	...	...	...	3	1	...	22	2
Poisonous fungi	...	...	...	...	...	...	...	...	...	1	...	...	...
Thorn-apple	...	...	...	...	...	...	1	...	...	...	...	...	...
Ptomaines	25	·84	5	...	1	...	1	...	...	2	...	...	...
Cantharides	1	·01	...	...	...	...	...	...	...	...	...	...	...
Vegetable not defined	...	...	...	...	...	...	...	...	4	1	...	...	...
Irritant drug not defined	...	...	...	...	...	...	...	...	...	...	...	3	3
POISONED WOUNDS:—													
Poisoned wound, not defined	2	·12	...	...	...	...	...	...	5	...	...	...	...
„ „ by snakes	...	...	...	...	...	...	...	...	10	2	1	19	...
„ „ by scorpions	2	·07	...	...	...	...	...	...	5	...	...	1	...
„ „ by centipedes	...	...	...	...	...	...	...	...	5	...	...	46	1
„ „ by hornets	2	·24	...	...	...	...	...	...	...	...	...	...	...
„ „ by tick	...		...	...	...	...	...	...	...	...	...	2	...
„ „ by stinging insects	...	...	...	...	...	...	...	...	10	...	...	4	...
„ „ by fish	...	...	...	...	...	...	...	...	...	...	...	5	...
„ „ by monkey	...	...	...	...	...	...	2	...	...	...	...	...	...
„ „ by horse	2	·05	...	...	...	...	...	...	...	...	...	...	...

Diseases.	Men.				Women.		Children.		Native Army of India.			Jail Population of India.	
	Admissions.	Constantly sick.	Deaths.	Invalids.	Admissions.	Deaths.	Admissions.	Deaths.	Admissions.	Deaths.	Invalids.	Admissions.	Deaths.
POISONED WOUNDS :—contd.													
Poisoned wound, by dog	76	19'02	...	...	1	...	9	...	11	...	...	5	...
„ „ by jackal	1	...	...	...	...	...	...	...	1	...	...	...	...
„ „ by dead animal matter	...	...	...	...	...	...	...	...	...	...	...	1	...
„ „ by tattooing	2	'03	...	...	...	...	...	...	...	...	...	...	...
„ „ by dirty clothing	1	'01	...	...	...	...	...	...	...	...	...	...	...
„ „ by septic matters	...	...	...	...	...	...	...	...	1	1	...	...	...
„ „ by vegetable substances	...	...	...	...	...	...	...	...	2	...	...	1	...
HOMICIDAL :—													
Cut-throat	...		...	...	...	...	...	...	...	...	...	...	1
Fracture of the skull	...	...	1	...	...	...	...	...	...	...	...	...	3
Gunshot wound	1	'01	3	...	...	...	...	...	...	6	...	3	...
Hanging	...		...	...	...	...	...	...	...	...	...	...	1
Probable homicide	...		...	...	...	...	...	...	...	...	...	...	1
Not defined	...		...	...	...	...	...	...	...	...	...	...	4
SUICIDAL :—													
Multiple injury	1	'01	1	...	...	...	...	...	...	...	...	...	...
Drowning	...	...	1	...	...	1	...	...	...	1	...	...	...
Hanging	...	...	1	...	...	...	...	...	...	3	...	...	13
Gunshot wound	1	...	11	...	...	...	...	...	...	9	...	...	...
Cut-throat	2	'01	4	...	...	...	...	...	...	...	...	4	1
Dislocation of spine	...		1	...	...	...	...	...	...	...	...	...	...
Fracture of base of the skull	...		2	...	...	...	...	...	...	...	...	...	...
Poison, oxalic acid	...		1	...	...	...	...	...	...	...	...	...	...
„ opium	...		...	...	...	...	...	...	...	1	...	...	...
Not defined	...		...	...	...	...	...	...	...	...	...	...	3
JUDICIAL :—													
Hanging	...	...	1	...	...	...	...	...	...	...	...	...	6
Punished	...	...	...	...	...	...	...	...	...	...	...	22	...
NOT DEFINED :—													
Cut-throat	...	...	...	...	...	...	...	...	...	...	...	12	2
No appreciable disease	209	11'95	...	...	41	...	55	...	17	...	...	35	...
Not yet diagnosed	...	...	...	...	...	...	...	...	...	...	...	35	...
Cause unknown	...	...	...	...	...	...	...	...	...	4	...	...	...
Hæmorrhage	...	...	...	...	...	...	...	...	...	...	...	...	1
Absent deaths	...	...	...	...	...	...	...	...	...	379	...	...	...
GRAND TOTAL	59,178	3,721'10	714	*1,508	2,183	37	2,092	200	73,739	1,363	1,113	86,688	2,327

```
  ┌ Northern Command .  374=20'47  per 1,000 of strength.
  │ Western      "    .  539=28'61      "    "    "
  │ Eastern      "    .  363=19'12      "    "    "
* ┤ Secunderabad Division  155=18'48    "    "    "
  │ Burma Division         57=14'95     "    "    "
  └ India        .    .  1,508=21'24    "
```

TABLE LIII—*concluded.*

DETAIL of DISEASES.

DISEASES.	EUROPEAN TROOPS. Aden Column Field Force. Average annual strength 421.		NATIVE TROOPS. Somaliland Field Force. Average annual strength 2,521.		Aden Column Field Force. Average annual strength 660.		Sikkim-Tibet Mission Force. Average annual strength 2,730.		China Garrison. Average annual strength 1,417.	
	Admissions.	Deaths.	Admissions.	Deaths.	Admissions.	Deaths.	Admissions.	Deaths.	Admissions.	Deaths.
Small-pox	...	...	3	...	...	...	...	...	...	...
Measles	...	...	...	...	...	...	...	...	1	...
Rubella	...	...	...	...	...	...	...	...	1	...
Influenza	...	...	...	...	...	...	...	...	77	...
Mumps	...	...	11	...	...	...	1	...	...	...
Simple continued fever	...	...	...	...	...	...	8	...	3	...
Dysentery	23	...	36	...	72	...	33	1	38	—
Intermittent fever	159	...	367	1	478	1	78	...	53	...
Remittent fever	...	...	...	...	2	...	39	7	...	—
Erysipelas	...	...	...	...	...	...	...	...	1	1
Tubercle of the lung	...	...	...	...	...	...	2	...	17	5
Syphilis	2	...	...	...	14	...	2	...	6	...
Gonorrhœa	6	...	1	...	5	...	9	...	22	...
Ascaris lumbricoides	...	...	...	...	...	...	1	...	...	...
Phthirius inguinalis	...	...	...	...	...	...	...	...	3	...
Scabies	...	...	...	...	...	...	2	...	15	...
Ringworm	...	...	...	...	...	...	...	...	1	...
Scurvy	...	...	1	...	2	...	...	...	...	—
Rheumatism	4		8	...	13	...	7	...	1	—
Anæmia	...	...	...	...	—	...	1	1	1	—
Debility	...	...	...	...	5	...	...	...	5	—
Other general diseases	18	...	3	—	40	...	...	...	...	—
Neuritis	...	...	...	...	...	...	3	...	—	—
Cerebral meningitis	...	...	...	...	...	...	...	...	1	1
Neuralgia	...	...	...	...	...	...	...	...	3	—
Other diseases of the nervous system	...	...	...	...	1	...	...	...	...	—
Conjunctivitis	4	...	4	...	6	...	2	...	...	...
Ecchymosis	...	...	...	...	...	...	...	...	1	—
Ulcerative keratitis	...	...	...	...	...	...	...	...	1	—
Atrophy of optic nerve	...	...	...	...	...	...	...	...	1	...
Amblyopia	...	...	...	...	...	...	...	...	2	—
Other diseases of the eye	2	...	1	...	4	...	...	...	...	...
Inflammation of the external ear	...	...	...	...	...	...	...	...	1	—
Deafness	...	...	...	...	...	...	...	...	3	...
Disordered action of the heart	5	...	...	...	2	...	...	...	...	—
Embolism	...	...	...	...	...	...	1	1	...	—
Other diseases of the circulatory system	2	...	...	...	2	...	...	...	...	...
Laryngitis	...	...	...	...	...	...	1	...	1	—
Bronchitis	3	...	5	...	9	...	12	...	23	...
Hæmoptysis	...	...	...	...	...	...	...	...	1	...
Pneumonia	2	1	3	...	2	1	1	...	20	—
Pleurisy	1	...	1	...	2	...	1	...	1	—
Other diseases of respiratory system	1	...	1	...	7	...	...	...	33	...
Caries of dentine	...	...	...	...	...	...	4	...	...	—
Sore throat	1	...	...	...	1	...	1	...	1	—
Tonsillitis	7	...	...	...	...	...	...	...	...	—
Tonsil itis follicular		...	...	...	1	...	1	...	1	—
Indigestion		...	...	...	...	...	...	...	3	—

Diseases.	European Troops — Aden Column Field Force. Average annual strength 421.		Native Troops — Somaliland Field Force. Average annual strength 2,521.		Native Troops — Aden Column Field Force. Average annual strength 660.		Native Troops — Sikkim-Tibet Mission Force. Average annual strength 2,720.		Native Troops — China Garrison. Average annual strength 1,417.	
	Admissions.	Deaths.	Admissions.	Deaths.	Admissions.	Deaths.	Admissions.	Deaths.	Admissions.	Deaths.
Inflammation of intestines	1	...	...	...	1	...	...	...	...	...
Obstruction of intestines	1	1	...	...	...	...	...	...	...	...
Diarrhœa	3	...	12	...	34	...	18	...	6	...
Abscess of the liver	1	1	...	...	...	...	1	...	...	...
Inflammation and congestion of liver	5	...	...	...	...	...	1	...	...	...
Jaundice	10	...	...	...	6	...	...	...	...	...
Other diseases of the digestive system	12	...	3	...	4	...	...	...	...	...
Inflammation of the lymph-glands	...	...	1	1	...	...	2	...	...	...
Diseases of the urinary system	4	...	...	...	...	...	...	...	...	...
Soft chancre	1	...	...	...	...	...	4	...	7	...
Hydrocele	...	...	...	...	...	...	2	...	...	...
Orchitis	...	...	...	...	...	...	2	...	...	...
Caries of bone	...	...	...	...	...	...	...	...	1	...
Synovitis	...	...	...	...	...	...	...	...	2	...
Myalgia	...	...	...	...	...	...	3	...	...	...
Inflammation of the connective tissue	...	...	...	...	...	...	...	...	2	...
Abscess of ,, ,, ,,	...	...	...	...	...	...	3	...	4	...
Eczema	...	...	...	...	...	...	...	...	1	...
Zona	...	...	...	...	...	...	1	...	...	...
Frost-bite	...	...	...	...	...	...	5	...	...	...
Ulcer	...	...	...	...	...	...	3	...	...	...
Cicatrices	...	...	...	...	...	...	1	...	...	...
Boil	...	...	...	...	...	...	...	...	5	...
Whitlow	...	...	...	...	...	...	...	...	2	...
Other local diseases	46	...	110	...	72	...	...	...	...	...
ACCIDENTAL :—										
Burns and scalds	...	...	...	...	...	...	2	...	...	...
Suffocation (by falling of roof)	...	...	...	...	...	...	...	1	...	...
Contusions	...	...	...	...	...	...	10	...	6	...
Effects of cold	...	...	...	...	...	...	...	...	...	1
Strains and sprains	...	...	...	...	...	...	3	...	2	...
Wounds	...	...	...	...	...	...	13	...	4	...
Abrasions	...	...	...	...	...	...	5	...	6	...
Fractures	...	...	...	...	...	...	2	...	...	...
Compression of brain (by falling of roof)	...	...	...	...	...	...	...	1	...	...
Other injuries	21	...	25	...	34	...	...	...	...	...
Poisoned wound by dog	...	...	...	...	...	...	2	...	...	...
All other causes	...	...	...	...	80	...	...	...	204	...
No appreciable disease	...	...	...	...	...	...	1	...	...	...
Total	345	3	596	2	908	2	293	12	594	8